Contemporary Human Geography

Fifth Edition

James M. Rubenstein

MIAMI UNIVERSITY, OXFORD, OHIO

Pearson

Content Development: Susan Teahan, Matt Walker
Content Management: Terry Haugen, Jeanne Zalesky
Content Production: Heidi Allgair, Ziki Dekel, Mike Early, Kassi Foley, Cheryl Chi, Wendy Romaniecki
Product Management: Aileen Pogran
Product Marketing: Candace Madden
Rights and Permissions: Felix Jun Sarte, Paul James, Tan, Ben Ferrini

Please contact https://support.pearson.com/getsupport/s/ with any queries on this content

Cover Image by triloks / E+ / Getty Images

Front Cover: A grandmother and grandson share a digital tablet. They live in one of the most remote places on Earth, the Spiti Mountains of northern India. The isolated location has helped them to preserve a unique identity, including a distinctive language, religion, and clothing. At the same time, they are aware of what's happening around the world, thanks to their tablet. *Contemporary Human Geography* explores the patterns of globalization and local cultural diversity.

Library of Congress Cataloging-in-Publication Data

Library of Congress Control Number: 2021950518

4 2023

Pearson

www.Savvas.com/Advanced
ISBN-13: 978-0-13-766210-4 High School Binding
ISBN-10: 0-13-766210-6 High School Binding

Pearson's Commitment to Diversity, Equity, and Inclusion

Pearson is dedicated to creating bias-free content that reflects the diversity, depth, and breadth of all learners' lived experiences.

We embrace the many dimensions of diversity, including but not limited to race, ethnicity, gender, sex, sexual orientation, socioeconomic status, ability, age, and religious or political beliefs.

Education is a powerful force for equity and change in our world. It has the potential to deliver opportunities that improve lives and enable economic mobility. As we work with authors to create content for every product and service, we acknowledge our responsibility to demonstrate inclusivity and incorporate diverse scholarship so that everyone can achieve their potential through learning. As the world's leading learning company, we have a duty to help drive change and live up to our purpose to help more people create a better life for themselves and to create a better world.

Our ambition is to purposefully contribute to a world where:

- Everyone has an equitable and lifelong opportunity to succeed through learning.
- Our educational content accurately reflects the histories and lived experiences of the learners we serve.
- Our educational products and services are inclusive and represent the rich diversity of learners.
- Our educational content prompts deeper discussions with students and motivates them to expand their own learning (and worldview).

Accessibility

We are also committed to providing products that are fully accessible to all learners. As per Pearson's guidelines for accessible educational Web media, we test and retest the capabilities of our products against the highest standards for every release, following the WCAG guidelines in developing new products for copyright year 2022 and beyond.

You can learn more about Pearson's commitment to accessibility at **https://www.pearson.com/us/accessibility.html**

Pearson

Contact Us

While we work hard to present unbiased, fully accessible content, we want to hear from you about any concerns or needs with this Pearson product so that we can investigate and address them.

Please contact us with concerns about any potential bias at **https://www.pearson.com/report-bias.html**

For accessibility-related issues, such as using assistive technology with Pearson products, alternative text requests, or accessibility documentation, email the Pearson Disability Support team at **disability.support@pearson.com**

Brief Contents

Contents

1 This is Geography

2 Population & Health

3 Migration

4 Culture & Social Media

5 Languages

6 Religions

7 Ethnicity & Race

8 Political Geography

11 Industry

12 Services & Settlements

13 Urban Patterns

14 Resource Issues

Preface

Welcome to a new kind of geography textbook! We live in a visual age, and geography is a highly visual discipline, so Pearson—the world's leading publisher of geography textbooks—invites you to study human geography as a visual subject. This book has been designed from the beginning to be equally readable in both print and digital formats.

The fifth edition of *Contemporary Human Geography* builds on the strengths of the first four editions while responding to user feedback to make important changes and improvements and incorporating innovative features, current data, and new information.

The main purpose of this book is to introduce you to the study of geography as a social science by emphasizing the relevance of geographic concepts to human problems. It is intended for use in college-level introductory human or cultural geography courses. The book is written for students who have not previously taken a college-level geography course.

What Makes This Book Contemporary?

Titling this book "contemporary" is a bold claim. All credible geography books—including this one—include up-to-date statistics, recent world events, and current geographic concepts. This book claims to be more contemporary—not merely up-to-date—for three reasons.

We live in a visual age. This book was written in the reverse order of traditional textbooks. For most books, the text is written first and the graphic material is added later. Instead of beginning with an author's complete manuscript, this book starts with a sketch of a visual concept for each one-page spread and two-page module. What would be the most important geographic idea presented on each spread, and what would be the most effective visual way to portray that idea? The maps, graphs, and photos are placed on the page first, and then the text is written around the graphics. The production of this book does not have a traditional manuscript; from the outset, the text is written to complement the graphics.

We live in a sound bite age. This book replaces the narrative style of traditional books. Each one-page spread and each two-page module of this book are self-contained. Material on a main concept is contained within a page. This places more of a premium on clear concise outlining as an important pedagogical feature. The captions under maps, graphs, and photos typically repeat material already presented in the text. Not so with this book—the graphic material stands on its own.

We live in an electronic age. This book has been designed to be equally usable—and attractive—in both print and digital formats. Most books are still composed in pages designed for print—as in the past—and converted to digital format after publishing of the print version. As a result, the conversion to digital format is frequently awkward. For example, maps and photos are often placed in the print version in positions that don't work well in digital format.

Contemporary Chapter Organization

This book has a clear, easy-to use organization and outline. Each chapter is organized into between 9 and 13 modules composed of two one-page sections that follow a consistent pattern:

Opening section. Each chapter opens with an outline of the four Key Issues that will be addressed in the chapter as well as a brief description of the material contained in each Key Issue. A Word Cloud introduces key words and phrases that appear in the chapter. A Locations map highlights some of the places that are illustrated in the chapter.

Key Issues. Each chapter is organized around four Key Issues that outline the main topics to be addressed. The four Key Issues are presented as questions to be addressed in that section of the chapter.

Self-contained modules. Each two-page module is titled and numbered to enhance the clarity of the outline. Each individual section is also titled and numbered.

Learning Objectives. Each two-page module begins with a Learning Objective that frames the main concept of that module.

Self-contained sections. Each page or section within the two-page module is also self-contained, with its own title and number. As a result, maps and photos appear next to where they are discussed in the text. No more going through a chapter to find a figure that is referenced on one page but actually appears on another page. This approach is especially critical for reading the eText on a tablet, computer, or smart phone.

Review, Analyze, & Apply. The final two-page module of each chapter (four pages for Chapter 1) reviews the main points of the chapter, organized around the four Key Issues. The end-of-chapter material also includes Key Terms as well as other features, including Thinking Geographically, Geospatial Analysis, and Explore activities.

Contemporary Content

This book discusses the following main topics:

What basic concepts do geographers use? Geographers employ several concepts to describe the distribution of people and activities across Earth, to explain reasons underlying the observed distribution, and to understand the significance of the arrangements. Chapter 1—This Is Geography—provides an introduction to ways that geographers think about the world.

Geographers have insights into the major issues of our time. COVID-19 is used to illustrate geography's five most basic concepts, five different ways to make maps, and features of distribution. Issues of cultural diversity and inequality include an introduction to Black Lives Matter (discussed in more detail in Chapter 7) as well as gender and LGBTQ distribution. Geographic perspectives on climate change are also introduced.

Where are people located in the world? Why do some places on Earth contain large numbers of people or attract newcomers whereas other places are sparsely inhabited? Chapters 2 and 3 examine the distribution and growth of the world's population as well as the movement of people from one place to another.

Since the last edition of the book, natural increase rates and birth rates have declined and death rates have increased, partly in response to COVID-19. Meanwhile, China has abandoned its long-standing one-child policy.

Migration has been an important issue for Americans. This edition expands its discussion of unauthorized immigrants and the spatial variations along the U.S. Mexico border. Coverage has been doubled from the previous edition concerning the main reasons that people migrate—environmental, economic, and political.

How are different cultural groups distributed? Geographers look for similarities and differences in the cultural features at different places, the reasons for their distribution, and the importance of these differences for world peace. Chapters 4 through 8 analyze the distribution of different cultural traits and beliefs and the political challenges that result from those spatial patterns. Important cultural traits discussed in Chapter 4 include food, clothing, shelter, and leisure activities. Chapters 5 through 7 examine three main elements of cultural identity: language, religion, and ethnicity. Chapter 8 looks at political problems that arise from cultural diversity.

The title of Chapter 4 has changed from Folk & Popular Culture to Culture & Social Media, a reflection of the growing importance of social media in understanding the geography of cultural patterns and preferences. A larger share of the chapter is devoted to distinctive spatial patterns of social media, including social media platforms, Internet freedom, cyberattacks, cyber espionage, and fake news. A spread has been added concerning the geography of travel.

Classification of the world's principal religions has been reorganized in accordance with recognized authorities on the geography of religions. The three universalizing religions with the largest number of adherents continue to be Christianity, Islam, and Buddhism. However, the ethnic religions with the largest number of adherents in addition to Hinduism are now considered by authorities to be Chinese Folk and Ethno-religions. Other religions are classified by number of adherents, with the most numerous in addition to Sikhism and Judaism now considered to be New Religions and Spiritism.

The title of Chapter 7 has been changed from Ethnicities to Ethnicity & Race to reflect the increasing importance of the term "race" in cultural geography. A key example of the increasing importance of race is the use of the term "Black" in most places instead of "African American." A new spread on Black Lives Matter examines the diffusion and patterns of racism on the landscape.

The content of Chapter 8 has been moved among the four key issues. The first key issue now includes internal structure of states (unitary and federal) and international cooperation such as the EU. The second key issue now explains the evolution of states from ancient times, through the development of nation-states, to recent creation of states especially in Eastern Europe. The third key issue focuses on the morphology of states, including boundaries and shapes. Coverage is expanded of gerrymandering of U.S. political boundaries, with specific examples of the two types of gerrymandering. The fourth key issue focuses on threats to states, including a new discussion of states with nuclear weapons.

How do people earn a living in different parts of the world? Human survival depends on acquiring an adequate food supply. One of the most significant distinctions among people globally is whether they produce their food directly from the land or buy it with money earned by performing other types of work. Chapters 9 through 12 look at the three main ways of earning a living: agriculture, manufacturing, and services. Chapter 13 discusses cities, where the world's economic and cultural activities are increasingly centered.

Chapter 9 (Food & Agriculture) has been reorganized to provide an increased emphasis on food. The chapter begins with the geography of food, including spatial patterns of what people eat and how much they eat. The remainder of the chapter is organized around the spatial patterns of where food comes from, especially as it relates to differences in climate.

The United Nations has created two new indexes of development that are now included in Chapter 10 (Development). They are Multidimensional Poverty Index (MPI) and Gender Social Norms Index (GSNI). The various indexes of development are presented to emphasize spatial differences in gender and economic inequalities. Immanuel Wallerstein's world-systems theory of core and periphery is now included as its own spread.

Material in Chapter 11 (Industry) is reorganized to place changes in site and situation factors alongside the rest of those subjects' material. The chapter concludes with a spread on changes in China's situation and site factors.

The title of Chapter 12 has been changed from Services to Services & Settlements. A discussion of current geographic patterns of settlements is now placed at the beginning of the chapter. The chapter contains a new spread on e-commerce, like Amazon, and a new spread on sharing services, like Airbnb and Uber.

Chapter 13 (Urban Patterns) includes a new case study of the geographic patterns within London.

What issues result from using Earth's resources? Geographers recognize that cultural problems result from the depletion, destruction, and inefficient use of the world's natural resources. Chapter 14 (Resource Issues) is devoted to a study of issues related to the use of Earth's natural resources.

The chapter includes expanded coverage of solar power and of the challenge of disposing of plastic waste. The chapter and the book conclude with a new spread that reviews geography's five key concepts by comparing Luxembourg and Timor-Leste—two small and little-known but very different countries.

Contemporary Relevance

Many speculated that geography would be irrelevant in the twenty-first century. Geography's future was thought to be grim because the diffusion of electronic communications and social media would make it easier for human activities to be conducted remotely. If any piece of information could be accessed from any place in the world (at least where electronic devices work), why live, shop, work, or establish a business in a crowded city or a harsh climate?

In reality, geography has become more, not less, important in people's lives and the conduct of business. Here are several ways that location matters more now than in the past, because of—not despite—the diffusion of electronic devices:

Geographical smart phone apps. Smart phones and other electronic devices match specific demand to supply in a particular locality. For example: Restaurant apps match hungry people to empty seats in a locality's restaurants. Real estate apps help people find housing for sale or for rent in a locality. Social apps let people know where their friends in a particular locality are hanging out that night. Transportation apps match vehicles with available seats to people trying to get to specific locations. These sorts of apps generate data on people's preferences in space, which in turn help even more location-based businesses get started and grow. Through an app, we can find a restaurant, get detailed directions to it via voice navigation, or order online for takeout delivery to our doorstep. Geographic apps, such as maps, navigation aids, travel, and transportation, rank among the most frequently used services on smart phones.

Navigation. Electronic devices are essential to the smooth movement of people and goods. For example: Turn-by-turn information can prevent you from getting lost or steer you back if you do get lost. Traffic jams on overcrowded roads can be avoided or minimized. Delivery services can find where to deliver packages. Drivers of car-sharing services can figure out how to get you to your destination. Instead of turning on a radio to hear traffic information, we look at the red and green traffic flow patterns on an electronic map. Vehicles in the future will be driverless, so you can spend driving time working, learning, or social networking.

Ideas. The people who make all of these new location-based apps are themselves highly clustered in a handful of places in the world, such as the San Francisco Bay area. Ideas—both brilliant and farfetched—are still easier to communicate face-to-face than across long distances. Living and working in places like Silicon Valley, despite high expenses and choking traffic jams, put people next to other like-minded innovators in the electronic-based geography of the twenty-first century.

Cultural diversity. Electronic devices also impact the changing geography of cultural diversity. What if you searched for an available restaurant table in a foreign language? Would you find the same places that you would if you searched in your native language? What if you conducted an internet search in a foreign country? Would you find the same information as in an internet search done at home?

Contemporary Tensions Between Globalization and Local Diversity

A central theme in this book is a tension between two important realities of the twenty-first-century world—

globalization and cultural diversity. In many respects we are living in a more unified world economically, culturally, and environmentally. The actions of a particular corporation or country affect people around the world. In the third decade of the twenty-first century, we continue to face wars in unfamiliar places and experience economic struggles unprecedented in the lifetimes of students or teachers. Geography's spatial perspectives help to relate economic change to the distributions of cultural features such as languages and religions, demographic patterns such as population growth and migration, and natural resources such as energy, water quality, and food supply.

This book argues that, after a period when globalization of the economy and culture has been a paramount concern in geographic analysis, local diversity now demands equal time. People are taking deliberate steps to retain distinctive cultural identities. They are preserving little-used languages, fighting fiercely to protect their religions, and questioning free trade agreements. Local diversity even extends to addressing issues, such as climate change, that at first glance are considered global. For example, the "greenest" cars for motorists to drive in Ohio are different than the "greenest" cars for Indiana.

Since 2013, I have written a weekly column for my local newspaper on behalf of my local cooperatively owned grocery store. The column has come to extol the virtues of local here in Midwestern USA: the local food, the local farmers, the local seasons, and the locally owned co-op grocery. I admire the farmers and the agriculture from far away, but our local food is more nutritious, consumes less energy, and tastes better. In a world where we feel anger and helplessness at the plight of people in other places, it is at the local scale that we all can make a difference.

A Contemporary Publishing Team

The steps involved in creating most traditional textbooks haven't changed much. The book passes from one to another like a baton in a relay race. The author writes a manuscript, which then passes in turn through development, editing, and production specialists on the way to the printing press. The preface typically includes a perfunctory litany of acknowledgments for the many fine people who contribute to the development, editing, and production of the book.

In contrast, this book starts as a genuine partnership among several development, editorial, and production specialists. The traditional separation of development, editorial, and production has been deliberately blurred.

Preparing this book demanded a contemporary approach to collaboration because it was created during the worst of the COVID-19 pandemic in the United States. The team, which is scattered around the country from New Hampshire to Oregon, met biweekly through Zoom. Key contributors include folks employed by Pearson and folks from elsewhere. The most important contributors from outside Pearson include the following:

Susan Teahan, Development Editor. A typical development editor provides useful information to the author early in the writing process and then fades from the scene. Not so with Susan. Susan's role actually expanded during the process of creating this book. We collaborated closely on tailoring the content to fit each and every page of the book. Increase or decrease the size of the artwork? Add or delete text? We were there together.

Erika Nelson, Research Assistant, has an MA in Geography from Miami University and currently teaches anthropology and geography courses at Miami. I have known Erika for many years as a student and friend and as a colleague on two books. Erika brings fresh perspectives to the book, such as her background living in Africa and her teaching online.

Kevin Lear, Senior Project Manager at International Mapping, and his team produced the outstanding maps for this book. Back in the 1980s, Kevin was the first cartographer to figure out how to produce computer-generated full-color maps that are more accurate and more attractive than hand-drawn ones. He has continued to stay well ahead of the pack in the quality of the maps. And Kevin has taken on additional tasks of creating all of the artwork in the book, not only the maps.

Katie Ostler and Sharon Cahill, Straive Project Managers. In a global economy, essential tasks in publishing a book are inevitably outsourced to specialists. In this case, the task of publishing the text and artwork into printed and digitized pages was in the hands of Straive, which is known as a leader in e-learning. When outsourcing, it is essential to be able to communicate with a single responsible leader. In this case, that person was first Katie and then Sharon.

Key contributors within Pearson include the following:

Heidi Allgair, Pearson Content Producer for Science. The smooth flow of pages, maps, photos, and so on requires someone who is able to stay on top of a complex project. Heidi perfectly performed this critical role of managing what each team member was doing at any given time.

Kassi Foley, Pearson Senior Learning Tools Strategist. Previous editions of the book have included several interactive features. We have shifted our priority to creating a handful of Geographic Interactive projects to help students understand some of the most challenging topics in the course, such as world agricultural regions and the diversity of languages. Kassi has played the key role at Pearson in creating the Geographic Interactive projects, in association with Social Explorer.

Ziki Dekel, Media Producer. In view of the fact that most students now read the book on an electronic device, the role of media producer has become more important. Most books are created in print version with little regard for their conversion to digital format. Thanks to folks at Pearson like Ziki, this book is hopefully more legible in digital version than most other textbooks.

Aileen Pogran, Pearson Product Manager Life Sciences, and Terry Haugen, Pearson Content Strategy Manager for Geosciences, Environmental Sciences, & Non-majors Biology. I've saved the most important team members for last. Aileen and Terry are the two editors for this project. When Pearson undertook a major reorganization of responsibilities a few years ago, Aileen and Terry both became my editors. In principle, Aileen has been responsible for ensuring that this book has had the support of Pearson's top management, and Terry has been responsible for ensuring that the content of this book meets Pearson's highest standards. A close relationship between author and editor in this case, editors is the heart of a successful book. I've been very fortunate that in my three decades of association with Pearson, I've had three outstanding editors: Paul Corey, Dan Kaveney, and Christian Botting. This first project with Aileen and Terry has continued my good fortune of outstanding editors. What a delight to work with them for the first time. I think this is the beginning of a beautiful friendship.

Digital & Print Resources

FOR TEACHERS:

This edition provides a complete human geography program for teachers.

Mastering Geography with Pearson eText for *Contemporary Human Geography*

The Mastering platform is the most widely used and effective online homework, tutorial, and assessment system for the sciences. It delivers self-paced coaching activities that provide individualized coaching, focus on course objectives, and are responsive to each student's progress. The Mastering system helps teachers maximize class time with customizable, easy-to-assign, and automatically graded assessments that motivate students to learn outside of class and arrive prepared for lecture. Mastering Geography offers:

- Assignable activities that include GIS-inspired MapMaster 2.0™ interactive maps, *Encounter Human Geography* Google Earth™ Explorations, GeoVideos, GeoTutors, Thinking Spatially & Data Analysis activities, end-of-chapter questions, reading quizzes, Test Bank questions, map labeling activities, and more.
- Student study area with GIS-inspired MapMaster 2.0 interactive maps, Geoscience Animations, web links, geography videos, glossary flash cards, "In the News" current articles, reference maps, an optional Pearson eText and more.

www.pearson.com/mastering/geography

PREVIEW AND ADOPTION PROCESS

Upon textbook purchase, students and teachers are granted access to Mastering Geography. High school teachers an obtain preview or adoption access to Mastering Geography in one of the following ways:

Preview Access

Teachers can request preview access by visiting Savvas .com/Access_Request. Select Initial Access then using Option 2, select your discipline and title from the drop-down menu and complete the online form. Preview Access information will be sent to the teacher via email.

Adoption Access

Upon purchase, teachers can request course adoption access by visiting Savvas.com/Access_Request. Select Initial Access, then, using Option 3, select your discipline and title from the drop-down menu and complete the online form. Access codes and registration instructions will be sent to the requester via email.

Students, ask your teacher about access.

Savvas, and any third party for which Savvas serves as the sales agent or distributor, reserve the right to change and/or update technology platforms, including possible edition updates to customers during the term of access. Customers will be notified of any change prior to the beginning of the new school year.

FOR TEACHERS

Instructor Resource Manual (Download Only)

Updated for the fourth edition, the *Instructor Resource Manual* is intended as a resource for both new and experienced instructors. It includes lecture outlines, additional source materials, teaching tips, advice about how to integrate online media, and various other ideas for the classroom.

www.pearson.com/mastering/geography

TestGen® Computerized Test Bank (Download Only)

TestGen is a computerized test generator that lets instructors view and edit *Test Bank* questions, transfer questions to tests, and print the test in a variety of customized formats. This *Test Bank* includes over 1,000 multiple choice and short answer/essay questions. Questions are correlated to the revised U.S. National Geography Standards and Bloom's Taxonomy to help instructors better map the assessments against both broad and specific teaching and learning objectives. The questions are also tagged to chapter specific learning outcomes.

www.pearson.com/mastering/geography

Instructor Resource Materials (Download Only)

The *Instructor Resource Materials* provides high-quality electronic versions of photos and illustrations from the book in JPEG, pdf, and PowerPoint formats, as well as customizable PowerPoint lecture presentations, Classroom Response System questions in PowerPoint, and the *Instructor Resource Manual* and *Test Bank* in MS. Word and TestGen formats. For easy reference and identification, all resources are organized by chapter.

About the Author

Dr. James M. Rubenstein received his B.A. from the University of Chicago in 1970, M.Sc. from the London School of Economics and Political Science in 1971, and Ph.D. from Johns Hopkins University in 1975. He was a Professor of Geography at Miami University for thirty-seven years, where he taught urban and human geography. Dr. Rubenstein is now a full-time writer. In addition to this book, Dr. Rubenstein is the author of *The Cultural Landscape* (13th edition) as well as co-author of *Introduction to Contemporary Geography,* both published by Pearson Education.

Dr. Rubenstein also conducts research in the automotive industry and has published four books on the subject *The Changing U.S. Auto Industry: A Geographical Analysis* (Routledge); *Making and Selling Cars: Innovation and Change in the U.S. Auto Industry* (The Johns Hopkins University Press); *A Profile of the Automobile and Motor Vehicle Industry: Innovation, Transformation, Globalization* (Business Expert Press); and *Who Really Made Your Car? Restructuring and Geographic Change in the Auto Industry* (W.E. Upjohn Institute, with Thomas Klier). He also writes a weekly column about local food for The Oxford Press. Winston, a lab-husky mix with one brown eye and one blue eye, takes Dr. Rubenstein for long walks in the woods most days, when he is up to it.

This book is dedicated to my wife
Bernadette Unger, the love of my life,
and my companion through life.

About our Sustainability Initiatives

Pearson recognizes the environmental challenges facing this planet, as well as acknowledges our responsibility in making a difference. This book is carefully crafted to minimize environmental impact. The binding, cover, and paper come from facilities that minimize waste, energy consumption, and the use of harmful chemicals. Pearson closes the loop by recycling every out-of-date text returned to our warehouse.

Along with developing and exploring digital solutions to our market's needs, Pearson has a strong commitment to achieving carbon-neutrality. As of 2009, Pearson became the first carbon- and climate-neutral publishing company, having reduced our absolute carbon footprint by 22% since then. Pearson has protected over 1,000 hectares of land in Columbia, Costa Rica, the United States, the UK and Canada.

In 2015, Pearson formally adopted *The Global Goals for Sustainable Development,* sponsoring an event at the United Nations General Assembly and other ongoing initiatives. Pearson sources 100% of the electricity we use from green power and invests in renewable energy resources in multiple cities where we have operations, helping make them more sustainable and limiting our environmental impact for local communities.

The future holds great promise for reducing our impact on Earth's environment, and Pearson is proud to be leading the way. We strive to publish the best books with the most up-to-date and accurate content, and to do so in ways that minimize our impact on Earth. To learn more about our initiatives, please visit **https://www.pearson.com/corporate/sustainability.html**

CHAPTER

1 This is Geography

▲ Floor Map, Google Office, Washington D.C.

Contemporary geography is the scientific study of where people and activities are found across Earth's surface and the reasons why they are found there. Geography is distinctive because it encompasses both social science and natural science. This book focuses on geography as a social science (human geography).

KEY ISSUES

1 Why Is Geography a Science?

Geography's most distinctive tool is the map. Prehistoric humans were the first people to make maps. Contemporary tools enable cartographers—and anyone else who has access to electronic devices—to make precise maps and to interpret their meaning.

2 Why Is Every Place Unique?

Geographers understand that each location on Earth is in some ways unique. Each specific place or larger region on Earth possesses a unique combination of features.

3 Why Are Different Places Similar?

Many geographical features are organized in a regular manner across space. Some regularities are global in scale, whereas others have distinctive local character.

4 Why Are Places Connected?

Distinctive to geography is the importance given to connections between human activities and the physical environment. Some human activities are sustainable, but others are not.

LOCATIONS IN THIS CHAPTER

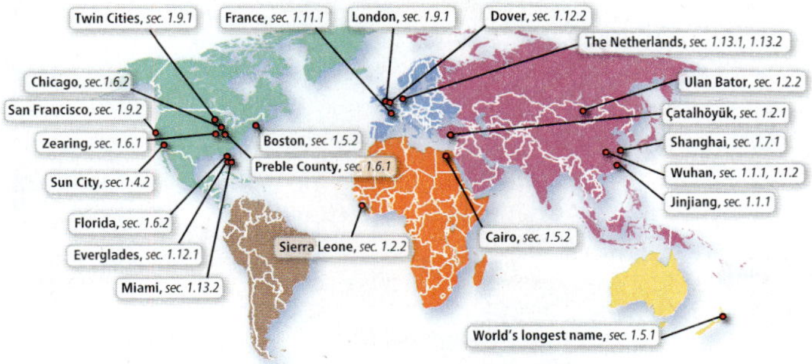

▲ **1.1.1 AN IMPORTANT PLACE: CHINA, FEBRUARY 2020**
During the lockdown to control the diffusion of COVID-19, people from Wuhan were not permitted to cross the Jiujiang Yangtze River Bridge, which is located 226 kilometers (140 miles) from Wuhan.

Welcome to Geography

- **Summarize geography's five most basic concepts.**

The word *geography*, invented by the ancient Greek scholar Eratosthenes (ca. 276–ca. 194 B.C.E.), is based on two Greek words: *geo* means "Earth," and *graphy* means "to write." Geography is the study of where things are found on Earth's surface and the reasons for their locations. Human geographers ask two questions: Where are people and activities found on Earth? Why are they there?

1.1.1 Comparing Geography & History

Human geography and history both rely on the collection of evidence about human activity. In his framework of all scientific knowledge, the German philosopher Immanuel Kant (1724–1804) compared geography and history:

Geographers . . .	Historians . . .
identify the location of important places (Figure 1.1.1).	identify the dates of important events.
explain why one human activity is found near another.	explain why one human activity follows another chronologically.
ask where and why.	ask when and why.
organize material spatially.	organize material chronologically.
recogn ze that an action at one location on Earth can be a reaction to actions taken at another point on Earth, which can, in turn, affect conditions elsewhere.	recognize that an action at one point in time can be in reaction to actions taken in the past and can, in turn, affect future actions.

Contrasting Geography & History

The capabilities of historians and geographers differ in one especially important aspect: A geographer can take a plane or car to another place on Earth (Figure 1.1.2), but a historian cannot travel back to another time in the past. The ability to reach other places lends excitement to the discipline of geography.

To introduce human geography, we concentrate on two main features of society: culture and economy. The first half of the book explains why the most important cultural features, such as languages, religions, and ethnicities, are arranged as they are across Earth. The second half of the book looks at the distribution of the most important economic activities, including agriculture, manufacturing, and services.

▼ **1.1.2 WUHAN, 2018**
Before the start of the COVID-19 pandemic.

1.1.2 Geographers Explain *Where & Why*

This chapter introduces basic concepts in geography and explains how geographers use a question approach—"where" and "why"—in their study. To explain where things are, one of geography's most important tools is a map. Geographers employ several basic concepts to explain why every place on Earth is in some ways unique and in some ways related to other locations. Two basic concepts relate to unique location:

- A **place** is a specific point on Earth distinguished by a particular characteristic. Every place occupies a unique location, or position, on Earth's surface.
- A **region** is an area of Earth defined by one or more distinctive characteristics.

Three basic concepts help to explain the interrelatedness of locations:

- **Scale** is the relationship between the portion of Earth being studied and Earth as a whole. Geographers are increasingly concerned with the global scale.
- **Space** refers to the physical gap, or interval, between two objects. Geographers observe that many objects are distributed across space in a regular manner and for discernible reasons.
- **Connection** refers to relationships among people and objects across the barrier of space. Geographers are concerned with the various means by which connections occur and are especially interested in connections between human activities and the physical environment.

Geography's Basic Concepts & COVID-19

Geography's basic concepts can help with our understanding of where and why the COVID-19 disease spread or was contained: place (Figure 1.1.3), region (Figure 1.1.4), scale (Figure 1.1.5), space (Figure 1.1.6), and connection (Figure 1.1.7).

◀ **1.1.3 PLACE** Wuhan, China, was the place where COVID-19 was first identified, in December 2019, according to the U.S. Centers for Disease Control and Prevention. During 2020, customers shopped in Wuhan's street market while wearing masks.

▼ **1.1.4 REGION** The Harbei region, which includes Wuhan, had more cases of COVID-19 per population than China's other regions during 2020.

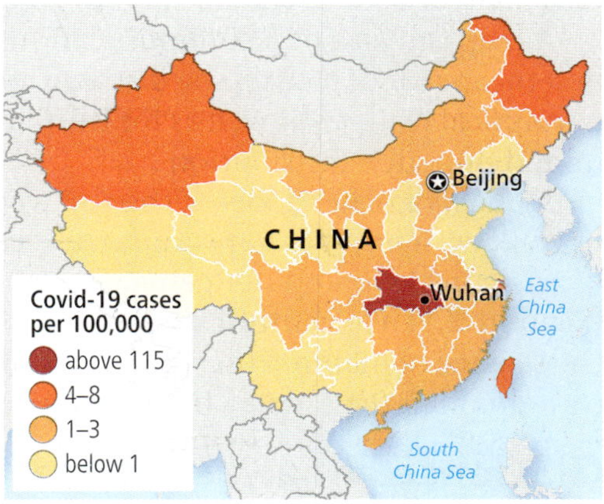

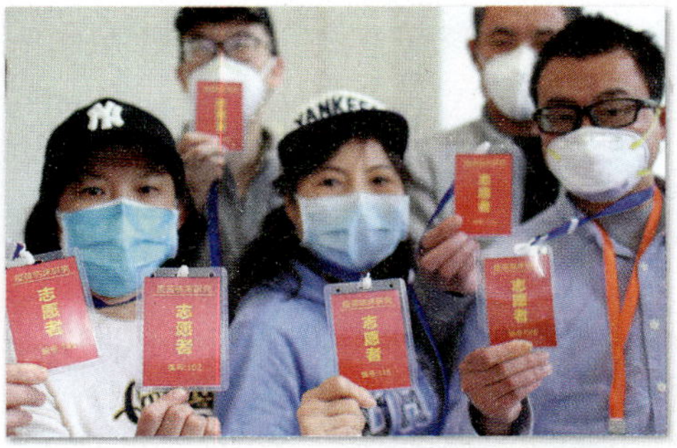

◀ **1.1.5 SCALE** These health-care workers in Wuhan were among the first in the world to receive COVID-19 vaccinations. China's vaccination cards look different than those issued in the United States by the Centers for Disease Control and Prevention.

◀ **1.1.6 SPACE** During COVID-19, a distance of 2 meters (6 feet) became the norm for safe social distance, as these workers eating lunch at a Honda factory in Wuhan demonstrate.

▼ **1.1.7 CONNECTION** (a) Australia quarantined people traveling to the country from Wuhan for 2 weeks in Christmas Island. (b) These Australian citizens returning from Wuhan arrive at Canberra Airport in Australia after their release from quarantine.

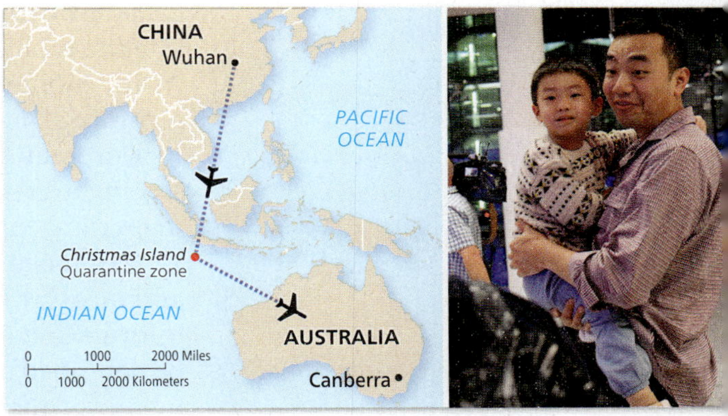

(a) (b)

Contemporary Geographic Tools

- Explain geography's contemporary analytic mapping tools.

Maps are not just paper references anymore. They have become an essential tool for online location-finding applications through smart phones, tablets, and computers.

1.2.1 What Is a Map?

Geography's most important tool for thinking spatially about the distribution of features across Earth is a map. A **map** is a two-dimensional or flat-scale model of Earth's surface or a portion of it. Geography is immediately distinguished from other disciplines by its reliance on maps to display and analyze information.

A map can serve as a reference tool, helping us find the shortest route between two places and avoid getting lost along the way. We consult maps to learn where in the world something is found, especially in relationship to a place we know, such as a town, body of water, or highway. The maps in an atlas or a road map are especially useful for this purpose.

A map can also be a communications tool. It is often the best means for depicting the distribution of human activities or physical features as well as for thinking about reasons underlying such distribution.

A map is a scale model of the real world, made small enough to work with on a desk or computer. It can be an elaborate work of art, a precise computer-generated product, or a hasty sketch (Figure 1.2.1). For centuries, geographers have worked to perfect the science of mapmaking, called **cartography**. Contemporary cartographers are assisted by computers and satellite imagery.

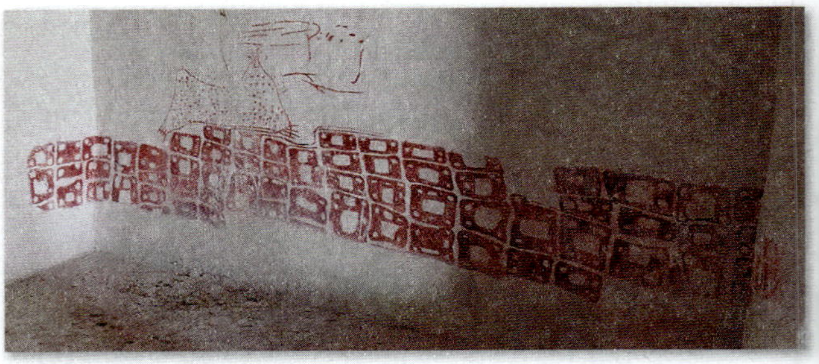

(a)

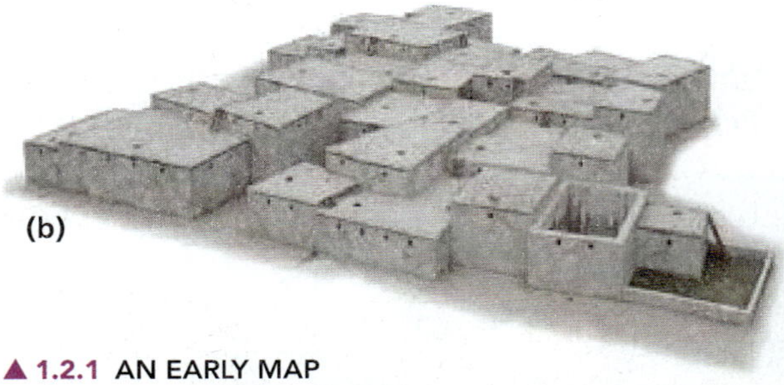

(b)

▲ 1.2.1 AN EARLY MAP
(a) This map, dating from 6200 B.C.E., depicts the town of Çatalhöyük, in present-day Turkey, and the eruption of the Hasan Dağı (Mount Hasan) twin-peaks volcano, which is actually located around 140 kilometers (87 miles) northeast of the town. Archaeological evidence indicates that the volcano did erupt around the time that the map was made. The map is now in the Konya Archaeology Museum. (b) Artist's impression of Çatalhöyük,created from details on the map.

GIScience: Analyzing Data

Geographic information science (GIScience) is analysis of data about Earth acquired through satellite and other electronic information technologies. A **geographic information system (GIS)** captures, stores, queries, and displays the geographic data. GIS produces maps (including those in this book) that are more accurate and attractive than those drawn by hand. Each type of information is stored in a layer. For example, separate layers could be created for boundaries of countries, bodies of water, roads, and names of places.

The science of taking measurements of Earth's surface from photographs is called **photogrammetry**. The acquisition of data about Earth's surface from a satellite orbiting Earth or from other long-distance methods is **remote sensing**. At any moment, an aerial sensor attached to a satellite, airplane, or drone may be recording the image of a tiny area on Earth's surface (Figure 1.2.2).

▼ 1.2.2 PHOTOGRAMMETRY
A drone flies a test course in Nevada to determine the precision of its mapping capabilities.

Corporations and government agencies use photogrammetry and remote sensing to create high-quality 3D virtual representations of portions of Earth. These maps can depict the distribution of a wide variety of urban and rural features.

GIScience helps geographers create more accurate and complex maps and measure changes over time in the characteristics of places. Layers of information acquired through remote sensing and produced through GIS can be described and analyzed. GIScience enables geographers to calculate whether relationships between objects on a map are significant or merely coincidental.

1.2.2 GPS: Pinpointing Locations

Our smart phones, tablets, and computers are equipped with **Global Positioning System (GPS)**, which is a system that determines the precise position of something on Earth. The GPS in use in the United States includes two dozen satellites placed in predetermined orbits; a series of tracking stations to monitor and control the satellites; and receivers that compute position, velocity, and time from the satellite signals.

GPS is most commonly used for navigation. Pilots of aircraft and ships stay on course with GPS. On land, GPS detects a vehicle's current position, the motorist programs the desired destination into a GPS device, and the device provides instructions on how to reach the destination.

Thanks to GPS, our electronic devices provide us with a wealth of information about the specific place on Earth we currently occupy. The locations to which we venture using mapping applications, as well as the photos we take with our electronic devices, are recorded through **geotagging**, which is identification and storage of a piece of information by its precise latitude and longitude coordinates (Figure 1.2.3).

▲ **1.2.3 GEOTAGGING: GOOGLE STREET MAPPING**
Ulan Bator, Mongolia.

VGI: Collecting & Sharing Data

Smart phones, tablets, and computers enable individuals to make maps and share them with others. **Volunteered geographic information (VGI)** is the creation and dissemination of geographic data contributed voluntarily and for free by individuals. VGI is part of the broader trend of **citizen science**, which is scientific research by amateur scientists, and **participatory GIS (PGIS)**, which is community-based mapping. Citizen science and PGIS collect and disseminate local knowledge and information through electronic devices.

OpenStreetMap is a collaborative project to create a free editable map of the world. The geodata underlying the map is considered the primary output of the project. Humanitarian OpenStreetMap Team (HOT) is a global community of volunteers, community leaders, and professionals who work together to create open map data to support sustainable development, such as where to build new housing for the least impact on the environment, and disaster response, such as where to send emergency medical teams during a disease outbreak (Figure 1.2.4).

▲ **1.2.4 HUMANITARIAN OPENSTREET MAPPING**
Assisting responders helping to reach Ebola victims in a rural area of Sierra Leone.

A **mashup** is a map that overlays data from one source on top of a map provided by a mapping service, such as Google Maps or Google Earth. The term *mashup* comes from the musical practice of mixing two or more songs.

Individuals can create mashups on their personal devices because mapping services provide access to the application programming interface (API), which is the computerized language that links a database such as an address list on your phone with software such as Google Maps.

Geocode System: What3words

What3words is a geocode system that has divided the entire world into 57 trillion squares of 3 meters (10 feet) per side. Each 9 square meter (100 square foot) square is identified by three English words. For example, the front door of the White House is identified by ///curve.empty.buzz. This app allows for finding the precise location of many places in the world that do not have traditional street addresses (Figure 1.2.5).

▶ **1.2.5 WHAT3WORDS**
Location of the Leaning Tower of Pisa, Italy.

Making Maps

- **Summarize the development of the science of geography.**

The science of geography developed in the ancient world, especially in the Eastern Mediterranean (mainly Greece and the Roman Empire), as well as in China. Little progress was made in development of geography until the first century C.E.

1.3.1 Geography in the Ancient World

Major contributors to geographic thought in the ancient Eastern Mediterranean included:

- Thales of Miletus (ca. 624–ca. 545 B.C.E.), who applied principles of geometry to measuring land area.
- Anaximander (610–ca. 546 B.C.E.), who made a world map shaped like a cylinder based on information from sailors.
- Pythagoras (ca. 570–ca. 495 B.C.E.), who may have been the first to propose a spherical world.
- Hecateus (ca. 550–ca. 476 B.C.E.), who may have produced the first geography book, called *Periodos Ges* (*Journey Around the Earth*).
- Aristotle (384–322 B.C.E.), who was the first to demonstrate the Earth was spherical.
- Eratosthenes (ca. 276–ca. 194 B.C.E.), the inventor of the word *geography*, who accepted that the Earth was round and calculated its circumference within 0.5 percent accuracy.
- Strabo (ca. 64 B.C.E.–ca. 24 C.E.), who described the known world in a 17-volume work titled *Geography*.
- Ptolemy (ca. 100–ca. 170 C.E.), who wrote the eight-volume *Guide to Geography* (Figure 1.3.1).

▲ **1.3.1 WORLD MAP BY PTOLEMY CA. 150 C.E.**
The map shows the known world at the height of the Roman Empire, surrounding the Mediterranean Sea and Indian Ocean.

Ancient Chinese geographic contributions included:

- "Yu Gong" ("Tribute of Yu"), by an unknown author, describing the economic resources of the country's different provinces.
- Pei Xiu (224–271 C.E.), the "father of Chinese cartography," who produced an elaborate map of the country.

Geography's Revival

After Ptolemy, little progress in mapmaking or geographic thought was made in Europe for several hundred years. Maps became less mathematical and more fanciful, showing Earth as a flat disk surrounded by fierce animals and monsters.

▲ **1.3.2 WORLD MAP BY WALDSEEMÜLLER, 1508**
The name *America* appears in very small print on the map.

Geographic inquiry continued, though, outside Europe. Contributors outside Europe included Muhammad al-Idrisi (1100–1165), a Muslim geographer who prepared a world map and geography text in 1154, and Ibn Battuta (1304–ca. 1368), a Moroccan scholar who wrote Rihla ("Travels") based on journeys through Africa, Asia, and Europe.

In Europe, making maps as reference tools revived during the Age of Exploration and Discovery. Columbus, Magellan, and other explorers who sailed across the oceans in search of trade routes and resources in the fifteenth and sixteenth centuries required accurate maps to reach desired destinations without wrecking their ships. Cartographers used information collected by the explorers to create more exact maps.

Influential European cartographers included Martin Waldseemüller (ca. 1470–1520, Figure 1.3.2), Abraham Ortelius (1527–1598), and Bernhardus Varenius (1622–1650).

1.3.2 Types of Maps

Maps can represent the distribution of data for one or more variables in a variety of ways. Figures 1.3.3 through 1.3.7 show five ways to display the distribution of data gathered on the COVID-19 pandemic in the United States.

A **choropleth map** is a map where recognizable areas are shaded or patterned in proportion to the data collected. For example, areas with darker colors may represent the highest range of the data (Figure 1.3.3).

This map is way more statistical than others

A **graduated symbol map** displays a graphic multiple times that changes in size according to the value of data. A higher value is typically represented by the graphic at a larger scale (Figure 1.3.4).

A **dot distribution map** depicts data as points and shows how those points are clustered together or spread out over an area. Each dot represents a predetermined number of observations, which could be one or many (Figure 1.3.5).

The more data there is, the bigger the map will get.

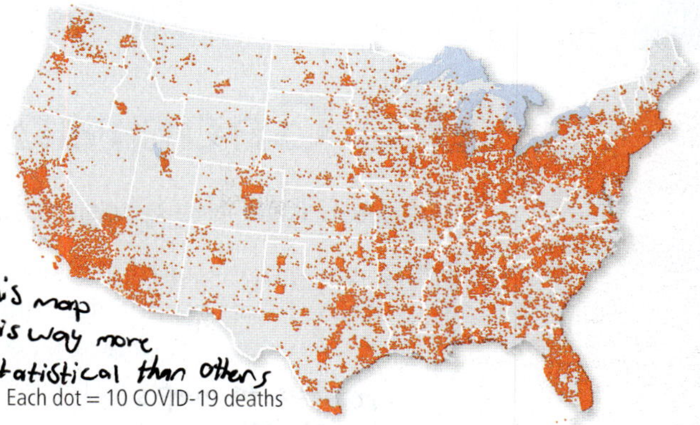

▲ **1.3.5 DOT DISTRIBUTION MAP**
More dots clustered in a region mean more deaths there attributed to COVID-19, March 2021.

An **isoline map** connects with lines all the places that have particular values (Figure 1.3.6).

A **cartogram** is a map in which the size of a territory, such as a U.S. state, is proportional to the value of particular data (Figure 1.3.7).

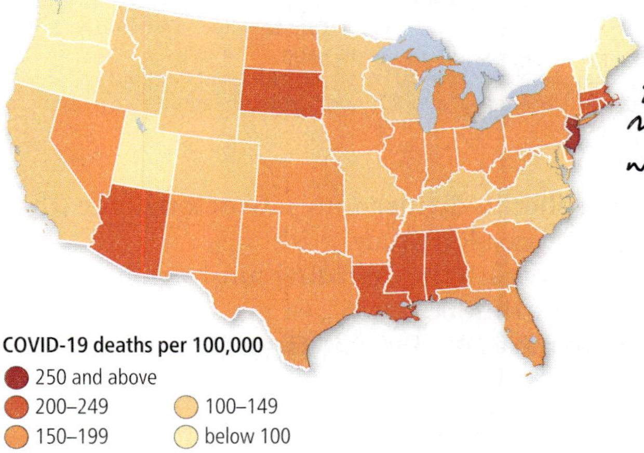

▲ **1.3.3 CHOROPLETH MAP**
The darker the state, the more deaths attributed to COVID-19 per capita, January 2020 through April 2021.

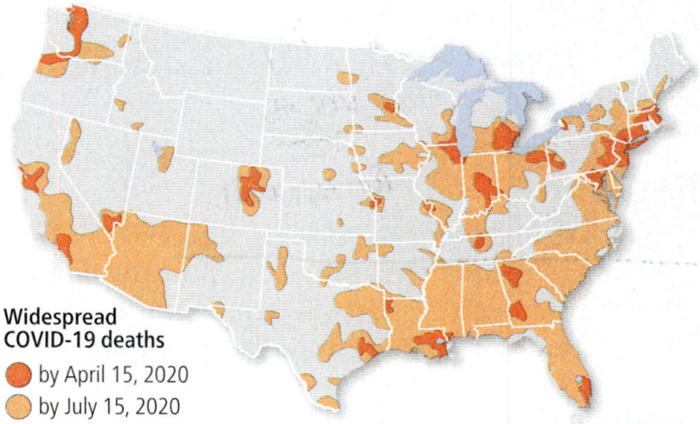

▲ **1.3.6 ISOLINE MAP**
The regions in the darker colors experienced COVID-19 deaths earlier than did the regions in the lighter colors.

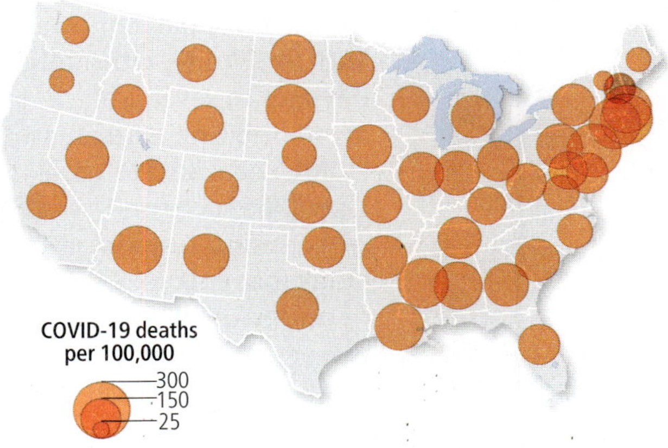

▲ **1.3.4 GRADUATED SYMBOL MAP**
The larger the circle, the more deaths attributed to COVID-19 per capita, January 2020 through April 2021.

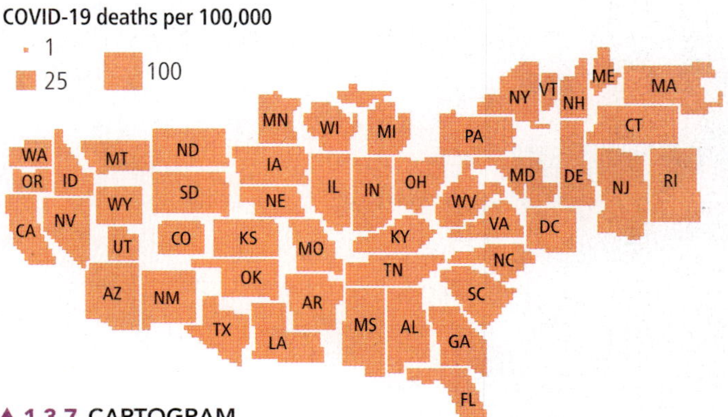

▲ **1.3.7 CARTOGRAM**
The larger the area of the state, the more deaths attributed to COVID-19 per capita, April 2021.

The Geographic Grid

- **Explain how the geographic grid determines location and time of day.**

The geographic grid is a system of arcs drawn in a pattern on a map of Earth's surface to represent the imaginary lines of latitude and longitude. The geographic grid plays an important role in telling time.

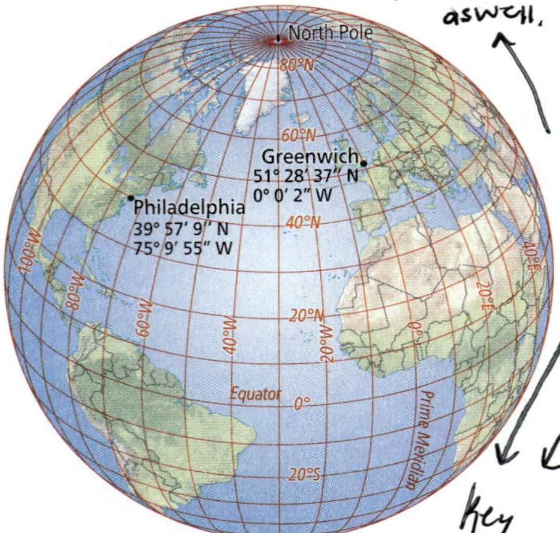

1.4.1 Longitude & Latitude

The location of any place on Earth's surface can be described by these human-imagined arcs and circles, known as meridians and parallels, respectively, that form a grid pattern on Earth's surface:

- A **meridian** is an arc drawn between the North and South poles. The location of each meridian is identified on Earth's surface according to a numbering system known as **longitude**.
- A **parallel** is a circle drawn around the globe parallel to the equator and at right angles to the meridians. The numbering system to indicate the location of a parallel is called **latitude** (Figure 1.4.1).

◀ 1.4.1 GEOGRAPHIC GRID
The meridian that passes through the Royal Observatory at Greenwich, England, is 0° longitude, and the meridian on the opposite side of the globe from 0° is 180° longitude. A meridian is numbered between 0° and 180° east or west longitude, depending on whether it is east or west of 0°. The equator is 0° latitude, the North Pole 90° north latitude, and the South Pole 90° south latitude. A parallel is numbered between 0° and 90° north or south latitude, depending on whether it is north or south of the equator.

Telling Time

Longitude is the basis for calculating time (Figure 1.4.2). Earth as a sphere is divided into 360° of longitude (the degrees from 0° to 180° west longitude plus the degrees from 0° to 180° east longitude).

As Earth rotates daily, these 360 imaginary arcs of longitude pass beneath the cascading sunshine. If we let every fifteenth degree of longitude represent one time zone, and divide the 360° by 15°, we get 24 time zones, or one for each hour of the day. By international agreement, **Coordinated Universal Time (UTC)**, known informally as Greenwich Mean Time (GMT), which is the time at the **prime meridian** (0° longitude), is the master reference time for all points on Earth.

Each 15° band of longitude is assigned to a standard time zone. The western United States, which is near 120° west longitude, is therefore 8 hours earlier than UTC (the 120° difference between the prime meridian and 120° west longitude, divided by 15° per hour, equals 8 hours).

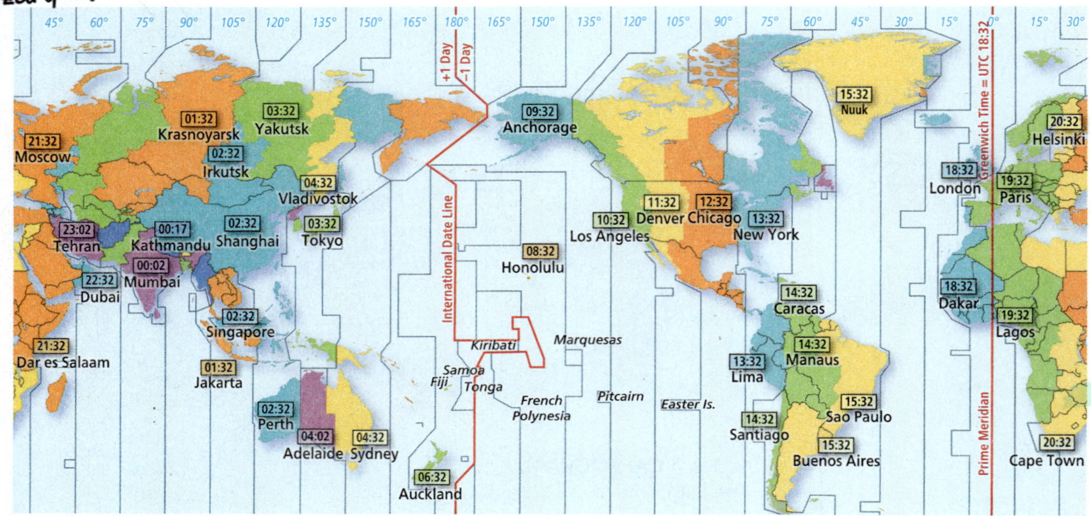

The **International Date Line** is an arc that for the most part follows 180° longitude, although it deviates in several places to avoid dividing land areas. When the International Date Line is crossed heading east (toward America), the clock moves back 24 hours, or one entire day. When it is crossed heading west (toward Asia), the calendar moves ahead one day.

◀ 1.4.2 TIME ZONES

1.4.2 Scale & Projection

To make a map, a cartographer must make two decisions: scale and projection.

Map Scale

Should a map show the entire globe, or a country, or a city? To make a scale model of the entire world, many details must be omitted because of lack of space. Conversely, a map showing only a small portion of Earth's surface can provide a wealth of detail about a particular place. The level of detail and the amount of area covered on a map depend on its **map scale**, which is the relationship of a feature's size on a map to its actual size on Earth. Map scale is presented in three ways (Figure 1.4.3):

▲ **1.4.3 MAP SCALE**
Sun City, Arizona.

- **Ratio.** A ratio or fraction shows a numerical relationship between distances on the map and Earth's surface. A scale of 1:1,000,000 means that 1 unit (such as inch or centimeter) on the map represents 1 million of the same unit on the ground.
- **Written.** A written scale describes the relationship between map and Earth distances in words. For example, in the statement "1 centimeter equals 10 kilometers," the first number refers to the map distance and the second to the actual distance on Earth's surface.
- **Graphic.** A graphic scale usually consists of a bar line on a map to represent distance on Earth's surface. To use a bar line, first determine with a ruler the distance on the map in inches or centimeters from one place to another. Then hold the ruler against the bar line and read the number on the bar line opposite the map distance on the ruler. The number on the bar line is the equivalent distance on Earth's surface.

Projection

Earth's spherical shape poses a challenge for cartographers because drawing Earth on a flat piece of paper unavoidably produces some distortion. Cartographers have invented hundreds of clever methods of producing flat maps, but none produces perfect results. The scientific method of transferring locations on Earth's surface to a flat map is called **projection**. Figures 1.4.2, 1.4.4, 1.4.5, and 1.4.6 display four projections. Four types of distortion can result with projection:

1. The ***shape*** of an area can appear more elongated or squat than in reality
2. The ***distance*** between two points may become increased or decreased.
3. The ***relative size*** of different areas may be altered, so that one area may appear larger than another on a map but is in reality smaller.
4. The ***direction*** from one place to another can be askew.

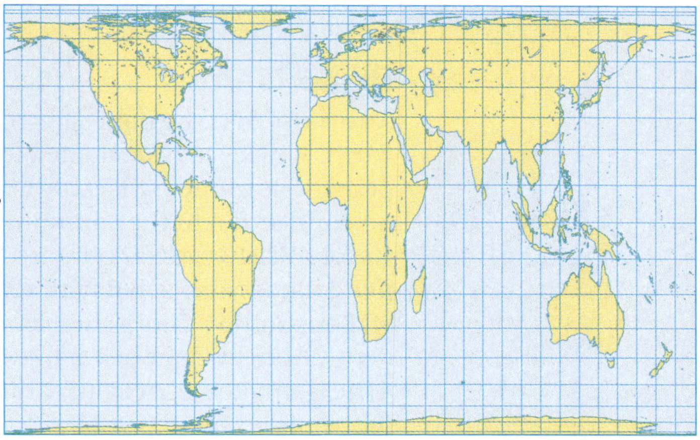

▲ **1.4.4 GALL-PETERS PROJECTION**
This projection distorts shape but not size. In contrast, the Mercator projection (Figure 1.4.2) has little distortion of shape but distorts size. In a Mercator projection, areas near the poles appear much larger than they actually are.

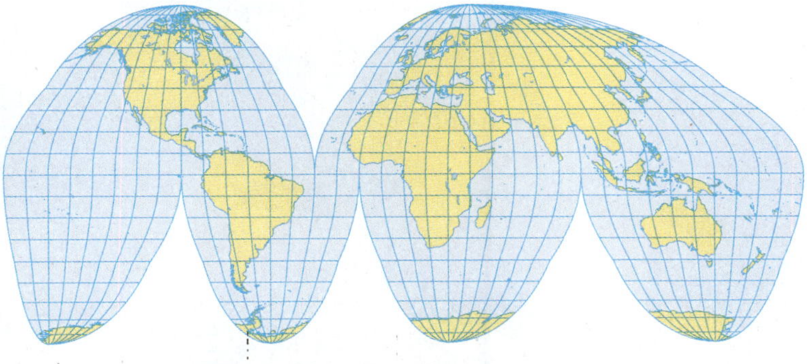

▲ **1.4.5 GOODE HOMOLOSINE PROJECTION**
This separates the hemispheres into two pieces. The meridians do not converge at the poles as they do in reality. Also, they do not form right angles with the parallels.

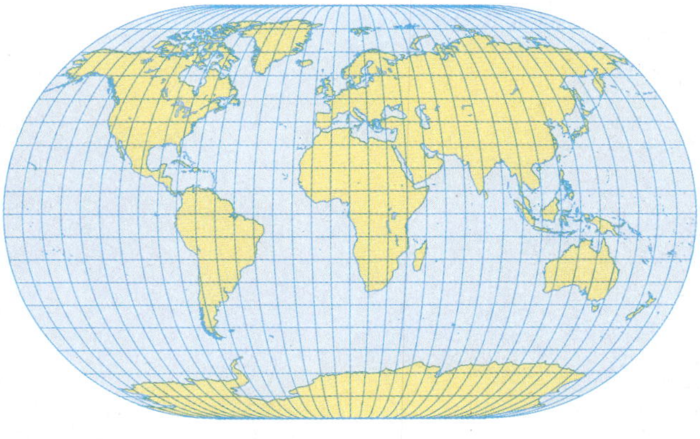

▲ **1.4.6 WINKEL PROJECTION**
The relative sizes of the land masses on the map are the same as in reality. However, by allocating space to the oceans, the land areas are much smaller than on interrupted maps of the same size.

Place: A Unique Location

- **Summarize the distinctive features of a place.**

Humans possess a strong sense of place—that is, a feeling for the features that contribute to the distinctiveness of a particular spot on Earth—perhaps a hometown, vacation destination, or college. Describing the features of a place is an essential building block for geographers to explain similarities, differences, and changes across Earth.

1.5.1 Location

Geographers describe a feature's place on Earth by identifying its **location**, which is the position that something occupies on Earth's surface. In doing so, they consider three ways to identify location: place name, site, and situation. Geographers analyze where particular places are located and the combination of features that make each place on Earth distinct.

Because all inhabited places on Earth's surface—and many uninhabited places—have been named, the most straightforward way to describe a particular location is often by referring to its place name.

Place Names

A **toponym** is the name given to a place on Earth. A place may be named for a person, perhaps its founder or a famous person with no connection to the community, such as George Washington. Some settlers selected place names associated with religion, such as St. Louis and St. Paul, whereas other names derive from ancient history, such as Athens, Attica, and Rome, or from earlier occupants of the place (Figure 1.5.1).

Names can be controversial. This book utilizes the name Czechia rather than Czech Republic in accordance with recent Czech government preferences. The Board of Geographical Names, operated by the U.S. Geological Survey, was established in the late nineteenth century to be the final arbiter of names on U.S. maps. In recent years, the board has been especially concerned with removing offensive place names, such as those with racial or ethnic connotations.

▼ 1.5.1 WORLD'S LONGEST PLACE NAME
This place in New Zealand is recognized as the world's longest one-word place name. It translates from the Maori language as "The summit where Tamatea, the man with the big knees, the climber of mountains, the land-swallower who travelled about, played his nose flute to his loved one."

1.5.2 Site & Situation

Geographers also describe the location of a place through the concepts of site and situation. **Site** is the physical character of a place. **Situation** is the location of a place relative to other places.

▲ **1.5.2 CHANGING SITE OF BOSTON**
The site of Boston has been altered by filling in much of Boston Harbor. Colonial Boston was a peninsula connected to the mainland by a very narrow neck. During the nineteenth century, a dozen major projects filled in most of the bays, coves, and marshes. A major twentieth-century landfill project created Logan Airport. Several landfill projects in the Boston area continue into the twenty-first century.

Site

Important site characteristics include climate, water sources, topography, soil, vegetation, latitude, and elevation. The combination of physical features gives each place a distinctive character.

Site factors have always been essential in selecting locations for settlements, although people have disagreed on the attributes of a good site, depending on cultural values. Some have preferred a hilltop site for easy defense from attack. Others have located settlements near convenient river-crossing points to facilitate communication with people in other places. Humans have the ability to modify the characteristics of a site. Central Boston is more than twice as large today as it was during colonial times (Figure 1.5.2).

Situation

Situation is a valuable way to indicate location, for two reasons:

- Situation helps us find an unfamiliar place by comparing its location with a familiar one. We give directions to people by referring to the situation of a place: "It's down past the courthouse, beside the large elm tree."
- Situation helps us understand the importance of a location. Many places are important because they are accessible to other places. For example, because of its situation, Cairo has become a center for the trading of goods and culture among Africa, Europe, and Asia (Figure 1.5.3).

▼▶ **1.5.3 SITUATION OF CAIRO, EGYPT**
Cairo is situated along the Nile River, the world's longest river, which flows north through much of Africa, and 100 kilometers (60 miles) west of the Suez Canal, which provides ships with passage between Europe and Asia. (a) Satellite image of the situation of Cairo. (b) The Nile River flows through central Cairo.

▲ **1.6.1 FORMAL REGION: U.S. CORN BELT**
Zearing, Illinois.

Region: A Unique Area

- **Identify three types of regions.**

The sense of place that humans possess may apply to a larger area of Earth rather than to a specific point. An area of Earth defined by one or more distinctive characteristics is a region. People, activities, and environment display similarities and regularities within a region and differ in some way from those of other regions.

1.6.1 Three Types of Regions

A region gains uniqueness from the **cultural landscape**, which is a combination of cultural features such as language and religion, economic features such as agriculture and industry, and physical features such as climate and vegetation, possessing not a single human or environmental characteristic but a combination of them. The designation of *region* can be applied to any area larger than a point and smaller than the entire planet.

Geographers most often define a region at one of two scales: (1) Several neighboring countries that share important features, such as the countries in Latin America that share languages and religions; (2) Localities within a country, such as southern California or the corn belt (Figure 1.6.1). Geographers identify three types of regions: formal, functional, and vernacular.

Formal Region

A **formal region**, also called a **uniform region**, is an area within which most people share one or more distinctive characteristics. The shared feature could be a cultural value such as a common language, an economic activity such as production of a particular crop, or an environmental feature such as climate. In a formal region, the selected characteristic is present throughout it.

Some formal regions are easy to identify, such as countries or U.S. states. Montana is an example of a formal region, characterized with equal intensity throughout the state by a government that passes laws, collects taxes, and issues license plates. The formal region of Montana has precise and legally recognized boundaries, and everyone living within them shares the status of being subject to a common set of laws.

Formal regions smaller than states can be found across the United States. For example, the U.S. Land Ordinance of 1785 divided much of the country into townships, 6 miles on each side. A township, in turn, was divided into 36 sections, 1 mile on each side (Figure 1.6.2 and refer ahead to Figure 1.8.1).

In identifying a formal region, it is important to recognize the diversity of cultural, economic, and environmental features, even while making a generalization. A minority of people in a region may speak a language, practice a religion, or use resources different from those of the majority. People in a region may play distinctive roles in the economy and hold different positions in society based on their gender or ethnicity.

▶ **1.6.2 FORMAL REGIONS**
(a) The State of Ohio,
(b) Preble County,
(c) Israel Township.

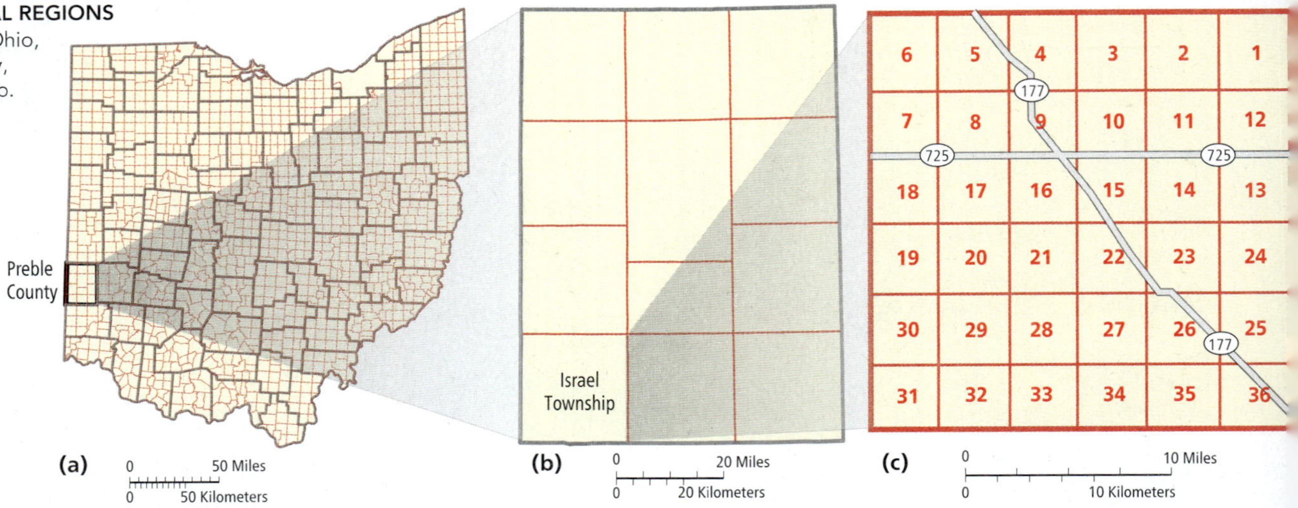

1.6.2 Functional & Vernacular Regions

In addition to formal regions, two other types of regions are functional and vernacular. A **functional region**, also called a **nodal region**, is an area organized around a node or focal point. A **vernacular region**, or **perceptual region**, is an area that people believe exists as part of their cultural identity.

Functional Region

The feature chosen to define a functional region dominates at a central focus or node and diminishes in importance outward. The region is tied to the central point by transportation or communications systems or by economic associations.

Geographers often use functional regions to relate information about the economics of an area. A region's node may be a shop or service, with the boundaries of the region marking the limits of its trading area. People and activities may be attracted to the node, and information may flow from the node to the surrounding area.

An example of a functional region is the area of reception of broadcast signals from a TV station (Figure 1.6.3). At the center of its service area, a TV station's over-the-air signal is strongest, and the percentage of cable and satellite dish receivers tuned to that station is highest. At some distance from the center, more people are watching a different station originating in another city. That place is the boundary between the nodal regions of the two TV market areas.

Social media offer numerous examples of functional regions. Although social media can be used to interact equally with someone nearby and someone far away, in reality most interaction is with people nearby and fewer with people far away. Nonetheless, new technology can break down some traditional functional regions. Through the internet, customers can shop at distant stores, and newspapers published in a particular city are delivered not only to local customers but also to people elsewhere.

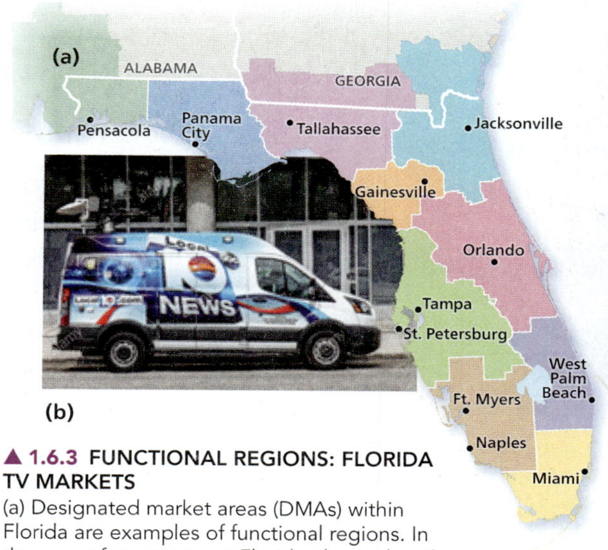

▲ **1.6.3 FUNCTIONAL REGIONS: FLORIDA TV MARKETS**
(a) Designated market areas (DMAs) within Florida are examples of functional regions. In the case of westernmost Florida, the node—the TV station—is in an adjacent state (Mobile, Alabama). The state of Florida is an example of a formal region. (b) The Miami ABC affiliate.

Vernacular Region

Vernacular regions emerge from people's informal sense of place rather than from strictly physical features. An example of a vernacular region is the U.S. Midwest (Figure 1.6.4). However, the actual location of the Midwest depends on the feature being mapped (Figure 1.6.5). A useful way to identify a perceptual region is to get a description of the region from the people living in it and to draw a **mental map**, which is based on a person's impressions of what is in the place.

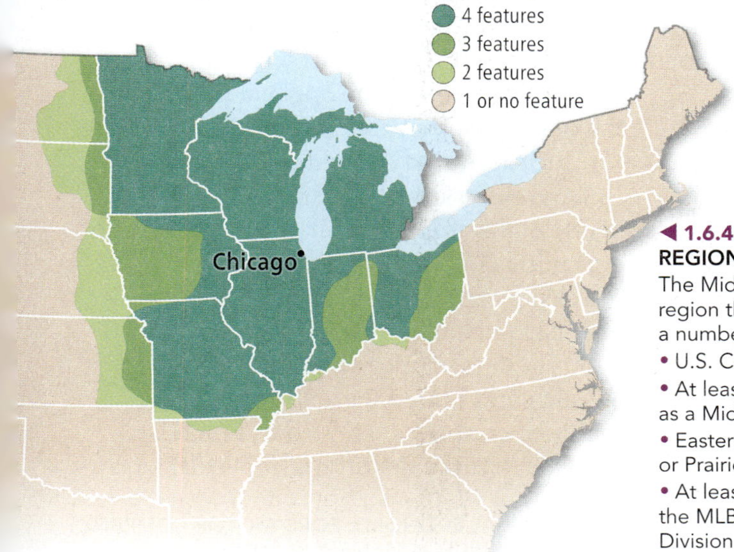

◄ **1.6.4 VERNACULAR REGION: THE MIDWEST**
The Midwest is a vernacular region that can be defined in a number of ways:
- U.S. Census definition
- At least 50% self-identify as a Midwesterner
- Eastern Broadleaf forest or Prairie parkland ecosystem
- At least 60% "like" a team in the MLB A.L. or N.L. Central Division

▼ **1.6.5 VERNACULAR REGION: "CHICAGOLAND"**
A feature of the Midwest vernacular region is a large number of fans of the Chicago Cubs.

▲ **1.7.1 GLOBAL CULTURE: JEANS AND MCDONALD'S** Shanghai, China.

Scale: From Global to Local

- **Compare local and global scales in economy and culture.**

Scale was defined at the beginning of the chapter as the relationship between the portion of Earth being studied and Earth as a whole. Geographers think about scale at many levels from local to global.

1.7.1 Globalization of Culture & the Economy

Although geographers study every scale from the individual to the entire Earth, increasingly they are concerned with global-scale patterns and processes. **Globalization** refers to actions or processes that involve the entire world and result in making something worldwide in scope. Geographers observe processes fostering globalization of the economy and globalization of culture (Figure 1.7.1).

Globalization of Culture

Geographers observe that increasingly uniform cultural preferences produce uniform "global" landscapes of material artifacts and of cultural values. Fast-food restaurants, service stations, and retail chains deliberately create a visual appearance that varies among locations as little as possible. That way, customers know what to expect, regardless of where in the world they happen to be.

As more people become aware of elements of global culture and aspire to possess them, local cultural beliefs, forms, and traits are threatened with extinction. The survival of a local culture's distinctive beliefs, forms, and traits may be threatened by interaction with such social customs as wearing jeans and Nike shoes, consuming Coca-Cola and McDonald's hamburgers, and communicating using cell phones and computers.

Globalization of the Economy

A few people living in very remote regions of the world may be able to obtain all of their daily necessities within that region. But most economic activities undertaken in one region are influenced by interaction with decision makers in other regions.

Globalization of the economy has been led primarily by transnational corporations, sometimes called multinational corporations (Figure 1.7.2). A **transnational corporation** conducts research, operates factories, and sells products in many countries, not just where its headquarters and principal shareholders are located.

Every place in the world is part of the global economy, but globalization has led to more specialization at the local level. Each place plays a distinctive role in the global economy, based on its local assets, as assessed by transnational corporations. A locality may be especially suitable for a transnational corporation to conduct research, to develop new engineering systems, to extract raw materials, to produce parts, to store finished products, to sell them, or to manage operations. In a global economy, transnational corporations remain competitive by correctly identifying the optimal location for each of these activities. Factories are closed in some locations and opened in others.

Geographers divide the world into regions of developed countries and regions of developing countries. Various shared features—such as per capita income, level of education, and life expectancy—distinguish developed regions and developing regions. Possession of wealth and material goods is higher in developed countries. These differences are discussed in more detail in Chapter 10.

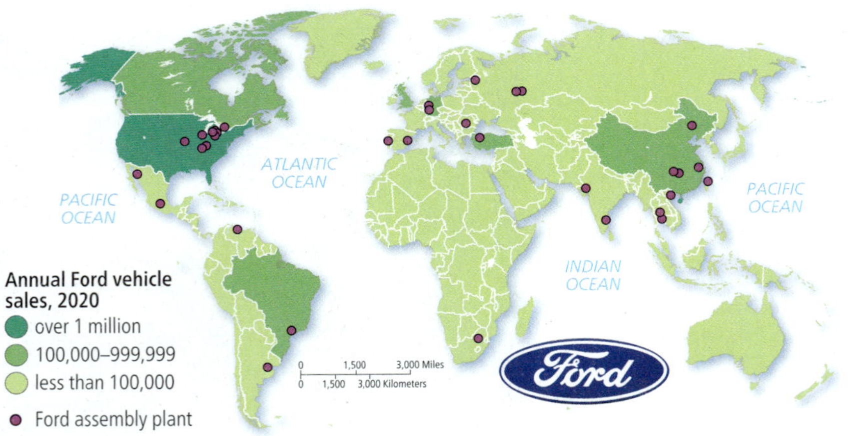

◀ **1.7.2 GLOBAL ECONOMY: FORD MOTOR COMPANY SALES AND VEHICLE ASSEMBLY PLANTS**

1.7.2 Local Diversity

▲ 1.7.3 TOGETHER APART
Each watching something different.

Globalization has not destroyed the uniqueness of local culture and economy. Human geographers understand that many contemporary social problems result from a tension between forces promoting global culture and global economy on the one hand and preservation of local economic autonomy and local cultural traditions on the other hand. Strong adverse reaction to globalization by many people in both developed and developing countries has inspired efforts to restore local cultural traditions and sever economic relations with other places. Lawmakers in countries and regions within countries disagree sharply on the appropriate policies to improve economic conditions.

Cultural differences within regions not only persist but actually flourish in many cases. The communications revolution that promotes globalization of culture also permits preservation of cultural diversity. TV, for example, was once limited to a handful of channels displaying one set of cultural values. With the distribution of programming through cable, satellite, the internet, and streaming services, people now can choose from an infinite number of programs in many languages. With the globalization of communications, people in two distant places can watch the same TV program. At the same time, with the fragmentation of the broadcasting market, two people in the same house can watch different programs (Figure 1.7.3). Groups of people on every continent may aspire to wear jeans, but unique clothing is still worn in regions of the world. In a global culture, companies can target groups of consumers with similar tastes in different parts of the world.

Spatial Association

A region gains meaning through its unique combination of features. The presence of some of these features may be coincidental, but others are related to each other. **Spatial association** occurs within a region if the distribution of one feature is related to the distribution of another feature. Spatial association is strong if two features have very similar distributions, and spatial association is weak if two features have very different distributions.

To illustrate spatial association, figures 1.7.4 and 1.7.5 display the distribution of two features related to the opioid epidemic crisis in the United States. Is the spatial association between the two maps strong or weak?

Figure 1.7.4 shows the distribution of overdose deaths related to opioids. Though most opioid-related deaths are in heavily populated urban areas, the highest per capita rates within the United States are in the Appalachians and the southwest. Figure 1.7.5 shows the distribution of prescriptions written for opioids. Prescription rates vary widely by state and county—three times higher in Alabama than in Hawaii, for example. The distribution of places with relatively high prescription rates is similar to the distribution of deaths from overdose. The two maps together show a spatial association between overdoses and prescriptions for opioids.

Health issues that cause people pain do not vary much from place to place and do not explain this variability in prescribing. So the variation may be related to regional differences among doctors concerning how much pain medication to prescribe.

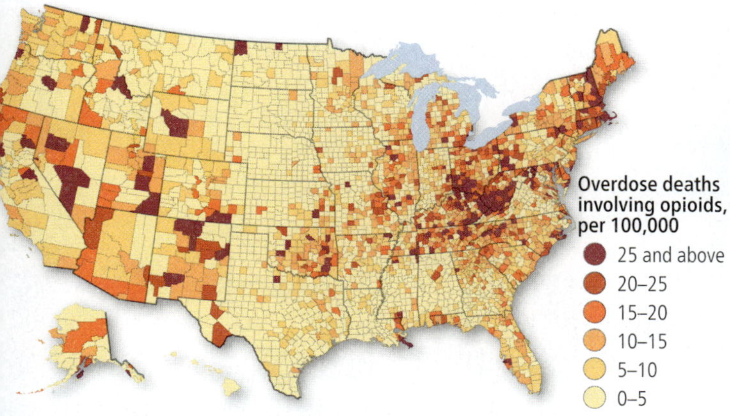

▲ 1.7.4 DEATHS INVOLVING OPIOIDS

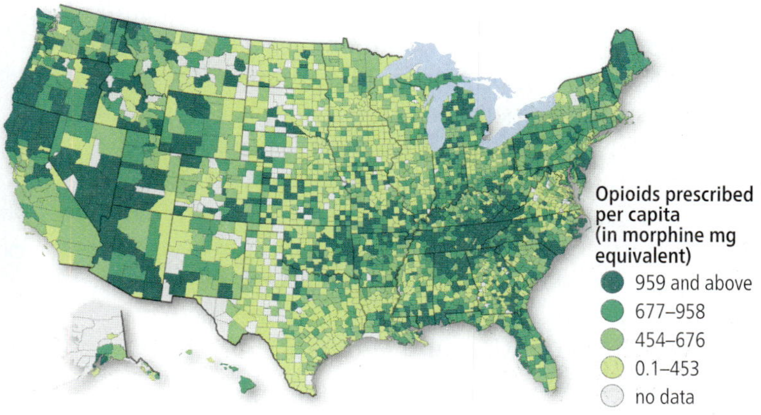

▲ 1.7.5 OPIOID PRESCRIPTIONS PER CAPITA

Space: Distribution of Features

- **Explain three properties of distribution.**

Space refers to the physical gap or interval between two objects. Geographers observe that many objects are distributed across space in a regular manner, for discernible reasons. Geographers think about the arrangement of people and activities found in space and try to understand why those people and activities are distributed across space as they are.

1.8.1 Distribution Properties: Density

Geographers explain how features such as buildings and communities are arranged across Earth. On Earth as a whole or within an area of Earth, features may be numerous or scarce, close together or far apart. The arrangement of a feature in space is known as its **distribution**. Geographers identify three main properties of distribution across Earth—density, concentration, and pattern.

Density is the frequency with which something occurs in space. The feature being measured could be people, houses, cars, trees, or anything else. The area could be measured in square kilometers, square miles, hectares, acres, or any other unit of area.

Remember that a large number of a feature does not necessarily lead to a high density. Density involves two measures—the number of a feature and the land area. China is the country with the largest number of people—approximately 1.4 billion—but it does not have the world's highest density. The Netherlands, for example, has only 17 million people, but its density of 508 persons per square kilometer is much higher than China's 144 persons per square kilometer. China may have a larger population, but it also has a much larger land area of 9.6 million square kilometers, compared to only 42,000 square kilometers for the Netherlands.

Distribution Properties: Concentration

The extent of a feature's spread over space is its **concentration**. If the objects in an area are close together, they are clustered; if relatively far apart, they are dispersed. To compare the level of concentration most clearly, two areas need to have the same number of objects and the same size area.

Geographers use concentration to describe changes in distribution. For example, the distribution of people across the United States is increasingly dispersed. The total number of people living in the United States is growing slowly—less than 1 percent per year—and the land area is essentially unchanged. But the population distribution is changing from relatively clustered in the Northeast to more evenly dispersed across the country.

Distribution Properties: Pattern

The third property of distribution is **pattern**, which is the geometric arrangement of objects in space. Some features are organized in a geometric pattern, whereas others are distributed irregularly. Geographers observe that many objects form a linear distribution, such as the arrangement of houses along a street or stations along a subway line.

Objects are frequently arranged in a square or rectangular pattern. Many American cities contain a regular pattern of streets, known as a grid pattern, which intersect at right angles at uniform intervals to form square or rectangular blocks. The system of townships, ranges, and sections established by the Land Ordinance of 1785 is another example of a square or grid pattern (Figure 1.8.1 and refer to Figure 1.6.2).

▼ 1.8.1 PATTERN: TOWNSHIP AND RANGE
The U.S. Land Ordinance of 1785 divided much of the United States into a checkerboard pattern, which is still visible in agricultural areas.

1.8.2 Distribution Properties

Concentration is not the same as density. Two neighborhoods could have the same density of housing but different concentrations. In a dispersed neighborhood, each house has a large private yard, whereas in a clustered neighborhood, the houses are close together, and the open space is shared as a community park.

Distribution: Social Distance

Understanding the difference between density and concentration became essential during the COVID-19 pandemic. Look back and remember a room you were in at a social gathering prior to the pandemic (Figure 1.8.2a). How many people were in the room? That number divided by the size of the room was the density. Were people spread evenly throughout the room, or were they huddled together in one part of the room? That was the concentration—either clustered or dispersed.

To stem the spread of a contagious, airborne disease such as COVID-19, the number of people in the same space must be limited—that is, the density reduced (Figure 1.8.2b). And to stop a COVID-19 outbreak from spreading, people must maintain a distance of 2 meters (6 feet) from each other—that is, the concentration must be dispersed (Figure 1.8.2c). In other words, people who practice social distancing, also called physical distancing, are keeping a safe space between each other in order to reduce the risk of spreading the highly contagious disease.

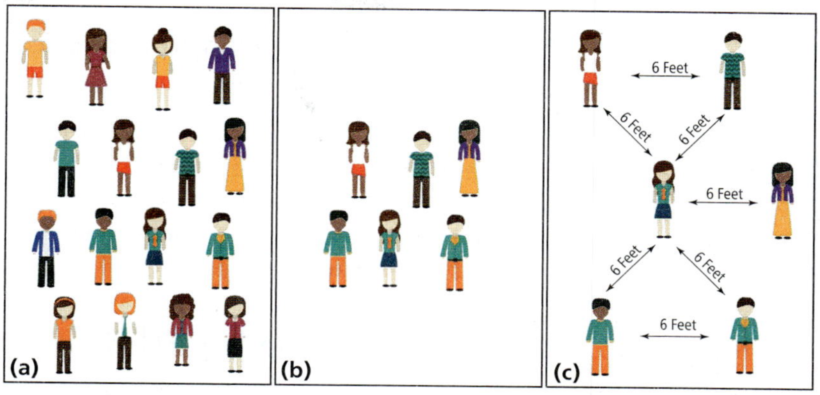

▲ **1.8.2 DISTRIBUTION OF PEOPLE DURING COVID-19 PANDEMIC**
(a) High density (many people in a room) and high clustering (people close together) was unsafe. (b) Lowering the density was safer, but the high clustering was still unsafe. (c) Lowering the density and dispersing the concentration was safer.

Distribution: Major League Baseball

The distribution of major league baseball teams illustrates the difference between density and concentration (Figure 1.8.3). After remaining unchanged during the first half of the twentieth century, the distribution of major league baseball teams changed during the second half of the twentieth century.

◀ **1.8.3 DISTRIBUTION OF BASEBALL TEAMS**
The changing distribution of North American baseball teams illustrates the difference between density and concentration.

These 6 teams moved to other cities during the 1950s and 1960s:

- Braves—Boston to Milwaukee in 1953, then to Atlanta in 1966
- Browns—St. Louis to Baltimore (Orioles) in 1954
- Athletics—Philadelphia to Kansas City in 1955, then to Oakland in 1968
- Dodgers—Brooklyn to Los Angeles in 1958
- Giants—New York to San Francisco in 1958
- Senators—Washington to Minneapolis (Minnesota Twins) in 1961

These 14 teams were added between the 1960s and 1990s:

- Angels—Los Angeles in 1961, then to Anaheim (California) in 1965
- Senators—Washington in 1961, then to Dallas (Texas Rangers) in 1971
- Mets—New York in 1962
- Astros—Houston (originally Colt .45s) in 1962
- Royals—Kansas City in 1969
- Padres—San Diego in 1969
- Expos—Montreal in 1969, then to Washington (Nationals) in 2005
- Pilots—Seattle in 1969, then to Milwaukee (Brewers) in 1970
- Blue Jays—Toronto in 1977
- Mariners—Seattle in 1977
- Marlins—Miami (originally Florida) in 1993
- Rockies—Denver (Colorado) in 1993
- Rays—Tampa Bay (originally Devil Rays) in 1998
- Diamondbacks—Phoenix (Arizona) in 1998

As a result of these relocations and additions, the density of teams increased, and the distribution became more dispersed.

Space: Identity

- **Discuss geographic approaches to cultural identity.**

Patterns in space vary according to gender, ethnicity, and sexuality. These are examples of social constructs (or construction), which are ideas that have been created and widely accepted by the people in a society. A social construct is an idea that would be seen as natural by the society but may or may not represent a reality shared by those outside the society. These attitudes vary from society to society and across time and can change.

1.9.1 Distribution by Ethnicity

The distribution of races and ethnicities in the United States varies considerably at all scales (refer to Chapter 7). Geographers are especially interested in ethnic identity because ethnicity is closely tied to the cultural traditions of a particular location on Earth, as discussed in more detail in Chapter 7. That chapter also discusses how the distribution of ethnicities in the United States varies considerably at all scales and how, for historic reasons, Blacks are clustered in the Southeast, whereas Hispanics are clustered in the Southwest.

Race, like ethnicity, is a social construct, but the two have different meanings. Ethnic identity is tied to a particular place, whereas racial identity is a social construct built around perception of a physiological trait, such as skin color. However, some cultures have further assigned racial identity a geographic feature through legalizing spatial segregation and other discriminatory practices.

▲ 1.9.1 BLACK LIVES MATTER

Black Lives Matter

Black Lives Matter is a movement that campaigns to stop violence and perceived discrimination against Blacks and educates others about the challenges that Black Americans face in the United States.

In Minneapolis during 2020, Black Lives Matter protests occurred in reaction to the death that year of George Floyd, a 46-year-old Black man (Figure 1.9.1). Floyd was killed during an arrest, after Minneapolis Police Department officer Derek Chauvin knelt on Floyd's neck for nearly eight minutes as three other officers looked on and prevented passersby from intervening.

The Twin Cities (Minneapolis and Saint Paul), like many U.S. urban areas, are characterized by intense degrees of clustering concentration based on racial identity. For example, Blacks are highly clustered in the northwest side of Minneapolis and the east side of St. Paul (Figure 1.9.2).

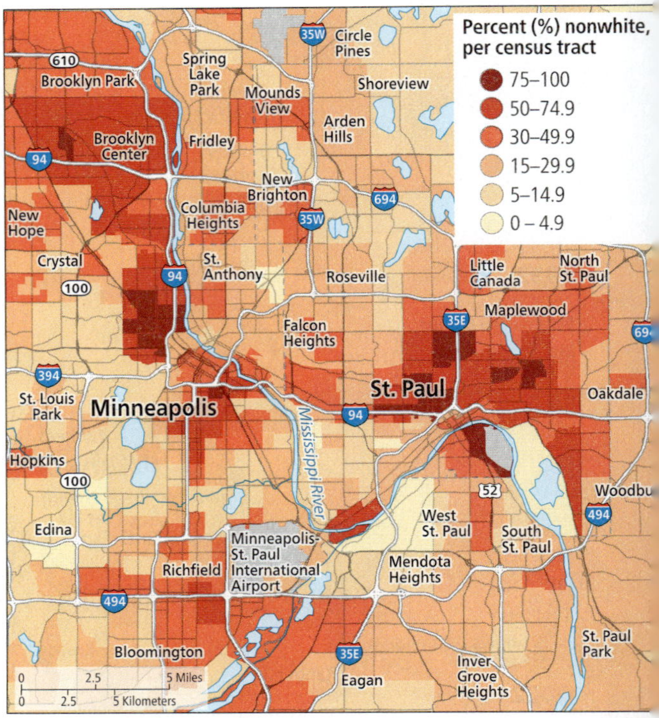

▲ 1.9.2 DISTRIBUTION OF RACES IN THE TWIN CITIES

1.9.2 Distribution by Sexual Orientation

The deliberate movement of goods, people, or information from one place to another place is referred to in geography as **spatial interaction**. With people, spatial interaction is exemplified by where they choose to live. Lesbian, gay, bisexual, transgender, and queer or questioning (LGBTQ) people may be attracted to some locations that reinforce their identity through spatial interaction with other LGBTQ people. Spatial interaction and a supporting environment for cultural identity can take place at many scales: neighborhood, city, state, and country.

Distribution: United States

Among the 50 largest U.S. cities, the percentage of LGBTQ people varies from 2.6 percent in Birmingham, Alabama, to 6.2 percent in the San Francisco metropolitan area. San Francisco, which has a reputation for being an especially hospitable city for LGBTQ people, has neighborhoods such as the Castro that have especially strong LGBTQ identities (Figure 1.9.3).

At the scale of the 50 U.S. states, the percentage of people who self-identify as LGBTQ also varies. According to a 2017 Gallup survey, the percentage varies among states from a high of 5.3 percent in Hawaii to a low of 2.0 percent in South Dakota. (The District of Columbia has 8.6 percent.) However, the percentage of people who supported same-sex marriage in 2015, a few months before the U.S. Supreme Court legalized it, varied much more widely among states, from a high of 75 percent in New Hampshire to a low of 32 percent in Mississippi.

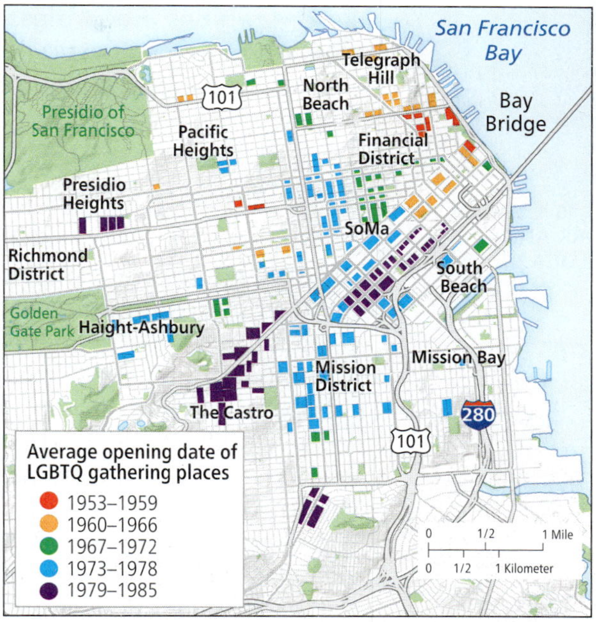

▲ 1.9.3 OPENING DATES, LGBTQ GATHERING PLACES, SAN FRANCISCO

Distribution: Worldwide

At the international scale, the International Lesbian, Gay, Bisexual, Trans and Intersex Association (ILGA) identifies laws that discriminate or protect based on sexual orientation or gender identity. Some countries protect LGBTQ people, whereas other countries criminalize being LGBTQ (Figure 1.9.4). In 2020, the ILGA classified 72 countries as criminalizing LGBTQ people, including eight that imposed the death penalty: Afghanistan, Iran, Iraq, Mauritania, Pakistan, Saudi Arabia, Sudan, and Yemen.

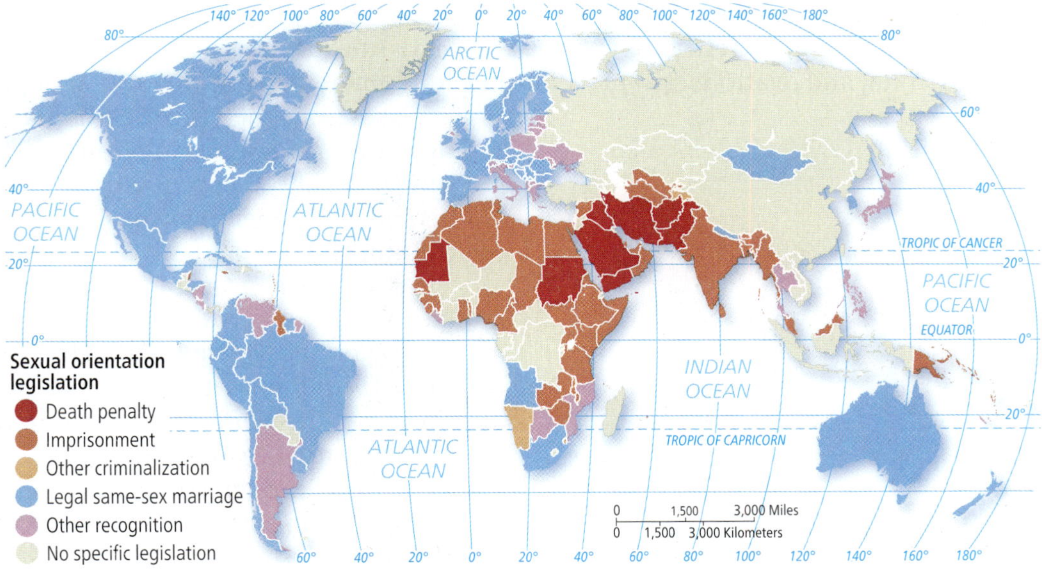

◀ 1.9.4 DISTRIBUTION OF SEXUAL ORIENTATION LAWS

▲ **1.10.1 BALANCING WORK AND CHILDCARE**

Space: Inequality

- **Discuss geographic approaches to inequality.**

Geographers focus on patterns of inequality within and among countries and regions. Regardless of location, women often face issues of inequality. At best, women achieved near-equality with men in some countries, specifically Canada, several developed countries in Europe, and a few developing countries with relatively low incomes.

1.10.1 Inequality: Gender

Children learn through a socialization process the gender roles that they are expected to follow in the home, community, and workplace of a given society. These gender norms are beliefs about males and females that are passed from generation to generation through processes of socialization. Gender relations refer to social relations between and among people that are based on gender norms and roles.

Expectations about gender relations often create hierarchies between and among groups that can lead to unequal power advantaging one group over another. For example, women may be expected to bear most of the responsibility for balancing work with childcare (Figure 1.10.1) and are more likely than men to do the shopping (Figure 1.10.2).

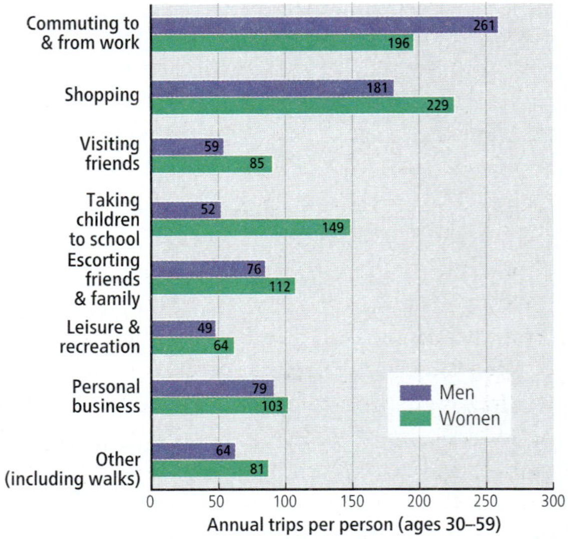

▲ **1.10.2 TRIPS BY PURPOSE AND GENDER**
The U.K. Department of Transport conducts an annual survey of why people travel in that country. According to the survey, women ages 30 to 59 on average take 17 percent more trips than do men of the same age.

Distribution by Gender

With gender, geographers focus on the distribution of inequality. Gender equality refers to equal chances or opportunities for females to access and control social, economic, and political resources. Gender equity refers to the different needs, preferences, and interests of women and men. This may mean that different treatment is needed to ensure equality of opportunity.

Any distinction, exclusion, or restriction (such as unfair or unequal treatment) made based on gender norms, roles, and relations that prevents people of different gender identity groups and ages from enjoying their human rights is considered a violation of gender equality. Such violations perpetuate gender inequality by legitimizing stereotypes.

As discussed in Chapter 10, gender inequality is reflected in numerous societal norms (Figure 1.10.3). The United Nations has not found a single country in the world where the average income earned by women exceeds that earned by men. Worldwide, the average income of women is around 50 percent that of men.

▼ **1.10.3 GENDER GAP**
The World Economic Forum Gender Gap combines four measures of gender parity: economic participation and opportunity, educational attainment, health and survival, and political empowerment. A lower number means less of a gap between women and men.

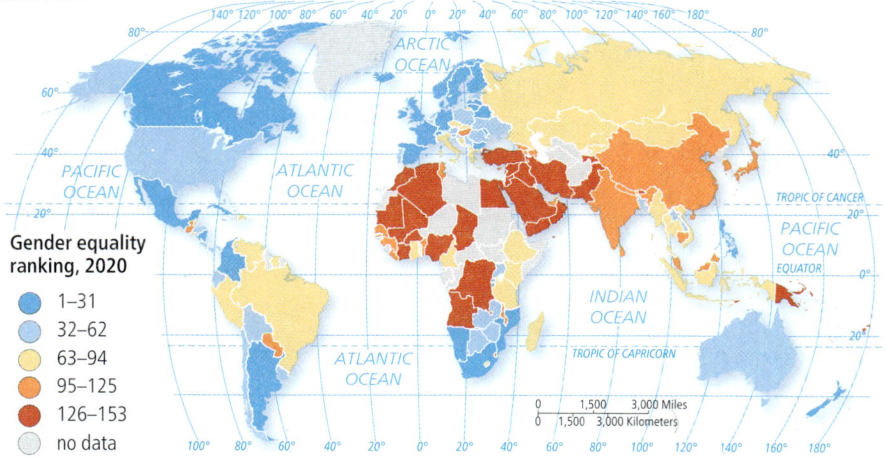

1.10.2 Uneven Development

Electronic communications play an especially important role in both removing barriers to interaction between people who are physically far from each other and erecting barriers through unequal access to electronics. Instantaneous electronic communications were once viewed as the "death" of geography because the ease of communications between distant places removed barriers to interaction and consequently the need to define borders. In reality, because of unequal access, geography matters even more than before.

Inequality Among Countries

Global culture and economy are increasingly centered on the three core, or hearth, regions of North America, Europe, and East Asia. These three regions have a large percentage of the world's advanced technology, capital to invest in new activities, and wealth to purchase goods and services. Surrounding the core are developing countries on the periphery with limited access to the world centers of consumption, communications, wealth, and power, which are clustered in the core. The increasing gap in economic conditions between regions in the core and periphery that results from the globalization of the economy is known as **uneven development**.

People in peripheral regions, who once toiled in isolated farm fields to produce food for their families, now produce crops for sale in core regions or have given up farm life altogether and migrated to cities in search of jobs in factories and shops. As a result, the global economy has produced greater disparities than in the past between the levels of wealth and well-being enjoyed by people in the core and in the periphery (Figure 1.10.4).

In a global culture and economy, every area of the world plays some role intertwined with the roles played by other regions. Workers and cultural groups that in the past were largely unaffected by events elsewhere in the world now share a single economic and cultural environment with other workers and cultural groups.

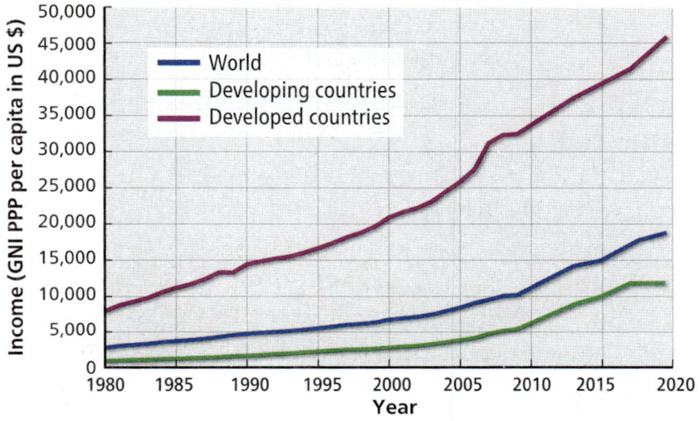

▲ 1.10.4 INEQUALITY: GAP BETWEEN HIGH INCOME AND LOW INCOME COUNTRIES

Inequality Within Countries

Unequal access and economic inequality have also increased within countries. In the United States, the share of the national income held by the wealthiest 1 percent increased from 24 percent in 1979 to around 40 percent in 2020, according to the Federal Reserve Bank (Figure 1.10.5). The share of Americans living in poverty has increased during that same period from 12 percent to around 18 percent.

Inequality in the distribution of wealth within the United States has been mapped in an innovative way by Sasha Trubetskoy as an undergraduate at the University of Chicago. One-half of the economic activity in the United States is clustered in a handful of small areas comprising well under 1 percent of the country's land area (Figure 1.10.6).

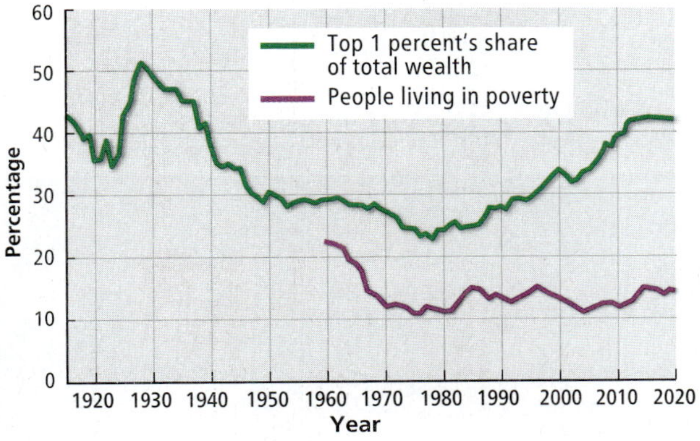

▲ 1.10.5 INEQUALITY: INCOME GAP WITHIN THE UNITED STATES

The share of wealth held in the United States by the wealthiest 1 percent has been rising since the 1980s. The percentage of people living in poverty has remained relatively unchanged since the 1960s.

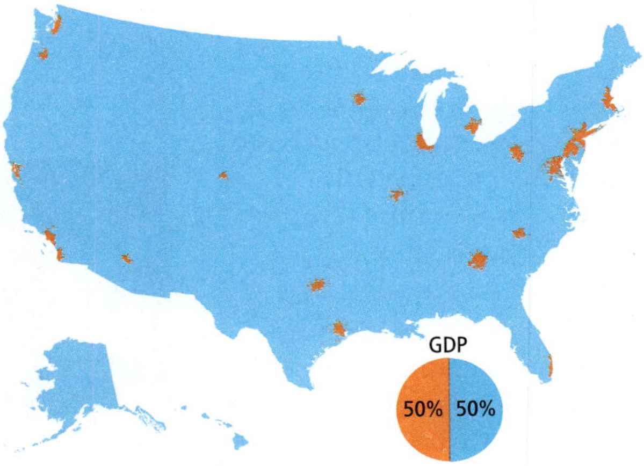

▲ 1.10.6 INEQUALITY WITHIN THE UNITED STATES

The map created by Sasha Trubetskoy shows that one-half of the economic activity in the United States is clustered in a handful of urban areas.

Connection: Diffusion & Interaction

- **Describe how features spread through diffusion.**

Connection refers to relationships between people, ideas, phenomena, and things across the barrier of space. Geographers are concerned with the various means by which connections occur, specifically diffusion and interaction.

1.11.1 Types of Diffusion

A new idea, phenomenon, or some other innovation originates at a node, known as a **hearth**, and spreads across space from one place to another through a process of **diffusion**. Geographers document the location of nodes of innovation and the processes by which diffusion carries that new idea, new phenomenon, or new thing elsewhere. Geographers distinguish between relocation diffusion and expansion diffusion.

Relocation Diffusion

The spread of an innovation through the physical movement of people from one place to another is termed **relocation diffusion**. We shall see in Chapter 3 that people migrate for a variety of cultural and environmental reasons. As discussed in Chapters 4 through 7, when people move, relocation diffusion occurs because they carry with them their culture, including language, religion, and ethnicity (Figure 1.11.1).

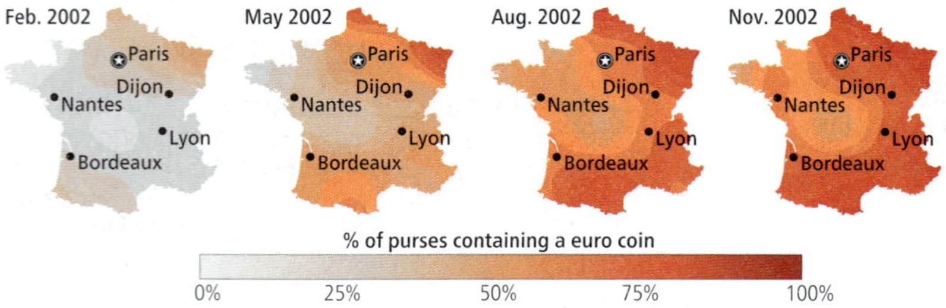

◀ 1.11.1 RELOCATION DIFFUSION
Introduction of a common currency, the euro, in 12 European countries in 2002 gave scientists an unusual opportunity to measure relocation diffusion from hearths. Each of the 12 countries minted its own euro coins in proportion to its share of the region's economy. A country's euro coins were initially distributed only inside its borders, although its euro coins could also be used in the other 11 countries. Scientists in France took month-to-month samples to monitor the proportion of euro coins from each of the other 11 countries. The percentage of euro coins from a particular country is a measure of the level of relocation diffusion to and from France.

Expansion Diffusion

The spread of an innovation from one place to another in an additive process is **expansion diffusion**. This expansion may result from one of three processes:

- **Hierarchical diffusion** is the spread of an idea from persons or nodes of authority or power to other persons or places. Hierarchical diffusion may result, for example, from the spread of ideas from political leaders, socially elite people, or other important persons to others in the community.
- **Contagious diffusion** is the rapid, widespread diffusion of an idea throughout a population. It occurs when something goes viral. For example, through widespread simultaneous diffusion, something is identified as cheugy (out of date or trying too hard). An example of something becoming cheugy through contagious diffusion, especially through TikTok, is watching reruns of the TV series *Friends* (Figure 1.11.2).
- **Stimulus diffusion** is the spread of an innovation but with only the underlying principles of it intact. For example, innovative features of Apple's iPhone and iPad have been adopted by competitors but altered for fit with each of those competitors' unique products.

▼ 1.11.2 CONTAGIOUS DIFFUSION
Watching reruns of the TV show *Friends* has been identified as cheugy.

1.11.2 Spatial Interaction

The farther away someone is from another, the less likely the two are to make connections. Geographers apply the term **space–time compression** to describe the reduction in the time it takes for something to reach another place (Figure 1.11.3). The diminishing in importance and eventual disappearance of a phenomenon is called **distance decay**. In the contemporary world, distance decay is much less severe because of space–time compression—that is, the connections between places takes much less time.

Connections & Culture

Connections between cultural groups can have several results:

- **Assimilation** is the process by which a group's cultural features are altered to resemble those of another group. The cultural features of one group may come to dominate the culture of the assimilated group.
- **Acculturation** is the process of changes in culture that result from the meeting of two groups. Changes may be experienced by both of the interacting cultural groups, but the two groups retain distinct cultures.
- **Syncretism** is the combination of elements of two groups into a new cultural feature. The two cultural groups come together to form a new culture.

To be connected with another place in the modern world, we may not need to travel at all. Ideas that originate in a hearth are now able to diffuse rapidly to other areas through communications networks. One example is the TV network (for example, BBC in the United Kingdom, CBC in Canada, NBC in the United States), which comprises a chain of stations simultaneously broadcasting to distant places the same program, such as a football game. Through a communications network, diffusion from one place to another is instantaneous in time, even if the physical distance between places—as measured in kilometers or miles—is large.

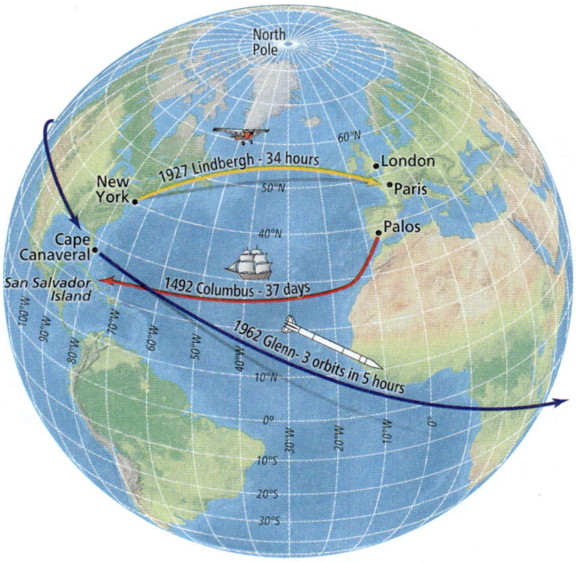

▲ **1.11.3 SPACE–TIME COMPRESSION**
Transportation improvements have shrunk the world. In 1492, Christopher Columbus took nearly 900 hours (37 days) to sail across the Atlantic Ocean. In 1927, Charles Lindbergh was the first to fly nonstop across the Atlantic, taking 33.5 hours. In 1962, John Glenn, the first American to orbit in space, crossed above the Atlantic in about 1/2 hour and circled the globe three times in 5 hours.

Networks

Interaction takes place through a **network**, which is a chain of communication that interconnects people, things, and places. An airline, for example, may have a network known as "hub-and-spokes" (Figure 1.11.4). With a hub-and-spokes network, an airline flies planes from a large number of places into one hub airport within a short period of time and then a short time later sends the planes to another set of places. In principle, travelers originating in relatively small towns can reach a wide variety of destinations by changing planes at the hub airport.

Modern networks make it possible for us to know more about what is happening elsewhere in the world, and space-time compression makes it possible for us to know it sooner. Distant places seem less remote and more accessible to us. With better connections between places, we are exposed to a constant barrage of cultural traits and economic initiatives from people in other regions, and perhaps we may adopt some of these cultural and economic elements.

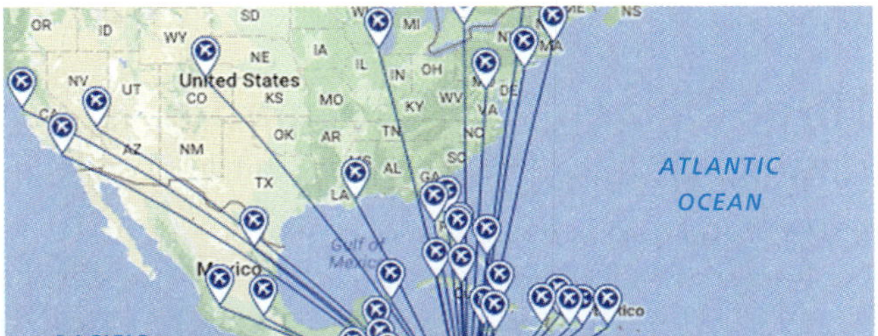

▲ **1.11.4 SPATIAL INTERACTION: HUB-AND-SPOKE NETWORK**
Copa Air flies from a large number of places into a hub airport in Panama, and a short time later sends the planes to another set of places.

Connection: Sustainability

- **Explain connections among the three pillars of sustainability.**

Geography is distinctive because it studies connections between social science (human geography) and natural science (physical geography). From the perspective of human geography, nature offers humans a large menu of resources. A **resource** is a substance in the environment that is useful to people, economically and technologically feasible to access, and socially acceptable to use.

1.12.1 Three Pillars of Sustainability

Sustainability is the use of Earth's resources in ways that ensure their availability in the future. The United Nations considers sustainability to rest on three interconnected pillars (Figure 1.12.1):

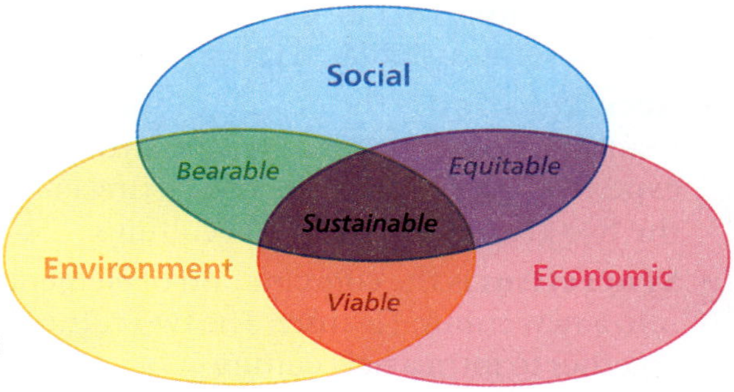

▲ 1.12.1 THREE PILLARS OF SUSTAINABILITY

Environment pillar. The sustainable use and management of Earth's natural resources to meet human needs such as food, medicine, and recreation is **conservation**. Resources such as trees are conserved if they are cut down at a slower rate than the time it takes for their replacement by new tree growth (Figure 1.12.2).

Social pillar. Humans need shelter, food, and clothing to survive, so they make use of resources to meet these needs. Consumer choices can support sustainability when people embrace sustainability as a value. For example, a consumer might prefer clothing made of sustainable resources such as cotton instead of clothing made of unsustainable resources such as polyester made from petroleum. A consumer might also choose products that benefit people living in a particular place (Figure 1.12.3).

Economic pillar. Natural resources acquire a monetary value through exchange in a marketplace. The price of a resource depends on the value placed on it by people and on people's ability to obtain it. The greater the supply of a resource, the lower the price, and the greater the demand for a resource by people, the higher the price (Figure 1.12.4).

Three Pillars: The Everglades

The three pillars can be illustrated in the Everglades.

▼ 1.12.2 ENVIRONMENT PILLAR
The Everglades was once a very wide and shallow freshwater river, slowly flowing south from Lake Okeechobee to the Gulf of Mexico. A sensitive ecosystem of plants and animals once thrived in this distinctive landscape, but much of it was destroyed by human actions to promote the economy.

◀ 1.12.3 SOCIAL PILLAR
The U.S. Army Corps of Engineers built levees and canals in the Everglades. These modifications opened up land for growing sugarcane and protected the growing South Florida population from flooding. However, the modifications also had unintended consequences, including polluted water and threats to native vegetation and animals.

▼ 1.12.4 ECONOMIC PILLAR
Everglades National Park attracts 1 million visitors a year. The National Park Service tries to provide an enjoyable visit while protecting the area's fragile environment.

1.12.2 Ecology & Ecosystems

Geographers classify natural resources as part of four interconnected systems classified as either biotic or abiotic (Figure 1.12.5). An **abiotic** system is composed of nonliving or inorganic matter. Three of Earth's four systems are abiotic. The **atmosphere** is a thin layer of gases surrounding Earth. The **hydrosphere** is water on and near Earth's surface. The **lithosphere** is Earth's crust and a portion of upper mantle directly below the crust.

A **biotic** system is composed of living organisms. One of Earth's four systems is biotic. The **biosphere** comprises all living organisms on Earth, including plants and animals, as well as microorganisms. The names of the four spheres are derived from the Greek words for air (*atmo*), water (*hydro*), stone (*litho*), and life (*bio*).

▼ **1.12.5 ECOSYSTEM: WHITE CLIFFS OF DOVER**

The city of Dover, in the United Kingdom, sits atop a lithosphere of chalky cliffs, beside the hydrosphere of the English Channel, beneath the atmosphere above, and surrounded by a green biosphere. Geographers are especially interested in the ecosystem of a city because more than half of Earth's humans live in urban areas. The lithosphere provides the ground and the materials to erect homes and businesses. The hydrosphere provides the water for urban dwellers to consume. The atmosphere is where urban dwellers emit pollutants. Some plants and other animals of the biosphere thrive along with humans in the cities, whereas others struggle.

ATMOSPHERE
A thin layer of gases surrounds Earth at an altitude up to 480 kilometers (300 miles). As atmospheric gases are held to Earth by gravity, pressure is created. Variations in air pressure from one location to another are responsible for producing such weather features as wind blowing, storms brewing, and rain falling.

HYDROSPHERE
Water exists in liquid form in the oceans, lakes, and rivers as well as groundwater in soil and rock. It can also exist as water vapor in the atmosphere and as ice in glaciers. Over 97 percent of the world's water is in the oceans. The oceans supply the atmosphere with water vapor, which returns to Earth's surface as precipitation, the most important source of freshwater.

LITHOSPHERE
Earth is composed of concentric spheres made of different materials. The core is a dense, metallic sphere about 3,500 kilometers (2,200 miles) in radius. Surrounding the core is a mantle about 2,900 kilometers (1,800 miles) thick. The mantle is mostly solid rock, but it is pliable and can flow slowly. The crust is a thin, brittle outer shell 8 to 40 kilometers (5 to 25 miles) thick. The lithosphere encompasses the crust and a portion of the mantle extending down to about 70 kilometers (45 miles). Powerful forces deep within Earth shape the crust to form mountains, continents, and ocean basins.

BIOSPHERE
The biosphere encompasses all of Earth's living organisms.

Earth's Connected Physical Systems

A group of living organisms and the abiotic spheres with which they interact is an **ecosystem**. The scientific study of ecosystems is **ecology**. Ecologists study interrelationships between living organisms and the three abiotic environments as well as interrelationships among the various living organisms in the biosphere.

For example, the principal cause of Earth's recent climate change has been human actions. The United Nations warns us that it is not too late to reverse climate change, but we are running out of time. Earth's temperature has risen an average of 1.5°C (2.8°F) during the past century. Why does this temperature rise matter to us? Earth's changing climate has increased the risks of various environmental disasters around the world (Figure 1.12.6). In some places because of climate change, food supply has declined and poverty increased. Climate change is inducing unprecedented numbers of humans to migrate in search of better living conditions in other places.

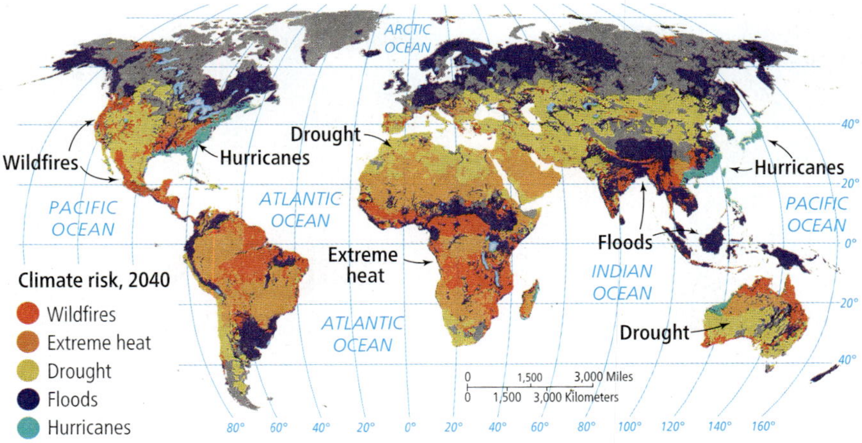

▶ **1.12.6 TOP RISKS BY 2040**

Humans & Their Environment

- **Compare and contrast sustainable and unsustainable ecosystems.**

The geographic study of human–environment relationships is known as **cultural ecology**. Geographers are interested in two main types of human–environment interaction: how people adjust to their environment and how they modify it.

1.13.1 Possibilism

Nineteenth-century geographers argued that the physical environment *caused* social development, an approach called **environmental determinism**. To explain connections between human activities and the physical environment, modern geographers embrace possibilism. According to **possibilism**, the physical environment may limit some human actions, but people have the ability to adjust to their environment. People can choose a course of action from many alternatives in the physical environment (Figure 1.13.1).

▲ **1.13.1 POSSIBILISM: ALTERNATIVE BEHAVIORS**
(a) Some people prefer to mow their lawns, whereas others (b) prefer to let wildflowers grow.

Sustainable Ecosystem: The Netherlands

The Dutch have modified their environment with two distinctive types of construction projects—polders and dikes (Figure 1.13.2). A **polder** is a piece of land that is created by draining water from an area. All together, the Netherlands has 6,500 square kilometers (2,600 square miles) of polders, comprising 16 percent of the country's land area. The Dutch government has reserved most of the polders for agriculture to reduce the country's dependence on imported food.

The Dutch have also constructed massive dikes to prevent the North Sea, an arm of the Atlantic Ocean, from flooding much of the country. The Zuider Zee project in the north of the Netherlands converted a saltwater sea to a freshwater lake called Lake IJssel. Some of the lake has been drained to create polders. The Delta Project in the southwest closed off several rivers with dams after a devastating flood killed nearly 2,000 people in 1953.

With these two massive projects completed, attitudes toward modifying the ecosystem have changed in the Netherlands. Plans have been scrapped to build more polders in Lake IJssel, and some dikes have been deliberately broken to flood fields. Still, climate change could threaten the Netherlands by raising the level of the sea. Rather than build new dikes and polders, the Dutch have become world leaders in advocating for the reduction in human actions that result in climate change.

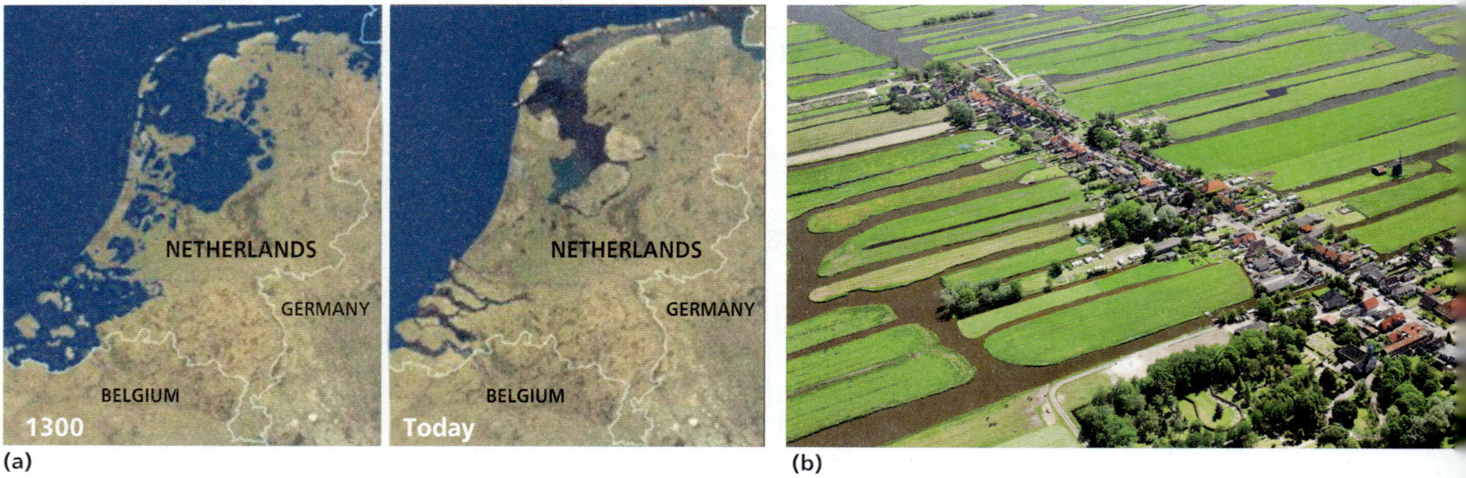

▲ **1.13.2 SUSTAINABLE ECOSYSTEM: THE NETHERLANDS**
(a) The Dutch have altered the site of the country through creation of polders and dikes. (b) Kalverpolder in North Holland has been created by pumping the water from the site into the canal, originally by means of windmills.

1.13.2 A Sustainability Doughnut

Climate change has raised the global sea level in the Netherlands about 19 centimeters (7.5 inches) since 1900. Should the sea further rise by 1 meter (36 inches), much of the Netherlands, including Amsterdam and the country's other largest cities, would be under water (Figure 1.13.3). The Dutch are trying to create a more sustainable ecosystem at the urban level.

The city of Amsterdam, for example, has adopted a sustainability model known as doughnut economics (Figure 1.13.4). The doughnut, developed by British economist Kate Raworth, depicts the sustainability model as two concentric rings. The inner ring shows what people need to lead a decent life, such as food, clean water, jobs, and education. People who don't have these necessities of life are falling through the doughnut's center hole.

The lines radiating out from the center represent the use of various resources; the further from the center, the less sustainable is the use of a resource. The outer ring of the doughnut represents ecological boundaries set by scientists. These are the boundaries that people should not cross because to do so would damage Earth's ecosystems, resulting in more pollution, biodiversity loss, and climate change.

To address its need for more affordable housing for its residents, Amsterdam could choose to simply build more houses. But the city's sustainability model suggests that meeting the need for affordable housing this way is not sustainable. For example, most building materials would have to be imported, utilizing a lot of energy. Instead, Amsterdam is regulating builders to maximize their use of recycled materials and to minimize their use of plastics and imported materials.

▲ **1.13.3 NETHERLANDS LAND PROJECTED TO BE BELOW ANNUAL FLOOD LEVEL IN 2050**
Source: Climate Central.

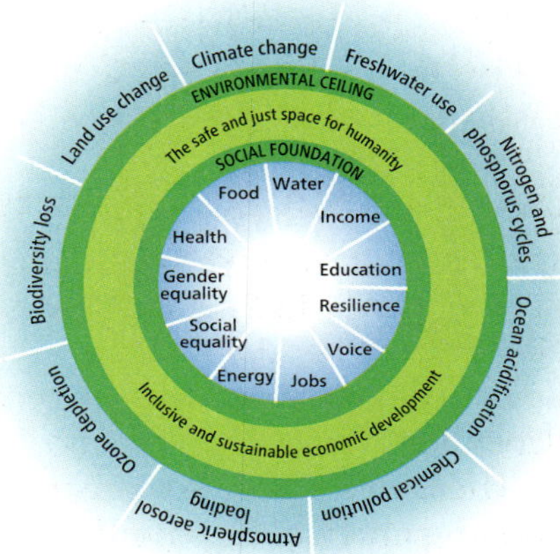

▲ **1.13.4 THE DOUGHNUT**
Meeting human needs (shown in the inner ring) can be achieved within the environment's constraints (shown in the outer ring).

Surging Seas

A model by Climate Central projects that 150 million people worldwide now live in land that will be below the high-tide line by 2050. Along the East Coast of the United States, sea level has risen by nearly 60 centimeters (2 feet) since 1880. Climate Central shows the likelihood of extensive flooding along U.S. coasts, including the Miami, Florida, area (Figure 1.13.5).

▲ **1.13.5 MIAMI LAND PROJECTED TO BE BELOW ANNUAL FLOOD LEVEL IN 2050**
Source: Climate Central.

Miami-Dade County officials assert that rising sea level is mostly manageable. Among the actions they have taken is building houses on stilts, elevating the ground with sand and rocks from elsewhere, and raising sidewalks and streets (Figure 1.13.6). Climate experts fear that local officials are much too upbeat. Efforts to mitigate the effects of rising sea level, they warn, are likely to prove much too inadequate.

▼ **1.13.6 SURGING SEAS**
Rising sea level threatens much of Miami, Florida. In response, local officials are raising sidewalks as well as taking other actions.

CHAPTER

1 Review, Analyze, & Apply

KEY ISSUE 1 Why Is Geography a Science?

- Geography has ancient and medieval roots.
- Maps are tools of reference and increasingly tools of communication.
- Reading a map requires recognizing its scale and projection.
- Contemporary mapping utilizes electronic technologies, such as GPS and GIS.

THINKING GEOGRAPHICALLY

Geography plays an important role in our electronic devices. What was the most recent geographic item that you consulted on your electronic device? Describe how this item illustrated each of geography's five basic concepts of place, region, scale, space, and connection.

▲ **1.CR.1** GPS MAPPING AND TRACKING APP

KEY ISSUE 2 Why Is Every Place Unique?

- Location is identified through name, site, and situation.
- Regions can be formal, functional, or vernacular.

THINKING GEOGRAPHICALLY

The South and the West are commonly cited vernacular U.S. regions. What are some features that you would expect to distinguish one of these regions?

▲ **1.CR.2** BIG SUR, CALIFORNIA

KEY ISSUE 3 Why Are Different Places Similar?

- Geographers examine features at all scales, though they are increasingly concerned with the global scale.
- Distribution has three properties: density, concentration, and pattern.
- Different cultural groups display different distributions in space.
- Differences in distributions can be associated with patterns of discrimination and inequality by gender, race, and sexual orientation.

THINKING GEOGRAPHICALLY

Describe elements of your daily life that derive from the globalization economy and culture and ways in which your actions and behavior represent distinctive displays of local diversity.

◀ **1.CR.3** TOUCH-SCREEN ORDERING, MCDONALD'S, SINGAPORE

KEY ISSUE 4 Why Are Places Connected?

- Features spread through diffusion, which can occur through relocation diffusion and expansion diffusion.
- Connections can result in spatial interaction, which could result in assimilation, acculturation, or syncretism.
- Interaction can take place by means of networks.
- Sustainability is the use of Earth's resources in ways that ensure their availability in the future.
- Sustainability combines environment, economy, and society.
- Earth's resources encompass three abiotic systems and one biotic system.

THINKING GEOGRAPHICALLY

How did your connections with other places change as a result of the COVID-19 pandemic? Can you identify a change in connections with other places that has been long-term or permanent since the onset of the pandemic? What might account for that permanent change in connections?

◀ **1.CR.4** VISITING GRANDPARENTS

Key Terms

Abiotic (1.12.2) Composed of nonliving or inorganic matter.

Acculturation (1.11.2) The process of changes in culture that result from the meeting of two groups, each of which retains distinct cultural features.

Assimilation (1.11.2) The process by which a group's cultural features are altered to resemble those of another group.

Atmosphere (1.12.2) The thin layer of gases surrounding Earth.

Biosphere (1.12.2) All living organisms on Earth, including plants and animals, as well as microorganisms.

Biotic (1.12.2) Composed of living organisms.

Black Lives Matter (1.9.1) Movement that campaigns against violence and perceived racism toward Blacks and educates others about the challenges that Blacks continue to face in the United States.

Cartogram (1.3.2) A map in which the projection and scale are distorted in order to convey the information of a variable.

Cartography (1.2.1) The science of making maps.

Choropleth map (1.3.2) A map where recognizable areas are shaded or patterned in proportion to the data collected.

Citizen science (1.2.2) Scientific research by amateur scientists.

Concentration (1.8.1) The extent of a feature's spread over a given area.

Connection (1.1.2) The relationships among people and objects across the barrier of space.

Conservation (1.12.1) The sustainable management of a natural resource to meet human needs.

Contagious diffusion (1.11.1) The rapid, widespread diffusion of a feature or trend throughout a population.

Coordinated Universal Time (UTC) (1.4.1) Informally Greenwich Mean Time (GMT). The time in the zone encompassing the prime meridian, or 0° longitude.

Cultural ecology (1.13) A geographic approach that emphasizes human–environment relationships.

Cultural landscape (1.6.1) An approach to geography that emphasizes the relationships among social and physical phenomena in a particular study area.

Density (1.8.1) The frequency with which something exists within a given unit of area.

Diffusion (1.11.1) The process of spread of a feature or trend from one place to another over time.

Distance decay (1.11.2) The diminished importance and eventual disappearance of a phenomenon with increasing distance from its origin.

Distribution (1.8.1) The arrangement of something across Earth's surface.

Dot distribution map (1.3.2) A map that depicts data consisting of discrete observations. Each dot represents a predetermined number of observations, which could be one or many.

Ecology (1.12.2) The scientific study of ecosystems.

Ecosystem (1.12.2) A group of living organisms and the abiotic spheres with which they interact.

Environmental determinism (1.13.1) A nineteenth- and early twentieth-century approach to the study of geography that argued that the general laws sought by human geographers could be found in the physical sciences. Geography was therefore the study of how the physical environment caused human activities.

Expansion diffusion (1.11.1) The spread of a feature or trend among people from one area to another in an additive process.

Formal region (or **uniform region**) (1.6.1) An area in which most people share in one or more distinctive characteristics.

Functional region (or **nodal region**) (1.6.2) An area organized around a node or focal point.

Geographic information science (GIScience) (1.2.1) Analysis of data about Earth acquired through satellite and other electronic information technologies.

Geographic information system (GIS) (1.2.1) A computer system that captures, stores, queries, and displays geographic data.

Geotagging (1.2.2) Identification and storage of a piece of information by its precise latitude and longitude coordinates.

Global Positioning System (GPS) (1.2.2) A system that determines the precise position of something on Earth through a series of satellites, tracking stations, and receivers.

Globalization (1.7.1) Actions or processes that involve the entire world and result in making something worldwide in scope.

Graduated symbol map (1.3.2) A map that displays symbols that change in size according to the value of the variable.

Hearth (1.11.1) A place from which an innovation originates.

Hierarchical diffusion (1.11.1) The spread of a feature or trend from one key person or node of authority or power to other persons or places.

Hydrosphere (1.12.2) All of the water on and near Earth's surface.

International Date Line (1.4.1) An arc that for the most part follows 180° longitude. When the International Date Line is crossed heading east (toward America), the clock moves back 24 hours, or one entire day. When it is crossed heading west (toward Asia), the calendar moves ahead one day.

Isoline map (1.3.2) A map that connects places of a particular value by lines.

Latitude (1.4.1) The numbering system used to indicate the location of parallels drawn on a globe and measuring distance north and south of the equator (0°).

Lithosphere (1.12.2) Earth's crust and a portion of upper mantle directly below the crust.

Location (1.5.1) The position of anything on Earth's surface.

Longitude (1.4.1) The numbering system used to indicate the location of meridians drawn on a globe and measuring distance east and west of the prime meridian (0°).

Map (1.2.1) A two-dimensional, or flat, representation of Earth's surface or a portion of it.

Map scale (1.4.2) The relationship between the size of an object on a map and the size of the actual feature on Earth's surface.

Mashup (1.2.2) A map that overlays data from one source on top of a map provided by a mapping service.

Mental map (1.6.2) A representation of a portion of Earth's surface based on what an individual knows about a place that contains personal impressions of what is in the place and where the place is located.

Meridian (1.4.1) An arc drawn on a map between the North and South poles.

Network (1.11.2) A chain of communication that connects places.

Parallel (1.4.1) A circle drawn around the globe parallel to the equator and at right angles to the meridians.

Participatory GIS (PGIS) (1.2.2) Community-based mapping, representing local knowledge and information.

Pattern (1.8.1) The geometric or regular arrangement of something in a particular area.

Photogrammetry (1.2.1) The science of taking measurements of Earth's surface from photographs.

Place (1.1.2) A specific point on Earth, distinguished by a particular characteristic.

Polder (1.13.1) Land that the Dutch have created by draining water from an area.

Possibilism (1.13.1) The theory that the physical environment may set limits on human actions, but people have the ability to adjust to the physical environment and choose a course of action from many alternatives.

Prime meridian (1.4.1) The meridian, designated as 0° longitude, that passes through the Royal Observatory at Greenwich, England.

Projection (1.4.2) A system used to transfer locations from Earth's surface to a flat map.

Region (1.1.2) An area distinguished by one or more distinctive characteristics.

Relocation diffusion (1.11.1) The spread of a feature or trend through bodily movement of people from one place to another.

Remote sensing (1.2.1) The acquisition of data about Earth's surface from a satellite orbiting the planet or from other longdistance methods.

Resource (1.12.1) A substance in the environment that is useful to people, is economically and technologically feasible to access, and is socially acceptable to use.

Scale (1.1.2) The relationship between the portion of Earth being studied and Earth as a whole.

Site (1.5.2) The physical character of a place.

Situation (1.5.2) The location of a place relative to another place.

Space (1.1.2) The physical gap or interval between two objects.

Space–time compression (1.11.2) The reduction in the time it takes to diffuse something to a distant place as a result of improved communications and transportation systems.

Spatial association (1.7.2) The relationship between the distribution of one feature and the distribution of another feature.

Spatial interaction (1.9.2) The deliberate movement of goods, people, or information from one place to another place.

Stimulus diffusion (1.11.1) The spread of an underlying principle even though a specific characteristic is rejected.

Sustainability (1.12.1) The use of Earth's renewable and nonrenewable natural resources in ways that do not constrain resource use in the future.

Syncretism (1.11.2) The combining of elements of two groups into a new cultural feature.

Toponym (1.5.1) The name given to a portion of Earth's surface.

Transnational corporation (1.7.1) A company that conducts research, operates factories, and sells products in many countries, not just where its headquarters or shareholders are located.

Uneven development (1.10.2) The increasing gap in economic conditions between core and peripheral regions as a result of the globalization of the economy.

Vernacular region (or **perceptual region**) (1.6.2) An area that people believe exists as part of their cultural identity.

Volunteered geographic information (VGI) (1.2.2) Creation and dissemination of geographic data contributed voluntarily and for free by individuals.

Geospatial Analysis

Log in to the Mastering Geography Study Area to access MapMaster 2.0.

Environmental Challenges for the Dutch

Add the ***World Elevations*** layer. Click the *Settings* icon in the *Legend*, and select *Show Political Labels*.

Zoom in as far as possible to center the map on the Netherlands (northwestern Europe between Belgium and United Kingdom). Add the ***Population Density*** layer, and select *Split map window.*

1. Hover over the Netherlands. What is the elevation of this area?
2. Compare the elevation map to the population density map. Describe the distribution of population within the Netherlands (density, concentration, and pattern).
3. What environmental challenges do people living in the Netherlands face? How are the Dutch addressing these challenges? Are these challenges likely to become more severe or less severe in the coming years? Explain your answer.

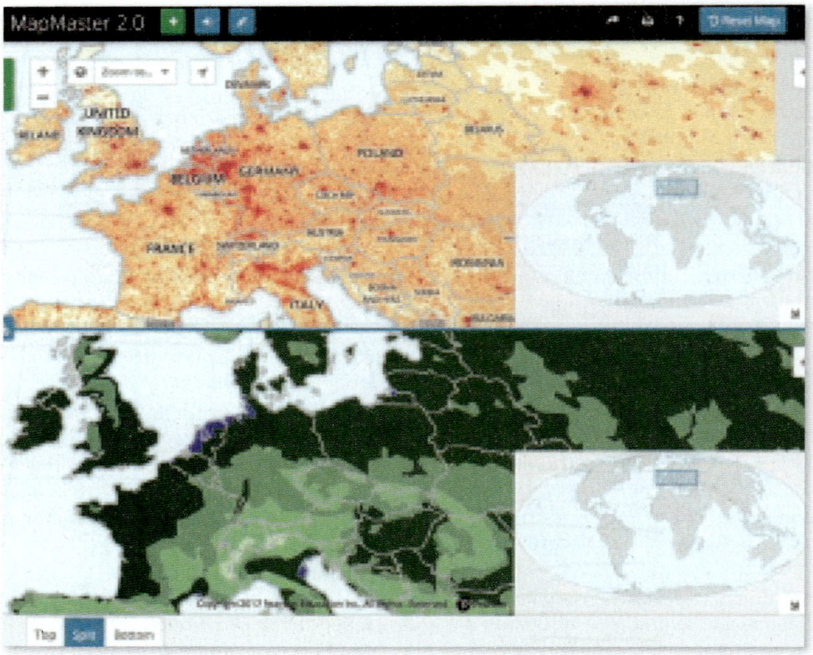

▲ **1.CR.5** ELEVATION OF THE NETHERLANDS

Explore

Dili

Use Google Earth to explore Dili, capital of Timor-Leste, one of the world's newest and least economically developed countries.
Fly to **Dili, Timor-Leste**.
Click the Street View icon. **Select** one of the blue dots along the waterfront.

▲ 1.CR.6 DILI, TIMOR-LESTE

1. What do you see on the water side of the street? Is there an attractive beach? Are people swimming in the water? Does this look attractive to you? Why or why not?
2. What activities are taking place on the land side? Is the area attractive? Why or why not?
3. Take a look at the road nearest the water. Is it busy with traffic or quiet? Why might that be?

▲ 1.CR.7 WATERFRONT, DILI, TIMOR-LESTE

CHAPTER 2

Population & Health

Waiting at a bus stop, Mumbai, India.

More people are alive at this time than at any other point in Earth's history, and most of the growth is concentrated in developing countries. Can Earth sustain nearly 8 billion people now, let alone the added billions in the future? Geographers have unique perspectives on the ability of people to live on Earth. South Asia, including India (shown in the chapter opener), is the world's most populous region and one of its fastest growing. So the region's future population growth will have a major impact on the future population of the world as a whole.

KEY ISSUES

1 Where Are People Distributed?

Humans are not distributed uniformly across Earth. They are highly clustered in particular places, like India, whereas other places are sparsely populated.

2 Why Is Population Increasing?

Population is growing at different rates in different regions of the world. Every country is at some stage in development that correlates to its population growth rate.

3 How Does Health Vary by Region?

Countries have distinctive health care challenges in part because of their share of the population that is elderly or young. Health and medical care have improved in the world as a whole, but less so in places like Uganda.

4 Why Might Population Change in the Future?

Geographers see strong connections between the size and growth of population and the adequacy of resources in some places, such as Cambodia and India.

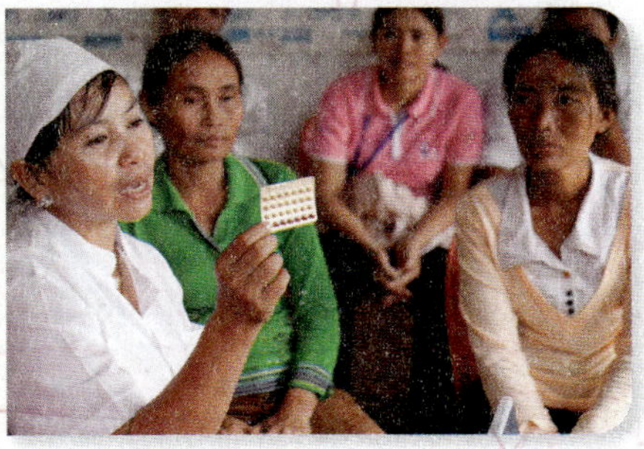

LOCATIONS IN THIS CHAPTER

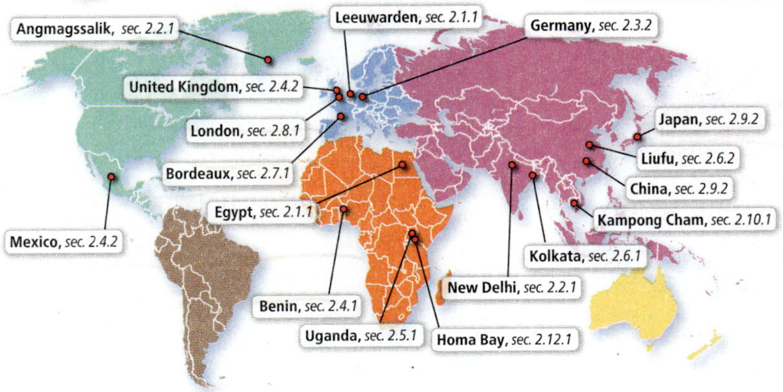

▲ **2.1.1 HIGH ARITHMETIC DENSITY**
Leeuwarden, The Netherlands.

Population Density

- **Define three types of density used in population geography.**

Geographers study population by first describing where Earth's nearly 8 billion people are distributed. They then turn to explaining why population grows or declines at different rates in different places. Geographers are increasingly concerned with differences in the health of people in different places, especially in the face of the spread of contagious diseases like the COVID-19 pandemic.

2.1.1 Density

Density was defined in Chapter 1 as the frequency with which something exists within a given unit of area. Population density is the number of humans living within an area, such as the city of Leeuwarden, in the Netherlands (Figure 2.1.1), which has a density of around 508 persons per square kilometer (1,340 persons per square mile). Three measures of population density frequently used by geographers are arithmetic density, physiological density, and agricultural density (Figure 2.1.2).

- **Arithmetic density** is the total number of objects in an area.
- **Physiological density** is the number of people per unit area of **arable land**, which is land suited for agriculture.
- **Agricultural density** is the ratio of the number of farmers to the amount of arable land.

▶ **2.1.2 DENSITIES OF FOUR COUNTRIES**
Density figures are in persons per square kilometer.

Country	Arithmetic Density	Physiological Density	Agricultural Density	Percent Farmers	Percent Arable Land
Canada	4	99	1	2	5
Egypt	104	3,514	565	28	3
The Netherlands	508	1,731	19	2	30
United States	36	211	1	1	17

Arithmetic Density

In population geography, arithmetic density refers to the total number of people divided by the total land area. To compute the arithmetic density, divide the population by the land area. Arithmetic density enables geographers to compare the number of people living in different regions of the world (Figure 2.1.3). Thus, arithmetic density answers the question, "Where are people distributed?" However, to explain why people are not uniformly distributed across Earth's surface, other density measures are more useful.

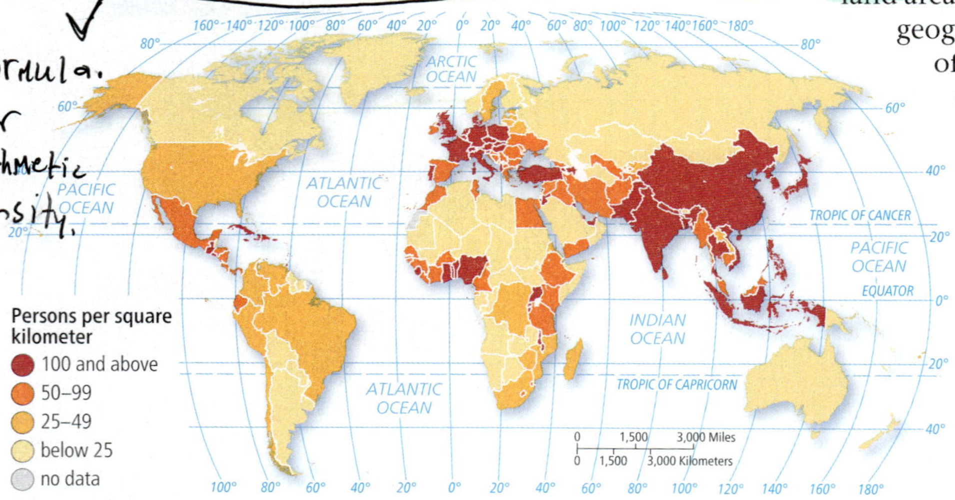

◀ **2.1.3 ARITHMETIC DENSITY**
Persons per square kilometer.

2.1.2 Physiological Density

Looking at the number of people per area of a certain type of land in a region can provide a more meaningful population measure than arithmetic density. For example, the physiological density helps geographers understand the capacity of the land to yield enough food for the needs of the people (Figure 2.1.4). In Egypt, for example, the large difference between the physiological density (3,514) and arithmetic density (104) indicates that most of the country's land is unsuitable for intensive agriculture (refer to Figure 2.1.2). In fact, all but 5 percent of Egyptians live in the Nile River valley and delta because it is the only area in the country that receives enough moisture (by irrigation from the river) to allow intensive cultivation of crops (refer to Figure 1.5.3).

"yield enough food" such as transporting goods? Limits farmers, as well as pushing out more and more goods for people

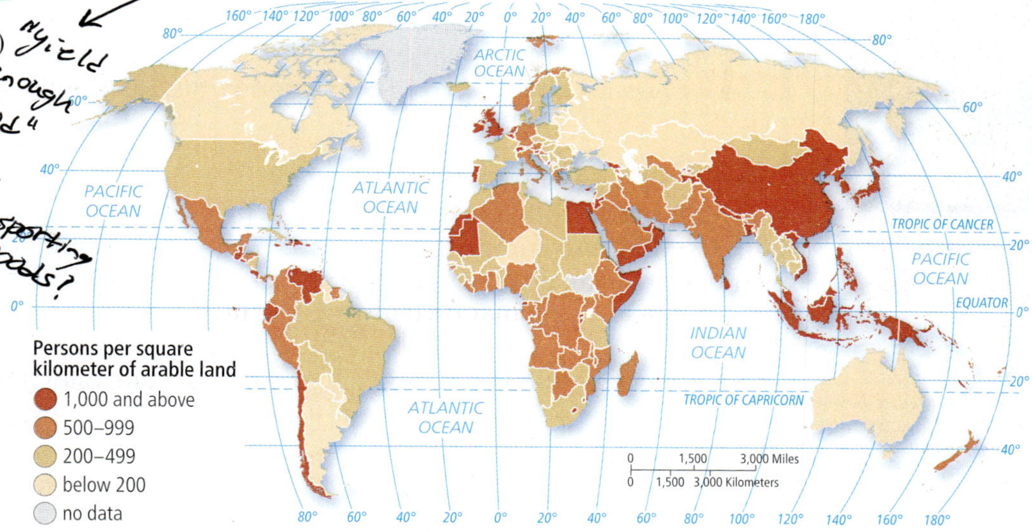

▲ **2.1.4 PHYSIOLOGICAL DENSITY**
Persons per square kilometer of arable land.

Agricultural Density

Agricultural density helps account for economic differences between countries. Developed countries have lower agricultural densities because technology and finance allow a few people to farm extensive land areas and feed many people (Figure 2.1.5).

To understand relationships between population and resources in a country, geographers examine a country's physiological and agricultural densities together (refer to Figure 2.1.2). For example, the physiological densities of both Egypt and the Netherlands are high, but the Dutch have a much lower agricultural density than the Egyptians. Geographers conclude that both the Dutch and Egyptians put heavy pressure on the land to produce food, but the more efficient Dutch agricultural system requires fewer farmers than does the Egyptian system.

does putting heavy pressure on ~~foot~~ land, disrupt the quality of food that is grown?

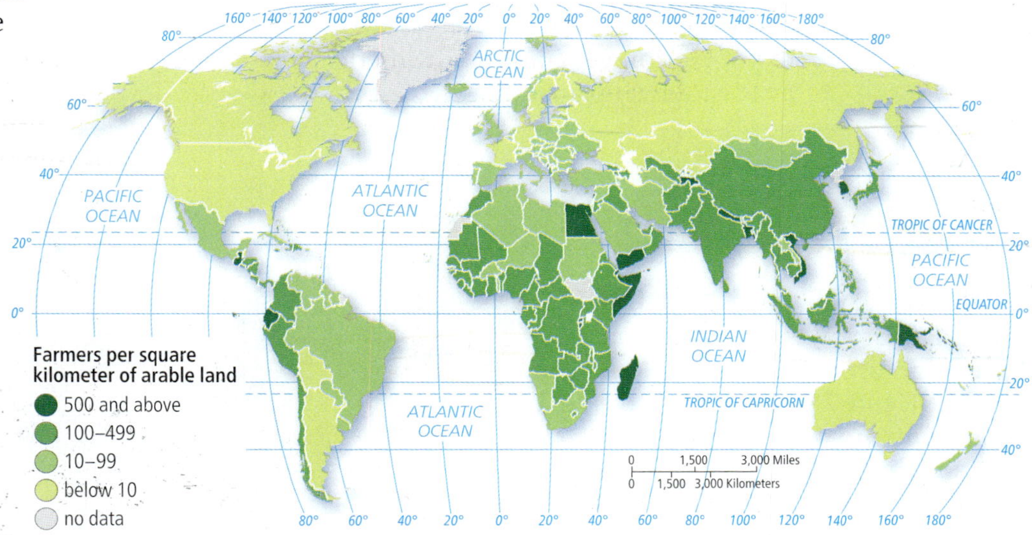

▲ **2.1.5 AGRICULTURAL DENSITY**
Farmers per square kilometer of arable land.

▲ **2.2.1 SOUTH ASIA POPULATION CLUSTER**
Filling containers with water supplied by a truck, New Delhi, India.

Population Concentrations

- **Explain reasons for the distribution of the world's peoples.**

Human beings are not distributed uniformly across Earth's surface. Geographers identify regions of Earth's surface where population is clustered and regions where it is sparse. World maps depict this distribution in several ways.

2.2.1 Population Clusters

Two-thirds of the world's inhabitants live in four regions—East Asia, South Asia (Figure 2.2.1), Southeast Asia, and Europe (Figure 2.2.2). The four population concentrations occupy generally low-lying areas, with temperate climate and soil suitable for agriculture.

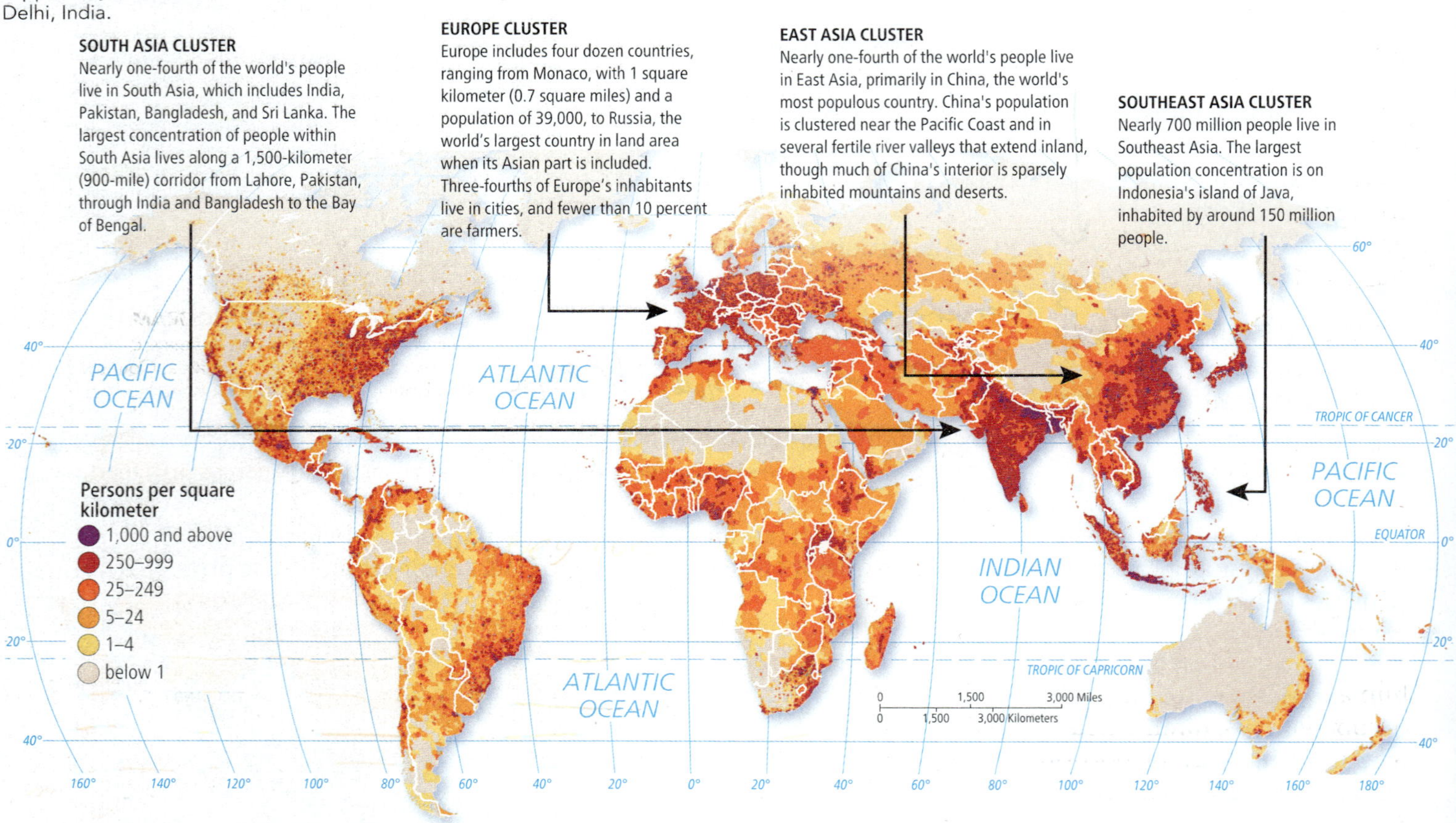

▲ **2.2.2 POPULATION CONCENTRATIONS**

Sparsely Populated Areas

Physical environments that are too dry, too mountainous, too wet, or too cold (Figure 2.2.3) have relatively few inhabitants. The areas of Earth that humans consider too harsh for occupancy have diminished over time.

▶ **2.2.3 SPARSELY POPULATED COLD LANDS**
Angmagssalik, Greenland.

2.2.2 Population Portions

The world can be divided into seven portions, each containing approximately 1 billion people (Figure 2.2.4). The combined population of North America, Latin America, the South Pacific, and Greenland (an area of 53 million square kilometers) is about the same as the population of India (3 million square kilometers) or eastern China (5 million square kilometers).

The small size of the Asia portions in the map of Figure 2.2.4 represents the large number of the world's inhabitants living there (a high concentration of people) compared with the large sizes of the Americas and the South Pacific portions that represent the small number of the world's inhabitants living there (low concentrations of people).

▲ **2.2.4** POPULATION PORTIONS
Each of the seven portions contains approximately 1 billion inhabitants.

Population Cartogram

Concentration can also be displayed on a cartogram (Figure 2.2.5). A cartogram was defined in Chapter 1 as a map that depicts the sizes of countries according to a feature such as population rather than land area.

▼ **2.2.5** POPULATION CARTOGRAM
In a cartogram, countries are displayed by size of population rather than land area. Countries with populations over 50 million are labeled.

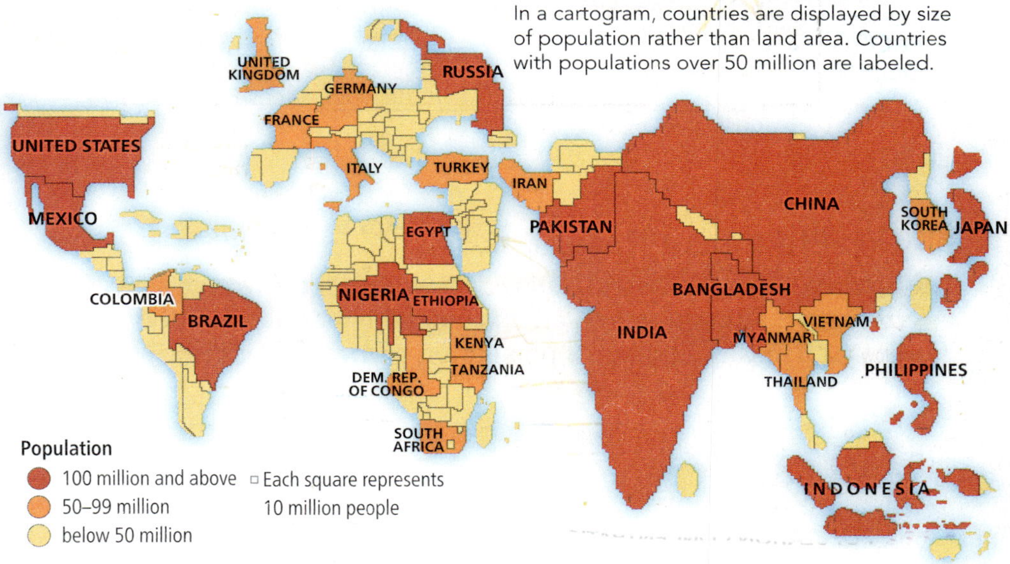

Ecumene

The areas of Earth that humans consider too harsh for occupancy have diminished over time, whereas the portion of Earth's surface occupied by permanent human settlement—called the **ecumene**—has increased (Figure 2.2.6).

◀ **2.2.6** ECUMENE

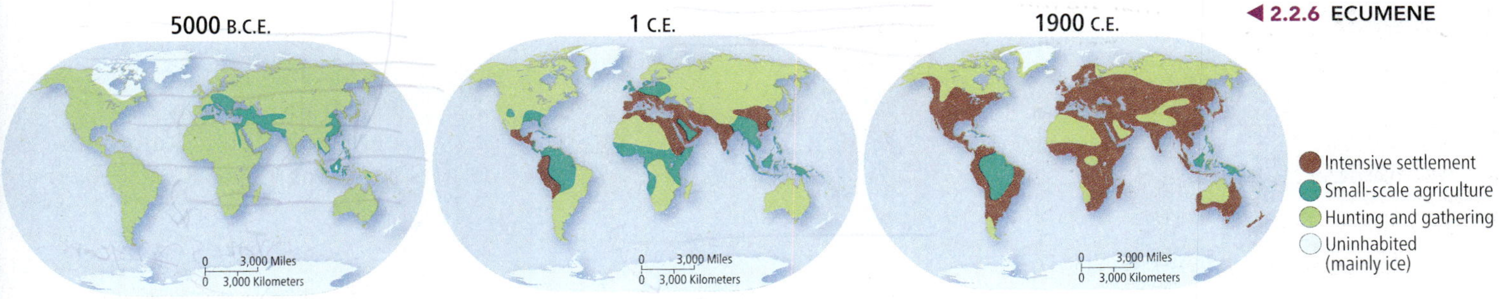

Population Change

- **Explain components contributing to a country's natural increase rate.**

Population increases rapidly in places where more people are born than die, and it declines in places where deaths outnumber births. Geographers most frequently measure population change in a country or the world as a whole by using three measures: natural increase rate (NIR), crude birth rate (CBR), and crude death rate (CDR).

2.3.1 Population Growth Through History

Through most of human history, population was virtually nil. Population increased rapidly beginning in the eighteenth century (Figure 2.3.1). Since 1980, developing countries have accounted for nearly all of the world's population growth (Figure 2.3.2).

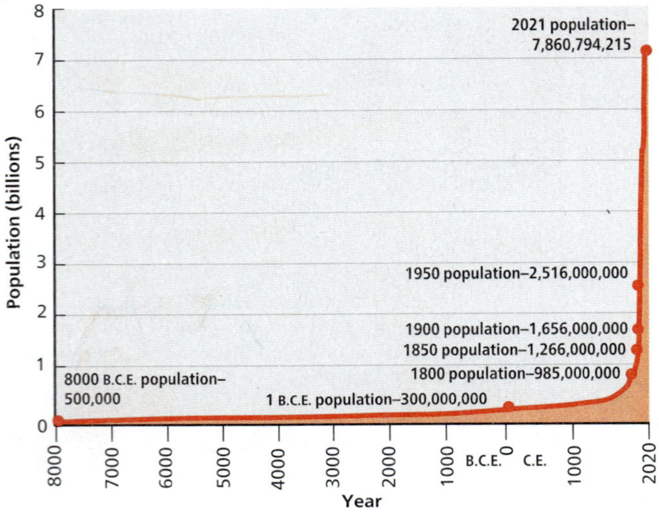

◀ 2.3.1 WORLD POPULATION GROWTH THROUGH THE PAST 10,000 YEARS

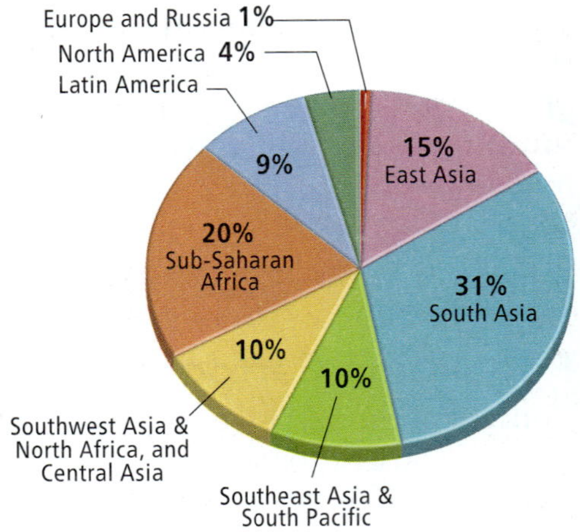

▶ 2.3.2 REGIONAL DISTRIBUTION OF POPULATION GROWTH

Natural Increase Rate

The **natural increase rate (NIR)** is the percentage by which a population grows in a year. The term *natural* means that a country's growth rate excludes migration. Migration is discussed in Chapter 3. The world NIR is currently 1.0, meaning that the population of the world is growing each year by 1.0 percent (Figure 2.3.3).

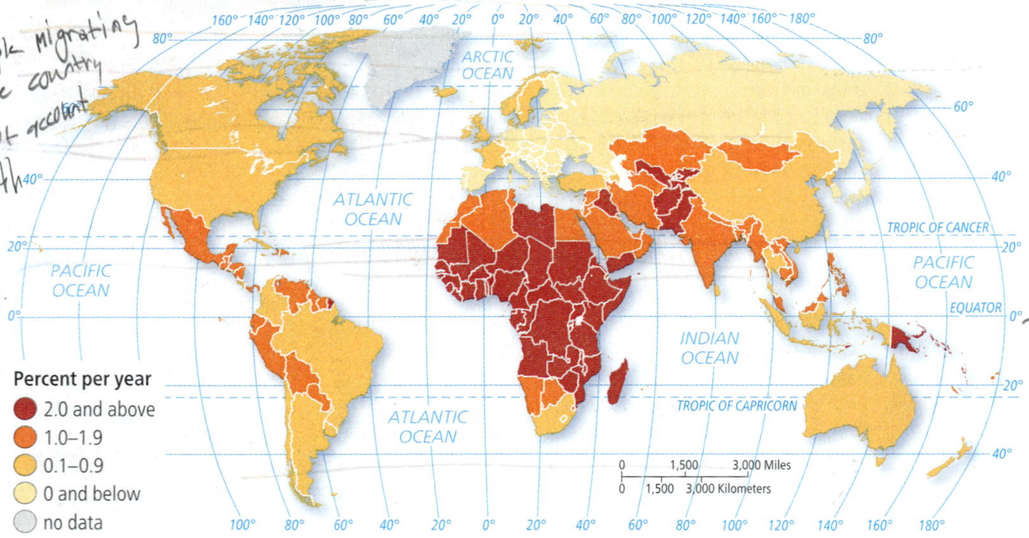

The rate of natural increase affects the **doubling time**, which is the number of years needed to double a population, assuming a constant rate of natural increase. At the current rate of 1.0 percent, world population would double in about 69 years. When the NIR was 2.1 percent in 1968, doubling time was 33 years.

◀ 2.3.3 NATURAL INCREASE RATE

2.3.2 Fertility

The **crude birth rate (CBR)** is the total number of live births in a year for every 1,000 people alive in the society. The current world CBR of 18 means that for every 1,000 people alive, 18 babies are born over a one-year period.

The world map of CBRs mirrors the distribution of NIRs. As was the case with NIRs, the highest CBRs are in sub-Saharan Africa, and the lowest are in Europe (Figure 2.3.4). Many sub-Saharan African countries have a CBR over 40, whereas many European countries have a CBR below 10.

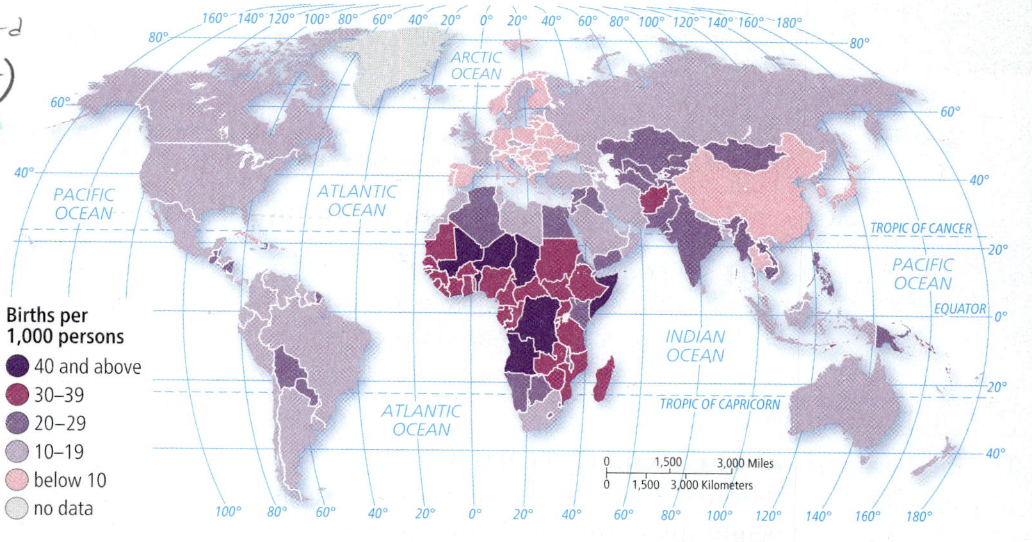

▲ 2.3.4 CRUDE BIRTH RATE

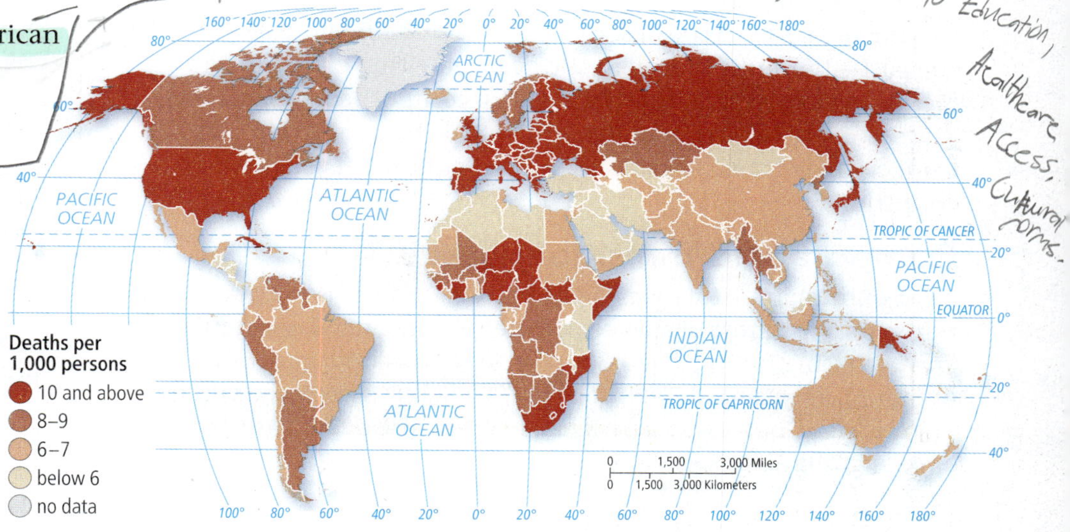

▲ 2.3.5 CRUDE DEATH RATE

Mortality

The **crude death rate (CDR)** is the total number of deaths in a year for every 1,000 people alive in the society. Comparable to the CBR, the CDR is expressed as the annual number of deaths per 1,000 people.

The CDR does not follow the same regional pattern as the NIR and CBR (Figure 2.3.5). The CDR is 7 in developing countries and 11 in developed countries, whereas the CBR is 20 in developing countries and 9 in developed countries. Thus, the combined CDR for all developing countries is actually lower than the combined rate for all developed countries. Furthermore, the variation among CDRs is much less extreme than the variation among CBRs.

Why does Germany, one of the world's wealthiest countries, with one of the world's best health-care systems, have one of the world's highest CDRs (Figure 2.3.6)? The answer is that the populations of different countries are at various stages in an important process known as the demographic transition, discussed on the next page.

▶ 2.3.6 AGING POPULATION
Germany.

The Demographic Transition

- **Describe the stages of the demographic transition.**

All countries have experienced some changes in NIR, CBR, and CDR but at different times and at different rates. The demographic transition helps us understand these differences.

2.4.1 Demographic Transition: Overview

The **demographic transition** is a process of change in a society's population from high crude birth and death rates and low natural increase rate to a condition of low crude birth and death rates, low natural increase rate, and higher total population.

The four-stage demographic transition is characterized by two big breaks with the past (Figure 2.4.1). The first break with the past—the sudden drop in the death rate that comes from technological innovation—has been accomplished everywhere. The second break—the sudden drop in the birth rate that comes from changing social customs—has yet to be achieved in many countries.

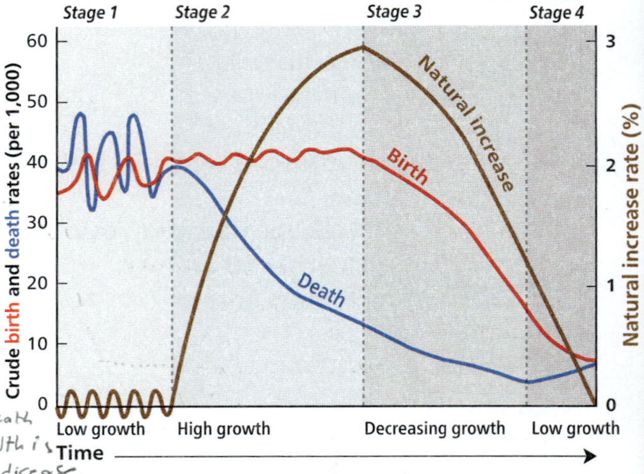

▲ 2.4.1 DEMOGRAPHIC TRANSITION MODEL

Demographic Transition: Stage 1

Most of human history was spent in stage 1 of the demographic transition. For most of this period, people depended on hunting and gathering for food (see Chapter 9). When food was easily obtained, a region's population increased, but it declined when people were unable to feed themselves because of inability to locate enough animals or vegetation.

▼ 2.4.2 DEMOGRAPHIC TRANSITION STAGE 2: BENIN

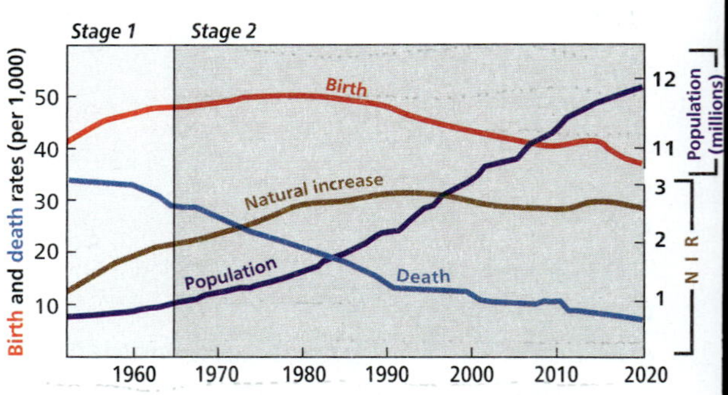

Demographic Transition: Stage 2

No country remains in stage 1. Every country has moved on to at least stage 2 and has experienced profound changes in population. In stage 2, rapidly declining death rates and very high birth rates produce a very high natural increase.

Europe and North America entered stage 2 of the demographic transition after 1750 as a result of the **Industrial Revolution** (see Chapter 11). An unprecedented level of wealth characterized the Industrial Revolution, some of which was invested to make communities healthier places to live.

Stage 2 of the demographic transition did not diffuse to Africa, Asia, and Latin America until after 1950. The push of developing countries into stage 2 was caused by the **medical revolution**. Development of new medicines and medical treatments and invention of advanced medical technologies in Europe and North America diffused to developing countries. Improved medical practices have eliminated many of the traditional causes of death in developing countries.

Benin is one of the least developed countries in Africa. Only 16 percent of women of reproductive age practice family planning. Benin's death rate declined rapidly beginning in the 1970s when the World Health Organization launched a program to immunize children in a number of countries, including Benin. This program helped reduce the CDR in Benin, transitioning the country from stage 1 to stage 2, where it is now (Figures 2.4.2 and 2.4.3).

▼ 2.4.3 BENIN: HIGH CRUDE BIRTH RATE

2.4.2 Demographic Transition: Stage 3

A country moves from stage 2 to stage 3 of the demographic transition when the CBR begins to drop sharply. During stage 3, the population continues to grow because the CBR is still greater than the CDR. But the rate of natural increase is more modest in countries in stage 3 than in those in stage 2 because the gap between the CBR and the CDR narrows (Figure 2.4.4).

A society enters stage 3 when people have fewer children. The decision people make to have fewer children is partly a delayed reaction to a decline in mortality. Economic changes during the transition from stage 2 to stage 3 also induce some people to have fewer offspring.

Most countries in Europe and North America (including the United States) moved from stage 2 to stage 3 of the demographic transition during the first half of the twentieth century. Many countries of Asia and Latin America, including Mexico, went through the transition into stage 3 during the second half of the twentieth century.

Mexico entered stage 2 of the demographic transition when the country experienced a combination of lower death rates and higher birth rates. The higher birth rates in Mexico occurred in part because the government proclaimed that higher birth rates would be good for the country's economic growth (Figure 2.4.5).

A dramatic decline in Mexico's birth rates came in the 1970s. In 1974, an amendment to Mexico's constitution guaranteed families the legal right to decide on the number of children they would have and the spacing of children. In 1979, Mexico established a National Population Council to promote family planning through education.

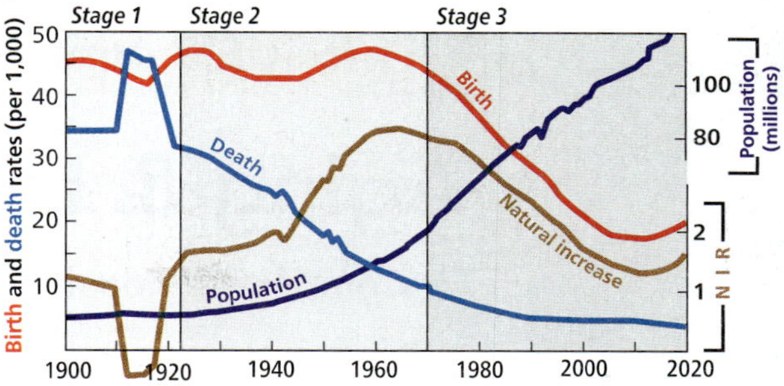

▲ **2.4.4** DEMOGRAPHIC TRANSITION STAGE 3: MEXICO

▲ **2.4.5** MEXICO: MODERATE NIR

Demographic Transition: Stage 4

A country reaches stage 4 when the CBR declines to the point where it equals the CDR and the NIR approaches zero (Figure 2.4.6). This condition is called **zero population growth (ZPG)**, a term often applied to stage 4 countries. ZPG may occur when the CBR is still slightly higher than the CDR because some females die before reaching childbearing years and because the number of females in their childbearing years can vary.

Social customs explain the movement to stage 4. Increasingly, women in stage 4 societies enter the labor force rather than remain at home as full-time homemakers. In stage 4, people have access to a wider variety of birth-control methods.

The United Kingdom entered stage 2 of the demographic transition in the late 1700s, as a result of the Industrial Revolution. Stage 3 occurred there in the twentieth century, and it entered stage 4 into the twenty-first century. As a result, elderly people now outnumber young people in the United Kingdom (Figure 2.4.7).

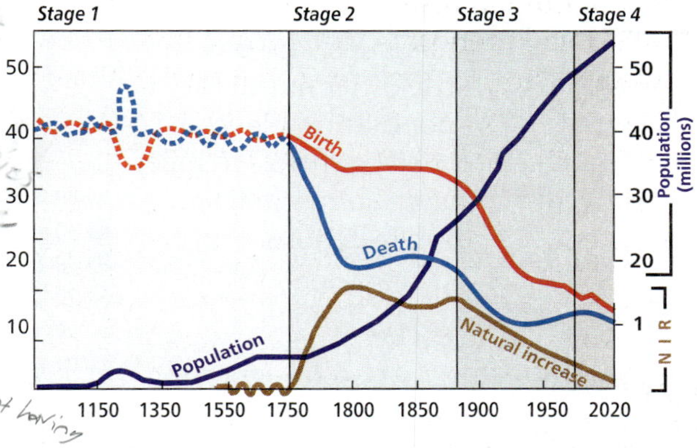

▲ **2.4.6** DEMOGRAPHIC TRANSITION STAGE 4: UNITED KINGDOM

▶ **2.4.7** UNITED KINGDOM: LOW NIR

Aging & Health

- **Compare regional differences in percentages of young and old.**

A country's stage of the demographic transition determines the percentage of people in different age groups. A country in stage 2 of the demographic transition typically has a relatively high percentage of young people, whereas a country in stage 4 has a relatively high percentage of elderly people. The percentage of different age groups helps to explain a country's distinctive health challenges.

▲ 2.5.1 MANY CHILDREN
Uganda.

2.5.1 Countries with Many Young People

The **dependency ratio** is the number of people who are too young or too old to work compared to the number of people in their productive years. Dependents are normally classified as people under age 15 and over age 65. The larger the dependency ratio, the greater the financial burden on those who are working to support those who do not. The dependency ratio is 54 percent in Europe, compared to 82 percent in sub-Saharan Africa.

Developing countries in stage 2 of the demographic transition have higher percentages of young people than do countries in stage 4 (Figure 2.5.1). In sub-Saharan Africa, 42 percent of the population is under age 15, whereas only 16 percent are under age 15 in Europe (Figure 2.5.2).

The large percentage of children in sub-Saharan Africa strains the ability of the developing countries in that region to provide needed services such as schools, hospitals, and day-care centers. On the other hand, the "graying" of the population places a burden on developed countries to meet their needs for income and medical care after they retire from jobs.

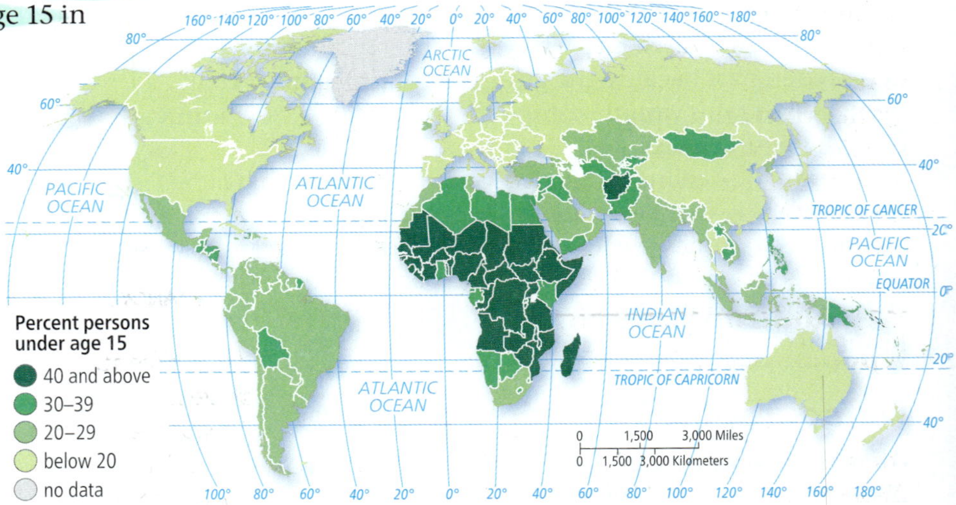

▲ 2.5.2 PERCENT UNDER AGE 15

Infant Mortality

The **infant mortality rate (IMR)** is the annual number of deaths of infants under 1 year of age, compared with the total live births (Figure 2.5.3). As is the case with the CBR and CDR, the IMR is expressed as the number of deaths among infants per 1,000 births rather than as a percentage (per 100). In general, the IMR reflects a country's health-care system.

The global distribution of IMRs follows a clear pattern: Lower IMRs are found in countries with well-trained doctors and nurses, modern hospitals, and large supplies of medicine. The IMR is 4 in European countries in stage 4, compared with an IMR of 50 in sub-Saharan Africa, where a number of countries are in stage 2.

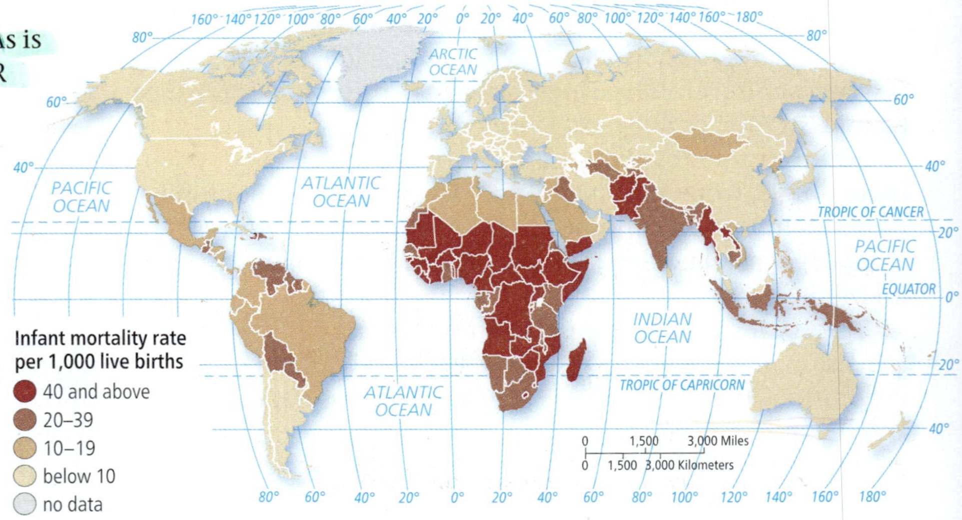

▲ 2.5.3 INFANT MORTALITY RATE

2.5.2 Countries with Many Older People

Life expectancy is the average number of years an individual can be expected to live, given current social, economic, and medical conditions. Life expectancy at birth measures the average number of years a newborn infant can expect to live at current mortality levels. People on average live much longer in developed countries than in developing ones. The life expectancy in Europe is 78, but only 62 in sub-Saharan Africa (Figure 2.5.4).

As countries progress through the demographic transition, they face increasing percentages of older people who need adequate levels of income and medical care after they retire from their jobs. In the United States, the percentage of the population over 65 varies widely; the highest percentages are in states to which many older people migrate and in states from which many younger people leave.

The "graying" of the population places a burden on governments in developed countries to meet these needs. The **potential support ratio** (also known as the **elderly support ratio**) is the number of working-age people (ages 15 to 64) divided by the number of persons 65 and older.

The world's potential support ratio is currently around 6, meaning that there are 6 people of working age for every elderly person. In 2050, the ratio is expected to decline to around 4, meaning that there will be only 4 people of working age available to support elderly people who have retired from work. Thus, as the potential support ratio gets smaller, fewer workers are available to contribute to the pensions, health care, and other support that older people need.

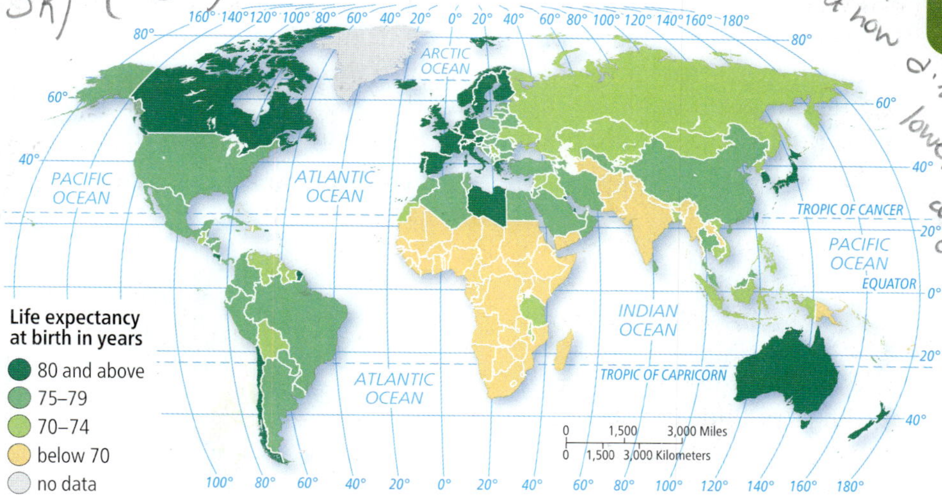

▲ 2.5.4 LIFE EXPECTANCY AT BIRTH

Population Pyramids

A **population pyramid** is a bar graph that displays the percentage of a place's population according to specified age groups and gender. A country that is in stage 2 of the demographic transition, such as Benin, has a pyramid with a broader base than that of a country in stage 4, such as the United Kingdom, indicating that there is a higher percentage of young people in Benin than in the United Kingdom. In contrast to Benin's gender distribution reflected by a more evenly tapering pyramid from bottom to top, the highest levels of the United Kingdom's gender distribution at the top of the pyramid are wider on the right than they are on the left because women live longer than men (Figure 2.5.5).

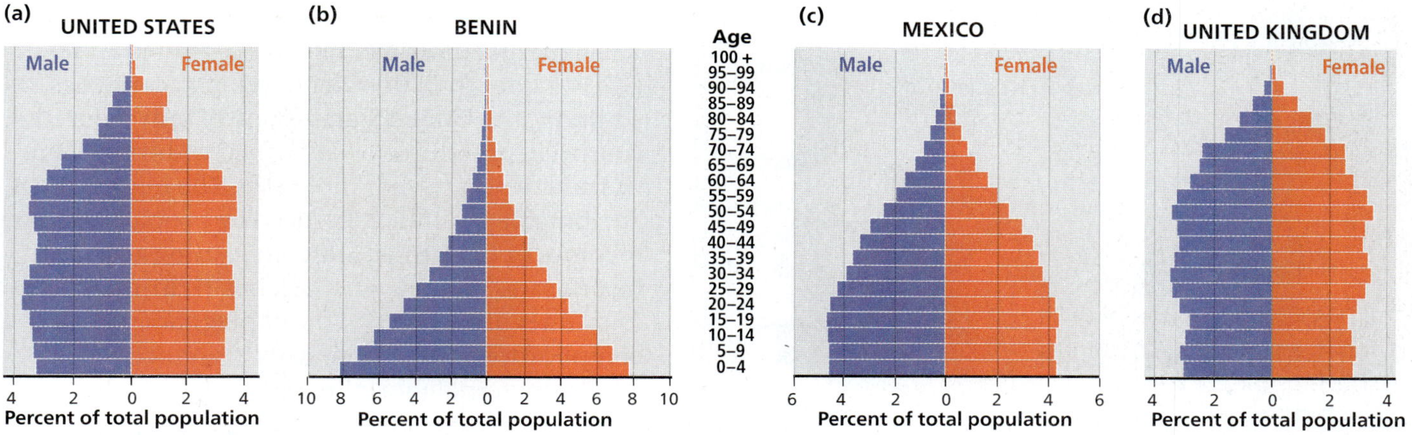

▲ 2.5.5 POPULATION PYRAMIDS
(a) United States, (b) Benin, (c) Mexico, (d) United Kingdom.

Gender & Health

- **Explain health risks faced by mothers and baby girls.**

As world NIR slows and more countries move into stage 3 or 4 of the demographic transition, geographers increasingly turn their attention to the health of the record number of people who are alive. Females face especially challenging health risks at childbirth. Both the mother and the baby girl are at risk.

2.6.1 Mothers at Risk

The **maternal mortality rate** is the annual number of female deaths per 100,000 live births from any cause related to or aggravated by pregnancy or its management (excluding accidental or incidental causes). The worldwide maternal mortality rate in 2018 was 211, amounting to more than 300,000 women. The rate exceeds 525 deaths per 100,000 live births in Africa, compared to 13 in Europe (Figure 2.6.1). Nonetheless, the world has made progress: the world maternal mortality rate was 385 in 1990.

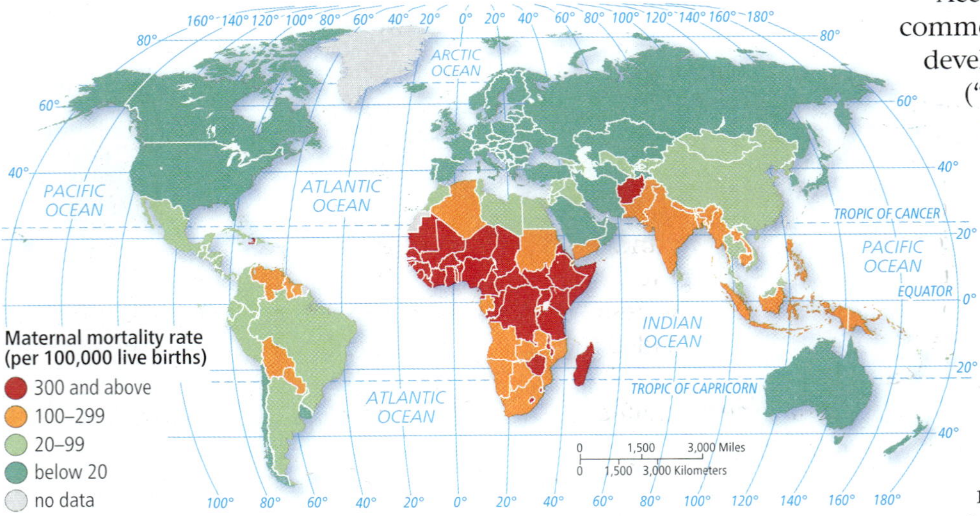

▲ **2.6.1** MATERNAL MORTALITY RATE

According to the U.N., the most common cause of maternal death in developing countries is heavy bleeding ("obstetrical hemorrhage"), followed by high blood pressure ("hypertensive disorders of pregnancy"). Developed countries have medical facilities, advanced technologies, and trained personnel to limit the incidence of life-threatening conditions during childbirth. Physicians in developed countries can monitor and treat conditions prior to delivery and manage complications during and after delivery (Figure 2.6.2).

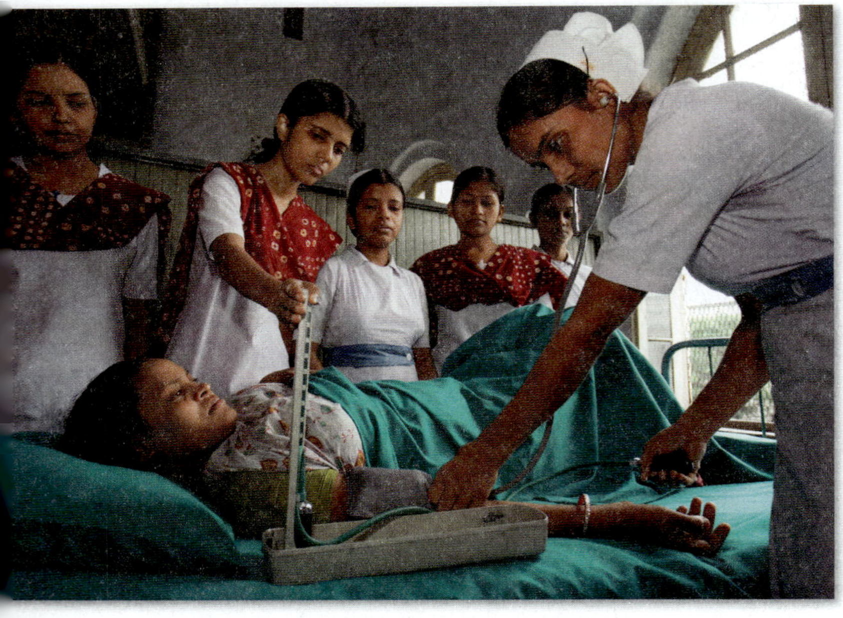

▲ **2.6.2** GIVING BIRTH, KOLKATA, INDIA
Training midwifes at Kolkata Nursing College.

Mothers at Risk: United States

Despite uniform advances in maternal health care throughout developed countries, maternal mortality rates vary among developed countries. The maternal mortality rate in the United States (17) is higher than in other developed countries. In addition, the United States is one of only a dozen countries, and the only developed country, where the rate has increased since 1990. The higher rate is attributable to difficulties faced by people with low incomes in the United States in gaining access to health care. Women who lack health insurance are four times more likely to die of pregnancy-related complications than those with insurance.

The maternal mortality rate varies widely among U.S. states. Several southern states have maternal mortality rates that are more than twice as high as the national average.

2.6.2 Baby Girls at Risk

Around 700,000 female babies are "missing" every year in China and India as a result of gender-based sex selection. The U.N. Population Fund estimates that, overall, 140 million more females should have been born in Asia over the past several decades. The likely explanations for the diminished female population in that region of the world are the abortion of female fetuses, female infanticide (intentional killing of a newborn female), and unreported births of females.

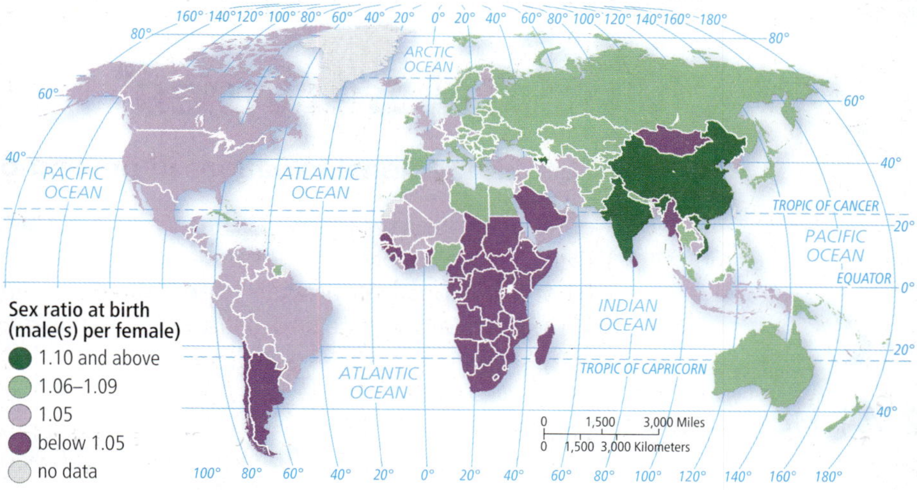

▲ **2.6.3** SEX RATIO AT BIRTH

We know about the large number of "missing" females in Asia because of the region's imbalanced **sex ratio**, which is the number of males per 100 females in the population. The standard sex ratio for human births is around 105 male babies for every 100 female babies. Scientists are not sure why naturally a few more males than females are born. The standard sex ratio is characteristic of the developed regions of North America and Europe as well as in the developing regions of Latin America and sub-Saharan Africa. However, the sex ratio at birth is 114:100 in China and 110:100 in India (Figure 2.6.3). The percentage of newborn females in the world's two most populous countries is much too low to have occurred by chance. The U.N. concludes that the "root cause" of the unequal sex ratio in Asia is gender inequality.

The extremely low percentage of female babies in China and India results from a cultural preference for sons over daughters. Sons are regarded as more likely than daughters to help the family economically. Government policies in China and India to promote smaller families have resulted in births of more male than female babies in those two countries. Many Chinese and Indians have decided that if they are going to have a small number of children, they want them to be boys (Figure 2.6.4).

Efforts to legally ban gender-based selection have been ineffective. An ultrasound reveals the gender of a fetus before birth, making gender-based selective abortion possible. If a baby girl is born, her parents may hide her from authorities in a remote rural area, and she may receive less food and health care than a baby boy.

Aside from ethical questions about the devaluation of female lives, the widespread practice of sex selection in China and India is creating a practical problem. As the babies grow to adulthood, these countries are left with an enormous surplus of men who are unable to find women to marry.

▼ **2.6.4** CHINA: MORE BOYS THAN GIRLS
Liufu, China.

Medical Services

- Understand regional variations in providing health care.

Health conditions vary around the world. Countries possess different resources to care for people who are sick. Developed countries use part of their wealth to provide publicly funded medical care to all citizens. The high expenditure on health care in developed countries is reflected in medical facilities.

2.7.1 Hospital Beds

Many countries in Europe have more than 50 hospital beds per 10,000 people, compared to fewer than 30 in sub-Saharan Africa and South and Southwest Asia (Figure 2.7.1).

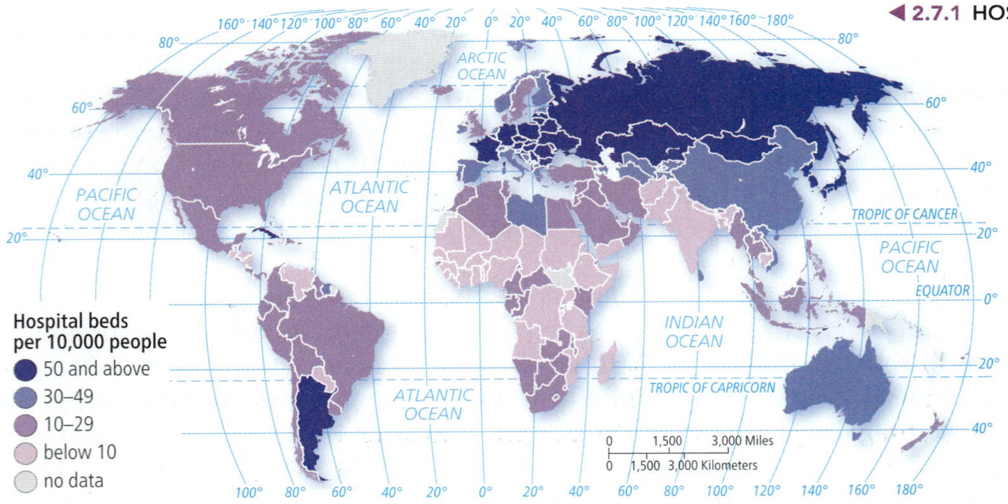

◀ **2.7.1 HOSPITAL BEDS PER 10,000 PEOPLE**

Physicians

Europe has more than 30 physicians per 10,000 people, compared to fewer than 5 in sub-Saharan Africa (Figures 2.7.2 and 2.7.3).

▶ **2.7.2 PHYSICIANS PER 10,000 PEOPLE**

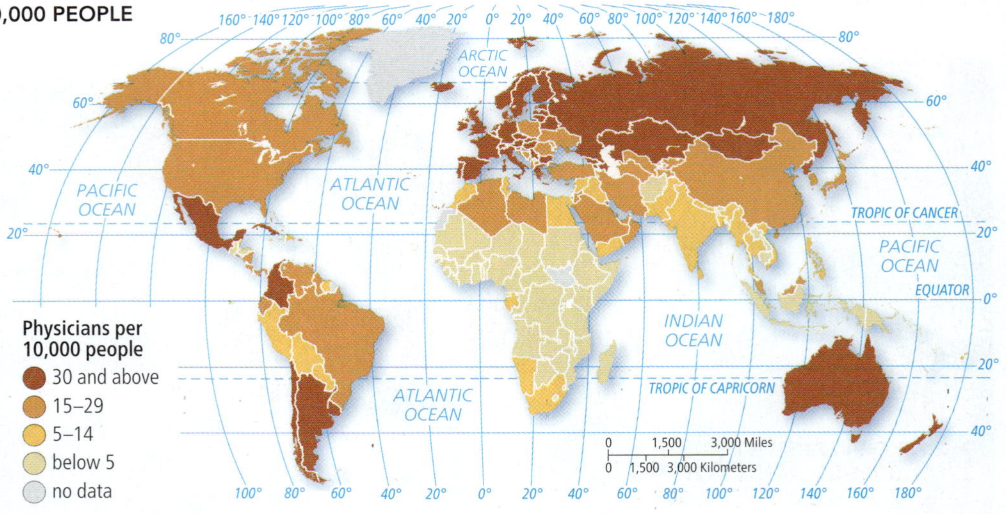

▼ **2.7.3 HOSPITAL, BORDEAUX, FRANCE**
A team of doctors perform brain surgery.

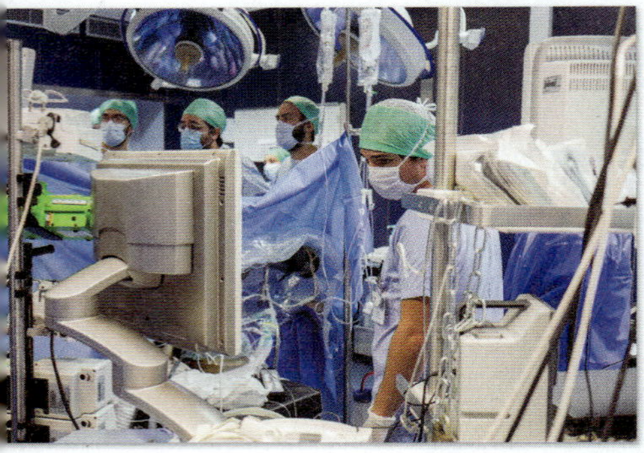

2.7.2 Health-Care Expenditures

Annual per capita expenditure on health care exceeds $1,000 in developed countries, compared to less than $100 in sub-Saharan Africa and South Asia (Figure 2.7.4). In most developed countries, health care is a public service that is available at little or no cost. Government programs pay more than 70 percent of health-care costs in most European countries, and private individuals pay less than 30 percent. In developing countries, private individuals must pay more than half the cost of health care. An exception to this pattern is the United States, a developed country where private individuals are required to pay an average of 55 percent of health care, more closely resembling the pattern in developing countries.

▶ **2.7.4 HEALTH-CARE EXPENDITURES PER CAPITA**

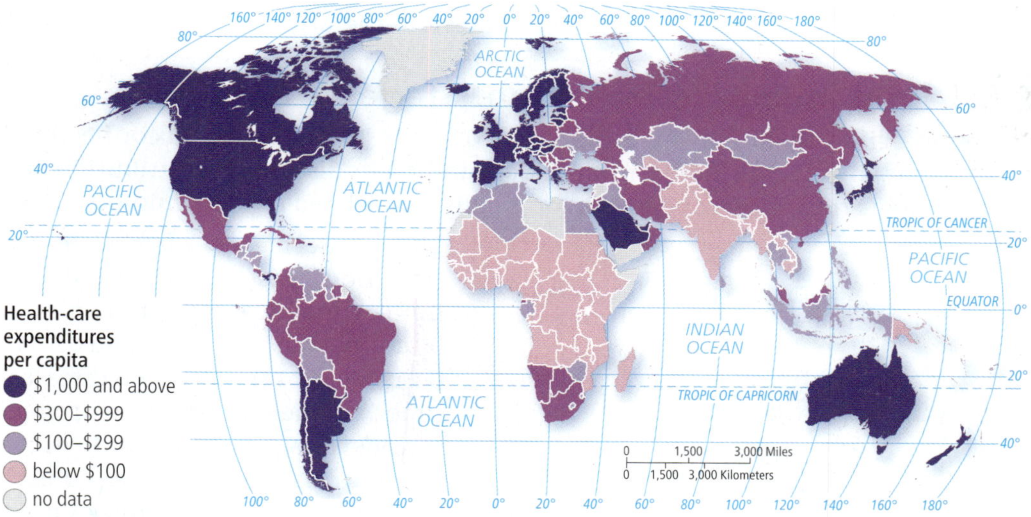

Government Expenditures on Health Care

The United States is the only populous country that does not have universal health care. Despite this, the U.S. government is spending a higher percentage of its GDP on health care than do other countries, even those with national health care systems (Figure 2.7.5).

Developed countries are hard-pressed to maintain their current levels of government expenditures for health care. In the past, rapid economic growth permitted these countries to finance generous health-care programs with little difficulty. But in recent years, economic growth has slowed while the percentage of people seeking government assistance to meet the cost of health care has increased. Governments have faced a choice between reducing financial assistance for health care and increasing taxes to pay for it. In some developing countries, threats to health and sustainability are not so much financial as environmental.

▶ **2.7.5 GOVERNMENT EXPENDITURES ON HEALTH CARE AS PERCENTAGE OF GDP**

Epidemiologic Transition

- **Summarize the first four stages of the epidemiologic transition.**

The **epidemiologic transition** focuses on distinctive health threats in each stage of the demographic transition. **Epidemiology** is the branch of medical science concerned with the incidence, distribution, and control of diseases that are prevalent among a population at a particular time and are produced by some special causes not generally present in the affected place.

2.8.1 Epidemiologic Transition, Stages 1 & 2

Epidemiologists rely heavily on geographic concepts such as scale and connection because measures to control and prevent an epidemic derive from understanding its distinctive distribution and method of diffusion. The epidemiologic transition concept was originally formulated by epidemiologist Abdel Omran in 1971.

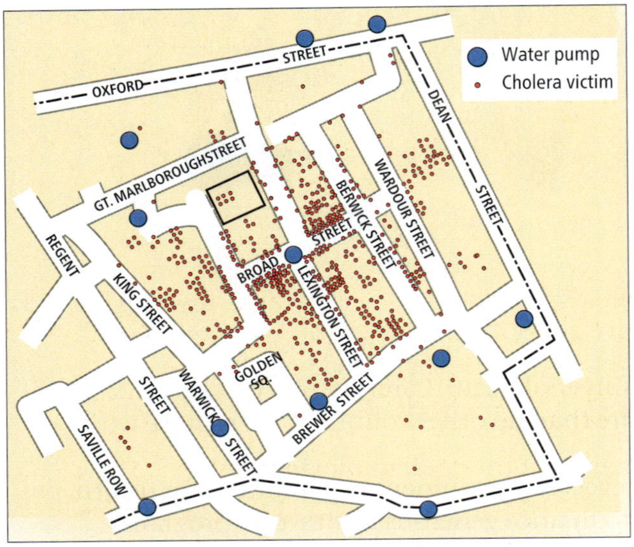

▲ 2.8.1 DR. JOHN SNOW'S 1854 CHOLERA MAP OF LONDON

Stage 1: Pestilence & famine. In stage 1 of the epidemiologic transition, epidemics and pandemics were principal causes of human deaths, along with famine and accidents and attacks by animals and other humans. An **epidemic** is a widespread occurrence of an infectious disease in a community at a particular time. An epidemic affects many persons at the same time and spreads from person to person in a locality where the disease is not permanently prevalent. **Pestilence** is the term for an especially virulent epidemic. A **pandemic** is an infectious disease that occurs over a wide geographic area and affects a very high proportion of the population at the same time. A disease is considered **endemic** when it is seasonal and/or more predictable.

History's most violent stage 1 pandemic was the Black Plague (bubonic plague), which was probably transmitted to humans by fleas from migrating infected rats. About 25 million Europeans—at least one-half of the continent's population—died between 1347 and 1350.

Stage 2: Receding pandemics. Stage 2 of the epidemiologic transition is the stage of receding pandemics. Improved sanitation, nutrition, and medicine during the Industrial Revolution reduced the spread of infectious diseases.

A century before the invention of computers, a mapping technique that anticipated GIS helped explain and battle the stage 2 pandemic of cholera in nineteenth-century London. Dr. John Snow (1813–1858), a British physician (not a geographer), created a hand-drawn GIS in 1854 (Figure 2.8.1). The overlay maps showed that cholera victims were not distributed uniformly through London's Soho neighborhood, but rather were clustered around one water pump on Broad Street. Tests at the Broad Street pump subsequently proved that the water there was contaminated by sewage. London's poor were disproportionately affected by cholera because they were more likely to have to use contaminated water.

Cholera persists to this day in several developing regions in stage 2 of the demographic transition, especially sub-Saharan Africa and South and Southeast Asia, where many people lack access to clean drinking water (Figure 2.8.2).

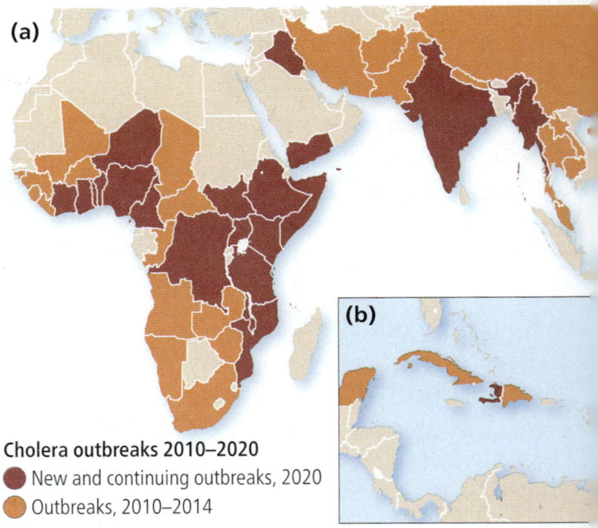

▲ 2.8.2 STAGE 2 DISEASE: CHOLERA
(a) Cholera in Africa and southern Asia, and (b) cholera in the Caribbean.

2.8.2 Epidemiologic Transition, Stages 3 & 4

Omran's epidemiologic transition included a third stage. It was extended by gerontologists S. Jay Olshansky and Brian Ault to stage 4.

Stage 3: Degenerative diseases. Stage 3 of the epidemiologic transition is characterized by a decrease in deaths from infectious diseases and an increase in chronic disorders and degenerative diseases associated with aging. The two especially important degenerative diseases in stage 3 are cardiovascular diseases, such as heart attacks, and various forms of cancer. The global pattern of cancer is the opposite of that for stage 2 diseases; sub-Saharan Africa and South Asia have the lowest incidence of cancer, primarily because of the relatively low life expectancy in those regions (Figure 2.8.3).

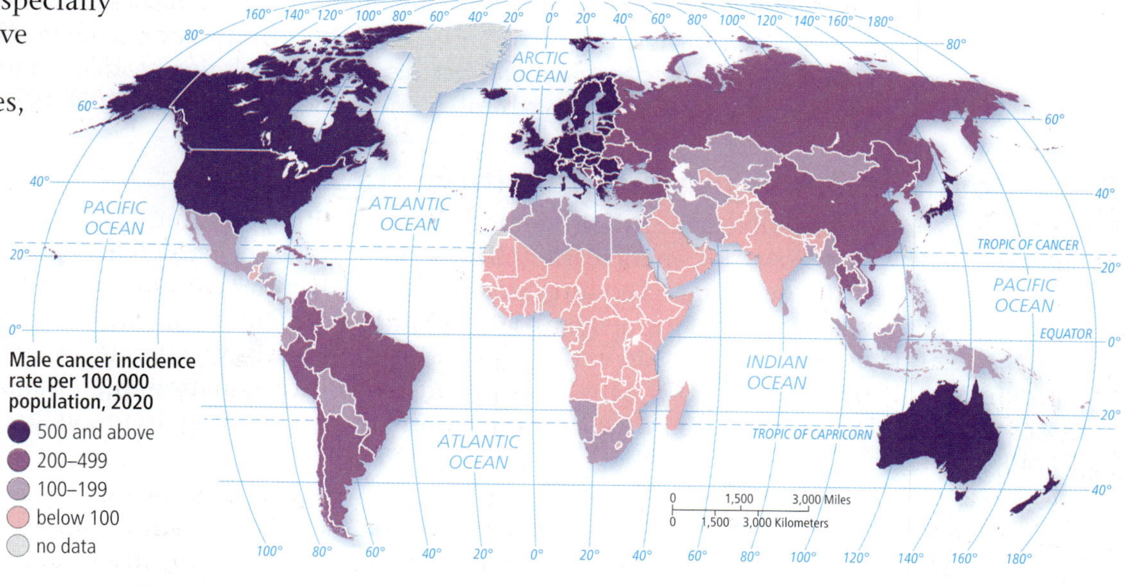

▲ **2.8.3 STAGE 3 DISEASE: MALE CANCER**

Stage 4: Delayed degenerative & lifestyle diseases. Stage 4 is characterized by delayed degenerative diseases. The major degenerative causes of death—cardiovascular diseases and cancers—exist through stage 4, but with stage 4 of the epidemiologic transition, the life expectancy is longer because of medical advances. Through medicine, cancers spread more slowly or are removed altogether. Operations such as bypasses repair deficiencies in the cardiovascular system.

Some behavioral changes have improved health in countries at stage 4 of the epidemiologic transition. These include better diet, reduced use of tobacco and alcohol, and more exercise. On the other hand, death rates have increased in stage 4 countries through use of prescription and illegal drugs, consumption of nonnutritious food, and sedentary behavior.

Increased deaths related to opioids have been especially challenging in stage 4 countries (Figure 2.8.4). As shown in Chapter 1, opioid-related deaths have not been distributed uniformly within the United States. Deaths are spatially associated with frequency of prescription of opioids (figures 1.7.4 and 1.7.5).

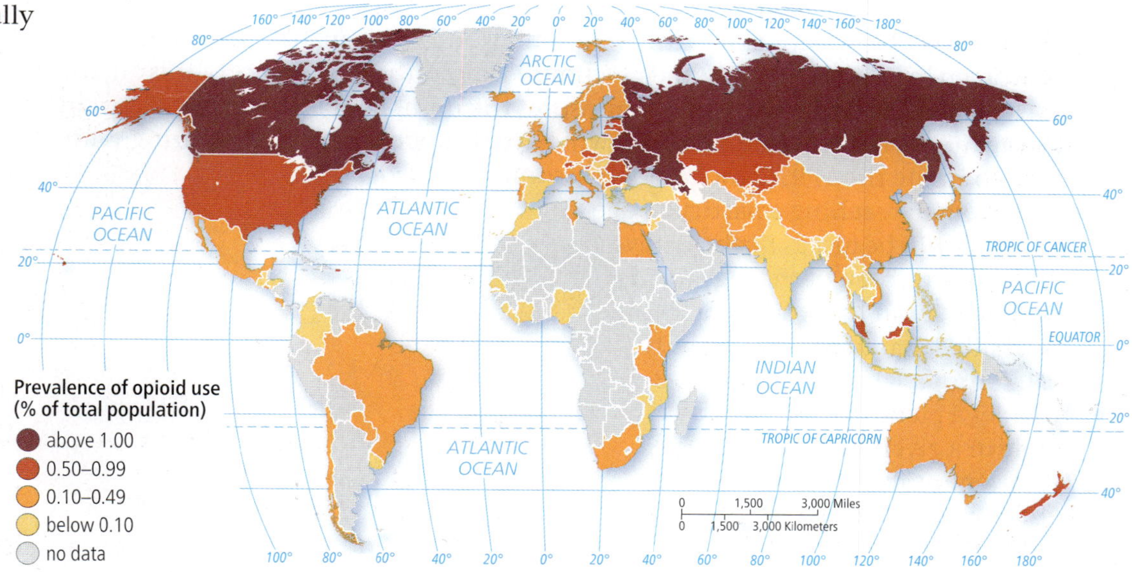

▲ **2.8.4 STAGE 4 DISEASE: OPIOIDS**

Population Futures

- **Describe a possible stage 5 of the demographic transition.**

Geographic concepts offer insights into future population and health trends. Considering the current size of Earth's population and the NIR, will there soon be too many of us? Or will Earth's total population stabilize and even decline in the future in the face of low birth rates and pandemics like COVID-19?

2.9.1 Possible Demographic Transition Stage 5

Demographers predict a possible stage 5 of the demographic transition for some developed countries. Stage 5 would be characterized by a very low CBR, an increasing CDR, and therefore a negative NIR (Figure 2.9.1). After several decades of very low birth rates, a stage 5 country would have relatively few young women aging into childbearing years. If those in the smaller pool of women have fewer children, the CBR would fall lower than in stage 4.

The U.N. estimates world population in the future. The moderate U.N. estimate for world population in 2100 is approximately 11.0 billion. The U.N.'s highest estimate for world population in 2100 is more than double the current world population, 15.6 billion. The low U.N. estimate for world population in 2021 is actually a decline to 7.3 billion (Figure 2.9.2).

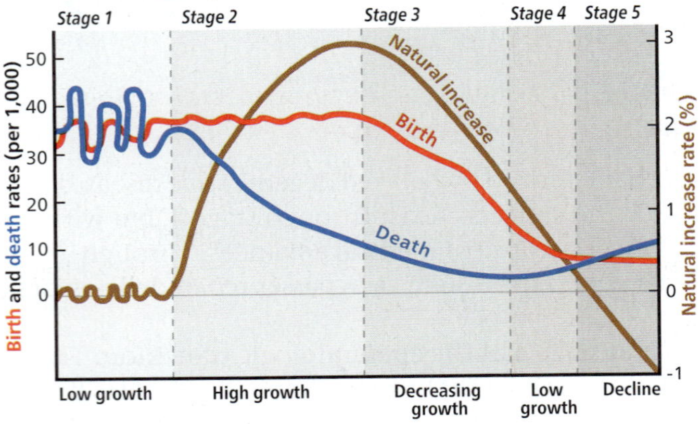

▲ 2.9.1 DEMOGRAPHIC TRANSITION WITH FIVE STAGES

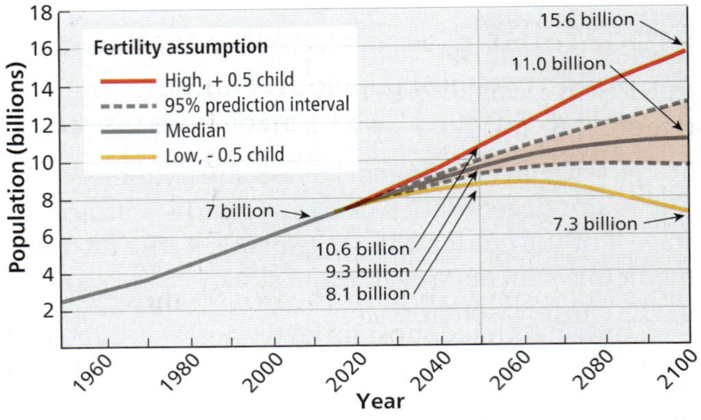

▲ 2.9.2 FUTURE POPULATION ESTIMATES

Total Fertility Rate

How many people will inhabit Earth in the future? Future population depends primarily on fertility: How many babies will women in the future bear in their lifetimes? Despite the importance of this question, the answer is unknowable.

The **total fertility rate (TFR)** measures the number of births in a society (Figure 2.9.3). The TFR is the average number of children a woman will have throughout her childbearing years (roughly ages 15 through 49).

To compute the TFR, demographers assume that a woman reaching a particular age in the future will be just as likely to have a child as are women of that age today. The TFR is 2.3 in the world as a whole, 4.7 in sub-Saharan Africa, and only 1.5 in Europe. The CBR (Figure 2.3.4) provides a picture of a society as a whole in a given year, whereas the TFR attempts to predict the future behavior of individual women in a world of rapid cultural change.

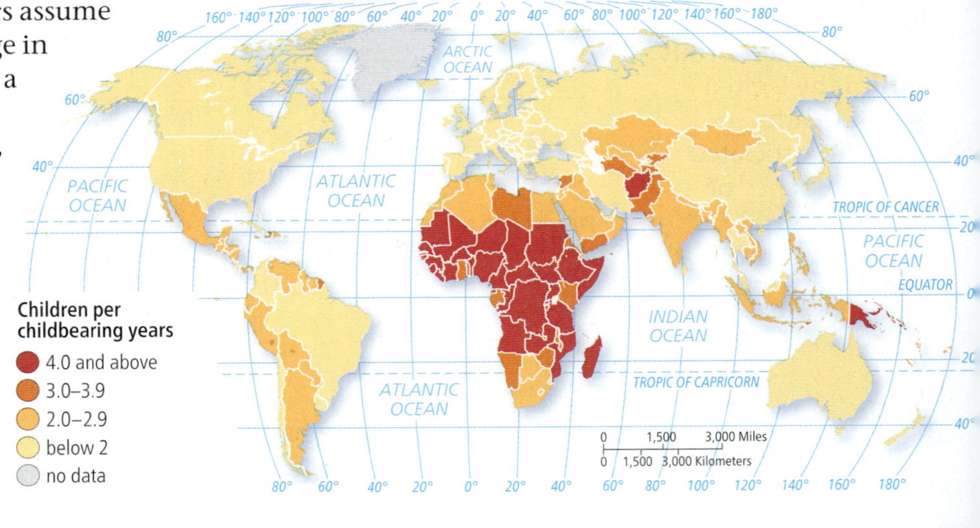

▶ 2.9.3 TOTAL FERTILITY RATE

2.9.2 Possible Stage 5 Countries

If the demographic transition is to include a stage 5, Japan and several European countries will be among the world's first countries to reach it.

Japan's population hit a historic high of 128 million in 2010 and is now starting to decline. The Statistics Bureau of Japan forecasts Japan's population to fall to 85 million in 2050. With the population decline will come an increasing percentage of elderly people (Figure 2.9.4).

Japan faces a shortage of workers. Instead of increasing immigration, Japan is addressing its shortage primarily by encouraging more Japanese people to work. Rather than combine work with child rearing, Japanese women are expected to make a stark choice: either marry and raise children or remain single and work. According to Japan's most recent census, the majority of women have chosen to work.

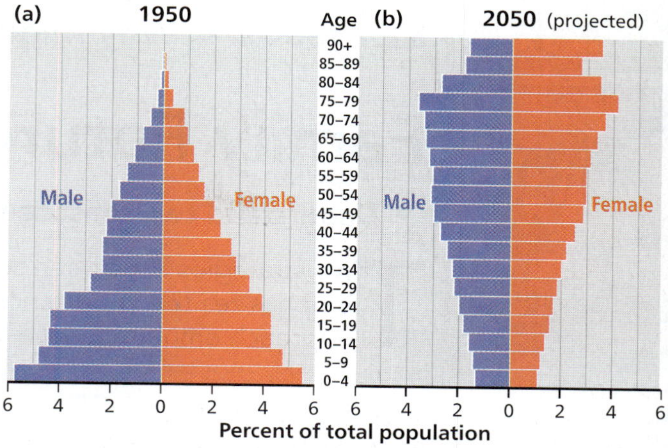

▲ 2.9.4 POSSIBLE STAGE 5: JAPAN'S POPULATION PYRAMIDS

Several European countries, notably Russia and other former Communist countries, already have negative NIRs. Russia's high CDR and low CBR are a legacy of a half-century of Communist rule. The low CBR in former Communist countries may stem from a long tradition of strong family planning programs and a deep-seated pessimism about having children in an uncertain world. The high CDR may be a legacy of inadequate pollution controls and inaccurate reporting of deaths by former Communist governments.

Pronatalist & Antinatalist Policies

The world's two most populous countries, China and India, will heavily influence future prospects for global population. A **pronatalist policy** is a government policy that supports higher birth rates, whereas an **antinatalist policy** is one that supports lower birth rates. In recent decades, China and India have both instituted antinatalist policies with varying success. These two countries—together encompassing more than one-third of the world's population—have adopted different family planning programs. As a result of less effective policies, India adds 13 million more people each year than does China. Current projections show that India will surpass China as the world's most populous country sometime in the 2020s.

China has made substantial progress in reducing its NIR. The core of the Chinese government's family planning program was the One Child Policy, adopted in 1980 and in effect through 2015. Under the One Child Policy, a couple needed a permit to have a child (Figure 2.9.5). Couples received financial subsidies, a long maternity leave, better housing, and (in rural areas) more land if they agreed to have just one child. To discourage births, people received free contraceptives, abortions, and sterilizations. Rules were enforced with penalties by a government agency.

Largely as a result of the One Child Policy, China's CBR declined from 18 in 1980 to 8 in 2021, and consequently, the NIR declined from 1.2 to 0.1. Since 2020, the CBR has been lower in China than in the United States. The number of people added to China's population each year dropped from 14 million to 1 million during the past four decades.

With the U.N. now forecasting China to lose population by 2035, China's government relaxed its antinatalist policies. China amended its One Child Policy to a Three Child Policy in 2021. But China's CBR is unlikely to increase much because, after three decades of intensive educational programs on population control, as well as coercion to limit births, most Chinese have accepted the benefits of family planning. Furthermore, as a result of preference for male children in China, there are fewer Chinese women of childbearing years.

▲ 2.9.5 CHINA: WALL ART PROMOTES FAMILY PLANNING

2.10 KEY ISSUE 4 Why might population change in the future?

Family Futures

- **Understand two reasons for declining birth rates.**

Two strategies have been successful in reducing birth rates. One approach emphasizes the importance of improving local economic conditions. The other approach emphasizes the importance of diffusing modern contraceptive methods.

▲ **2.10.1** POSTER PROMOTES FAMILY PLANNING IN INDIA

2.10.1 Declining CBR

The CBR declined rapidly between 1990 and 2020, from 27 to 18 in the world as a whole and from 31 to 20 in developing countries. Declines have been most rapid in Latin America, Southwest Asia, and South Asia (Figure 2.10.1). The CBR has declined in all but a handful of countries (Figure 2.10.2).

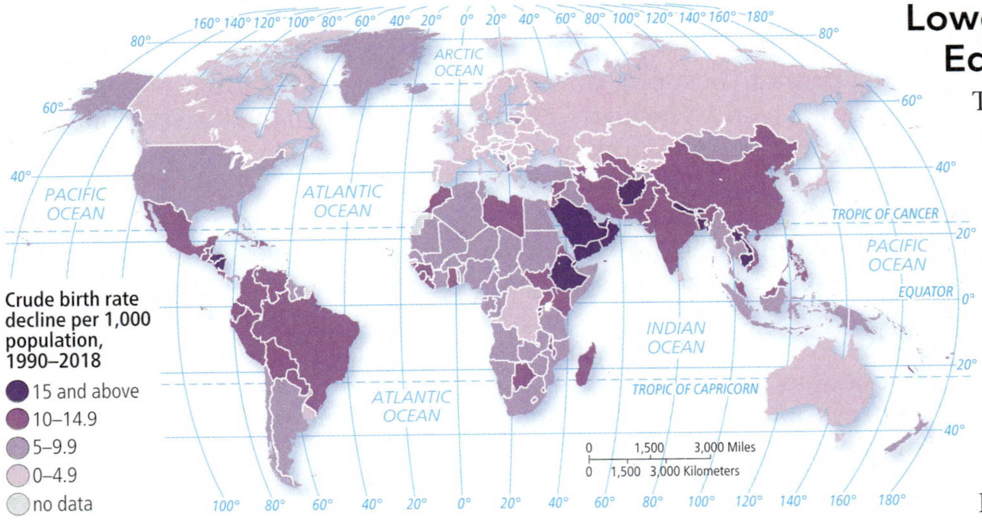

▲ **2.10.2** CBR CHANGE

Lowering CBR Through Education & Health Care

The U.N. Sustainable Development Goals, adopted in 2015, recognize that women's access to reproductive health care and education plays a key role in the overall quality of life for families (see Chapter 10). An economically developed country has money to spend on education and health-care programs that promote lower birth rates.

According to the approach of lowering CBR through education and health care, the following outcomes would be expected:

- With more women able to attend school and to remain in school longer, they will be more likely to learn employment skills and gain more economic control over their lives.
- With better education, women would better understand their reproductive rights, make more informed reproductive choices, and select more effective methods of contraception (Figure 2.10.3).
- With improved health-care programs such as prenatal care, counseling about sexually transmitted diseases, and child immunization, IMRs will decline.
- With the survival of more infants ensured, women will be more likely to use contraceptives for family planning.

◀ **2.10.3** FAMILY PLANNING EDUCATION
Women talk about birth control at a reproductive health clinic in Kampong Cham, Cambodia.

2.10.2 Lowering CBR Through Contraception

More than two-thirds of couples in developed countries use a family planning method (Figure 2.10.4). Family planning varies widely in developing countries. China reports the world's highest rate of family planning, whereas the lowest rates are in sub-Saharan Africa.

In developing countries, demand for contraceptive devices is greater than the available supply. Therefore, the most effective way to increase their use is to distribute more of them cheaply and quickly. Even though economic development may promote lower birth rates in the long run, the world cannot wait around for that alternative to take effect. Allocating resources into family planning programs can reduce birth rates much more rapidly.

Rapid growth in the acceptance of family planning in a number of developing countries is evidence that in the modern world, ideas can diffuse rapidly, even to places where people have limited access to education and modern communications. Bangladesh is an example of a country that has had little improvement in the wealth and literacy of its people, but 63 percent of married women in the country used contraceptives in 2021 compared to 6 percent in 1980. Similar growth in the use of contraceptives has occurred in other developing countries, including Colombia, Morocco, and Thailand. The percentage of women using contraceptives is especially low in sub-Saharan Africa, in part because of reluctance of men to use them. Therefore, the alternative of distributing contraceptives could have an especially strong impact there.

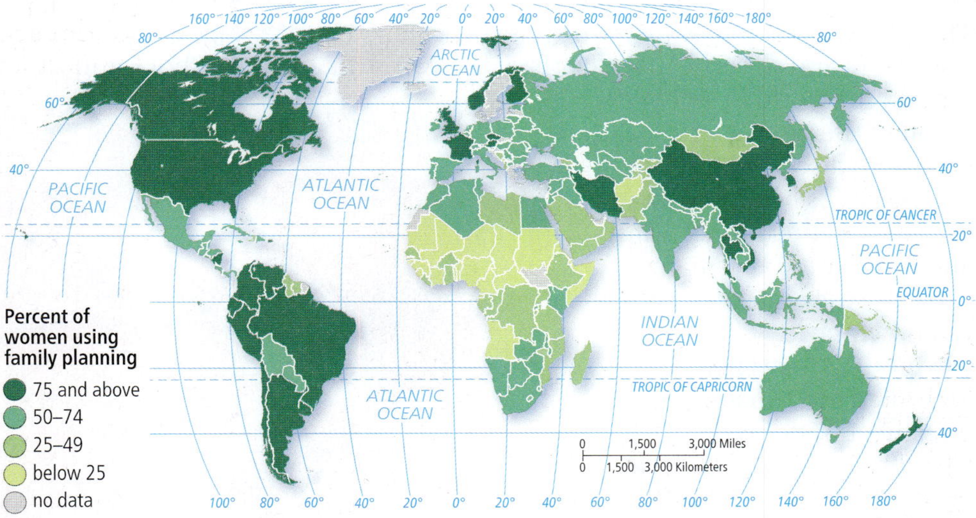

▲ **2.10.4** WOMEN USING FAMILY PLANNING

Family Planning Methods

Family planning methods vary widely (Figure 2.10.5). The principal family planning methods in developed countries are condoms and birth-control pills. The principal methods in China are intrauterine devices (IUDs) and condoms. People in sub-Saharan African countries such as Nigeria make minimal use of family planning.

Many oppose birth-control programs for religious and political reasons. Adherents of several religions have convictions that prevent them from using some or all birth-control methods. Opposition to terminating pregnancy by abortion is especially strong in some countries.

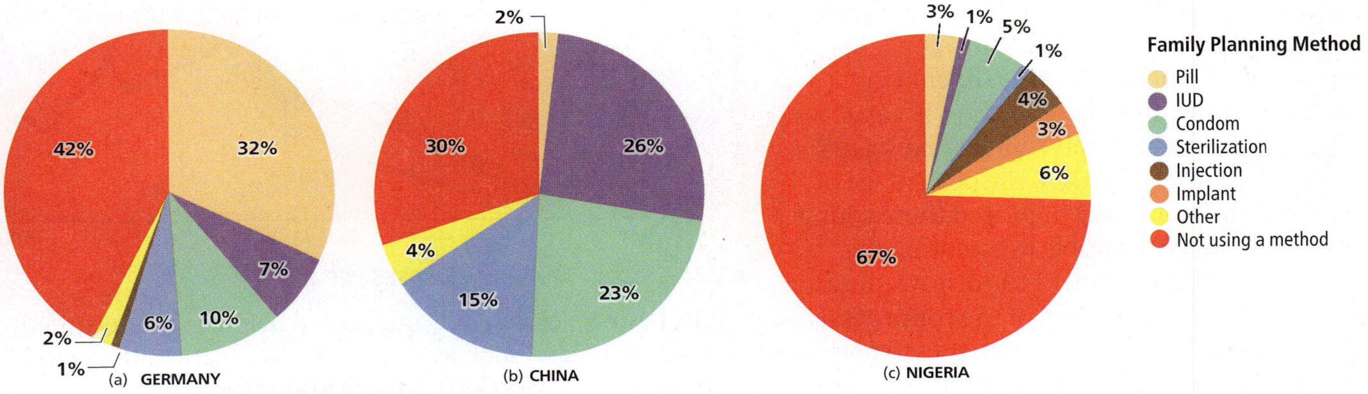

▲ **2.10.5** FAMILY PLANNING METHODS
(a) Germany, (b) China, (c) Nigeria.

Epidemiologic Futures

- **Understand geographic factors underlying recent pandemics.**

Recall that in the possible stage 5 of the demographic transition, CDR rises because more of the population is elderly. Some medical analysts argue that the world is moving into stage 5 of the epidemiologic transition, brought about by a reemergence of infectious and parasitic diseases. Infectious diseases thought to have been eradicated or controlled have returned, and new ones have emerged.

2.11.1 Reason for Possible Stage 5: Diffusion

Three reasons help to explain the possible emergence of a stage 5 of the epidemiologic transition: diffusion, evolution, and poverty. AIDS, cholera, the swine flu, measles, Ebola, and COVID-19 have been able to spread around the world in recent decades through the process of relocation diffusion, discussed in Chapter 1. As people are on the go, traveling domestically and internationally, many carry disease with them, often unknowingly.

COVID-19 was first identified in China in December 2019 (see Chapter 1). From there, the virus was carried through movement of people to other countries. Worldwide, around 240 million people had COVID-19 as of October 2021, and 4.9 million had died of the disease as of that date (Figure 2.11.1).

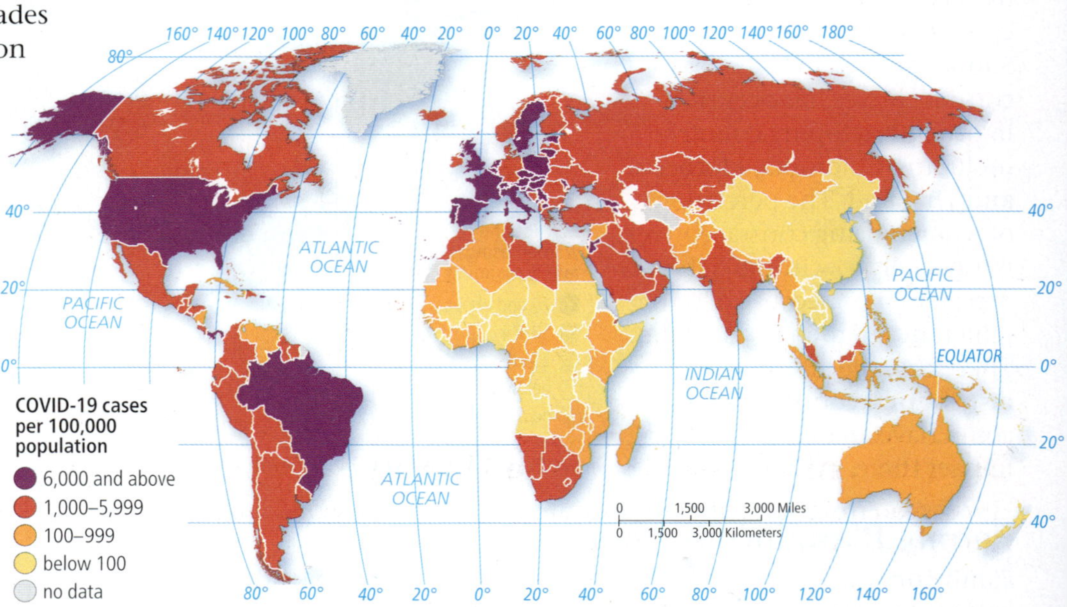

▲ 2.11.1 COVID-19 CASES PER CAPITA

Global Distribution

The COVID-19 pandemic has been especially severe in developed countries (Figure 2.11.2). Europe and North America, with only 14 percent of the world's population, have 50 percent of the world's COVID-19 cases.

Several explanations have been offered for the relatively small number of COVID-19 cases in Africa. With many African countries in stage 2 of the demographic transition, the percentage of younger people is relatively high, and these younger people may be less susceptible to catching the disease. To prevent diffusion of the disease, drastic measures were taken in several of the countries. For example, Lesotho closed its borders even before it had recorded a single case. Compliance with measures to prevent spread of the disease, such as wearing masks and avoiding large gatherings, may have been higher in African countries than in Europe and North America. On the other hand, some epidemiologists contend that Africa has had rates of COVID-19 comparable to developed countries, but poor testing and inadequate health-care systems in developing countries may have kept many cases from being recorded.

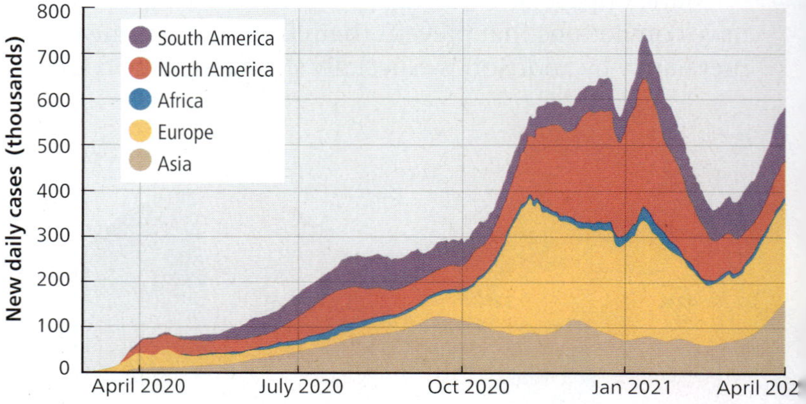

▲ 2.11.2 COVID-19 AVERAGE CASES PER DAY

2.11.2 Reason for Possible Stage 5: Increased Connections

The most lethal pandemic in recent years—even greater than COVID-19—has been acquired immunodeficiency syndrome (AIDS). Worldwide, 37 million people died from AIDS-related illnesses from the start of the epidemic in 1981 through 2021, and 38 million were living with human immunodeficiency virus (HIV, the cause of AIDS) in 2021. The impact of AIDS has been felt most strongly in sub-Saharan Africa, home to 26 million of the world's 37 million HIV-positive people (Figure 2.11.3).

▶ 2.11.3 PERCENT OF POPULATION DIAGNOSED AS HIV-POSITIVE

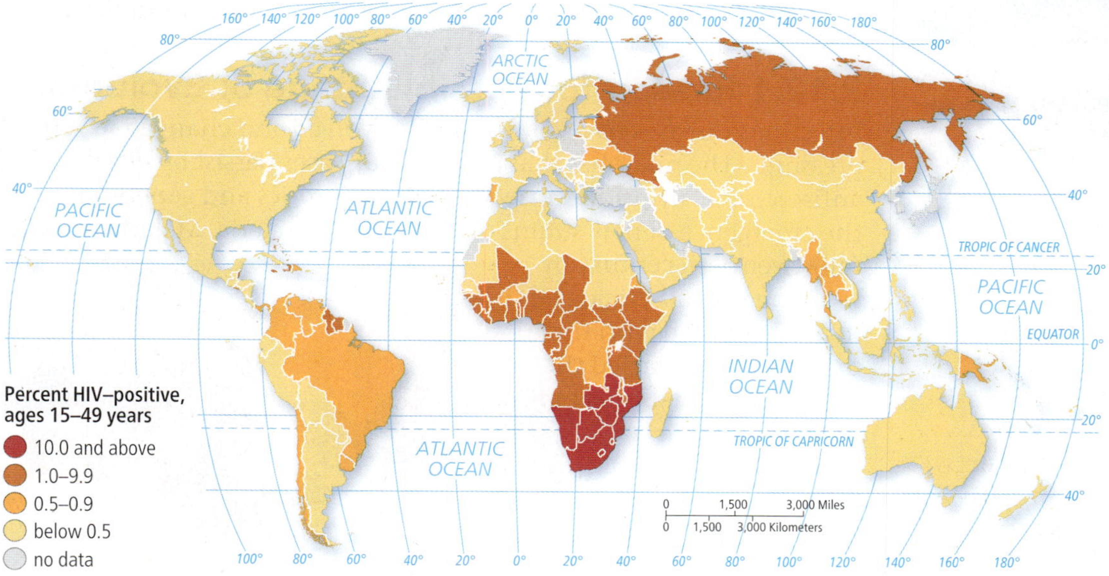

Diffusion of HIV/AIDS in the United States

HIV spread from sub-Saharan Africa to the rest of the world through relocation diffusion, both by Africans and by visitors to Africa returning to their home countries. HIV entered the United States during the early 1980s through New York, California, and Florida. Not by coincidence, the three leading U.S. airports for international arrivals are in these three states. Though HIV diffused to every state during the 1980s, these three states, plus Texas (a major port of entry by motor vehicle), accounted for half of the country's new HIV diagnoses in the peak year of 1993.

The number of people newly diagnosed with HIV has dropped sharply since the late 1990s because of the rapid diffusion of preventive methods and medicines. The rapid diffusion of these innovations is an example of expansion diffusion rather than relocation diffusion.

Most people newly diagnosed as HIV-positive are found in large urban areas (Figure 2.11.4). However, as a percentage of total population in the United States, southern states, led by Louisiana and Georgia, now have the largest rate of new HIV-positive diagnoses. Male-to-male sexual contact remains the principal form of diffusion in the United States.

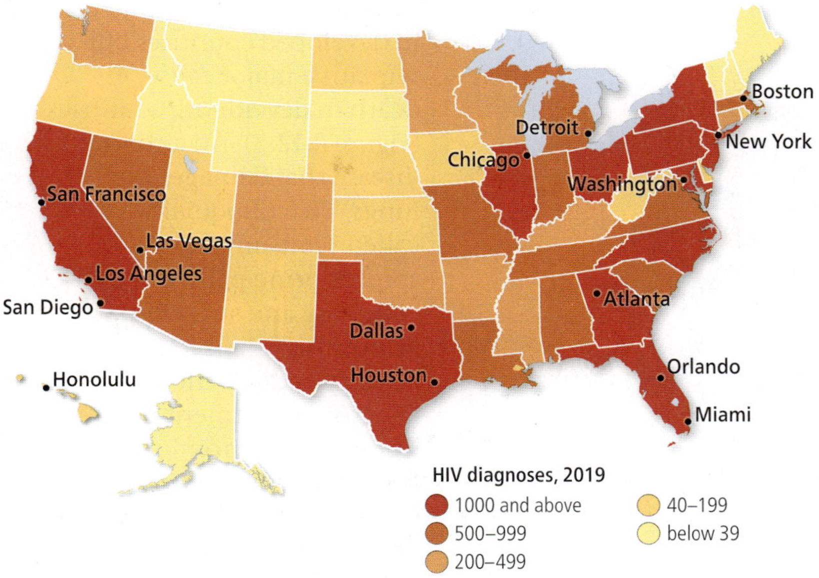

▲ 2.11.4 U.S. HIV DIAGNOSES

Population & Resources

▲ **2.12.1 VACCINATION TO PREVENT MALARIA**
Homa Bay, Kenya.

- **Summarize arguments supporting and opposing Malthus's theory.**

Overpopulation is a condition in which the number of people in an area exceeds the capacity of the environment to support them at a decent standard of living. British economist Thomas Malthus (1766–1834) was one of the first to argue that the world's rate of population growth was far outrunning the production of food supplies.

2.12.1 Reason for Possible Stage 5: Evolution

Infectious diseases have continuously evolved and changed in response to environmental pressures. The microbes that cause the diseases persist by developing resistance to antimicrobial drugs and insecticides. Antibiotics and genetic engineering contribute to the emergence of new strains of viruses and bacteria. Malaria was nearly eradicated in the mid-twentieth century by spraying the insecticide DDT in areas infested with the mosquito that carried the parasite. The first vaccination to prevent malaria was developed in 2021 (Figure 2.12.1). However, malaria still causes around a half-million deaths worldwide annually, nearly all in sub-Saharan Africa (Figure 2.12.2). A major reason is the evolution of DDT-resistant mosquitoes.

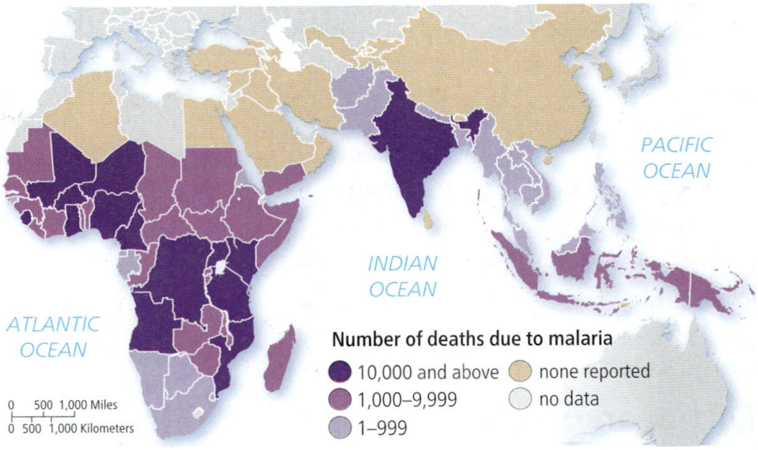

▲ **2.12.2 EVOLUTION: MALARIA DEATHS**

Reason for Possible Stage 5: Poverty

Infectious diseases are more prevalent in areas of poverty than other places because of unsanitary conditions and because residents can't afford the drugs needed for treatment. Tuberculosis (TB) is an example of an infectious disease that has been largely controlled in developed countries but remains a major cause of death in developing countries (Figure 2.12.3). An airborne disease that damages the lungs, TB (colloquially called "consumption") spreads principally through coughing and sneezing. TB is more prevalent in areas of poverty because the long, expensive treatment poses a significant economic burden. People with immune systems compromised by poor nutrition or HIV infection are especially susceptible to TB.

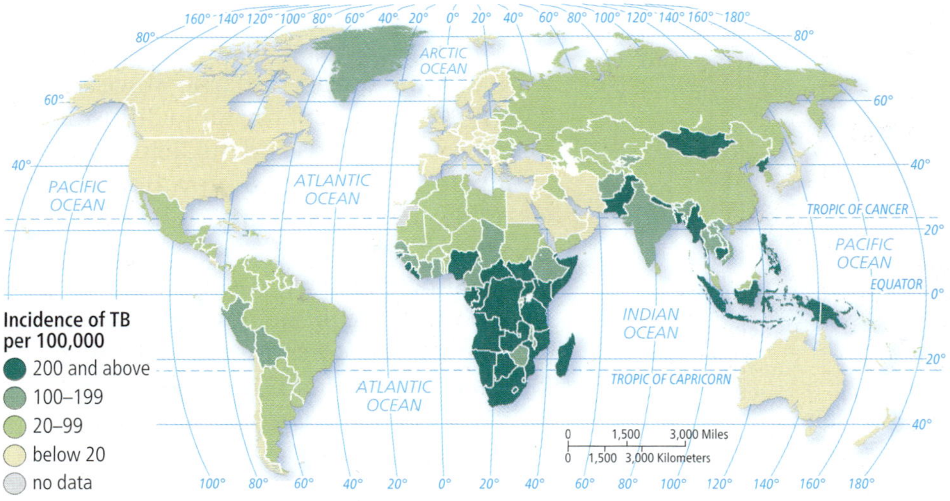

▲ **2.12.3 POVERTY: INCIDENCE OF TB**

2.12.2 Malthus's Theory

In *An Essay on the Principle of Population*, published in 1798, Thomas Malthus claimed that the population was growing much more rapidly than Earth's food supply was increasing because population grows geometrically, whereas food supply increases arithmetically (Figure 2.12.4).

Malthus made these conclusions several decades after the United Kingdom had become the first country to enter stage 2 of the demographic transition, in association with the Industrial Revolution. He concluded that population growth would press against available food supply in every country unless "moral restraint" produced lower CBRs or unless disease, famine, war, or other disasters produced higher CDRs.

According to Malthus, failure to act would produce the following ratios of people to food in the future:

Today: 1 person, 1 unit of food.
25 years from now: 2 persons, 2 units of food.
50 years from now: 4 persons, 3 units of food.
75 years from now: 8 persons, 4 units of food.
100 years from now: 16 persons, 5 units of food.

Malthus's views remain influential today. In fact, contemporary geographers and other analysts are taking another look at Malthus's views because world population growth is outstripping a wide variety of resources, if not food, as well as contributing to frightening pandemics.

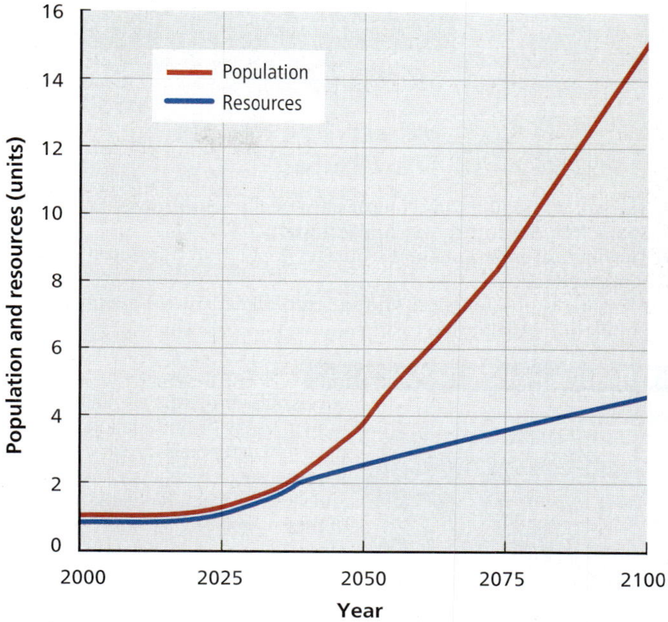

▲ 2.12.4 MALTHUS'S THEORY

Malthus's Critics

Criticism has been leveled at both the population and food supply sides of Malthus's theory. Many geographers consider Malthus's theory unrealistically pessimistic because it is based on a belief that the world's supply of resources is fixed rather than expanding.

Critics also argue that population growth is not a problem. To the contrary, they think that a larger population could stimulate economic growth and therefore production of more food. Marxist economists argue that poverty and hunger are a result of an unjust society and economic inequality, not population growth, and that the world possesses sufficient resources to eliminate hunger and poverty if these resources are shared equally.

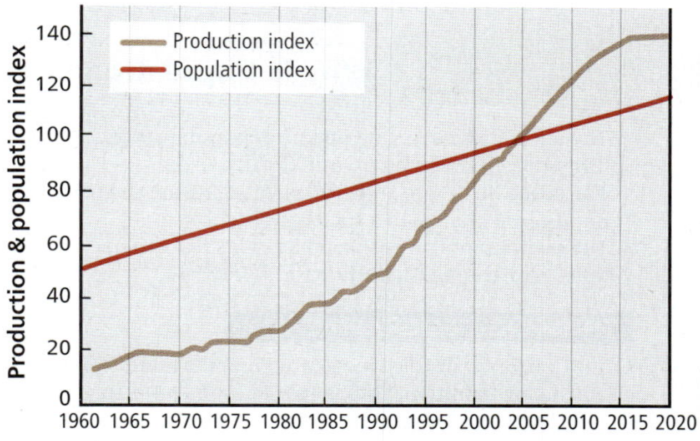

▲ 2.12.5 MALTHUS'S THEORY APPLIED TO CHINA
Food production and population are set to 100 in 2005. Food production has increased much more rapidly than population.

Malthus's Theory & Reality

Evidence from the past half-century lends support to both Neo-Malthusians and their critics. Malthus was fairly close to the mark on food supply but much too pessimistic on population growth. In contrast with Malthus's theory, food production has increased during the last half-century in China more rapidly than its population (Figure 2.12.5). Better growing techniques, higher-yielding seeds, and cultivation of more land have contributed to the expansion in the food supply (see Chapter 9). On the other hand, the food supply has barely kept up with population growth in Africa (Figure 2.12.6).

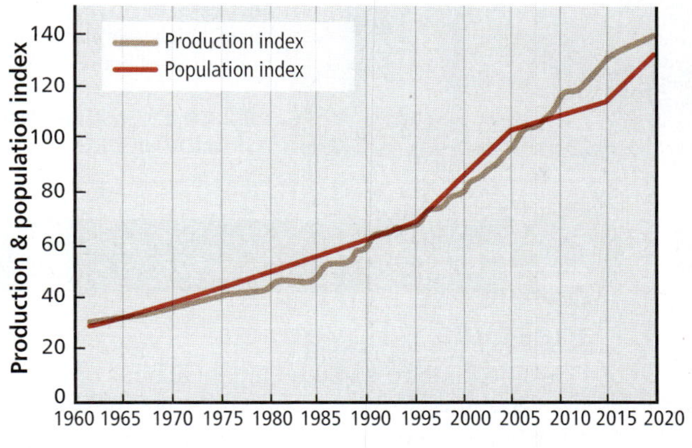

▲ 2.12.6 MALTHUS'S THEORY APPLIED TO AFRICA
Food production and population are set to 100 in 2005. Food production is barely keeping up with population growth.

CHAPTER 2

Review, Analyze, & Apply

KEY ISSUE 1 Where Are People Distributed?

- The world's population is highly clustered in four regions: East Asia, South Asia, Southeast Asia, and Europe.
- The physical environment discourages population concentrations in some regions.
- Arithmetic, physiological, and agricultural densities represent population distribution in different ways.

THINKING GEOGRAPHICALLY

Scientists disagree about the effects of high-density population on human behavior. Some laboratory tests have shown that rats display evidence of increased aggressiveness, competition, and violence when very large numbers of them are placed in a box. Does living in very high-density places, such as São Paulo, cause humans to behave especially aggressively or violently? Why or why not?

◀ 2.CR.1 SÃO PAULO, BRAZIL

KEY ISSUE 2 Why Is Population Increasing?

- The natural increase rate measures population growth as the difference between births and deaths.
- The crude birth rate and crude death rate are the principal measures of population change in a society as a whole.
- The demographic transition has four stages characterized by varying rates of births, deaths, and natural increase.

THINKING GEOGRAPHICALLY

Countries display distinctive age structures as a result of their stage in the demographic transition. The United States and other stage 3 and stage 4 countries have relatively large numbers of boomers (born 1946–64). Compared with their parents' generation, boomers received more education and women were more likely to work outside the home. Boomers married later (if at all), and were more likely to cohabit. They had fewer children and more children while unmarried. In what ways do you think the Gen Z generation (born after 1995) might display demographic characteristics similar to, or different from, those of boomers? Why?

◀ 2.CR.2 GRANDFATHER AND GRANDCHILDREN GO FOR A HIKE

KEY ISSUE 3 How Does Health Vary by Region?

- Baby girls and mothers face challenging health risks, especially in developing countries.
- The percentage of younger and older people varies among countries and impacts the dependency ratio.
- Health care and medical facilities vary sharply between developed and developing countries.

THINKING GEOGRAPHICALLY

▲ 2.CR.3 FREE EYE CARE CLINIC, MIAMI, FLORIDA

Health care varies widely around the world. Health-care indicators for the United States do not always match those of other developed countries. What reasons might explain these differences? Why might there be a need to set up a free medical clinic in a van such as in Miami?

KEY ISSUE 4 Why Might Population Change in the Future?

- Neo-Malthusians argue that population is outstripping resources, but others do not agree.
- The demographic transition may display a possible stage 5 of population decline.
- Birth rates have declined through the use of two strategies—education and contraception.

THINKING GEOGRAPHICALLY

▲ 2.CR.4 ENJOYING CHERRY BLOSSOMS, TOKYO, JAPAN

Authorities on demography such as the Population Reference Bureau (PRB) and the U.N. Population Fund still prefer the four-stage model of the demographic transition, but some geographers believe that the demographic transition now has a fifth stage in places such as Japan. What sort of evidence might help to determine the existence of a possible stage 5?

Explore

Nagel-Gulab, Egypt

Use Google Earth to explore Nagel-Gulab, a small town in Egypt along the banks of the Nile River. *Search* for **Nagel-Gulab, Egypt**.

1. What color is most of the land immediately to the west of the town? Does this indicate that the land is used for agriculture, or is it desert?
2. How wide is the green strip? Why is the green strip closer to the river and the tan area farther away?

▶ 2.CR.5 NILE RIVER NEAR NAGEL-GULAB, EGYPT

Key Terms

Agricultural density (2.1.1) The ratio of the number of farmers to the total amount of arable land (land suitable for agriculture).

Antinatalist policy (2.9.2) Government policy that supports lower birth rates.

Arable land (2.1.1) Land suited for agriculture.

Arithmetic density (2.1.1) The total number of people divided by the total land area.

Crude birth rate (CBR) (2.3.2) The total number of live births in a year for every 1,000 people alive in the society.

Crude death rate (CDR) (2.3.2) The total number of deaths in a year for every 1,000 people alive in the society.

Demographic transition (2.4.1) The process of change in a society's population from a condition of high crude birth and death rates and low natural increase rate to a condition of low crude birth and death rates, low natural increase rate, and higher total population.

Dependency ratio (2.5.1) The number of people under age 15 and over age 64 compared to the number of people active in the labor force.

Doubling time (2.3.1) The number of years needed to double a population, assuming a constant natural increase rate.

Ecumene (2.2.2) The portion of Earth's surface occupied by permanent human settlement.

Endemic (2.8.1) A widespread occurrence of an infectious disease that is considered seasonal and/or predictable.

Epidemic (2.8.1) A widespread occurrence of an infectious disease in a community at a particular time.

Epidemiologic transition (2.8.1) The process of change in the distinctive causes of death in each stage of the demographic transition.

Epidemiology (2.8.1) The branch of medical science concerned with the incidence, distribution, and control of diseases that are prevalent among a population at a certain time and are produced by some special causes not generally present in the affected locality.

Industrial Revolution (2.4.1) A series of improvements in industrial technology that transformed the process of manufacturing goods.

Infant mortality rate (IMR) (2.5.1) The total number of deaths in a year among infants under 1 year of age for every 1,000 live births in a society.

Life expectancy (2.5.2) The average number of years an individual can be expected to live, given current social, economic, and medical conditions.

Maternal mortality rate (2.6.1) The annual number of female deaths per 100,000 live births from any cause related to or aggravated by pregnancy or its management (excluding accidental or incidental causes).

Medical revolution (2.4.1) Medical technology invented in Europe and North America that has diffused to the developing countries in Latin America, Asia, and Africa. Improved medical practices have eliminated many of the traditional causes of death in developing countries and enabled more people to live longer and healthier lives.

Natural increase rate (NIR) (2.3.1) The percentage growth of a population in a year, computed as the crude birth rate minus the crude death rate.

Overpopulation (2.12.1) A situation in which the number of people in an area exceeds the capacity of the environment to support them at a decent standard of living.

Pandemic (2.8.1) Disease that occurs over a wide geographic area and affects a very high proportion of the population.

Pestilence (2.8.1) An especially virulent epidemic.

Physiological density (2.1.1) The number of people per unit area of arable land, which is land suitable for agriculture.

Population pyramid (2.5.2) A bar graph that represents the distribution of population by age and sex.

Potential support ratio (or **elderly support ratio**) (2.5.2) The number of working-age people (ages 15 to 64) divided by the number of persons 65 and older.

Pronatalist policy (2.9.2) Government policy that supports higher birth rates.

Sex ratio (2.6.2) The number of males per 100 females in the population.

Total fertility rate (TFR) (2.9.1) The average number of children a woman will have throughout her childbearing years.

Zero population growth (ZPG) (2.4.2) A decline of the total fertility rate to the point where the natural increase rate equals zero.

Geospatial Analysis

Log in to the Mastering Geography Study Area to access MapMaster 2.0.

Obesity & Income

Chapter 1 defined spatial association. Let's investigate possible spatial association among demography and health features. Add ***Gross National Income per capita***. In the *Legend*, deselect all income classes below $20,000. Only $20,000–$49,999 and $50,000–$103,630 are to be displayed. Add the ***Adult Obesity Rate*** layer and select *Join with data layer*. In the *Legend*, deselect the two lowest classes for ***Gross National Income per capita***.

1. Hover over the United States. What are the figures for income and obesity for the United States?
2. Is the obesity rate in the United States relatively high or low compared with other countries with relatively high incomes? What might account for this?

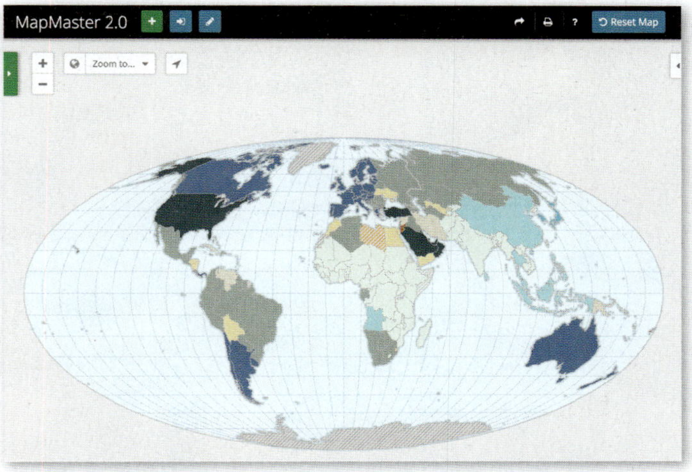

▲ **2.CR.6** OBESITY AND GROSS NATIONAL INCOME PER CAPITA

CHAPTER

3 Migration

▲ Migrants from Honduras and Guatemala cross Mexico in hopes of ultimately reaching the United States.

Migration is the permanent move to a new location. Geographers study migration in part because it is increasingly important in explaining changes in population in various places and regions. Migration is also important because when people migrate, they take with them to their new home cultural values and economic practices. At the same time, they become connected with the cultural and economic patterns of their new place of residence.

KEY ISSUES

1 Where Are Migrants Distributed?

People emigrate from some places and immigrate to other places. The United States has long been an especially important destination for migrants.

2 Where Do People Migrate Within Countries?

Large countries may have extensive migration between regions. Migration is also important within regions of countries, especially from rural areas into urban areas.

3 Why Do People Migrate?

People migrate for cultural, environmental, and economic reasons. Women and children comprise increasing shares of migrants.

4 What Challenges Do Migrants Face?

People often face legal restrictions on their ability to immigrate into another country. In some places, they also face hostility from people already residing there.

LOCATIONS IN THIS CHAPTER

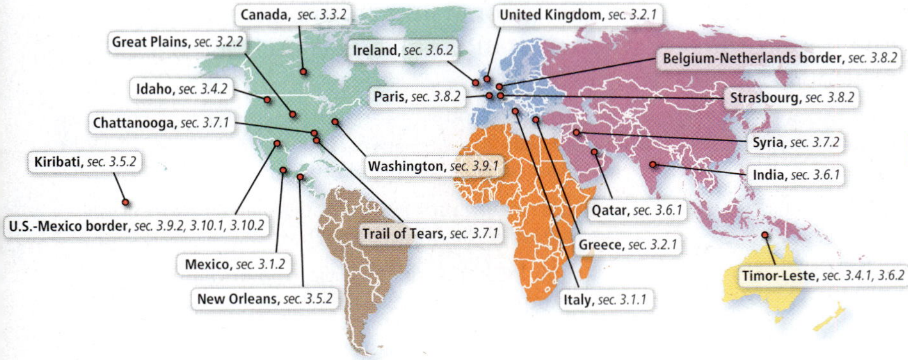

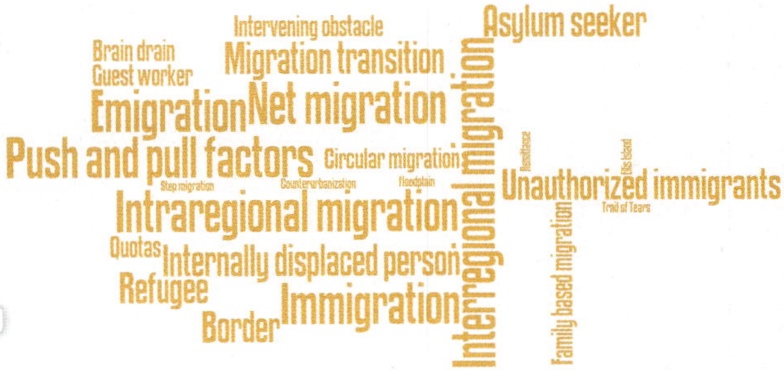

Migration Principles

- **Introduce reasons for migration.**

The diffusion of human settlement from a small portion of Earth's land area to the current distribution, across most of it, is the result of migration. To accomplish their expansion across Earth, humans have continually changed their place of residence—where they sleep, store their possessions, and practice their customs with family and friends.

▲ 3.1.1 MIGRATION Immigrants from Eritrea arrive in Italy.

3.1.1 Migration "Laws"

Migration is a permanent move to a new location (Figure 3.1.1). Geographers document the migration of people across Earth and the reasons for that migration. Migration is a specific type of relocation diffusion, which was defined in Chapter 1 as the spread of an innovation through the physical movement of people from one location to another.

Geographers study migration in part to explain changes in population in various places and regions. Migration causes innovations in societies because when people migrate, their cultural values and economic practices come with them.

Migration is a form of **mobility**, which is a more general term covering all types of movements from one place to another. People display mobility in a variety of ways, such as by journeying every weekday from their homes to places of work or school and once a week to shops, places of worship, or recreation areas. These types of short-term, repetitive, or cyclical movements that recur on a regular basis, such as daily, monthly, or annually, are called **circulation**. Another form of mobility is seasonal mobility, movement based on the time of year. College students engage in seasonal mobility when they move to a residence hall each fall and return home the following spring (Figure 3.1.2).

▲ 3.1.2 MOBILITY Move-in day, university residence hall.

Migration Transition

Geography has no comprehensive theory of migration, although an outline of migration principles written by nineteenth-century geographer E. G. Ravenstein is the basis for contemporary geographic migration studies. To understand where and why migration occurs, Ravenstein's principles can be organized into three groups:

- The distance that migrants typically move (discussed in sections 3.1.2, 3.2.1, 3.2.2, 3.3.1, 3.3.2, 3.4.1, and 3.4.2).
- The characteristics of migrants (discussed in section 3.5.1).
- The reasons why migrants move (discussed in sections 3.5.2, 3.6.1, 3.6.2, 3.7.1, and 3.7.2).

Geographer Wilbur Zelinsky identified a migration transition, which consists of changes in a society comparable to those in the demographic transition (Figure 3.1.3). The **migration transition** is a change in the migration pattern in a society that results from the social and economic changes that also produce the demographic transition. According to the migration transition, international migration is primarily a phenomenon of countries in stage 2 of the demographic transition, whereas internal migration is often a phenomenon of countries in stages 3 and 4 of the demographic transition.

▶ 3.1.3 MIGRATION TRANSITION AND DEMOGRAPHIC TRANSITION

Stage	Demographic Transition	Migration Transition
1	Low NIR, high CBR, high CDR	High daily or seasonal mobility in search of food
2	High NIR, high CBR, rapidly falling CDR	High international emigration and interregional migration from rural to urban areas
3	Declining NIR, rapidly declining CBR, still declining CDR	High international immigration and intraregional migration from cities to suburbs
4	Low NIR, low CBR, low CDR	Same as stage 3

3.1.2 International & Internal Migration

Migration occurs at two principal scales: international migration and internal migration.

International Migration

A permanent move from one country to another is **international migration** (Figure 3.1.4). About 4 percent of the world's people are international migrants—that is, they currently live in countries other than those in which they were born. International migration is further divided into two types:

Voluntary migration means that the migrant has chosen to move, usually for economic reasons, though sometimes for environmental reasons. **Forced migration** means that the migrant has been compelled to move because of political conflict or adverse environmental conditions.

The distinction between forced migration and voluntary migration is not clear-cut. For example, a person who migrates for economic reasons may seek a better life through improved access to jobs and food but could also be compelled to migrate because of violence in their country of origin.

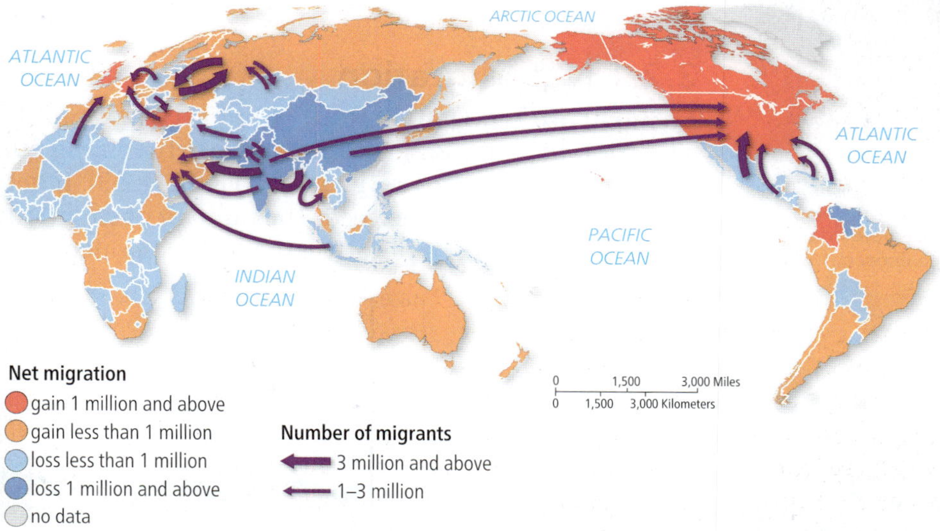

▲ 3.1.4 GLOBAL MIGRATION PATTERNS

Internal Migration

A permanent move within the same country is **internal migration** (Figure 3.1.5). Consistent with the distance-decay principle presented in Chapter 1, the farther away a place is located, the less likely it is that people will migrate to it. Furthermore, in contrast with international migration, most countries permit their citizens to migrate internally. Thus, internal migrants are much more numerous than international migrants.

Internal migration can be divided into two types: **Interregional migration** is movement from one region of a country to another in that same country. Historically, the main type of interregional migration has been from rural to urban areas in search of jobs. **Intraregional migration** is movement within one region of a country. The main type of intraregional migration has been within urban areas, from older cities to newer suburbs.

Most people find internal migration less traumatic than international migration because they find familiar language, foods, broadcasts, literature, music, and other social customs after they move. Internal migration also generally involves much shorter distances than does international migration. However, internal migration can involve long-distance moves in large countries, such as in the United States and Russia.

▲ 3.1.5 INTERNAL MIGRATION

Mexico's two principal interregional migration flows are net migration from the south to the north and from the center to the north. Mexico's principal intraregional migration flow is from Mexico City to outer states in the Mexico City metropolitan area. Also shown are Mexico's two principal patterns of international migration—arrivals from Central America and departures to the United States.

Global Migration Distribution

- **Explain the difference between emigration and immigration.**

The flow of migration always involves two-way connections. Given two locations, A and B, some people migrate from A to B, while at the same time, others migrate from B to A.

▲ 3.2.1 EMIGRATION People trying to emigrate from Turkey to Greece.

3.2.1 Net Migration

Emigration is migration from a location (Figure 3.2.1), and **immigration** is migration to a location (Figure 3.2.2). The difference between the number of immigrants and the number of emigrants is the **net migration** (Figure 3.2.3). If net migration is positive because the number of immigrants exceeds the number of emigrants, the region has net in-migration, and if net migration is negative, the region has net out-migration.

▲ 3.2.2 REFUGEES FROM SYRIA Arriving in Greece.

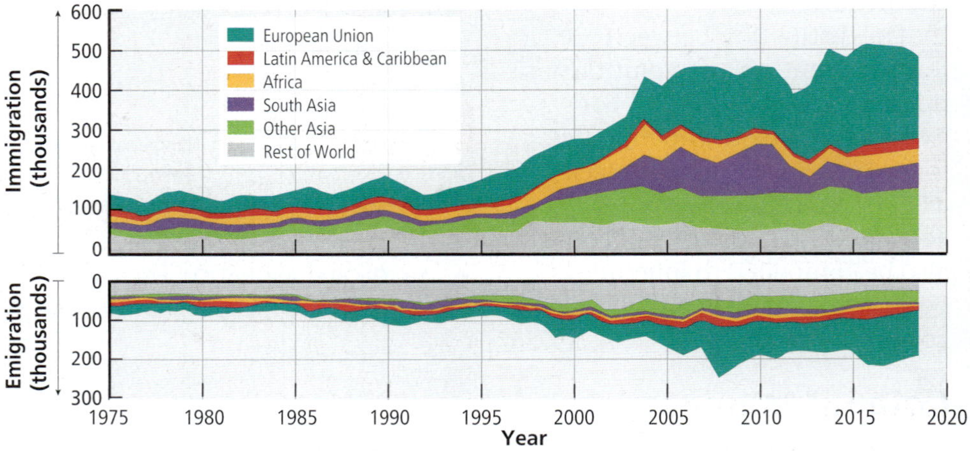

▲ 3.2.3 NET MIGRATION, UNITED KINGDOM

Origin & Destination

Figure 3.2.4 summarizes the origin and destination of international migrants between 1990 and 2020. In 2020, 281 million of the world's 7.8 billion people were living outside their country of birth. The largest number of emigrants came from Asia, and the largest number of immigrants went to Asia or Europe. North America, Europe, and the South Pacific have more immigrants than emigrants, whereas Africa, Asia, and Latin America have more emigrants than immigrants.

▶ 3.2.4 ORIGIN AND DESTINATION OF MIGRANTS, 1990–2020

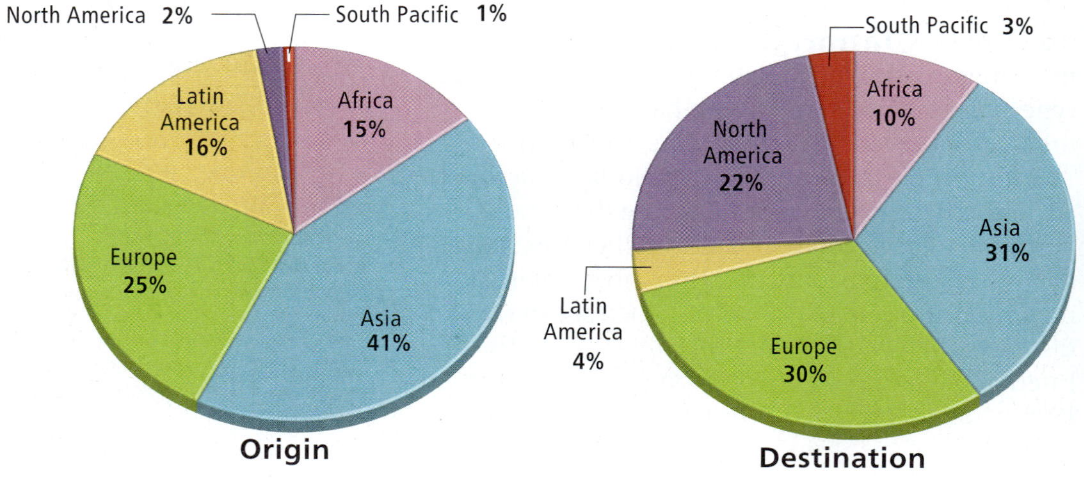

3.2.2 Changing U.S. Immigration

The United States plays a special role in the study of international migration. As the world's third most populous country, the United States is inhabited overwhelmingly by direct descendants of immigrants. About 80 million people migrated to the United States between 1820 and 2020, including 45 million who were alive in 2020 (Figure 3.2.5).

The United States has had three main eras of immigration:

- Immigration from Europe and Africa in the seventeenth and eighteenth centuries, during the colonial era.
- Mass immigration from Europe in the late nineteenth and early twentieth centuries.
- Immigration from Asia and Latin America in the late twentieth and early twenty-first centuries.

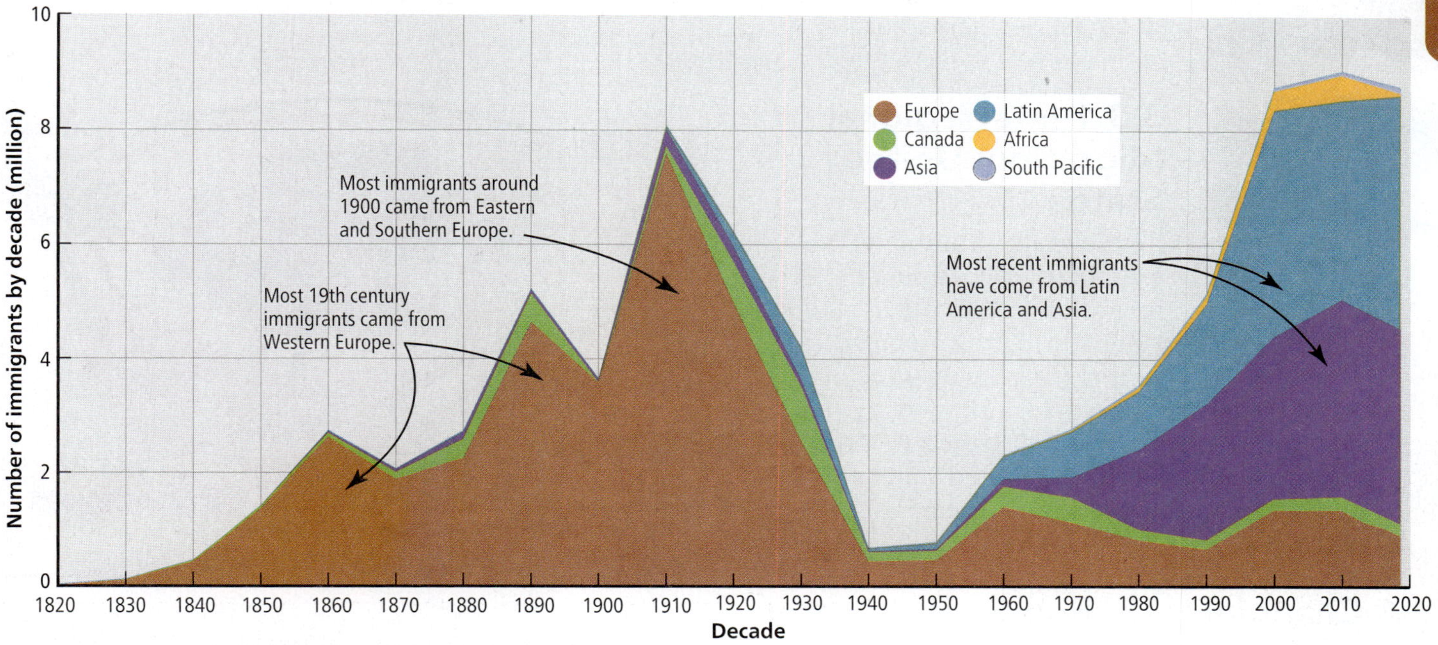

▲ 3.2.5 TWO CENTURIES OF IMMIGRATION TO THE UNITED STATES

Colonial Immigration

Immigration to the American colonies and the newly independent United States came from Europe and Africa. Most Europeans were voluntary immigrants. However, harsh economic conditions and religious persecution in Europe at the time blurred the distinction between forced and voluntary migration. Most African immigrants had been captured in Africa and forcibly transported to the United States, where they were sold as property.

Nineteenth-Century Immigration

During the nineteenth century, nearly 90 percent of immigrants to the United States emigrated from Europe. For many emigrants from Europe, the United States offered a great opportunity for economic success. Migration from Europe to the United States peaked at several points during the nineteenth and early twentieth centuries.

Desperate economic factors in their native countries compelled Irish and Germans to cross the Atlantic during the 1840s and 1850s. Germans also emigrated to escape political unrest. Emigration from Ireland and Germany resumed during the 1870s following a temporary decline during the U.S. Civil War (1861–1865). The Industrial Revolution diffused to Scandinavia during the 1880s, triggering a rapid population increase and inducing many people from there to emigrate to the United States in search of jobs. The shift in the primary source of immigrants coincided with the diffusion of the Industrial Revolution to Southern and Eastern Europe during the early twentieth century, along with rapid population growth in those regions.

Recent Immigration

More than three-fourths of recent U.S. immigrants have emigrated from Latin America and Asia. Although the pattern of immigration to the United States has changed over the country's history, the reasons for immigration to the country remain virtually unchanged: rapid population growth and limited prospects for economic advancement at home combined with hopes for economic opportunity and social advancement in the United States. Europeans left when their countries entered stage 2 of the demographic transition in the nineteenth century, and Latin Americans and Asians have responded to similar motivations in recent years.

▲ 3.3.1 MIGRATING ACROSS THE GREAT PLAINS, APPROXIMATELY 1890

Interregional Migration

- Describe interregional migration within North America and Europe.

In the past, people migrated from one region of a country to another in search of better farmland. Today, the principal type of interregional migration is from rural areas to urban areas.

3.3.1 Interregional Migration: United States

An especially prominent example of interregional migration in the United States during the nineteenth century was the opening of the West (Figure 3.3.1). People pushed west, where land to farm was abundant. The largest flows of interregional migrants in the United States are now heading South from the Northeast and West (Figure 3.3.2).

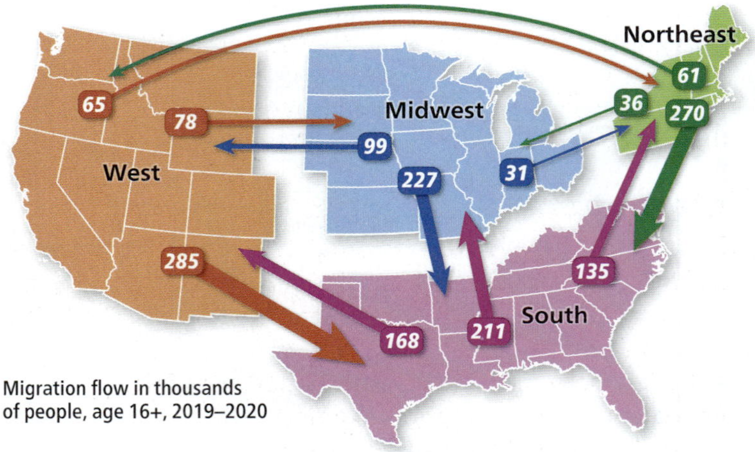

▲ 3.3.2 INTERREGIONAL MIGRATION IN THE UNITED STATES

A Changing U.S. Population Center

The U.S. Census computes the country's population and its population center every decade in "0"- ending years, such as 2020. The population center is the average location of where people live when considering everyone in the country, or the "center of population gravity." If the United States were a flat plane placed on top of a pin and each individual weighed the same, the population center would be the point where the population distribution causes the flat plane to balance on the head of a pin. The changing location of the population center graphically demonstrates the march of the American people across the North American continent (Figure 3.3.3).

▼ 3.3.3 CHANGING CENTER OF U.S. POPULATION

1790: Hugging the Coast. Virtually all colonial-era settlements were near the Atlantic Coast. The Appalachian Mountains initially hindered western development, and Native Americans still occupied large areas to the west.
1800–1840: Crossing the Appalachians. Transportation improvements, especially canals, helped open the interior.
1850–1890: Rushing to the Gold. The population center shifted more rapidly, as pioneers kept going all the way to California, lured by the Gold Rush.
1900–1940: Filling In the Great Plains. Westward movement slowed, as international migration from Europe to the East Coast offset most of the internal migration from the East to the West. Immigrants began to fill in the Great Plains that earlier generations had bypassed on their way to California.
1950–2020: Moving South. Interregional migration has slowed in recent years. The South has had net in-migration, a reversal of past patterns.

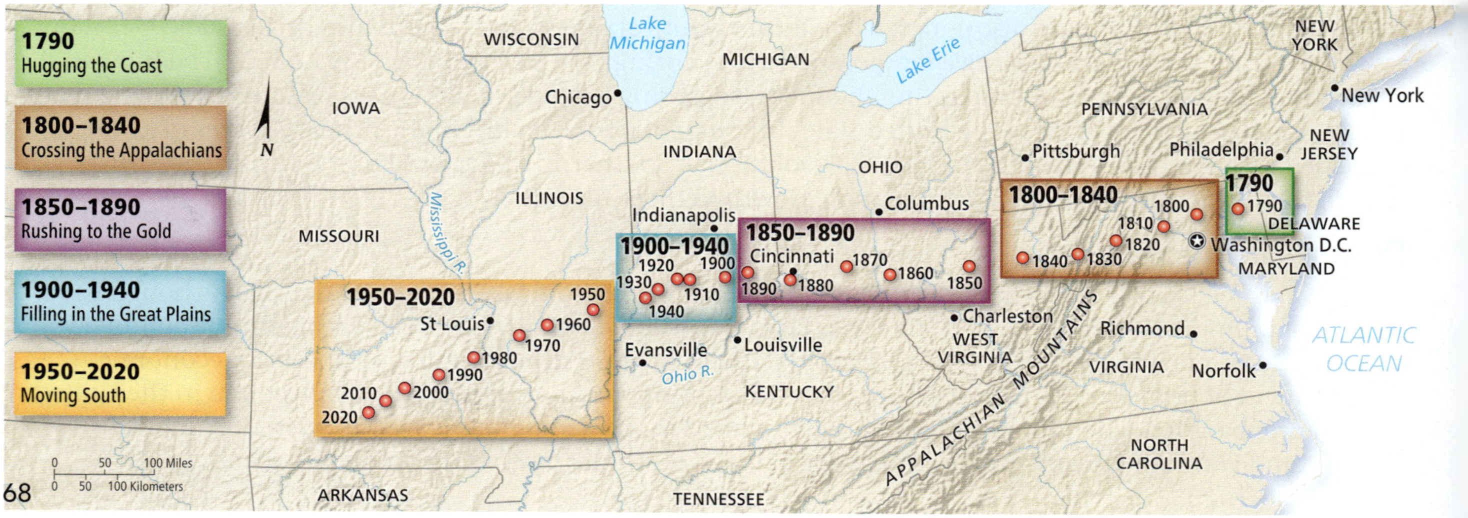

3.3.2 Interregional Migration: Developed Countries

As in the United States, long-distance interregional migration has been an established pattern in other countries.

Interregional Migration: Canada

Canada, like the United States, has had interregional migration primarily from east to west for more than a century (Figure 3.3.4). Alberta has had the largest net in-migration and Quebec the largest net out-migration. The three largest interprovincial flows in Canada are from Ontario to Alberta, from British Columbia to Alberta, and from Alberta to British Columbia.

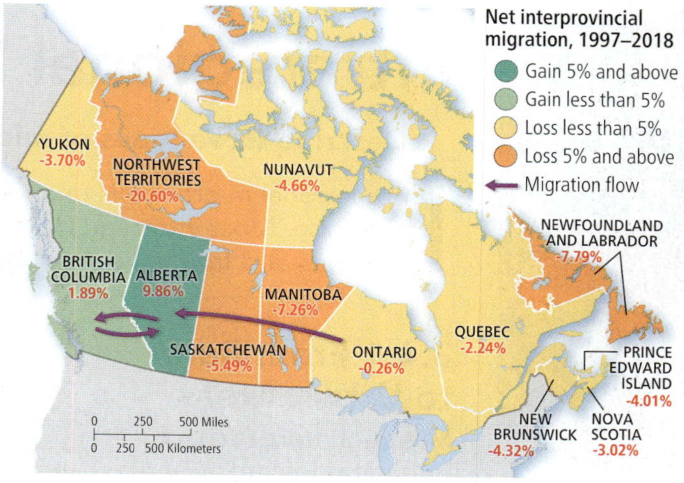

▲ **3.3.4 INTERREGIONAL MIGRATION: CANADA**

Interregional Migration: Europe

Europe (including Russia) is home to 87 million of the world's 281 million immigrants as of 2020. In other words, Europe has around 10 percent of the world's total population and around 30 percent of the world's migrants. Germany has 16 million immigrants, Russia has 12 million, the United Kingdom has 9 million, France has 9 million, Spain has 7 million, and Italy has 6 million.

In Europe as a whole, around 10 percent of the population consists of immigrants. The foreign-born population is 48 percent in Luxembourg and 29 percent in Switzerland. Germany, the most populous country in Europe (other than Russia), not only has the largest number of immigrants, it also has one of the largest percentages of immigrants.

Europe's 87 million immigrants include 44 million originating in another European country and 43 million from other regions of the world. The general pattern of movement of migrants in Europe is from south and east to north and west (Figure 3.3.5). This pattern holds both for migrants within Europe and for those originating in other regions of the world, primarily from North Africa and Southwest Asia as well as Asian countries once part of the former Soviet Union.

Russia is the country of origin in Europe for the largest number of emigrants, 11 million, although overall it has net in-migration. Other leading sources of emigrants include 5 million from Ukraine, 4 million each from Kazakhstan and Poland, and 3 million each from Morocco, Romania, and Turkey. The arrival of large numbers of refugees from outside the region has altered the patterns of migration within Europe (see section 3.8.2).

▲ **3.3.5 NET MIGRATION, EUROPEAN UNION**

Most portions of southern and eastern Europe have net out-migration. Most of these emigrants are heading for northern and western Europe.

▲ 3.4.1 RURAL TO URBAN MIGRANTS: TIMOR-LESTE

Intraregional Migration

- Describe three types of migration occurring within a region.

Intraregional migration is much more common than interregional or international migration. Most intraregional migration is from rural to urban areas in developing countries and from cities to suburbs in developed countries (Figure 3.4.1).

3.4.1 From Rural to Urban Areas

Migration from rural (or nonmetropolitan) areas to urban (or metropolitan) areas began in the 1800s in Europe and North America as part of the Industrial Revolution (see Chapter 11). The percentage of people living in urban areas in the United States, for example, increased from 5 percent in 1800 to 50 percent in 1920 and 82 percent in 2021. In recent years, urbanization has diffused to developing countries in Asia, Latin America, and Africa (see Chapter 12). Between 1950 and 2021, the percentage living in urban areas increased from 40 percent to 79 percent in Latin America, 10 percent to 41 percent in sub-Saharan Africa, and from 15 percent to 52 percent in Asia (Figure 3.4.2). As with interregional migrants, most people who move from rural to urban areas seek economic advancement. They are pushed from rural areas by declining opportunities in agriculture, such as having to pay higher rents for farmland, and are pulled to the cities by the prospect of work in factories or in service industries.

▼ 3.4.2 RURAL TO URBAN MIGRATION: TIMOR-LESTE
Intraregional migration has been from rural areas into Dili, the capital and largest city.

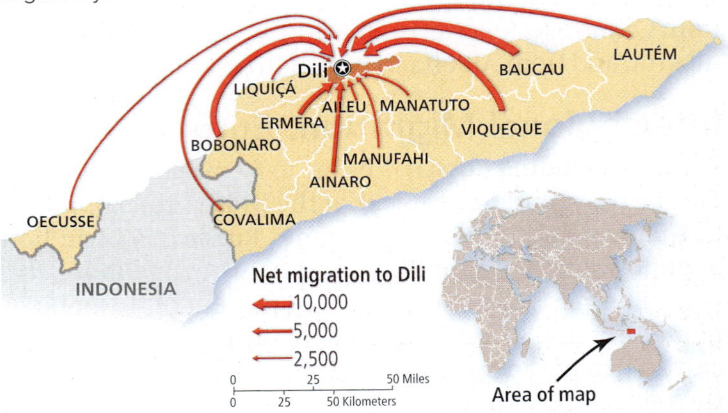

From Urban to Suburban Areas

Most intraregional migration in developed countries is from cities to surrounding suburbs. The population of most cities in developed countries has declined since the mid-twentieth century while suburbs have grown rapidly. Nearly twice as many Americans migrate from cities to suburbs each year as migrate from suburbs to cities (Figure 3.4.3). Comparable patterns are found in Canada and Europe.

The major reason for the large-scale migration to the suburbs is not related to employment, as is the case with other forms of migration. For most people, migration to suburbs does not coincide with changing jobs. Instead, people are attracted by a suburban lifestyle.

Suburbs offer the opportunity to live in a detached house rather than an apartment, surrounded by a private yard where children can play safely. A garage or driveway on the property guarantees space to park cars at no extra charge. In the United States, suburban schools tend to be more modern, better equipped, and less crowded than those in cities. Cars and trains enable people to live in suburbs yet have access to jobs, shops, and recreational facilities throughout the urban area (see Chapter 13).

As a result of suburbanization, the territory occupied by urban areas has rapidly expanded. To accommodate suburban growth, farms on the periphery of urban areas are converted to housing and commercial developments, where new roads, sewers, and other services must be built.

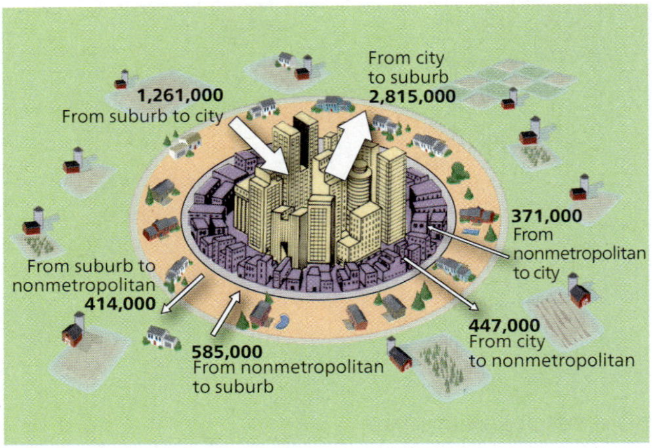

▲ 3.4.3 INTRAURBAN MIGRATION: UNITED STATES, 2019–2020

3.4.2 From Urban to Rural Areas

Developed countries witnessed a new migration trend beginning in the late twentieth century. For the first time, more people immigrated into rural areas than emigrated out of them in some years. Net migration from urban to rural counties is called **counterurbanization**.

Counterurbanization results in part from very rapid expansion of suburbs. The boundary where suburbs end and the countryside begins cannot be precisely defined. But some counterurbanization represents genuine migration from cities and suburbs to small towns and rural communities (Figure 3.4.4).

As with suburbanization, people move from urban to rural areas for lifestyle reasons. Some are lured to rural areas by the prospect of swapping the frantic pace of urban life for the opportunity to live on a farm, where they can own horses or grow vegetables. Others move to farms but do not earn their living from agriculture; instead, they work in nearby offices, small town shops, or other services. In the United States, evidence of counterurbanization can be seen primarily in the Rocky Mountain states. Some rural counties in states such as Colorado, Idaho, Utah, and Wyoming have experienced net in-migration (Figure 3.4.5).

Counterurbanization is not a phenomenon every year. Before COVID-19, cities had become more attractive to those wanting to relocate, especially younger people (see Chapter 13).

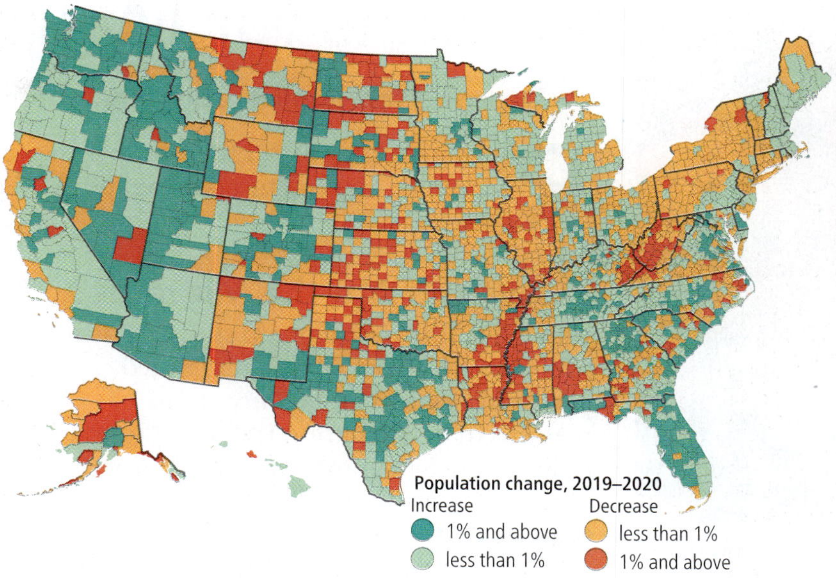

▲ 3.4.4 POPULATION CHANGE BY COUNTY

▲ 3.4.5 COUNTERURBANIZATION: IDAHO

Reasons for Decreasing Intraregional Migration

With modern communication and transportation systems, no location in a developed country is truly isolated, either economically or socially. Computers, tablets, and smart phones enable us to work anywhere and still have access to an international network. We can buy most products online and have them delivered within a few days. We can follow the fortunes of our favorite sports teams anywhere in the country, thanks to cable, satellite dishes, and webcasts.

Intraregional migration in the United States has slowed considerably since the 1980s (Figure 3.4.6). Most intraregional migration in the United States continues to be between cities and suburbs. Since 2010, the number of Americans moving from cities to suburbs has decreased, whereas the number moving from suburbs to cities has increased.

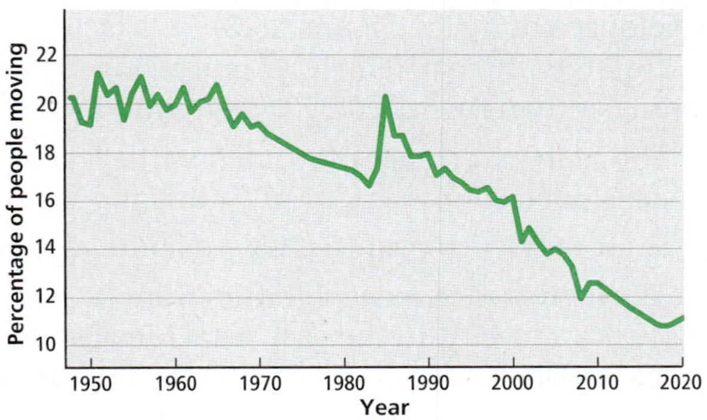

▲ 3.4.6 PERCENT OF AMERICANS WHO MOVE IN A YEAR

Reasons for Migrating

- **Describe the demographic characteristics of migrants.**

Ravenstein suggested that people migrate for three reasons: environmental, economic, and cultural (or political). Migration may involve a combination of these reasons.

▲ 3.5.2 TEENAGE MIGRANTS
Teenagers from Central America trying to migrate to the United States.

3.5.1 Who Are the Migrants?

When forming his theory in the nineteenth century, Ravenstein noted distinctive gender and family-status patterns in his migration theories. Most long-distance migrants were male rather than female and adult individuals rather than families with children. Recent U.S. migration patterns match the theory in some respects but not in others.

▼ 3.5.1 AGE OF IMMIGRANTS TO THE UNITED STATES
Compare to Figure 2.5.5.

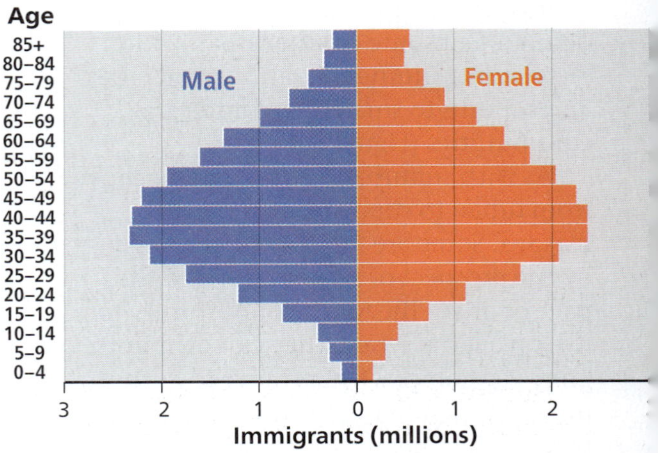

Age of Migrants

As Ravenstein expected, a relatively high share of U.S. immigrants are young adults, and U.S. immigrants are less likely to be elderly people (Figure 3.5.1). However, the number of unaccompanied minors trying to cross into the United States without proper documentation has increased sharply in recent years (Figure 3.5.2).

As with other migration flows, the large increase in teenage boys trying to reach the United States stems from a mix of push and pull factors. Many are pushed out of Honduras and El Salvador because of increased gang violence there and are pulled to the United States because of rumors that they won't be deported if they are caught crossing the U.S. border without authorization.

▼ 3.5.3 FEMALE IMMIGRANTS AS A PERCENTAGE OF ALL IMMIGRANTS

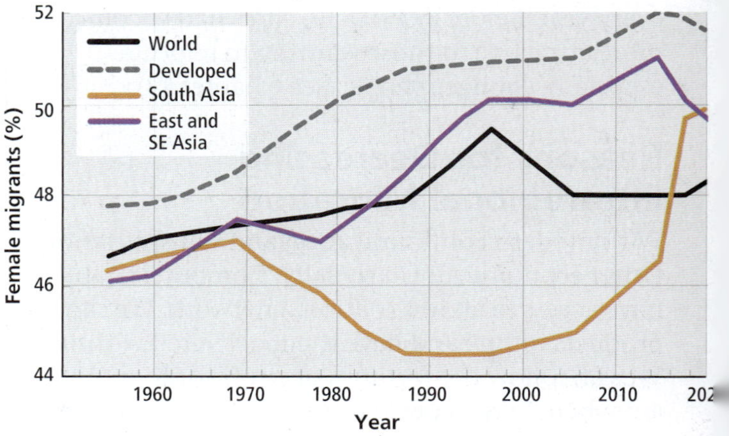

Female & Male Migrants

Ravenstein theorized that males were more likely than females to migrate long distances to other countries because searching for work was the main reason for international migration and males were much more likely than females to be employed. This held true for U.S. immigrants during the nineteenth and much of the twentieth centuries, but female immigrants to the United States began to outnumber male immigrants around 1970. Female immigrants also outnumber males in other developed countries (Figure 3.5.3). Two factors contribute to the larger share of females immigrating to developed countries than theorized by Ravenstein:

- The high percentage of women in the labor force in developed countries logically attracts a high percentage of female immigrants.
- Some developed countries have made it possible for wives to join husbands who have already immigrated.

The increased female migration from Mexico to the United States partly reflects the changing role of women in Mexican society. In the past, rural Mexican women were obliged to marry at a young age and to remain in the village to care for children. Now some Mexican women are migrating to the United States to join husbands or brothers already in the United States, but most immigrating Mexican women are seeking jobs because of poor economic conditions in Mexico.

3.5.2 Environmental Reasons for Migrating

People migrate because of push factors and pull factors. A **push factor** induces people to move out of their present location. A **pull factor** induces people to move to a new location. People sometimes migrate for environmental reasons. They may be pulled toward beautiful places and pushed from hazardous ones.

Attractive Environments

Attractive environments for migrants include mountains, seasides, and warm climates. Proximity to the Rocky Mountains lures Americans to the state of Colorado, and the Alps pull French people to eastern France. The highlands of Scotland attract migrants from southern areas of the United Kingdom. Regions with warm winters, such as southern Spain and the southwestern United States, attract migrants from harsher climates. Of all elderly people who migrate from one U.S. state to another, one-third select "the Sunshine State" of Florida as their destination.

Flooding & Migration

Water—either too much or too little—poses the most common environmental threat. Many people are forced to move by water-related disasters because they live in a vulnerable area, such as a floodplain. A **floodplain** is the area subject to flooding during a specific number of years, based on historical trends. People living in the "100-year floodplain," for example, can expect flooding on average once every century. Many people are unaware that they live in a floodplain, and even people who do know often choose to live there anyway.

Rising waters during hurricanes can force people to migrate. Large-scale forced migration has occurred in recent years in the United States because of hurricanes Katrina, Harvey, and Maria.

Hurricane Katrina hit New Orleans and neighboring communities in August 2005. It killed 1,836 people and temporarily displaced more than 1 million people, including between 80 and 90 percent of the residents of New Orleans.

Most of the New Orleans residents who were forced to emigrate in the aftermath of Katrina were Black. Most vulnerable to the flooding were the lowest-lying areas of New Orleans, such as the Lower Ninth Ward, where rental housing with a high density of Black residents was located. Areas in the city spared the worst of the flooding were those on slightly higher ground, such as the Vieux Carre (the French Quarter), where expensive homes with a high density of White residents were located (Figure 3.5.4).

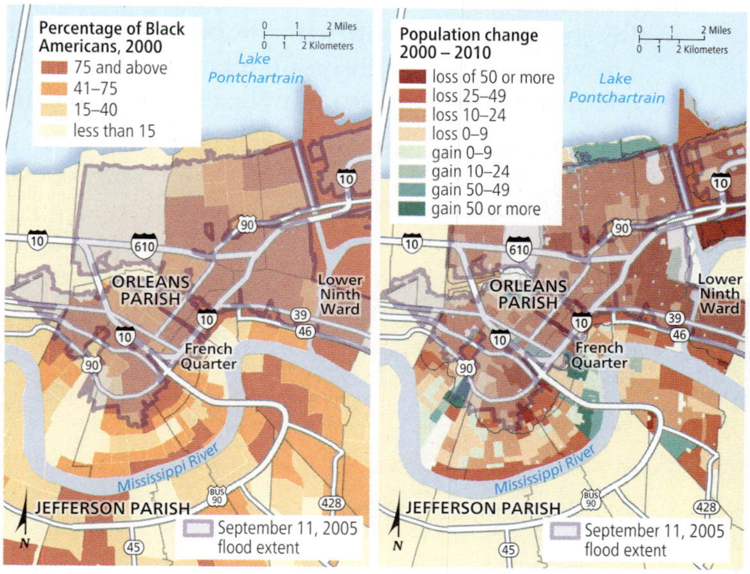

▲ 3.5.4 IMPACT OF HURRICANE KATRINA ON NEW ORLEANS
(a) Distribution of Blacks compared with extent of flooding;
(b) Population change after Katrina.

Rising Seas & Migration

Rising sea levels threaten to submerge in whole or part many of the world's islands. Sea levels rose around 17 centimeters (6.7 inches) during the twentieth century (see Section 1.13.2). The country of Kiribati, a collection of 33 small islands, in the Pacific Ocean, has already witnessed the disappearance of two of its islands under rising sea levels (Figure 3.5.5). Anticipating that the remaining islands will be gone by 2050, Kiribati purchased 2,500 hectares (6,000 acres) of land in Fiji in 2014, as a place to relocate all of the country's 120,000 inhabitants. Kiribati President Anote Tong called the project "migration with dignity."

President Tong's successor, Taneti Maamau, has rejected the migration program. Instead, he has announced a plan to raise the elevation of Kiribati's islands, with financial support from other countries, primarily China.

▼ 3.5.5 FLOODING IN KIRIBATI

Economic Reasons for Migrating

- Identify economic reasons for migration.

Most people migrate for economic reasons. People often emigrate from places that have few job opportunities and immigrate to places where jobs seem to be available. Because of economic restructuring, such as a decline in manufacturing and an increase in services, job prospects often vary from one country to another and within regions of the same country.

3.6.1 Economic Destinations

The United States and Canada have been especially prominent destinations for economic migrants. Many European immigrants to North America in the nineteenth century truly expected to find streets paved with gold. Although not literally so gilded, the United States and Canada did offer Europeans prospects for economic advancement. This same perception of economic plenty now lures people to the United States and Canada from Latin America and Asia.

People unable to migrate permanently to a new country for employment opportunities may be allowed to migrate temporarily. Examples of temporary work are found in Asia (discussed here) and in Europe (discussed later in this chapter). Asia is both a major source of and a major destination for migrants in search of work (Figure 3.6.1).

The world's largest influxes of migrants in search of work occur in South and East Asia. More than 2 million people annually migrate from India, Bangladesh, China, and Pakistan. An estimated 50 million Chinese and 27 million Indians live in other countries. The United States is a leading receiving country of migrants from South and East Asia, although most migrants from that region move to other countries in Asia.

The largest numbers of Chinese migrants are in Thailand, Malaysia, Indonesia, and Myanmar as well as the United States. Chinese people comprise one-half of the population of Singapore, one-fourth of Malaysia, and one-sixth of Thailand. The largest numbers of Indian migrants are in Nepal, Myanmar, and Malaysia as well as the United States.

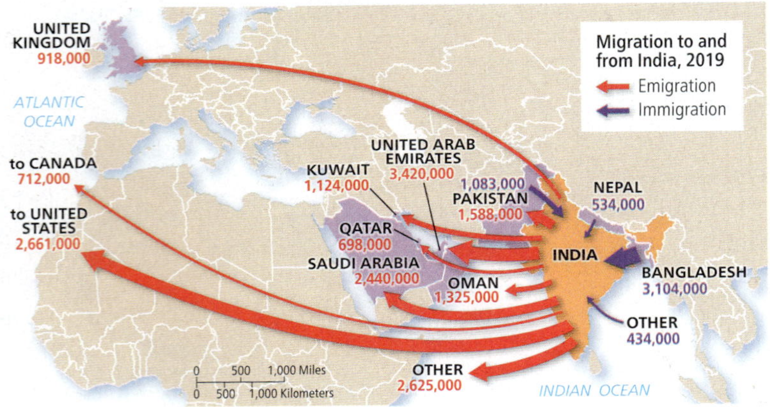

▲ 3.6.1 MIGRATION TO AND FROM INDIA

Working Conditions

The oil-producing countries of Southwest Asia with very high per capita incomes have been major destinations for emigrants from countries in Southeast Asia. In addition, citizens of countries in Southwest Asia where oil is not produced have emigrated to the region's oil-producing countries. Immigrants comprise 84 percent of the population of the United Arab Emirates (UAE), 74 percent of Qatar (Figure 3.6.2), 60 percent of Kuwait, and 55 percent of Bahrain. Saudi Arabia and the UAE have the largest numbers of immigrants in Southwest Asia.

Working conditions for immigrants have been considered poor in some of these countries. The Philippine government determined in 2011 that only two countries in Southwest Asia— Israel and Oman—were "safe" for their Filipino migrants, and the others lacked adequate protection for workers' rights. For their part, oil-producing countries fear that the increasing numbers of immigrants will spark political unrest and abandonment of traditional Islamic customs.

◀ 3.6.2 IMMIGRANTS IN QATAR
Allowed to work in Qatar to build soccer stadiums.

3.6.2 Remittances

Migrants who find work in another country frequently send a portion of the wages they have earned to relatives back home. The transfer of money by migrant workers to people in the country from which they migrated is a **remittance**.

The total amount of remittances worldwide was $715 billion in 2019. The figure had been increasing by nearly 10 percent annually until declining sharply during the COVID-19 pandemic, when fewer migrants were able to find work in other countries. Remittances are a significant portion of basic income for people in developing countries, especially following cutbacks in official assistance from foreign governments and international aid agencies (Figure 3.6.3).

People in India received the most remittances in 2020 ($83 billion), followed by people in China ($60 billion). Nearly one-third of the GDP of Tajikistan and Kyrgyzstan comprised remittances, primarily from migrants working in Russia.

The cost of transferring money is high in many places. Banks and firms such as Western Union that specialize in money transfers charge high fees for the service, an average of 9 percent worldwide (Figure 3.6.4). To transfer $200 from the United States, it costs an average of $6 to Mexico and $12 to Haiti; it costs around $20 to transfer $200 between many African countries.

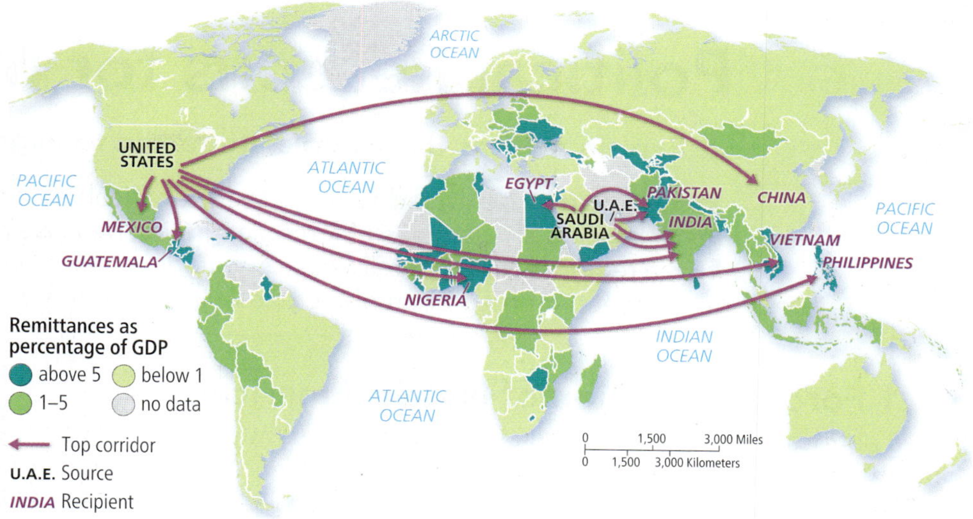

▲ 3.6.3 FLOW OF REMITTANCES

▲ 3.6.4 COLLECTING REMITTANCES
Dili, Timor-Leste.

Changing Attraction

The relative attractiveness of a region to migrants can shift with economic change. Ireland was a place of net out-migration through most of the nineteenth and twentieth centuries. Dire economic conditions in that country produced net out-migration in excess of 200,000 a year during the 1850s. The pattern reversed during the 1990s, as economic prosperity made Ireland a destination for immigrants, especially from Eastern Europe. The collapse of Ireland's economy as part of the severe global recession in 2008–2009 brought a return in that country of a net out-migration. However, an improving economy in Ireland has brought net in-migration in recent years (Figure 3.6.5).

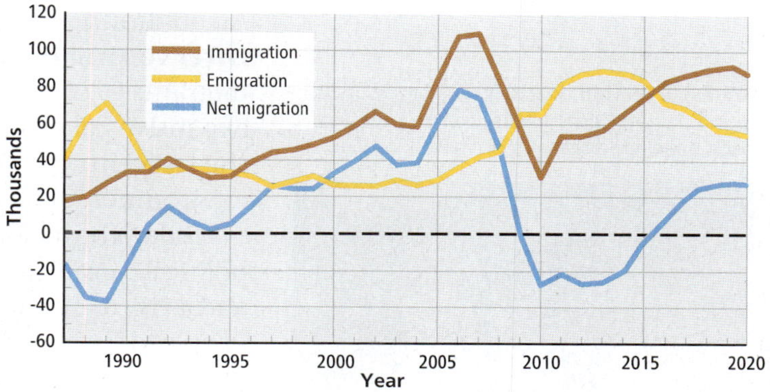

▲ 3.6.5 CHANGING MIGRATION TO AND FROM IRELAND

Political Reasons for Migrating

- **Explain reasons that people are forced to migrate.**

People may be forced to migrate for political reasons. Political reasons are often classified as examples of forced migration because people may be compelled to leave a region because of religious or cultural persecution, civil war, genocide, an unstable government, disintegration of the rule of law, and similar dangerous circumstances.

3.7.1 Forced Migration of Native Americans

Forced political migration occurred extensively in the United States during the nineteenth century. The victims were Native Americans. Like other people in North America, many Native Americans also migrated west in the nineteenth century. But their migration was forced rather than voluntary. This inequality was written in law, when the Indian Removal Act of 1830 authorized the U.S. Army to remove five Native American tribes from their land in the southeastern United States and move them to an area that is now part of the state of Oklahoma. The Choctaw were forced to emigrate from Mississippi in 1831, the Seminole from Florida in 1832, the Creek from Alabama in 1834, the Chickasaw from Mississippi in 1837, and the Cherokee from Georgia in 1838 (Figure 3.7.1).

The Indian Removal Act of 1830 opened up 100,000 square kilometers (25 million acres) of land for Whites to settle and relocated tribes to places that were too dry to sustain their traditional ways of obtaining food. Approximately 46,000 Native Americans were estimated to have been uprooted, and many of them died in the long trek west. The route became known as the Trail of Tears; parts of it are preserved as a National Historic Trail (Figure 3.7.2).

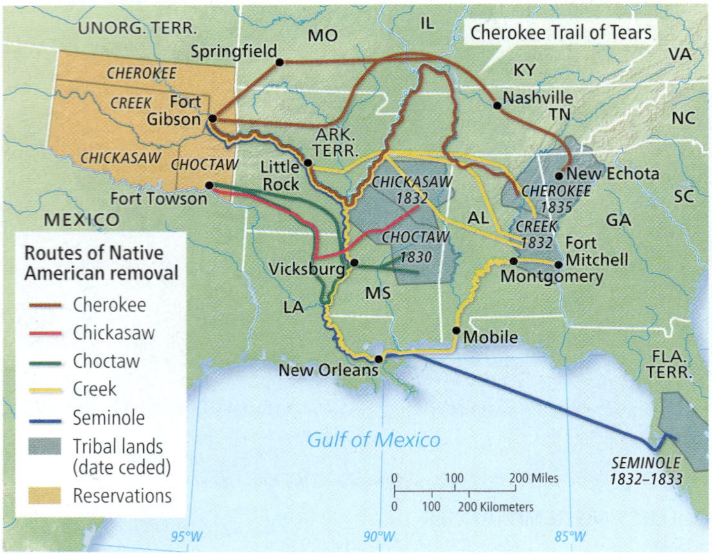

▲ 3.7.1 TRAIL OF TEARS ROUTES

▲ 3.7.2 TRAIL OF TEARS MEMORIAL
The Memorial Passage sculpture in Chattanooga, Tennessee, marks the beginning of the forced journey of Cherokee Tribes west to Oklahoma.

Intervening Obstacles

An environmental or political feature that hinders migration is an **intervening obstacle**. Historically, the principal obstacle faced by migrants to other countries was environmental: the long, arduous, and expensive passage over land or by sea. Transportation improvements over the last century—roads, canals, trains, cars, and airplanes—diminished environmental obstacles.

Today, the major intervening obstacles faced by most immigrants are political. Legally, a migrant needs a passport to emigrate from a country and a visa to immigrate to a new country. Intervening obstacles often cause people to engage in **step migration**, which is migration that follows a path of a series of stages or steps toward a final destination (Figure 3.7.3).

▼ 3.7.3 STEP MIGRATION
Migrants from Central America hoping to reach the United States stop in Mexico City.

3.7.2 Forced Political Migration

The United Nations High Commission for Refugees (UNHCR) recognizes three groups of people who are forced to migrate for political reasons:

- A **refugee** is forced to migrate to another country to avoid armed conflict, situations of generalized violence, violations of human rights, or other threats to their well-being. Often a refugee is an emigrant who cannot return to their country of origin for fear of persecution because of race, religion, nationality, membership in a social group, or political opinion.
- An **internally displaced person (IDP)** has been forced to migrate for reasons similar to those of a refugee, but an IDP has not migrated across an international border.
- An **asylum seeker** is someone who has migrated to another country in the hope of being recognized as a refugee.

The U.N. counted 26.4 million refugees, 48.0 million IDPs, and 4.1 million asylum seekers worldwide in 2020 (Figure 3.7.4).

▲ 3.7.4 REFUGEES AND IDPS

The Largest Numbers of Refugees

The largest numbers of refugees and IDPs in recent years are from Syria, Afghanistan, South Sudan, Myanmar, and Somalia. Ongoing civil wars in those countries cause this forced political migration. Neighboring countries received the most refugees. Refugees in Pakistan and Iran are primarily from Afghanistan, in Lebanon and Turkey primarily from Syria (Figures 3.7.5 and 3.7.6), in Uganda and Ethiopia primarily from South Sudan, in Bangladesh primarily from Myanmar, and in Ethiopia primarily from Somalia.

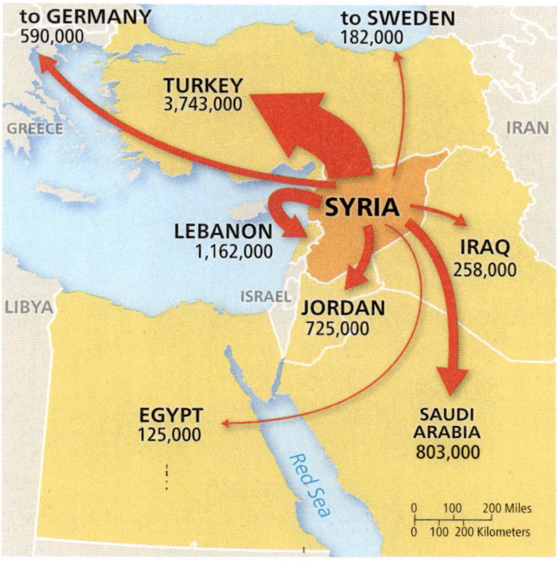

▲ 3.7.5 REFUGEES FROM SYRIA

▼ 3.7.6 REFUGEES FROM SYRIA ARRIVE IN JORDAN

▲ 3.8.1 PASSPORT CONTROL, LONDON HEATHROW AIRPORT

Government Migration Policies

- **Describe government policies that restrict immigration.**

Most countries, including the United States, have adopted legal restrictions and selective immigration policies that admit some groups of immigrants but not others (Figure 3.8.1). Policies toward immigrants and refugees are influenced by a number of factors: public opinion (for or against admitting migrants), economic (positive and negative impacts of newcomers on the local economy), resources (ability of the country to accommodate newcomers), and international relations (perception of humanitarian responsibility among the family of nations). The weight given to each of these factors can determine how countries address immigration.

3.8.1 Classifying Immigration & Emigration Policies

The U.N. classifies countries' immigration policies according to four criteria: Does a country wish to (1) maintain a steady level of immigration, (2) increase the level of immigration, (3) reduce the level of immigration, or (4) have no policy? Emigration policies are classified using the same four criteria as immigration.

Distribution of Immigration & Emigration Policies

According to the U.N., 24 countries seek more immigrants, 25 want fewer immigrants, 120 wish to maintain their current level of immigrants, and 27 do not have an immigration policy (Figure 3.8.2). Fourteen of the 24 countries with policies to encourage more immigration are in Europe, including most of the former communist countries of Central and Eastern Europe. The 25 countries with policies to reduce immigration include 11 in Asia and 7 in Africa.

The distribution of emigration policies is different. The U.N. found policies to increase emigration in 18 countries, to decrease emigration in 46 countries, to maintain the current level of emigration in 43 countries, and counted 88 countries with no emigration policy. The 18 countries wishing to increase emigration include 5 each in South Asia, Southeast Asia, and the South Pacific. Sub-Saharan Africa had the most countries seeking to lower emigration.

▶ 3.8.2 IMMIGRATION POLICIES

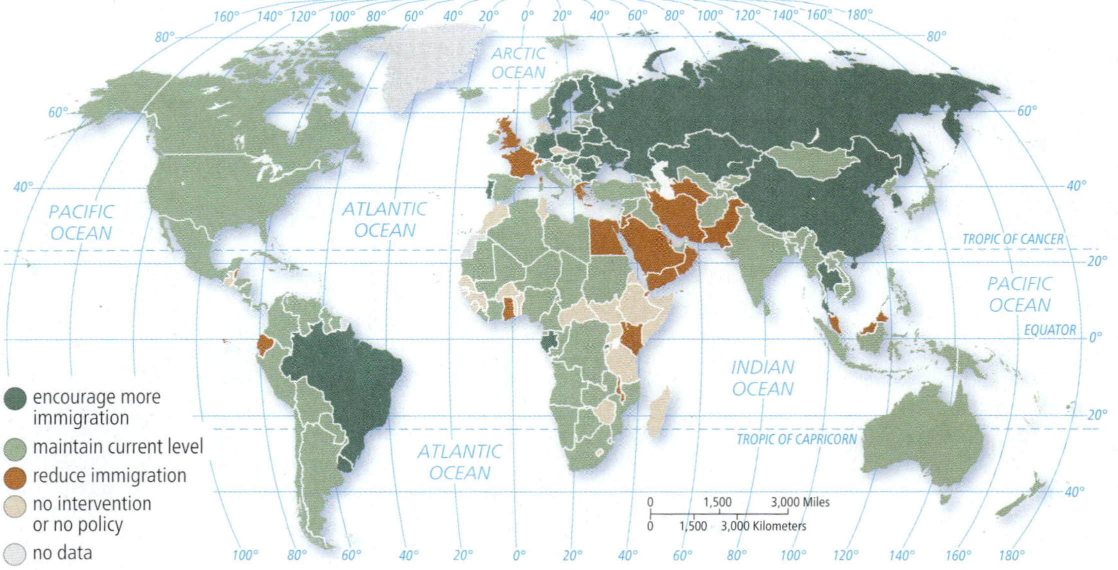

3.8.2 Europe's Immigration Issues

Migration issues have been especially challenging for Europeans in recent years. On the one hand, it is easier than ever for a European to migrate from one country to another within Europe. At the same time, it is harder for non-Europeans to migrate into Europe.

Open Borders Within Europe

Most borders within Europe are open for Europeans to cross without having to show a passport. Europeans can freely travel across borders by foot, airplane, car, or train. For example, people can simply cross by foot, bicycle, or streetcar over a bridge that spans the Rhine River connecting Strasbourg, France, and Kehl, Germany (Figure 3.8.3). Similarly, the border between Belgium and the Netherlands runs through the middle of a town, known as Baarle-Nassau on the Netherlands side and Baarle-Hertog on the Belgium side (Figure 3.8.4).

▼ **3.8.3 BEATUS RHEANUS BRIDGE**
No passport control between Strasbourg, France, and Kehl, Germany.

▼ **3.8.4 BELGIUM-NETHERLANDS BORDER**

Guest Workers

Until recently, immigrants were regarded as useful additions to the labor force in Europe. With most European countries in stage 4 or 5 of the demographic transition, characterized by stagnant or declining population, immigrants took low-status and low-skill jobs that local residents didn't want to do. In cities such as Berlin, Brussels, Paris, and Zurich, immigrants provide essential services, such as driving buses, collecting garbage, repairing streets, and washing dishes (Figure 3.8.5).

Germany and other developed European countries operated **guest worker** programs, mainly during the 1960s and 1970s, in which people from developing countries were allowed to immigrate temporarily to obtain jobs. Although relatively low paid by European standards, immigrants could earn far more than they would in their home country. The guest workers were expected to be examples of **circular migration**, which is the temporary movement of a migrant worker between their home and host countries to seek employment. However, many immigrants who arrived originally under the guest worker program have remained permanently in Europe.

In recent years, immigrants trying to reach Europe from Africa and Asia have been less welcome. Many Europeans fear that long-standing cultural traditions of the host country are threatened by immigrants who adhere to different religions and have different cultural traditions. Many African immigrants try to reach Europe by dangerous boat passages across the Mediterranean Sea (Figure 3.8.6).

▼ **3.8.5 GUEST WORKER CLEANS THE STREETS IN PARIS**

▼ **3.8.6 MIGRANTS FROM NORTH AFRICA APPREHENDED BY AUTHORITIES TRYING TO REACH ITALY BY BOAT**

U.S. Immigration Policies

- **Describe U.S. policies concerning immigration.**

To regulate migration, a **quota** is a law establishing a maximum limit on the number of people who can immigrate into a country. Until the twentieth century, the United States did not have quotas.

3.9.1 U.S. Quotas

The Immigration and Nationality Act (INA) allows the United States to grant up to 675,000 permanent immigrant visas each year. In addition to those visas, the INA sets no limit on the annual admission of U.S. citizens' spouses, parents, and children under the age of 21. In addition, each year the president is required to consult with Congress to set an annual number of refugees to be admitted to the United States through the U.S. Refugee Resettlement Process.

A person who obtains immigrant visas and comes to the United States becomes a lawful permanent resident (LPR). In some circumstances, a noncitizen already inside the United States can obtain LPR status. LPRs are eligible to apply for nearly all jobs and can remain in the country permanently.

After residing in the United States for five years (or three years in some circumstances), an LPR is eligible to apply for U.S. citizenship. Only LPRs may apply for citizenship.

Quota Preferences

Because the number of applicants for admission to the United States far exceeds the quotas, Congress has set preferences:

- **Family reunification.** Approximately three-fourths of immigrants are admitted to reunify families, primarily spouses or unmarried children of lawful permanent residents of the United States (Figure 3.9.1).
- **Skilled workers.** Exceptionally talented professionals receive most of the remainder of the quota.
- **Diversity lottery.** A few immigrants are admitted by lottery under a diversity category for people from countries that historically sent few people to the United States.

The quota does not apply to refugees, who may be admitted once their refugee status is verified. The number of refugees has varied sharply in recent years because of changes in policy. For example, under President Trump, the number of refugees admitted to the United States declined from 110,000 in 2017 to 18,000 in 2020.

Other countries charge that by giving preference to skilled workers, immigration policies in the United States as well as other developed countries contribute to a **brain drain**, which is a large-scale emigration by talented people. When scientists, researchers, and doctors immigrate to countries like the United States, countries from which they emigrated could then be left with shortages of skilled workers.

◀ 3.9.1 FAMILY REUNIFICATION
Recently arrived immigrants are learning English.

3.9.2 Unauthorized Immigration

The number of people allowed to immigrate into the United States is at a historically high level, but the number who wish to come is even higher. Many who cannot legally enter the United States immigrate illegally. Those who do so are entering without proper documents and thus are called **unauthorized immigrants**. The principal reason for unauthorized immigration to the United States is to seek a job.

Unauthorized immigrant is a neutral term preferred by academic observers, including the authoritative Pew Hispanic Center. *Undocumented immigrant* is the term preferred by some of the groups that advocate for the rights of those who immigrate without authorization. In 2020, around 250,000 people were apprehended trying to enter the United States from Mexico without authorization (Figure 3.9.2).

▲ 3.9.2 UNAUTHORIZED IMMIGRANT CHILDREN
Mother and children in U.S. Border Patrol custody after trying to cross into the United States.

The U.S. Department of Homeland Security estimated that 11.3 million unauthorized immigrants lived in the United States in 2018. The number increased rapidly between 1990 and 2005 (Figure 3.9.3). After reaching nearly 12 million in 2007, the number declined because of reduced job opportunities in the United States during the severe recession in 2008–2009. The number of unauthorized immigrants entering the United States has remained steady at around 11 million per year.

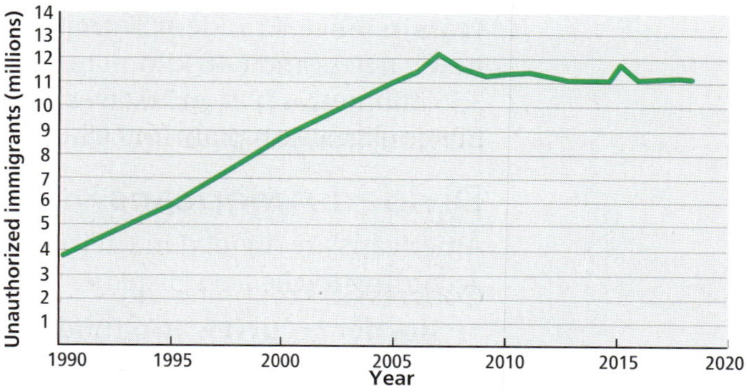

▲ 3.9.3 UNAUTHORIZED IMMIGRANTS IN THE UNITED STATES

Who Are the Unauthorized Immigrants?

The following are statistics on the 11 million unauthorized immigrants living in the United States as of 2018, according to surveys by the Migration Policy Institute, Pew Hispanic Center, and the U.S. Department of Homeland Security.

Origin & destination. Around 51 percent of the 11 million unauthorized immigrants came from Mexico. The share of unauthorized immigrants coming from Mexico has decreased in recent years. The remainder were about evenly divided between other Latin American countries and other regions of the world. Around 24 percent lived in California and 16 percent in Texas.

Children. The 11 million unauthorized immigrants included 1 million children. In addition, while living in the United States, unauthorized immigrants have given birth to approximately 4.4 million babies, who are legal citizens of the United States. The Deferred Action for Childhood Arrivals (DACA), a policy that President Obama established by Executive Order in 2012, President Trump rescinded in 2017, and President Biden restored in 2021, permitted 650,000 children brought to the United States as unauthorized immigrants to remain in the United States legally, as of 2021.

Years in the United States. Around 40 percent of unauthorized immigrants have lived in the United States for less than 10 years and 37 percent for at least 15 years.

Labor force. Approximately 7 million unauthorized immigrants were employed in the United States, accounting for around 4 percent of the total U.S. civilian labor force. Unauthorized immigrants were much more likely than the average American to be employed in construction and hospitality (food service and lodging) jobs and less likely to be in white-collar jobs such as education, health care, and finance.

U.S.–Mexico Border

- **Summarize the diversity of conditions along the U.S.–Mexico border.**

The U.S.–Mexico border is 3,145 kilometers (1,954 miles) long. The border runs through diverse landscapes, from major metropolitan areas to rugged mountains.

3.10.1 Along the U.S.–Mexico Border

A joint U.S.–Mexican International Boundary and Water Commission is responsible for keeping official maps of the border region on the basis of a series of nineteenth-century treaties. The commission is also responsible for marking the border by maintaining 276 6-foot-tall iron monuments erected in the late nineteenth century as well as 440 15-inch-tall markers added in the 1970s. Actually locating the border is difficult in some remote areas.

The Secure Fence Act of 2006 funded construction of 1,125 kilometers (700 miles) of barriers along the U.S.–Mexico border. By 2011, 1,044 kilometers (649 miles) had been built. President Trump signed an executive order in 2017 to construct a border wall. During its four years in office, President Trump's administration constructed 79 kilometers (49 miles) of new barriers and 648 kilometers (403 miles) of replacements for existing barriers. Another 356 kilometers (221 miles) were under construction and 253 kilometers (157 miles) were in pre-construction planning when President Trump left office in January 2021 (Figure 3.10.1).

Divided Americans

Americans are divided in their opinions on border policies and unauthorized immigration, according to the Pew Hispanic Center, for example:

- **Border security.** Two-thirds of Americans would like to see an increase in security—surveillance towers, air and boat patrols, and so forth—along the border. However, Americans are evenly divided on whether that security should include expansion of the existing barrier.
- **Legal status.** Two-thirds of Americans favor establishing a way for unauthorized immigrants to remain legally in the United States.
- **DACA immigrants.** Three-fourths of Americans favor granting permanent immigration status to DACA immigrants, who were brought to the United States as children by parents who were unauthorized immigrants.

◀ 3.10.1 **BORDER WALL, CIUDAD JUAREZ, MEXICO** A section of the 356 kilometers under construction in January 2021, the month that President Trump left office.

3.10.2 Diverse Border Crossings

The 49 vehicle and pedestrian border crossings between the United States and Mexico look very different depending on where one is trying to cross (Figure 3.10.2):

- Driving across the border in urban areas, such as between San Diego and Tijuana, can have heavy traffic (Figure 3.10.3).
- Crossings through adjacent small towns, such as between Columbus, New Mexico, and Palomas, Mexico, rarely have the heavy traffic of large urban areas (Figure 3.10.4).
- Crossing the border on foot is possible in several places, such as between Progreso, Texas, and Nuevo Progreso, Mexico (Figure 3.10.5).
- The Rio Grande River is easy for migrants to cross in places such as from Ciudad Juarez, Mexico, to El Paso, Texas, if they are able to evade the authorities (Figure 3.10.6).

▲ 3.10.5 U.S.–MEXICO BORDER: PROGRESO AND NUEVO PROGRESO

▼ 3.10.2 U.S.–MEXICO BORDER: DAILY ENTRIES PER CROSSING

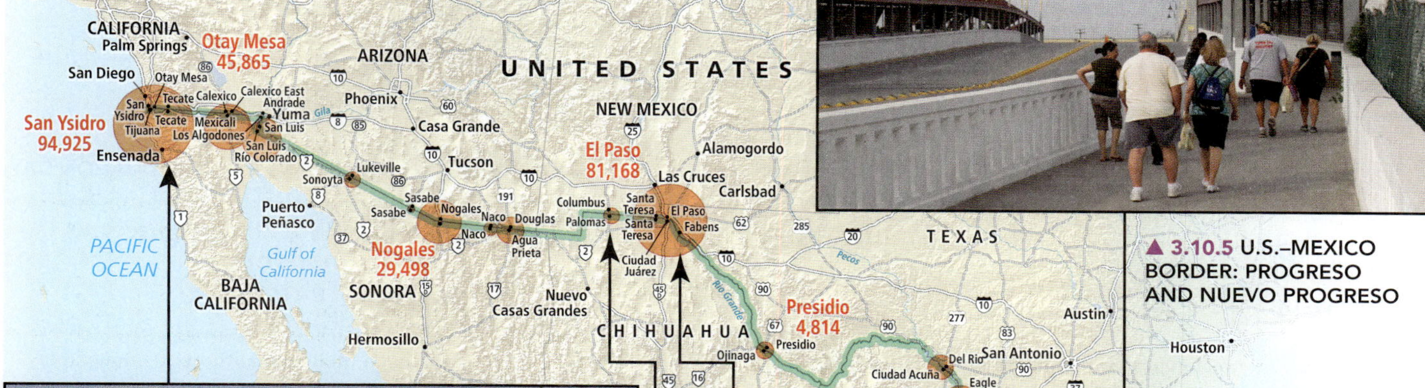

▲ 3.10.3 U.S.–MEXICO BORDER: SAN DIEGO AND TIJUANA

▲ 3.10.4 U.S.–MEXICO BORDER: COLUMBUS AND PUERTO PALOMAS

▲ 3.10.6 U.S.–MEXICO BORDER: EL PASO AND CIUDAD JUAREZ

CHAPTER 3

Review, Analyze, & Apply

KEY ISSUE 1 Where Are Migrants Distributed?

- Migration is the permanent move to a new location.
- Migration can be international (voluntary or forced) or internal (interregional or intraregional).
- The number and places of origin of immigrants to the United States have varied over time.

THINKING GEOGRAPHICALLY What might be the impact of emigration on the country from which migrants departed?

◀ 3.CR.1 DESERTED VILLAGE, MEXICO

KEY ISSUE 2 Where Do People Migrate Within Countries?

- Large-scale interregional migration has resulted in the movement of the U.S. population center to the west and south.
- Other large countries have experienced substantial interregional migration.
- Intraregional migration has been primarily from rural areas to urban areas in developing countries and from cities to suburbs in developed countries.

THINKING GEOGRAPHICALLY Has your community experienced net in-migration or net out-migration in recent years? What factors might explain your community's net migration?

◀ 3.CR.2 MOVING DAY

KEY ISSUE 3 Why Do People Migrate?

- People migrate for a combination of environmental, economic, and political push and pull factors.
- Most people migrate in search of work.
- Most migrants are young adults.

THINKING GEOGRAPHICALLY At some point in the future, do you expect to migrate? If so, do you expect to undertake international, interregional, or intraregional migration? Why?

◀ 3.CR.3 GRADUATION DAY

KEY ISSUE 4 What Challenges Do Migrants Face?

- Immigration is tightly controlled by most countries.
- The United States has around 11 million unauthorized immigrants, mostly from Mexico.
- Americans and Europeans are divided on attitudes toward immigrants.

THINKING GEOGRAPHICALLY The number of unauthorized immigrants from Mexico to the United States has been declining since 2007. Given the push and pull factors underlying people's reasons for migrating, what might account for this decline?

◀ 3.CR.4 BARRIER BETWEEN SAN DIEGO, CALIFORNIA, AND TIJUANA, MEXICO

Explore

Mexico's Other Border

Use Google Earth to explore Mexico's southern border (Figure 3.CR.6). Search for Talisman, Mexico, border crossing.

1. Use Street View for a ground-level view of the border crossing. Which if any of Mexico's border crossings with the United States in Section 3.10.2 does the Talisman one resemble? Why?
2. Click the X in the upper left to return to aerial view. Zoom out to around 3 km.
3. What is the country bordering Mexico? The border mostly follows a river. What is the river's name?
4. Follow the river to the south. Do you see any other official border crossings? Does the border look easy or difficult to cross? Why?

▶ 3.CR.5 MEXICO'S SOUTHERN BORDER WITH GUATEMALA

Key Terms

Asylum seeker (3.7.2) Someone who has migrated to another country in the hope of being recognized as a refugee.

Brain drain (3.9.1) Large-scale emigration by talented people.

Circular migration (3.8.2) The temporary movement of a migrant worker between home and host countries to seek employment.

Circulation (3.1.1) Short-term, repetitive, or cyclical movements that recur on a regular basis.

Counterurbanization (3.4.2) Net migration from urban to rural areas in developed countries.

Emigration (3.2.1) Migration from a location.

Floodplain (3.5.2) The area subject to flooding during a given number of years, according to historical trends.

Forced migration (3.1.2) Permanent movement, compelled by cultural or environmental factors.

Guest worker (3.8.2) A term once used for a worker who migrated to the developed countries of Northern and Western Europe, usually from Southern and Eastern Europe or from North Africa, in search of a higher-paying job.

Immigration (3.2.1) Migration to a new location.

Internal migration (3.1.2) Permanent movement within a particular country.

Internally displaced person (IDP) (3.7.2) Someone who has been forced to migrate for similar political reasons as a refugee but has not migrated across an international border.

International migration (3.1.2) Permanent movement from one country to another.

Interregional migration (3.1.2) Permanent movement from one region of a country to another.

Intervening obstacle (3.7.1) An environmental or cultural feature of the landscape that hinders migration.

Intraregional migration (3.1.2) Permanent movement within one region of a country.

Migration (3.1.1) A form of relocation diffusion involving a permanent move to a new location.

Migration transition (3.1.1) A change in the migration pattern in a society that results from industrialization, population growth, and other social and economic changes that also produce the demographic transition.

Mobility (3.1.1) All types of movements between locations.

Net migration (3.2.1) The difference between the level of immigration and the level of emigration.

Pull factor (3.5.2) A factor that induces people to move to a new location.

Push factor (3.5.2) A factor that induces people to leave old locations.

Quota (3.9.1) In reference to migration, a law that places maximum limits on the number of people who can immigrate to a country each year.

Refugees (3.7.2) People who are forced to migrate from their home country and cannot return for fear of persecution because of their race, religion, nationality, membership in a social group, or political opinion.

Remittance (3.6.2) Transfer of money by workers to people in the country from which they emigrated.

Step migration (3.7.1) Migration that follows a path of a series of stages or steps toward a final destination.

Unauthorized immigrant (3.9.2) A person who enters a country without proper documents to do so.

Voluntary migration (3.1.2) Permanent movement undertaken by choice.

Geospatial Analysis

Log in to the Mastering Geography Study Area to access MapMaster 2.0.

Europe's Immigrants & Emigrants

Substantial immigration is occurring within Europe. Let's see where people are coming and going. Add the ***Net Migration*** data layer. Select the *Settings* icon from the *Legend*, and select *Show Political Labels*. Add the ***Gross National Income per capita*** layer, and select *Split Map Window*. Zoom to Europe.

1. Europe can be divided between east and west. Which of the two has net outmigration, and which has net in-migration?
2. Which half of Europe has higher gross national income per capita?
3. In what way do you think the patterns on the two maps are associated with each other?

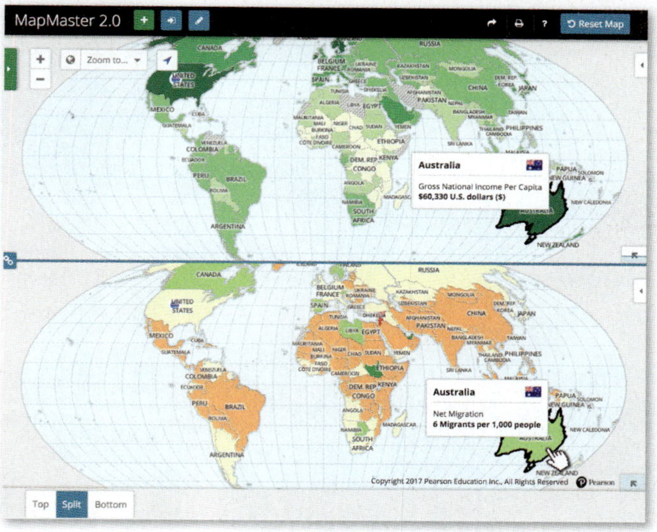

▶ **3.CR.6** NET MIGRATION AND GROSS NATIONAL INCOME PER CAPITA EUROPE

CHAPTER

4 Culture & Social Media

Children operating a drone, Vietnam.

Geographers distinguish groups of people according to important cultural features, describe where particular cultural groups are distributed, and offer reasons to explain the observed distribution. A consideration of culture follows logically from the discussion of migration in Chapter 3. Material culture diffuses from its place of origin to new locations because people bring along their culture when they migrate. Modern technology and interconnectivity allow ideas, images, and cultural values to spread almost instantaneously, sometimes bringing rapid change to once isolated regions of the world.

KEY ISSUES

1 How Are Leisure Activities Distributed?

Compared to folk culture, popular culture activities, including leisure and recreation, are more likely to originate at a specific time and place and to diffuse over a wider region.

2 How Does Culture Diffuse?

Material folk culture, including food, shelter, and clothing, is more likely to diffuse through relocation diffusion, whereas material popular culture is more likely to diffuse through expansion diffusion.

3 Why Is Access to Culture Unequal?

Popular culture is increasingly likely to diffuse through electronic media. People in some regions lack access to electronic media.

4 Why Do Cultures Face Sustainability Challenges?

Although popular culture tends to produce cultural uniformity throughout the world, differences in folk culture preferences persist among people in some regions.

LOCATIONS IN THIS CHAPTER

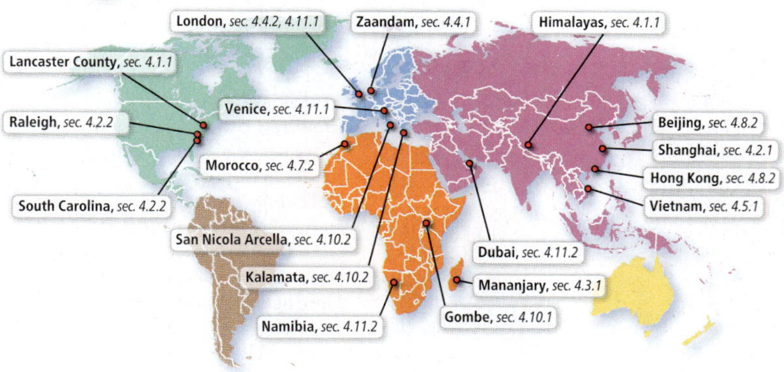

Material Culture & Geography

- **Introduce main categories of cultural geography.**

Geographers define **culture** as the body of material possessions, customary beliefs, and social norms that together constitute the distinct tradition of a group of people. This chapter focuses on material culture—the possessions and leisure activities that define a group of people and that they preserve for the future.

4.1.1 Scale & Types of Culture

Culture falls into two basic categories that differ according to scale. **Folk culture** is traditionally practiced primarily by small, homogeneous groups living in relative isolation from other groups. **Popular culture** is found in large, heterogeneous societies that share certain customs (such as wearing jeans) despite differences in other personal characteristics. Culture can be distinguished from a habit or a custom. A **habit** is a repetitive act that a particular individual performs, such as wearing jeans to class every day. A **custom** is a repetitive act of a group, performed to the extent that it becomes characteristic of the group, such as many students typically wearing jeans to class.

Origin

Culture originates at a hearth, which was defined in section 1.11.1 as a node of innovation. Folk culture often has anonymous hearths, originating from anonymous sources, at unknown dates, through unidentified originators. Popular culture is typically traceable to a specific person or corporation in a particular place. It is most often a product of developed countries, especially in North America and Europe.

Distribution

Popular culture and folk culture have different distributions. Popular culture is distributed widely across many countries, with little regard for physical factors. The distribution is influenced by the ability of people to access the material. The distribution of folk culture is influenced by a combination of local physical and cultural factors. For example, geographers P. Karan and Cotton Mather found distinctive artistic representations of the physical environment among neighboring cultural groups that are isolated from each other in the rugged Himalaya Mountains (Figure 4.1.1).

4.1.1 DISTRIBUTION OF FOLK CULTURE: ART IN THE HIMALAYAS

Muslims. Islamic folk art is inspired by the area's beautiful plants and flowers.

Buddhists. Buddhists paint idealized divine figures, such as monks and saints, often set in the local environment.

Folk religionists. Animists paint symbols and designs that derive from their religion rather than from their surroundings.

Hindus. Hindus create scenes from everyday life and familiar local scenes. Their paintings sometimes include a deity in a domestic scene.

4.1.2 Diffusion of U.S. Houses

Folk and popular cultures have different processes of diffusion:

- Folk culture is transmitted from one location to another relatively slowly and on a small scale, primarily through relocation diffusion (migration).
- Popular culture typically spreads through a process of expansion diffusion, diffusing rapidly and extensively from hearths or nodes of innovation with the help of modern communications.

The differences between the relocation diffusion of folk culture and the expansion diffusion of popular culture are illustrated through U.S. housing.

Relocation Diffusion of U.S. Folk Houses

The distribution of older houses built in the United States displays evidence of relocation diffusion as it applies to folk culture. In contrast, housing built more recently in the United States demonstrates expansion diffusion as it applies to popular culture.

Geographer Fred Kniffen identified three major hearths, or nodes, of folk house architecture in the United States (Figure 4.1.2). When families migrated westward in the 1700s and 1800s, they cut trees to clear fields for planting and used the wood to build houses, barns, and fences. The style of the homes reflected that of the most popular house style at the time on the East Coast region from which they migrated.

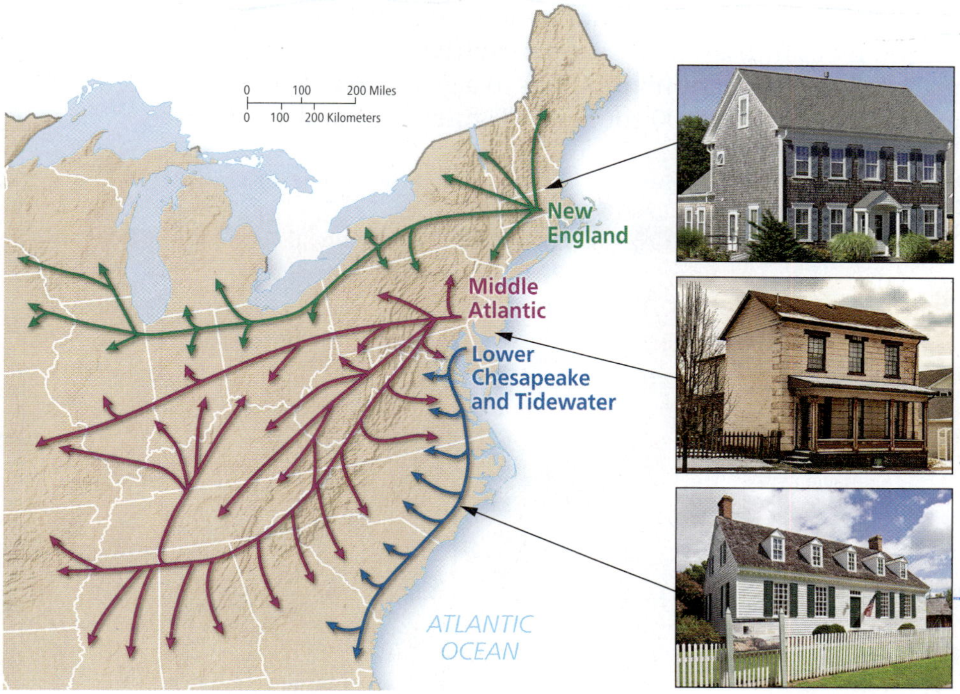

New England. The distinctive New England house style was box shaped with a central hall. The house style can be found throughout the Great Lakes area as far west as Wisconsin because that region was settled primarily by migrants from New England.

Middle Atlantic. The principal house style in the Middle Atlantic region was known as the "I house" and typically featured two full stories in height, was one room deep, and at least two rooms wide. Middle Atlantic migrants carried their house type westward across the Ohio Valley and southwestward along the Appalachian trails.

Lower Chesapeake/Tidewater. The house style that originated in the Lower Chesapeake Bay & Tidewater Virginia area typically comprised one full story, with a steep roof and chimneys at either end. Migrants spread this style along the Southeast Coast.

▲ **4.1.2 FOLK HOUSING: U.S. HEARTHS**
U.S. folk houses originated in three East Coast regions and diffused west and south along different paths. These paths coincided with predominant routes taken by migrants from the East Coast toward the interior of the country.

Expansion Diffusion of U.S. Popular Housing

Houses built in the United States since the mid-twentieth century display popular culture influences. The degree of regional distinctiveness in housing style has diminished because rapid communication and transportation systems provide people throughout the country with knowledge of alternative styles.

Furthermore, most people do not build the houses in which they live. Instead, houses are usually mass-produced by construction companies. Houses show the influence of shapes, materials, detailing, and other features of architectural style in vogue at any one point in time. In the years immediately after World War II, which ended in 1945, most U.S. houses were built in a modern style. Since the 1960s, styles that architects call neo-eclectic have predominated.

▲ 4.2.1 MCDONALD'S IN CHINA
Shanghai.

Popular & Folk Food Customs

- Identify differences between popular and folk food preferences.

According to the nineteenth-century cultural geographer Paul Vidal de la Blache, "Among the connections that tie [people] to a certain environment, one of the most tenacious is food supply; clothing and weapons are more subject to modification than the dietary regime, which experience has shown to be best suited to human needs in a given climate."

4.2.1 Popular Food Culture

Folk food customs derive in part from environmental features, as suggested by Vidal de la Blache a century ago. However, folk food customs are also strongly influenced by cultural traditions, especially in developed countries. What is eaten suggests one's social, religious, and ethnic memberships. The surest way to identify a family's ethnic origins is to look in its kitchen.

However, in the popular culture of the twenty-first century, food preferences seem far removed from folk traditions. The popular food culture is influenced more by changing cultural values and economic conditions than by the local environment. For example, McDonald's, founded in 1940 at a single location in San Bernardino, California, evolved in the 1950s into a franchise. A **franchise** is an agreement between a corporation and businesspeople to market that corporation's products in a given area. The corporation now has around 39,000 McDonald's franchises in more than 100 countries (Figures 4.2.1 and 4.2.2).

Regional Differences in Popular Food Culture

Regional differences in popular food culture can be observed between and within countries. For example, Subway restaurants outnumber McDonald's restaurants in some countries, whereas in others McDonald's are more numerous (Figures 4.2.3 and 4.2.4).

▶ 4.2.2 NUMBER OF MCDONALD'S RESTAURANTS

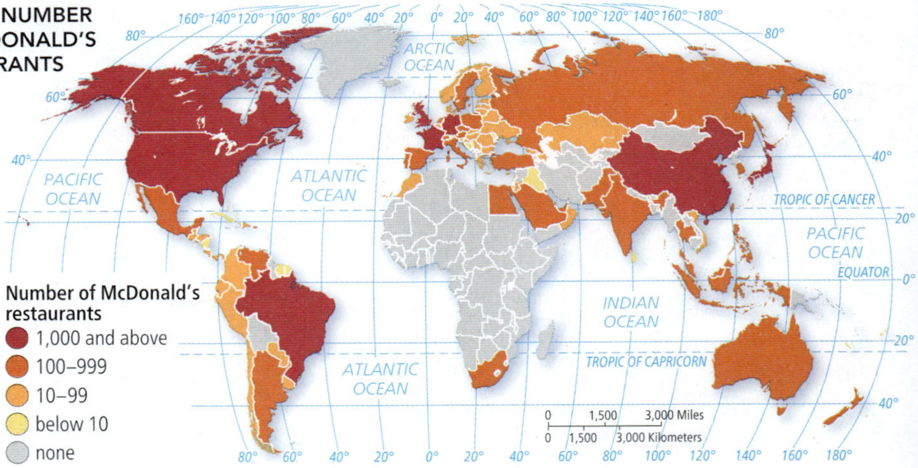

▼ 4.2.3 SUBWAY
Chennai, India.

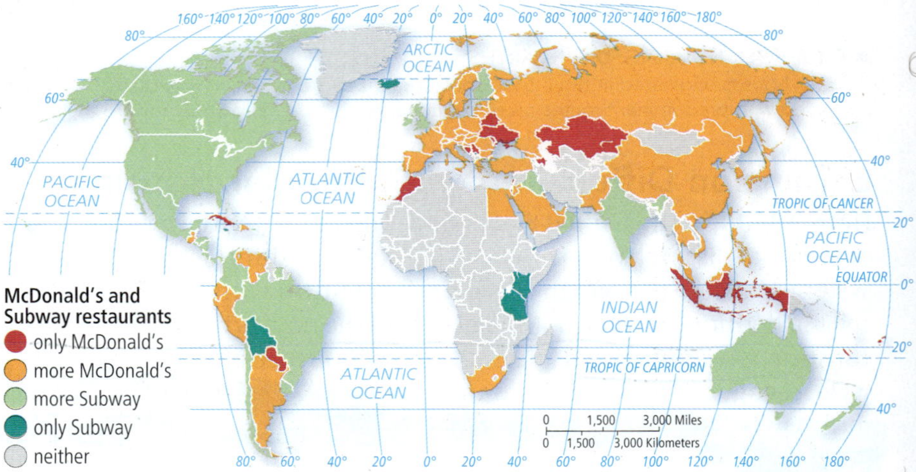

▶ 4.2.4 SUBWAY VS. MCDONALD'S
McDonald's has more outlets than Subway in Europe and China.

4.2.2 Folk Food Customs

Folk food customs differ significantly within and among regions. For example, barbecue styles vary widely within the United States and at varying scales. At the national scale, four major barbecue styles include two relatively old styles—Carolina and Memphis—which use pork, and two more recently evolved styles—Kansas City and Texas—which use beef as well as pork (Figure 4.2.5). Meats other than pork and beef can be found elsewhere, such as lamb and mutton in Kentucky (Figure 4.2.6). Variations in barbecue style can be found at an even more local scale. For example, the preferred barbecue sauce varies widely within South Carolina (Figure 4.2.7).

▲ **4.2.5 PORK BARBECUE STAND**
Raleigh, North Carolina.

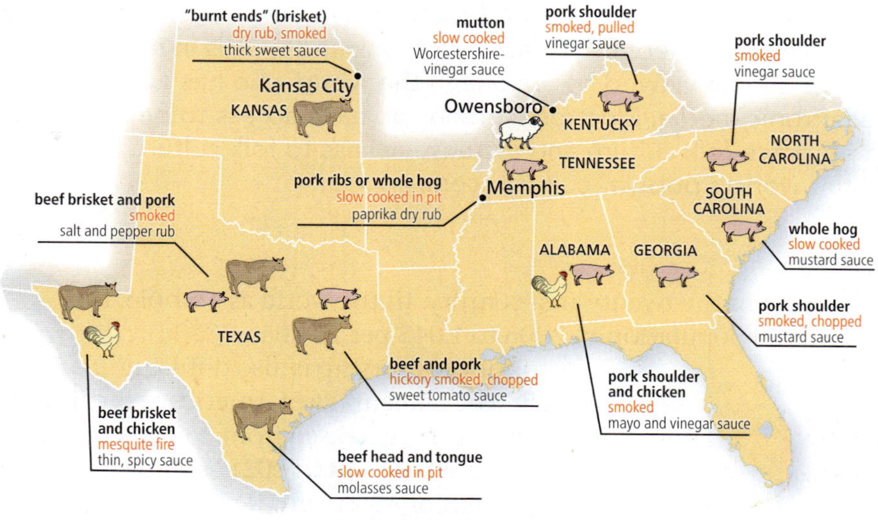

▲ **4.2.6 REGIONAL VARIATIONS IN BARBECUE: SOUTHEAST UNITED STATES**

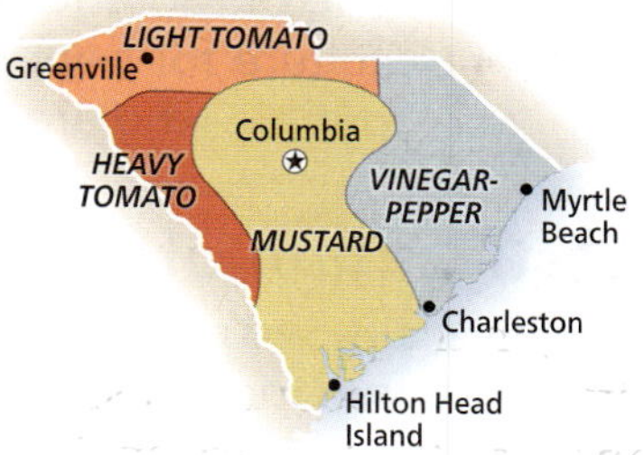

▲ **4.2.7 VARIATIONS IN BARBECUE SAUCE: SOUTH CAROLINA**

Food Taboos

According to many folk customs, everything in nature carries a signature, or distinctive characteristic, based on its appearance and natural properties. Consequently, people may desire or avoid certain foods in response to perceived beneficial or harmful natural traits.

People refuse to eat particular plants or animals for a variety of reasons. Most of the time, they refuse because they don't like the taste or texture of the food, but at times, the reason for refusing to eat a food is cultural. A restriction on behavior imposed by religious law or social custom is a **taboo**. Other behaviors may be prohibited, such as certain practices related to sex and marriage, but taboos are especially strong in the area of food.

Some folk cultures may establish food taboos because of concern for the natural environment. Such food taboos may help protect endangered animals or conserve scarce natural resources. For example, to preserve scarce animal species, only a few high-ranking people in some tropical regions are permitted to hunt.

Relatively well-known food taboos are found in the Jewish Bible. The ancient Hebrews were prohibited from eating a wide variety of foods, including animals that do not chew their cud or that have cloven feet and fish lacking fins or scales. These biblical taboos were developed through oral tradition and by rabbis to become the kosher laws observed today by some Jews.

Muslims embrace a taboo against pork in part because pigs are unsuited for the dry lands of the Arabian Peninsula. Pigs would compete with humans for food and water, without offering compensating benefits, such as pulling a plow, carrying loads, or providing milk and wool. Widespread raising of pigs would be an ecological disaster in Islam's hearth.

Hindu taboos against eating beef are partly explained by India's environment. Cows are the source of oxen (castrated male bovine), the traditional choice for pulling plows as well as carts. A large supply of oxen must be maintained in India because every field has to be plowed at approximately the same time—when the monsoon winds bring rain. Religious sanctions have kept India's cattle population large as a form of insurance against the loss of oxen and an increasing human population.

But the taboo against consumption of meat among many people, including Muslims, Hindus, and Jews, cannot be explained primarily by environmental factors. Social values also influence diet because people in similar climates and with similar levels of income consume different foods. Biblical food taboos help the Jewish people maintain their identity and communal affiliation. That Christians ignore Biblical food taboos is consistent with Christianity, as a religion that seeks to attract adherents from many places (see Chapter 6).

Distribution of Electronic Communications

- **Identify the distribution of social media devices.**

Geographic concepts help us understand the distribution and diffusion of popular culture. Popular culture is distributed around the world in the twenty-first century primarily through electronic media. The latest fashions and trends in material culture and leisure activities can be viewed by anyone in the world who has access to one or more forms of electronic media. Electronic media increase access to popular culture for people who embrace folk culture and increase access to folk culture for people who are part of the world's popular culture scene.

4.3.1 Distribution of Smartphones

Cell phones have diffused rapidly around the world in the twenty-first century. In the world as a whole, the number of cell phones increased from 338 per 1,000 population in 2005 to 1,043 per 1,000 in 2020.

Smartphone ownership is also diffusing rapidly. The percentage of Americans owning smartphones increased from 35 percent in 2011 to 90 percent in 2021. More than 95 percent of Americans between 18 and 29 years old have smartphones, but ownership is also increasing rapidly among older Americans. Internationally, the rate of ownership of smartphones remains higher among developed countries than among developing ones, but the gap is narrowing (Figure 4.3.1).

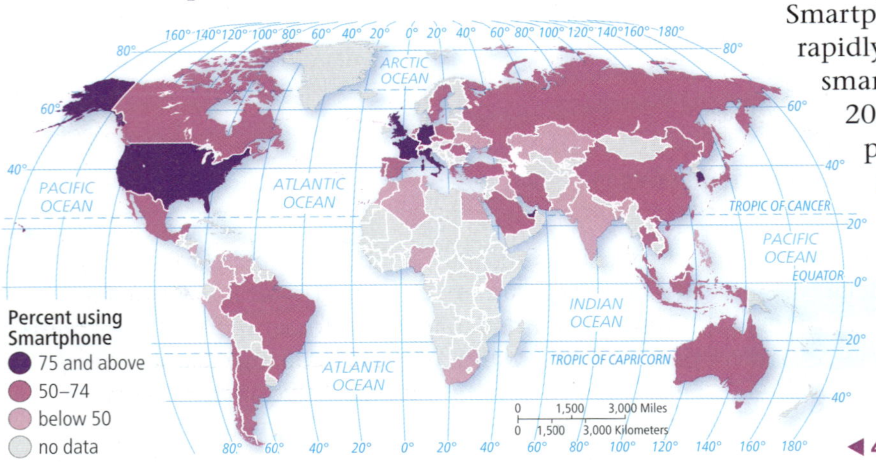

◀ 4.3.1 SMARTPHONE OWNERSHIP

Cell Phones in Africa

Cell phones and smartphones have become common in Africa (Figure 4.3.2). Traditional land line services are scarce in Africa, and access to electricity is limited. Such obstacles to telephone communication on the continent are largely bypassed because cell phones do not require the costly investment of connecting wires to each individual building, and thousands of individuals can obtain cell phone service from a single tower or satellite (Figure 4.3.3).

◀ 4.3.2 SMARTPHONES IN AFRICA
Mananjary, Madagascar.

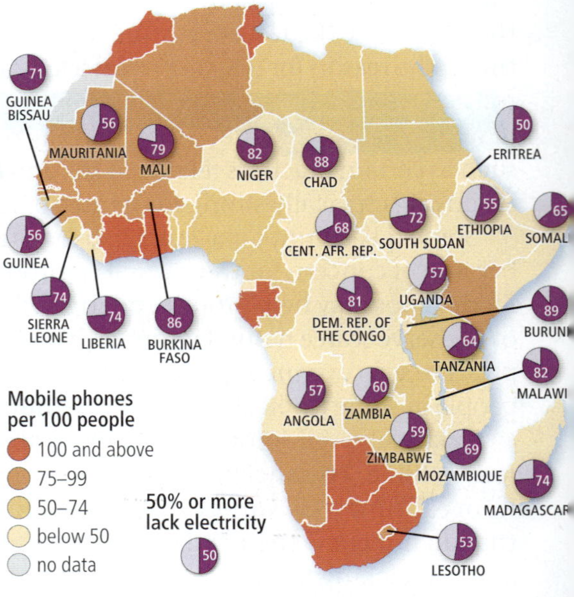

▲ 4.3.3 CELL PHONE SERVICE, AFRICA

4.3.2 Distribution of Social Media Platforms

The world's most popular social media platforms are Facebook, Twitter, and Instagram. Around two-thirds of the world's social media users are on Facebook, and Instagram and Twitter together account for nearly one-fourth (Figure 4.3.4). In the first years of social media, many distinct networks were popular. In many places, competing social networks were quickly supplanted by Facebook, especially in Latin America (Figure 4.3.5).

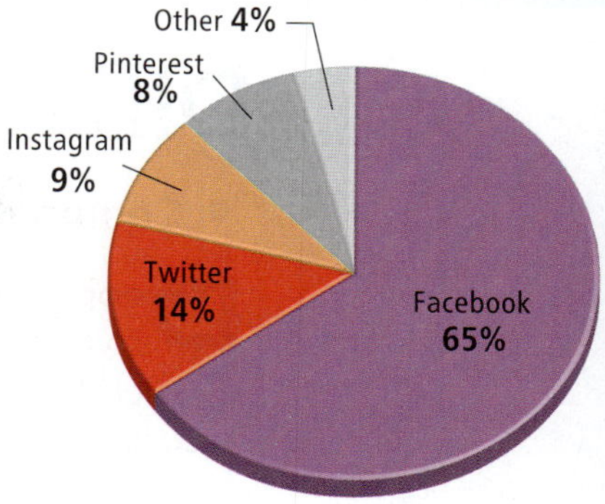

▲ **4.3.4 MOST POPULAR SOCIAL MEDIA BY MARKET SHARE**

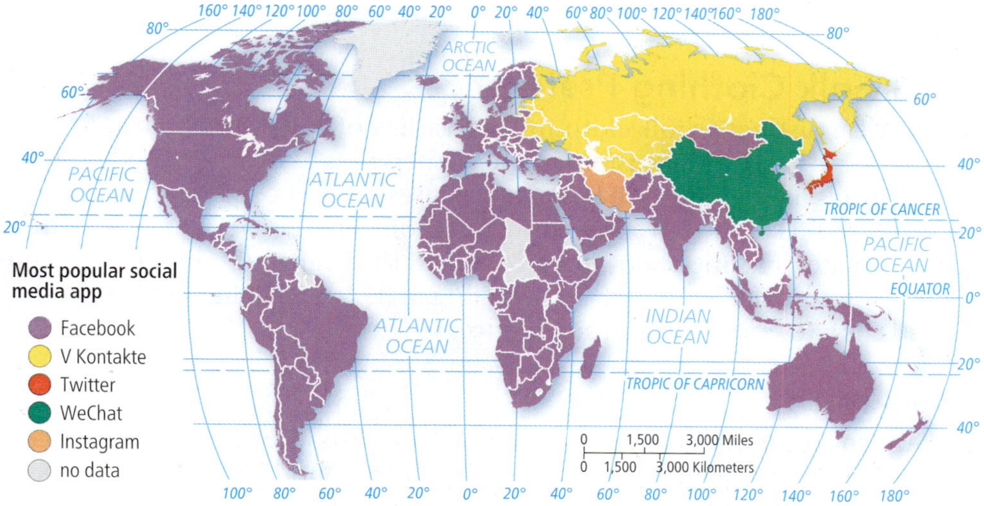

◀ **4.3.5 MOST POPULAR SOCIAL MEDIA BY COUNTRY**

Regional Differences in Popularity

Regional differences persist in the use of social media. The United States is the leading source of Twitter and Instagram messages (Figures 4.3.6 and 4.3.7). Notably absent from the list of leading users is the world's most populous country, China. China's government has limited the ability of the Chinese people to use several of the most popular social media developed in the United States and other countries, including the three most popular: Facebook, Twitter, and Instagram.

Regional variations of social media can also be seen in messaging applications. Six messaging applications are dominant worldwide (Figure 4.3.8). However, two of them are used primarily in China (WeChat and QQ), whereas the other four are not available for most people in China.

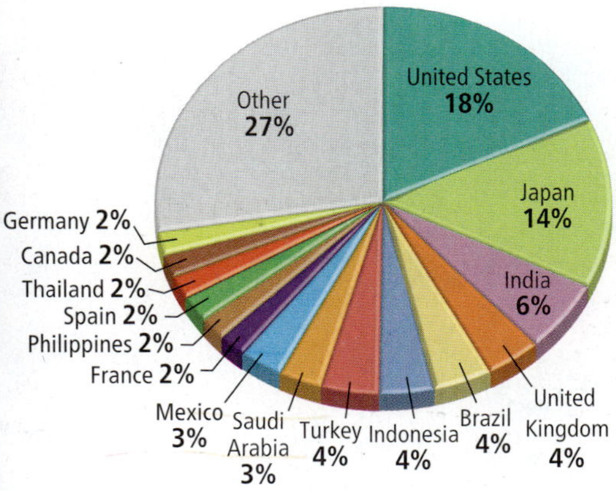

▲ **4.3.6 DISTRIBUTION OF TWITTER USERS**

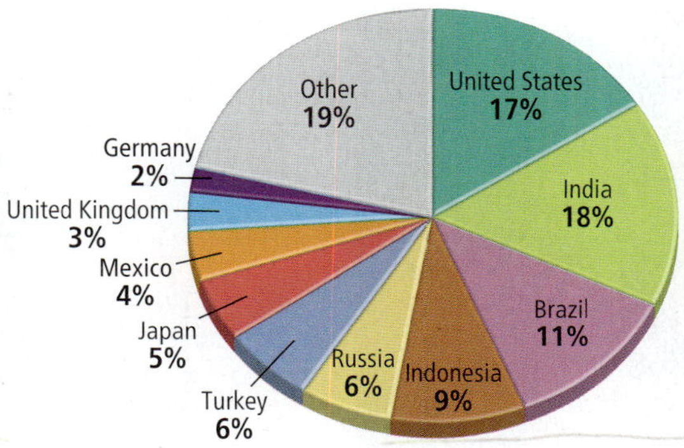

▲ **4.3.7 DISTRIBUTION OF INSTAGRAM USERS**

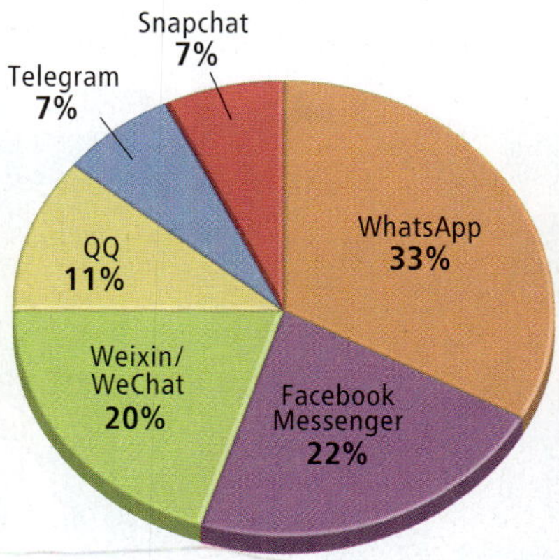

▲ **4.3.8 MOST POPULAR MESSAGING APPLICATIONS**

Diffusion of Clothing

- **Compare distribution of clothing styles in folk and popular culture.**

People living in folk cultures have traditionally worn clothing partly in response to distinctive agricultural practices and climatic conditions. In popular culture, clothing preferences generally express class, occupation, and other elements of cultural identity rather than particular environments.

▲ **4.4.1 ARTISAN MAKING WOODEN SHOES, USING TRADITIONAL METHODS**
Zaandam, The Netherlands.

4.4.1 Folk Clothing Preferences

People wear distinctive folk clothing for a variety of environmental and cultural reasons. The folk custom in the Netherlands of wearing wooden shoes may appear quaint, but it still has practical uses in a wet climate. Artisans continue to make wooden shoes using traditional methods (Figure 4.4.1). In arctic climates, fur-lined boots protect against the cold, and snowshoes permit walking on soft, deep snow without sinking in it. People living in warm and humid climates may not need any footwear.

The Amish

The Amish are identifiable by their distinctive clothing and farming practices (Figure 4.4.2). Although the Amish number only around 350,000, their folk culture remains visible on the landscape in at least 19 U.S. states (Figure 4.4.3). The distribution of Amish folk culture across a major portion of the U.S. landscape is explained by examining the diffusion of that culture through migration.

Several hundred Amish families migrated to North America in two waves. The first wave, primarily from Bern, Switzerland, and the Palatinate region of southwestern Germany, settled in Pennsylvania in the early 1700s, enticed by English colonial proprietor William Penn's offer of low-priced land. Around 30,000 Amish currently live in Lancaster County, Pennsylvania. Because of lower land prices in the 1800s, the second wave, from Alsace in northeastern France, settled in Ohio, Illinois, and Iowa in the United States and Ontario, Canada. From these U.S. states and Canadian province, groups of Amish migrated to other locations in both countries where inexpensive land was available.

▼ **4.4.2 AMISH**
Amish children wear traditional clothing, Lancaster County, Pennsylvania.

▼ **4.4.3 DISTRIBUTION OF AMISH**

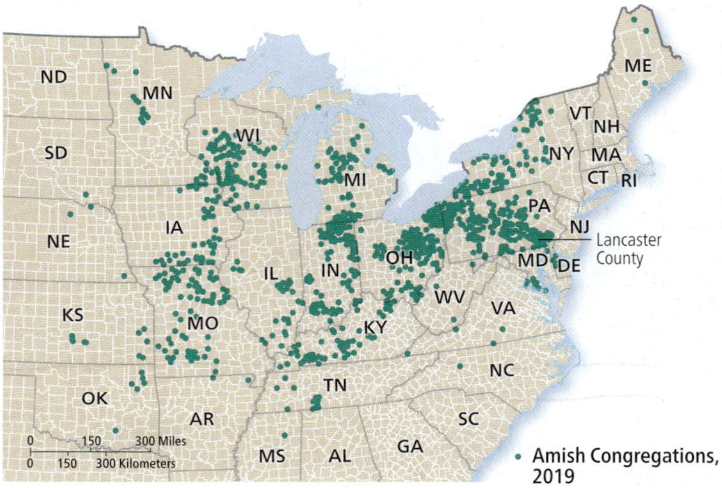

4.4.2 Distribution of Popular Clothing Styles

Changes in contemporary fashion and popular clothing take place in an interconnected world. Fashionable clothes, which constantly feature new styles, are among the most globally distributed products available to consumers today. What an individual chooses to wear reveals that popular culture can be distributed across the landscape with little regard for a region's distinctive environment and climate.

▶ 4.4.4 COUNTRIES WITH HOLLISTER STORES

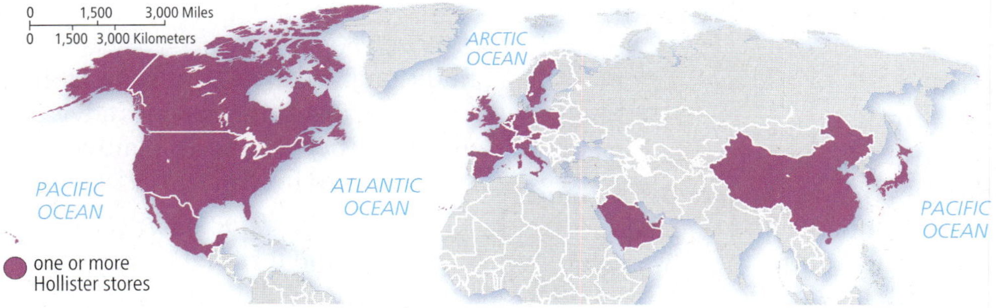

Instead, popular clothing is more likely to reflect occupation and income. A lawyer in Mexico is more likely to dress like a lawyer in the United States than like a factory worker in Mexico.

Most people wear clothing that fits the cultural traditions where they live, but the diffusion of popular clothing is eroding that norm. A shopper who wants to buy a trendy, international-retail brand of clothing such as Hollister will find stores that carry that brand around the world (Figures 4.4.4 and 4.4.5). And the brand of clothing may have been made in and shipped from another region of the world (see Chapter 11).

▲ 4.4.5 HOLLISTER STORE, LONDON

Jeans

Jeans have become popular clothing in many parts of the world (Figure 4.4.6). However, consumers in two places who choose the same brand of jeans might wear them in different cultural contexts. Clothing considered commonplace and traditional in one part of the world may be viewed as bold and controversial in another.

▼ 4.4.6 UNIVERSITY STUDENTS IN EUROPE WEARING JEANS, UNITED KINGDOM

Diffusion of Music

- **Compare differences in geographic features of folk and popular music.**

Every culture in human history has had some tradition of music, asserts music researcher Daniel Levitan. As music is a part of both folk and popular culture traditions, it can be used to illustrate the differences in the origin, diffusion, and distribution of folk and popular culture.

4.5.1 Folk Music

According to a Chinese legend, music was invented in 2697 B.C.E., when the Emperor Huang Ti sent Ling Lun to cut bamboo poles that would produce a sound matching the call of the phoenix bird. In reality, folk songs typically originate anonymously and are transmitted orally. A song may be modified from one generation to the next as cultural conditions change, but the content of a song is most often derived from events in daily life that are familiar to the majority of the people at any given time. As people migrate, folk music travels with them as part of the diffusion of folk culture.

Folk Songs: Vietnam

Folk songs may tell a story or convey information about life-cycle events, such as birth, death, and marriage, or environmental features, such as agriculture and climate. For example, in Vietnam, where many people are subsistence farmers, information about agricultural technology was traditionally conveyed through folk songs. The following folk song provides advice about the difference between seeds planted in summer and seeds planted in winter:

Ma chiêm ba tháng không già
Ma mùa tháng ruôi át là không non.[1]

This song can be translated as follows:

While seedlings for the summer crop are not old when they are three months of age,
Seedlings for the winter crop are certainly not young when they are one-and-a-half months old.

English-language folk songs often draw upon similar themes, even if the specific information conveyed about the environment differs.

Festivals throughout Vietnam feature music in locally meaningful environmental settings, such as hillsides or on water. Singers in traditional clothes sing about elements of daily life in the local village, such as surrounding trees, flowers, and water sources (Figure 4.5.1).

◀ **4.5.1 FOLK MUSIC: QUAN HO, VIETNAM**
Vietnamese singers perform Quan Ho folk songs as part of the Lim Festival, which is held annually on the 13th to the 15th day of the first lunar month. Quan Ho folk music dates back more than 500 years. It is included by UNESCO on its Lists of Intangible Cultural Heritage, which have the goals of protecting and increasing awareness of the world's most important cultural features.

[1]From John Blacking and Joann W. Kealiinohomoku, eds., *The Performing Arts: Music and Dance* (The Hague: Mouton, 1979), 144. Reprinted by permission of the publisher.

4.5.2 Popular Music

In contrast to folk music, popular music is written by specific individuals for the purpose of being sold to the general public and performed in front of a large number of people. The performance of popular music frequently displays a high degree of technical skill through manipulation of sophisticated electronic equipment.

For example, popular music as we know it today originated around 1900. At that time, the main popular musical entertainment in North America and Europe was the variety show, called the music hall in the United Kingdom and vaudeville in the United States. To provide songs for music halls and vaudeville, a music industry was developed in a district of New York that became known as Tin Pan Alley. The diffusion of American popular music worldwide began in earnest during the 1940s, when the Armed Forces Radio Network broadcast music to American soldiers abroad and to citizens of countries where American forces were stationed or fighting during World War II.

Popular music is increasingly delivered through streaming services (Figure 4.5.2). These services require customers to pay a monthly fee for access to a music library. Worldwide, the leading music streaming service is Spotify, which has around one-third of the market.

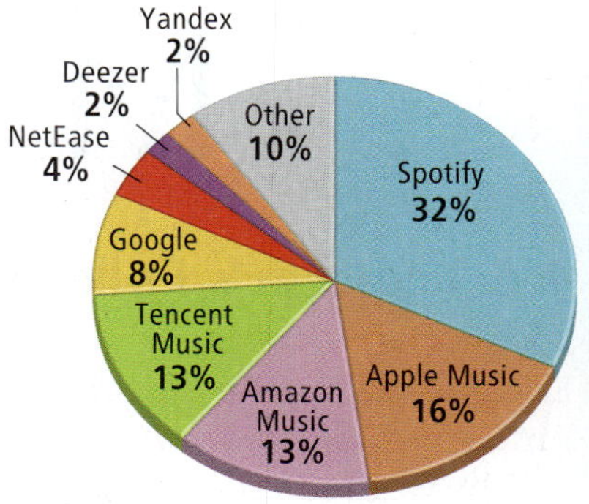

▲ 4.5.2 WORLDWIDE MUSIC STREAMING SERVICES

Popular Music Clusters

Popular musicians cluster in communities where other creative artists reside, regardless of the particular style (Figure 4.5.3). Nashville has the highest concentration of popular musicians, especially those performing country and gospel. New York, Los Angeles, and San Francisco, which are much larger metropolitan areas than Nashville, have relatively high total numbers of musicians, as well as high concentrations when compared to the country as a whole. Musicians cluster in these cities so they can be near sources of employment and cultural activities that appeal to a wide variety of artists, not just performers of a specific type of music. Popular musicians are also attracted to these cities for better access to agencies that book live performances, an important component of the popular music industry, even with the growing importance of streaming services.

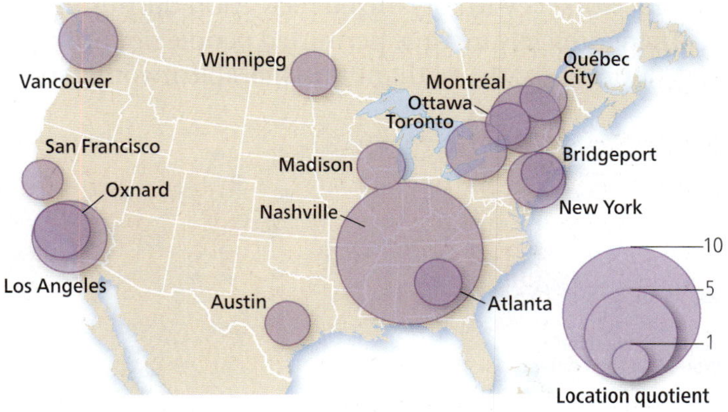

▲ 4.5.3 POPULAR MUSIC CLUSTERS
The location quotient is the percentage of employees who work in the music industry in a city compared with the percentage of employees who work in the music industry throughout Canada and the United States.

Regional Variations

Regional variations can be observed in popular music preferences, both within and between countries. For example, within the United States, hip-hop (including rap and R&B) is relatively popular in large urban areas like New York and Los Angeles, whereas Latin music is in Miami and south Texas, and country music is in many rural areas (Figure 4.5.4).

Figure 4.5.2 points to international variations in music streaming services. Spotify and Apple Music are leading services in the United States, whereas Tencent is the leader in China, and Gaana and JioSaavn are the leaders in India.

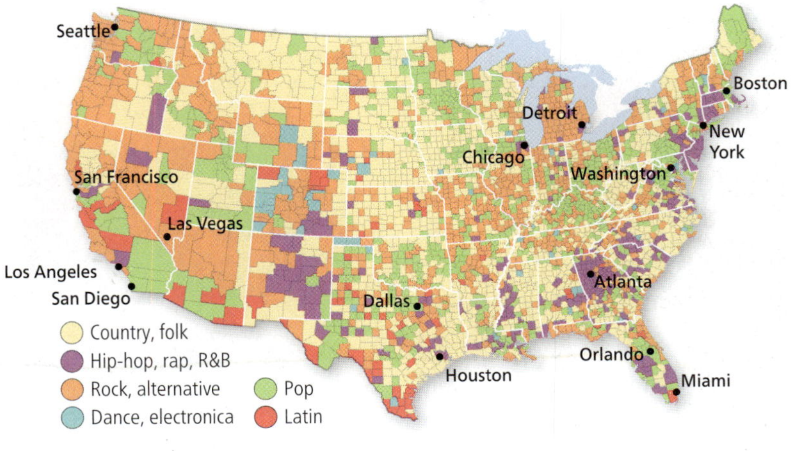

▲ 4.5.4 REGIONAL PREFERENCES IN MUSIC GENRES

Diffusion of Sports

- **Describe the transformation of sports from folk to popular culture.**

Many sports originated as isolated folk customs and were diffused like other folk culture through the migration of individuals. The contemporary diffusion of organized sports, however, displays the characteristics of popular culture.

4.6.1 Soccer's Folk Culture Origins

Soccer, the world's most popular sport—known in most of the world as football—originated as a folk custom in England during the eleventh century. It was transformed into a part of global popular culture beginning in the nineteenth century.

The word *soccer* originated after 1863, when supporters of the game formed the Football Association. *Association* was shortened to *assoc*, which ultimately became twisted around into the word *soccer*.

Soccer as Popular Culture

The transformation of football from an English folk custom to global popular culture began in the 1800s. Football and other recreation clubs were founded in the United Kingdom, frequently by churches, to provide factory workers with organized recreation during leisure hours. Sport became a subject that was taught in school.

Increasing leisure time permitted people not only to participate in sporting events but also to become spectators of sports. With more disposable income, spectators paid to see first-class sporting events. To meet public demand, football clubs began to hire professional players. Several British football clubs formed an association in 1863 to standardize the rules of the game and to organize professional leagues. Organization of the sport into a formal structure in the United Kingdom marks the transition of football from folk to popular culture.

Beginning in the late 1800s, the British exported association football around the world, first to continental Europe and then to other countries. In the twentieth century, soccer, like other sports, was further diffused by new communication systems, especially radio and TV.

The global popularity of soccer is seen in the World Cup, in which national soccer teams compete every four years (Figure 4.6.1).

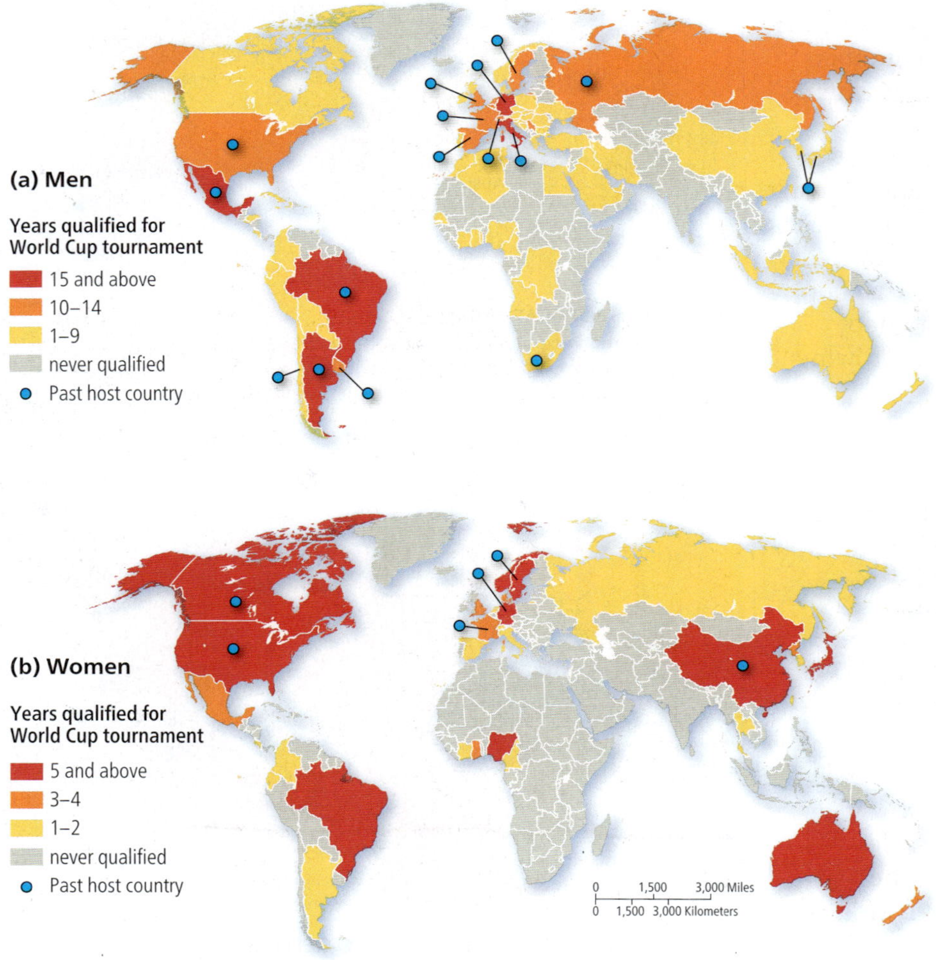

◀ **4.6.1 POPULAR SPORTS: WORLD CUP SOCCER**
Participating countries and hosts for (a) men and (b) women.

4.6.2 Regionally Distinctive Sports

Sports can be a strong force for cultural and regional identity. Lacrosse has fostered cultural identity among the Iroquois Confederation of Six Nations (Cayugas, Mohawks, Oneidas, Onondagas, Senecas, and Tuscaroras) because they have been invited by the International Lacrosse Federation to participate in the World Lacrosse Championships, along with teams from sovereign states such as Australia, Canada, and the United States (Figure 4.6.2). Support for Major League Baseball teams has fostered regional identities (Figure 4.6.3).

Most sports have diffused less widely than soccer. Cultural groups still have their own preferred sports, which are often unintelligible to people elsewhere. For example, cricket is popular primarily in the United Kingdom and former British colonies. Wushu, martial arts that combine forms such as kicking and jumping with combat such as striking and wrestling, are especially popular in China. Australian football is a sport distinct from soccer and the football played in North America. Distinctive forms of football developed in Australia, as well as the United States and Canada, as a result of a lack of interaction among sporting nations during the nineteenth century.

◀ **4.6.2 IROQUOIS NATIONALS LACROSSE**
Gewas Schindler, a member of the Onondaga Nation, has played lacrosse and served as general manager for the Iroquois Nationals, the only team composed entirely of Native American (First Nations) members sanctioned to compete in international sports. Schindler is holding his passport, which is issued by the Iroquois (known in the Nation's language as Haudenosaunee Confederacy). The Iroquois Nationals were unable to travel to the 2010 World Lacrosse Championship in the United Kingdom, because U.K. immigration officials would not accept their Haudenosaunee Confederacy passports.

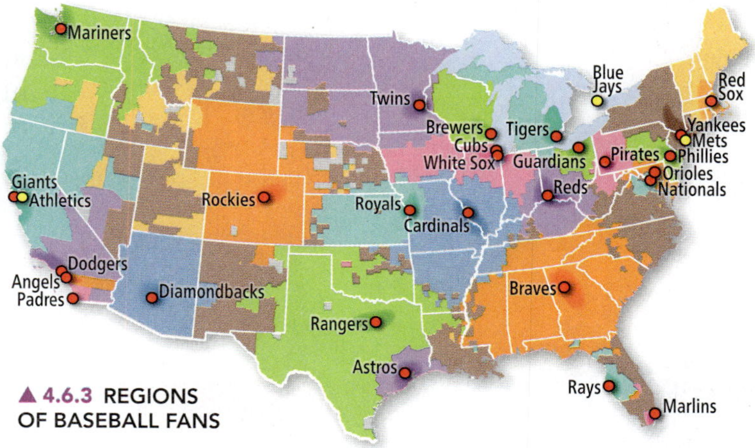

▲ **4.6.3 REGIONS OF BASEBALL FANS**

Olympic Sports

Despite the diversity in the distribution of sports, organized spectator sports today are part of popular culture. The common element in professional sports is the willingness of people throughout the world to pay for the privilege of viewing professional athletes in person, on TV, or through streaming.

To be included in the Summer Olympics, a sport must be widely practiced in at least 75 countries and on four continents (50 countries for women). Under consideration for future Olympics are **esports**, which are competitions using video games. Esports first became popular in Asia. South Korea was the first country to license professional esports players.

Once reserved for amateurs, esports are now played by professional teams, with spectators watching through live streaming. Online streaming media platforms, especially YouTube and Twitch, have become key to the growth of esports. For example, the Overwatch League, launched in 2018, includes 20 esports teams that attack or defend against opponents while competing for an objective (Figure 4.6.4).

▲ **4.6.4 ESPORTS**
Watching an Overwatch game, China.

Electronic Diffusion of Culture

- **Explain how TV and the internet have altered diffusion of popular culture.**

Culture diffuses rapidly around the world, primarily through electronic media. The latest in material culture and leisure activities can be viewed by anyone with access to electronic media. Diffusion of popular culture across the world was facilitated by TV during the twentieth century and increasingly by video programming and the internet in the twenty-first century.

4.7.1 Diffusion of TV: Mid-Twentieth Century

The principal obstacle to diffusion of popular culture is lack of access to electronic media. Access is limited primarily by lack of income. In some developing countries, access is also limited by lack of electricity and phone service.

The world's most important electronic media format by far is TV. During the twentieth century, TV supplanted other formats, notably radio and telegraph, and diffused from the United States to Europe and other developed countries and then to developing countries (Figure 4.7.1). Into the twenty-first century, other formats have become popular, but they have not yet supplanted TV worldwide. TV remains especially important for popular culture for two reasons. First, watching TV is the most popular leisure activity in the world. Second, TV is the most important mechanism by which popular culture rapidly diffuses.

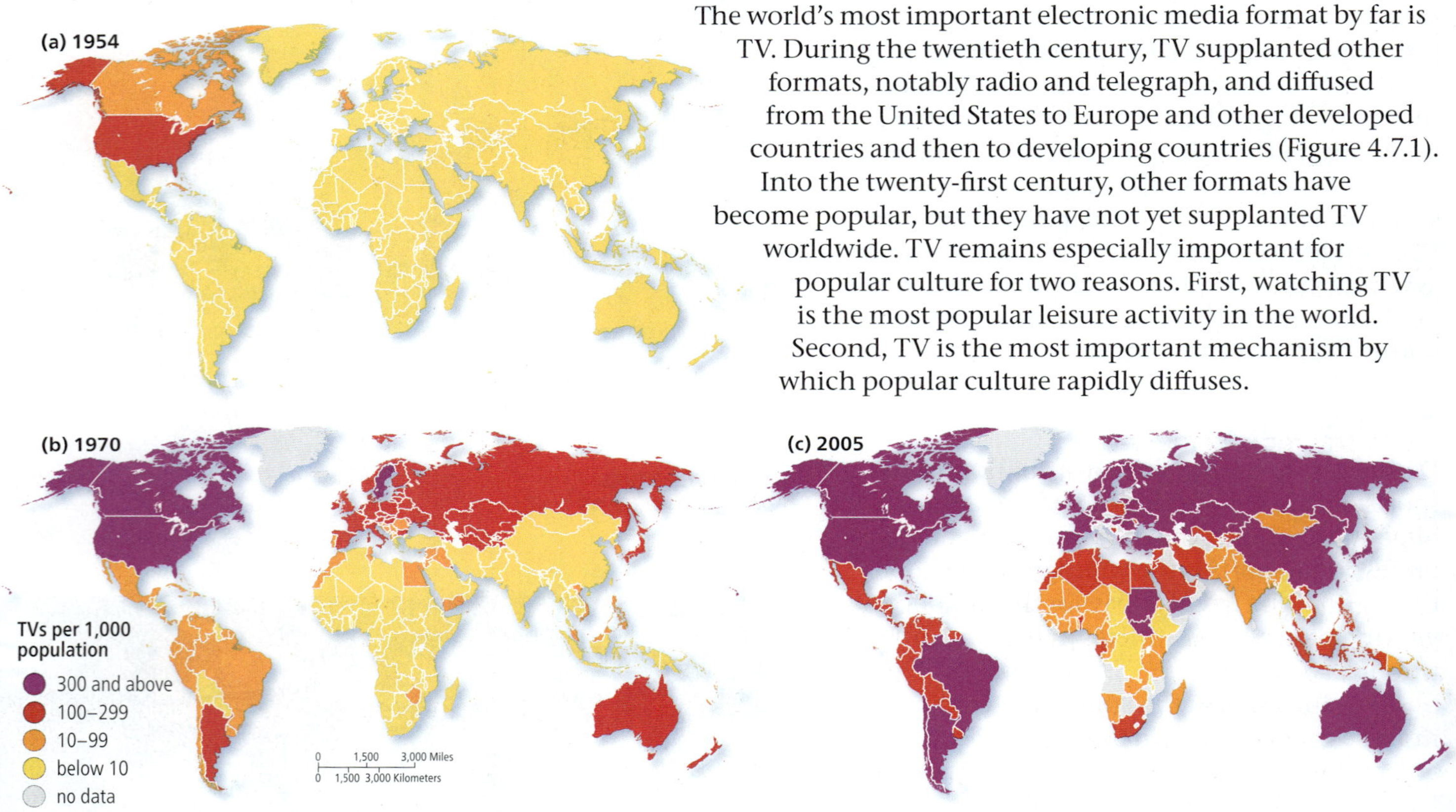

▲ **4.7.1 DIFFUSION OF TV**
TVs per 1,000 population in (a) 1954, (b) 1970, and (c) 2005.

Streaming Services

Video programming is increasingly delivered through streaming services, and regional differences can be seen in the choice of services. The leading streaming services in the United States are totally different than those in China (Figure 4.7.2). As with other social media platforms, the government of China prefers that its citizens utilize domestic streaming services rather than those originating in the United States and other Western countries.

▼ **4.7.2 VIDEO STREAMING MARKET SHARE**
(a) United States, (b) China.

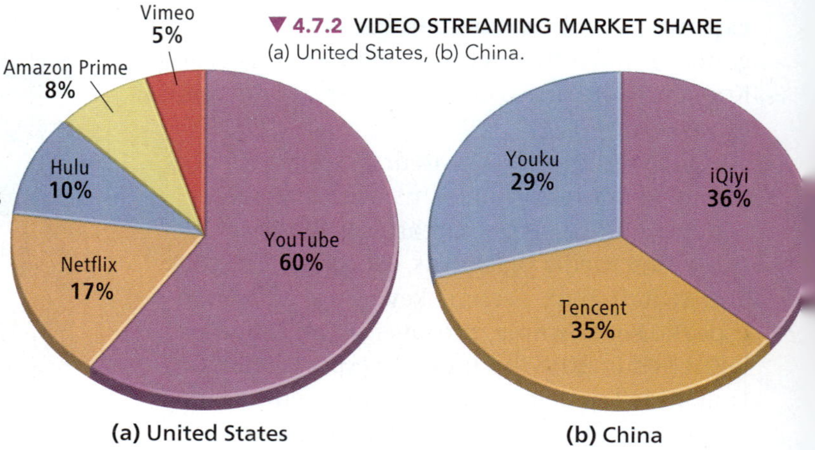

4.7.2 Diffusion of the Internet: Late Twentieth Century

The average person spent 2 hours and 24 minutes per day on the internet in 2020. In 1995, most countries did not have internet service, and the United States had two-thirds of the world's users. Between 1995 and 2000, internet usage increased rapidly in the United States, but worldwide increase was even greater. As internet usage diffused rapidly, the share of the world's internet users clustered in the United States declined from two-thirds to one-third (Figure 4.7.3).

Between 2000 and 2019, internet usage continued to increase rapidly in the United States, to around 90 percent of the country's population. Again, the U.S. increase was more modest than in the rest of the world, and the share of the world's internet users found in the United States continued to decline, to 7 percent in 2019.

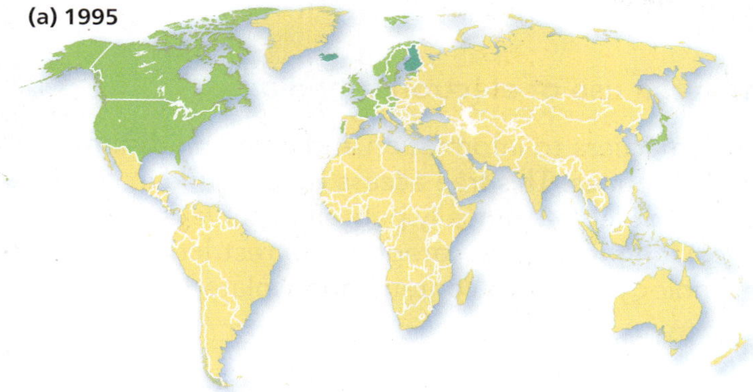

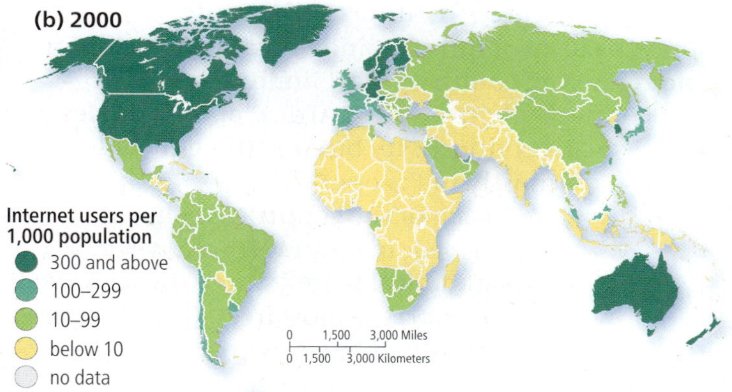

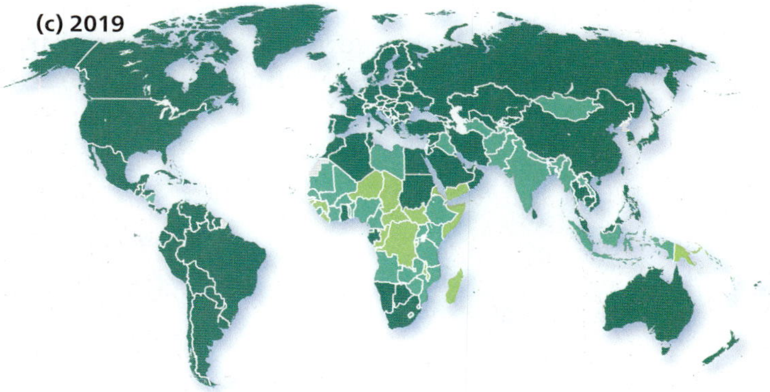

▲ **4.7.3 DIFFUSION OF THE INTERNET**
Internet users per 1,000 population in (a) 1995, (b) 2000, and (c) 2019.

Comparing the Diffusion of TV & the Internet

The diffusion of TV in the twentieth century took much longer than the diffusion of the internet in the twenty-first century, both within the United States and from the United States to other countries. Within the United States, it took around 25 years for the number of TV sets to grow to around 500 per 1,000 population and around 60 years to reach 900 per 1,000 population. In comparison, it took only 10 years for the number of internet users within the United States to reach 500 per 1,000 population and only 30 years for the country to reach 900 internet users per 1,000 population.

Worldwide, note that all six maps in Figures 4.7.1 and 4.7.3 use the same intervals. For example, the highest class in all maps is 300 or more per 1,000. What is different is the time interval period. The diffusion of TV from the United States to the rest of the world took a half-century, whereas the diffusion of the internet took only a decade (Figure 4.7.4).

Given the history of TV and the internet, other forms of social media may diffuse even more rapidly within the United States and worldwide.

▶ **4.7.4 INTERNET CAFE, MOROCCO**

Challenges in Accessing Electronic Media

- **Evaluate threats to freedom in the use of electronic media.**

Most Americans take for granted access to information and communications through the internet and smartphones. This free access is not found in many other countries. Freedom on the Net (an annually published survey and report on internet policies in select countries) identifies three categories of restrictions on the free use of the internet: banned technology, blocked content, and violated user rights.

4.8.1 Freedom on the Net

Freedom on the Net, published by Freedom House, a U.S.-based research institute for the study of democracy, political freedom, and human rights, measures the level of internet and digital media freedom each year in 65 countries. Excluded from their survey and report are countries with limited internet connectivity (mostly in Africa) and countries with connectivity but insufficient evidence (mostly in Europe).

Each of the countries included in Freedom on the Net receives a numerical score from 0 (the least free) to 100 (the most free). Countries are classified "free" with a score of 70 or above, "partly free" with a score between 40 and 69, and "not free" with a score below 40.

Only 15 of the 65 countries included in Freedom on the Net were classified as "free," 28 were "partly free," and 22 were "not free" (Figure 4.8.1).

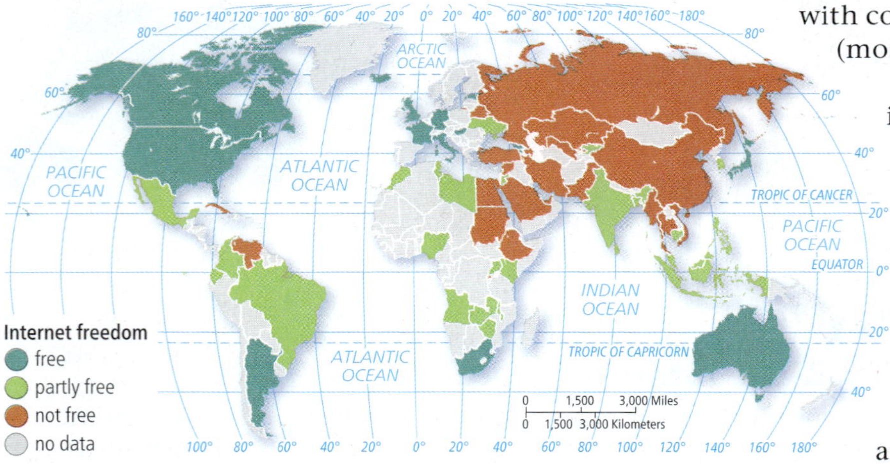

▲ 4.8.1 INTERNET FREEDOM

Banned Technology

In countries lacking complete internet freedom, some models of phones, tablets, and computers may not be sold. Devices that are sold must be configured to exclude certain applications and technologies.

Governments can effectively prevent unwanted electronic technology by regulating the most commonly used technology platforms that are supported by the infrastructure in the country. Even international travelers moving through free countries, such as to and from the United States and Western European countries, find that their electronic devices do not always operate because of incompatible cellular infrastructure.

China is especially aggressive at restricting foreign applications. The small number of Facebook, Twitter, and Instagram users in China, displayed in Figures 4.3.5, 4.3.6, and 4.3.7, is evidence of those restrictions. Instead, electronic interaction in China is undertaken primarily through home-grown apps, such as WeChat, Weibo, and QQ (Figure 4.8.2).

▲ 4.8.2 QQ, CHINA'S MOST POPULAR MESSAGE SERVICE

4.8.2 Blocked Content

Some websites are censored or prevented from being seen altogether on devices in a particular country. As with technology, blocking internet content is a continuation of a widespread practice relating to TV.

Three developed countries—Japan, the United Kingdom, and the United States—have dominated TV in developing countries, including supplying much of the programming. Leaders of many developing countries view American control of much of the world's TV programming as a new method of economic and cultural imperialism. American TV programs present characteristically American beliefs and social norms, such as upward social mobility, relative freedom for women, glorification of youth, and stylized violence. Such emerging popular culture may conflict with and drive out traditional folk culture.

To show sensitivity to traditional folk culture, many TV and smartphone providers in developing countries block offending networks such as MTV and censor unacceptable programs and apps (Figure 4.8.3). The entertainment programs that are substituted may emphasize local values and avoid controversial or edgy cultural, economic, and political content. In the twenty-first century, concern with American-produced media content has spilled over into the internet.

▲ **4.8.3 APPLE STORE, BEIJING, CHINA**
Apple removed 94,000 apps from the phones it sold in China in 2020.

Violated User Rights

Individuals in certain countries are harassed, physically attacked, or jailed for online activity deemed unacceptable by authorities. Governments are finding it increasingly difficult to stop the diffusion of technology. Their citizens are finding ways to circumvent government restrictions on ownership of hardware, use of software, and viewing of online content.

In some countries, governments have enacted laws that provide a justification for their restrictions on technology. For example, a law in The Gambia makes it a criminal offense to use the internet to criticize public officials. A law in Ethiopia permits the government to search computers, internet sites, and other social media of its citizens for anything that it considers damaging to the country.

A number of countries require transnational corporations to maintain a local server in order to do business there. The government has the right to access the data that is stored on the local server.

Freedom House determined that women and LGBTQ communities have been targeted in a number of countries for their online activities. A woman was stoned to death in Pakistan in 2013 for possessing a cell phone. Gay men in Russia have been lured through social media to in-person meetings, where they are assaulted by hate groups.

The three worst offending countries, according to Freedom House, are Iran, Syria, and China. In Iran, especially harsh punishments are imposed on people who promote causes opposed by the government. In Syria, government supporters launched malware cyberattacks against the computers of thousands of antigovernment protestors. In China, individuals posting antigovernment messages on blogs and websites have been arrested.

Russia and Turkey are considered the two countries where internet freedom has suffered the most severe declines in recent years. In Russia, several laws have been enacted since 2012 to block online content critical of the government, and these laws have been used to arrest antigovernment leaders (Figure 4.8.4). In Turkey, government censorship of internet content has increased, including shut down of YouTube and Twitter, and in that country, government protestors and journalists have been arrested for posting online content critical of the Turkish government.

▲ **4.8.4 SOCIAL MEDIA AND POLITICAL PROTEST**
A policeman confronts a woman and tries to stop her from filming other officers as they restrain a pro-democracy protester in Hong Kong, China.

Spatial Dimensions to Cyberattacks

- **Describe geographic features of cyberattacks and other misuse of social media.**

Social media and electronic devices make it easier for people to interact across the distance of space with friends and businesses. At the same time, people have learned to misuse social media for criminal activities or to diffuse false information.

4.9.1 Cyberattacks

Criminals have learned to take advantage of hardware and software for launching attacks against individuals and communities. These cyberattacks successfully compromise individual privacy, penetrate national and corporate security, and disseminate deliberately false information. Through cyberattacks, perpetrators are able to steal information and make money through ransom. The United States is the most frequent country of origin for attacks (Figure 4.9.1).

Countries are not equally prepared to withstand cyberattacks. Rapid7, a private company, assesses readiness of countries to defend against cyberattacks through measures that determine the number of unsecured server ports and the amount of out-of-date encryption software (Figure 4.9.2).

The Global Cybersecurity Index (GCI), constructed by the U.N. International Telecommunication Union (ITU), incorporates five measures of cybersecurity: strength of legislation to fight cybercriminals, the presence of a response team in case of cyberattack, a strategy to respond to a cyberattack, a program to increase public awareness of the cyber threat, and cybersecurity structures that foster cooperation among countries.

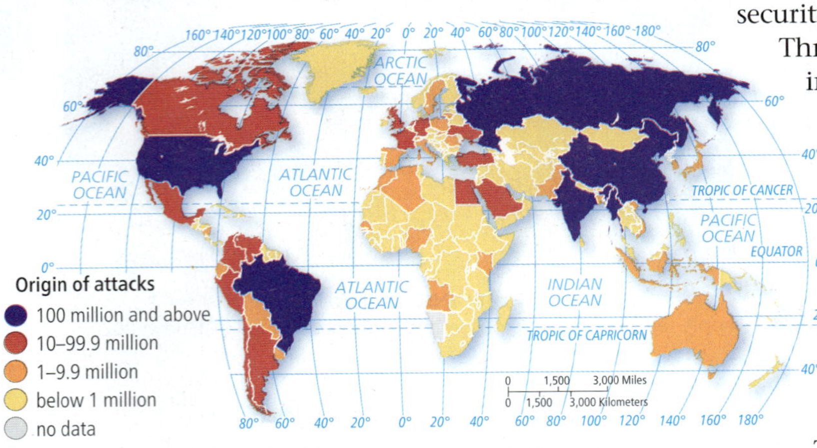

▲ **4.9.1 COUNTRY OF ORIGIN FOR ATTACKS ON WEB APPLICATIONS**

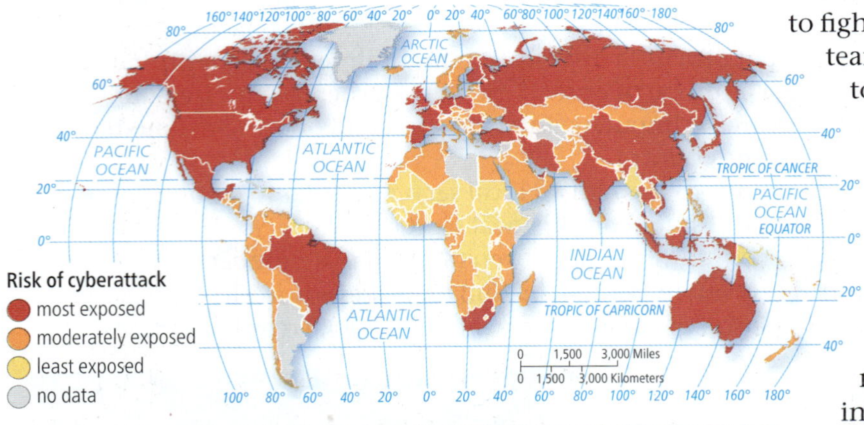

▲ **4.9.2 COUNTRIES LEAST AND MOST EXPOSED TO CYBERATTACKS**

Fake News & Alternative Facts

Fake news is a false, often sensational report disseminated under the guise of an authentic news report. Creators of fake news disseminate information that they know to be untruthful, in order to maliciously spread misinformation and mislead consumers of the content. Elected officials and other political leaders may use the term *fake news* to describe press coverage unfavorable to them.

The rise of fake news has generated other practices that misuse social media. An **alternative fact**, or **alt fact**, is a statement that can be proved to be false. The phrase is attributed to Kellyanne Conway, a counselor to President Trump. She labeled as "alternative fact" an incorrect statement made by Press Secretary Sean Spicer that President Trump's inauguration attracted a higher attendance than did President Obama's. **Trolling** is the practice of posting deliberately inflammatory, extraneous, or off-topic messages in social media in order to provoke quarrels or otherwise agitate people.

The lessons from fake news and alt facts are many but principally are the following: If we are motivated to like or dislike someone or something, we are more likely to accept fake news about that person or subject. If information is shared from someone we know and trust, such as a friend on Facebook, we are more likely to believe it and not check its source. In a world with more fake news and alt facts, individuals and communities have a great burden to verify disseminated information through fact-checking.

4.9.2 Victims of Cyberattacks

Most victims of cyberattacks are clustered in a handful of countries. The most adversely affected countries vary depending on the nature of the attack.

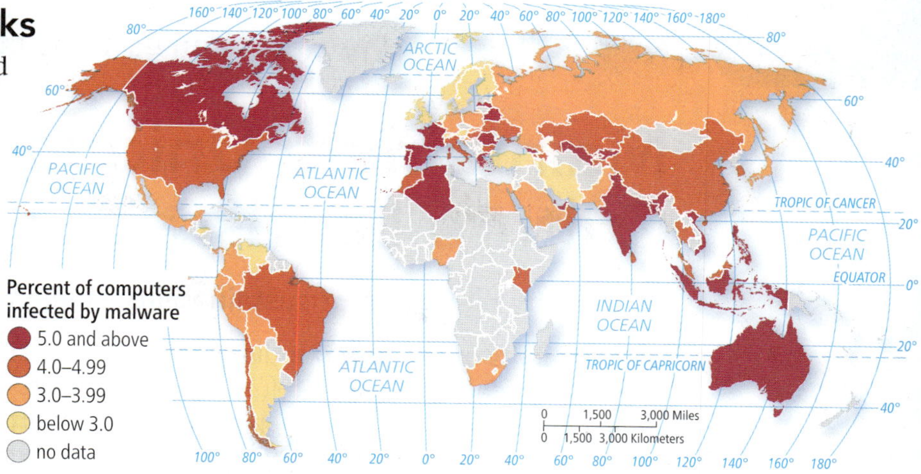

▲ **4.9.3 PERCENTAGE OF COMPUTERS WITH MALWARE INFECTION**

Malware

Malware, or **malicious software**, is a virus inserted into a computer system, often downloaded by accident. Malware takes control of a victim's computers in order to steal personal, financial, and business information. Figure 4.9.3 shows the countries that have the lowest and highest rates of malware infection. Countries in Western Europe and South Asia have relatively high rates of malware infection of computers.

Cyber Espionage

Cyber espionage is the unauthorized and clandestine deployment of a computer virus—malicious software, computer worm, spyware, trojan horse, and so on—to observe or destroy data in the computer systems of government agencies and large corporations, especially those involved in defense and high-tech work. Countries that have been most affected by cyber espionage are displayed in Figure 4.9.4. Government agencies and corporations in the United States have been by far the most affected by cyber espionage.

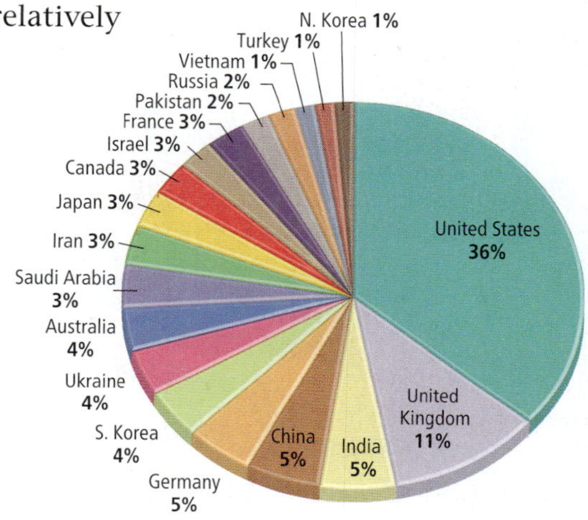

▲ **4.9.4 CYBER ESPIONAGE ATTACKS BY COUNTRY, 2006–2020**

Ransomware

Ransomware, or **cryptoviral extortion**, is a form of malware that encrypts the victim's files, making them inaccessible, until a ransom is paid to decrypt them (Figure 4.9.5). There were 1,001 cases just in the United States in 2020 (Figure 4.9.6).

Using Our Personal Data

The billions of people who communicate by social media have accepted a tradeoff: the convenience of free services in exchange for allowing advertisers and other third parties access to personal information. When companies use our personal data to advertise products that we may wish to purchase, the tradeoff seems acceptable. But when organizations use information about us to influence how we vote or make other important decisions, the tradeoff is likely sinister.

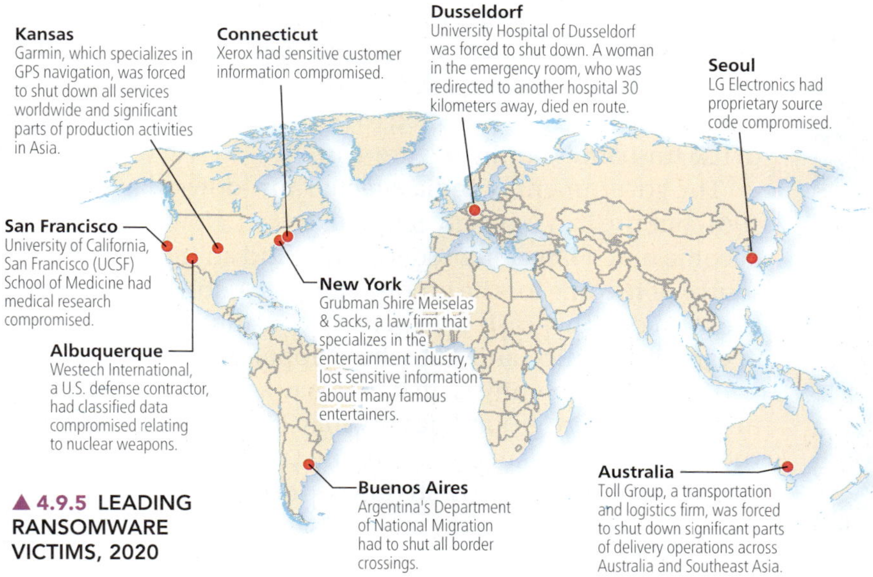

▲ **4.9.5 LEADING RANSOMWARE VICTIMS, 2020**

▼ **4.9.6 MACGYVER TV SHOW: DRAMATIZATION OF A RANSOMWARE ATTACK AND SUBSEQUENT BLACKOUT IN LOS ANGELES**

Maintaining Unique Folk Culture Landscapes

- **Identify examples of sustainability challenges for folk and popular cultures.**

Elements of folk and popular culture face challenges in maintaining identities that are sustainable into the future. For folk culture, the challenges are to maintain unique local landscapes in an age of globalization. For popular culture, the challenges derive from unsustainable practices designed to promote uniform landscapes.

4.10.1 Maintaining Unique Folk Culture

▲ 4.10.1 HOUSES MADE OF MUD BRICK
Gombe, Uganda.

Distinctive environmental and cultural features influence the provision of housing in folk cultures. The style of construction can be influenced by the environment. For example, windows may face in the direction of the Sun in temperate climates to take advantage of the Sun's heat and light. In hot climates, window openings may be smaller to protect the interior from the full heat of the Sun.

The type of building materials used to construct folk houses is influenced partly by the resources available in the immediate environment. Stone, grass, sod, and animal skins may be used, but the two most common building materials in the world are wood and brick. For example, in much of Africa bricks are made of mud. To make bricks, Africans mix soil with water, pour the mixture into molds, and dry the mixture in the mold for several days. A soil and water mix is also used as mortar between the bricks (Figure 4.10.1).

Marriage in India: Challenging Cultural Values

The global diffusion of popular culture has challenged the folk culture in some regions of the world that holds women subservient to men. In folk culture, traditionally women are relegated to performing household chores, such as cooking and cleaning, and to bearing and raising children, and women who worked outside the home are likely to perform agricultural work or trade handicrafts.

Global diffusion of popular culture has had a specific unintended negative impact for women in India: an increase in demand for dowries. Traditionally, a dowry was a "gift" from one family to another, as a sign of respect. In the past, the local custom in much of India was for the groom to provide a small dowry to the bride's family. In the twentieth century, the custom reversed, and the family of a bride was expected to provide a substantial dowry to the husband's family (Figure 4.10.2).

▼ 4.10.2 RALLY AGAINST DOWRY, INDIA

The government of India enacted anti-dowry laws in 1961, but the ban is widely ignored. In fact, dowries have become much larger in modern India and an important source of income for the groom's family. A dowry can take the form of either cash or expensive consumer goods, such as cars, electronics, and household appliances.

If the bride's family is unable to pay a promised dowry or installments, the groom's family may cast the bride out on the street, and her family may refuse to take her back. The government has tried to ban dowries because of the adverse impact on women.

4.10.2 Maintaining Unique Food Customs

Folk food customs are embedded especially strongly in the environment. Humans eat mostly plants and animals—living things that spring from the soil and water of a region. Inhabitants of a region must consider the soil, climate, terrain, vegetation, and other characteristics of the environment to determine which particular foods that environment can produce.

▲ 4.10.3 TERROIR: SHAKING TREES TO HARVEST KALAMATA OLIVES, GREECE

Terroir

The contribution of a location's distinctive physical features to the way food tastes is known by the French term **terroir**. The word comes from the same root as *terre* (the French word for *land* or *earth*) and is definitely not derived from the word *terror*. *Terroir* does not translate precisely into English; it has a similar meaning to the English expressions "grounded" and "sense of place." Terroir is the sum effects of soil, climate, and other features of the local environment on a particular food. For example, a special type of olive is grown only around the village of Kalamata, Greece (Figure 4.10.3). The European Union designates Kalamata olives as "protected designation of origin" to ensure the olive's quality and safeguard against Kalamata olive counterfeiters.

Wine Geography

The spatial distribution of wine production shows the influence of both environmental and cultural elements. The distinctive character of a wine derives from a vineyard's unique combination of soil, climate, and other physical features at the place where the grapes are grown. Most grapes used for wine are grown near the Mediterranean Sea or in areas of similar climate. Vineyards are planted on hillsides, if possible, to maximize exposure to sunlight and to facilitate drainage (Figure 4.10.4).

Wine is produced today primarily in locations that have a tradition of excellence in winemaking and where there is an abundance of people who like to drink it and can afford to purchase it (Figure 4.10.5). Income, preferences, and other social customs influence the distribution of wine consumption, as seen in the lower production levels of predominantly Islamic countries south of the Mediterranean, where little wine is consumed.

▲ 4.10.4 GROWING GRAPES ON HILLSIDES OVERLOOKING THE MEDITERRANEAN SEA
San Nicola Arcella, Italy.

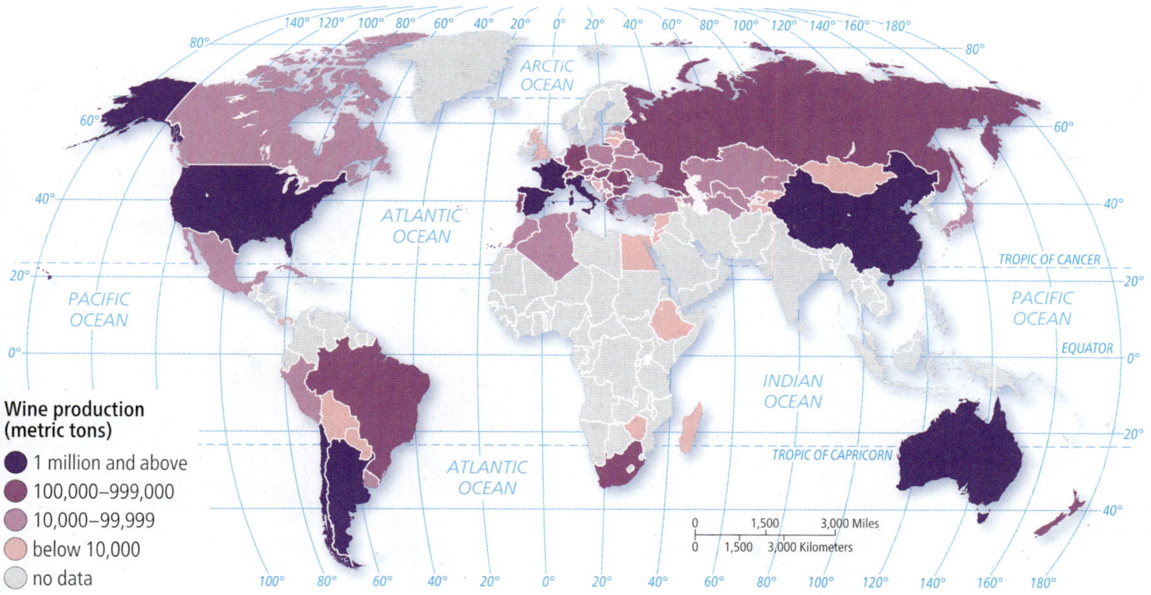

◀ 4.10.5 WINE PRODUCTION

Uniform Landscapes of Popular Culture

- **Identify examples of sustainability challenges for popular culture.**

Popular culture faces challenges that derive from unsustainable practices and experiences. It can significantly modify or control the environment in ways that promote uniform landscapes. Whereas folk culture tends to spring from the surrounding environment, popular culture may be imposed on it.

▲ 4.11.1 STUDY ABROAD PROGRAM London.

4.11.1 Travel

Travel plays an increasingly important role in refashioning the landscape in the service of popular culture. At its best, travel offers people the opportunity to learn about other cultures firsthand. For example, many students benefit from study abroad programs (Figure 4.11.1).

Most international tourists travel from one developed country to another, such as between the United States and Western Europe (Figure 4.11.2). The hospitality industry, including transportation as well as hotels and restaurants, has become the world's largest industry in many respects, such as employment and spending. The service sector of the economies of these countries is increasingly organized around travel and tourism.

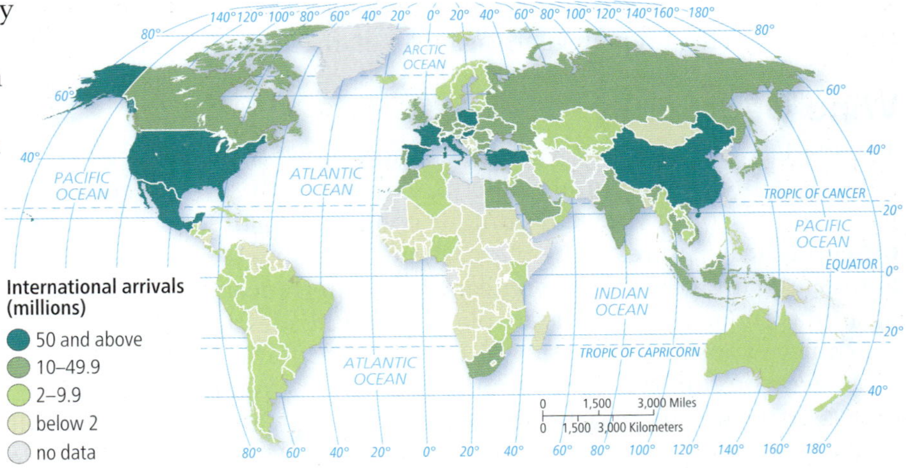

▲ 4.11.2 INTERNATIONAL TOURIST ARRIVALS Data are for 2019, before the COVID-19 lockdown.

Travel has also become more important for developing countries in Africa, Asia, and Latin America. These regions face challenges in maintaining traditional folk cultural features that attract tourists while at the same time catering to the comforts of popular culture that these tourists expect.

Overtourism

Many cities in Europe have struggled with overtourism (too many people visiting). Cities such as Venice are searching for ways to limit the numbers of tourists through such measures as keeping out very large cruise ships and reducing the number of Airbnb lodgings (Figure 4.11.3 and 4.11.4).

▶ 4.11.3 HUGE CRUISE SHIP DOCKS IN THE CENTER OF VENICE

▼ 4.11.4 TOURISTS CROWD THE NARROW STREETS OF VENICE

4.11.2 Unsustainable Landscapes

▲ 4.11.5 SKI DUBAI

In order for people in a society to enjoy popular culture, often their surrounding environment must be modified, sometimes dramatically.

The popular culture of shopping malls, smartphones, the internet, sports arenas, and so forth requires roads, buildings, power lines, cell towers, and the like. In the diffusion of popular culture, land is cleared, waterways are occupied, trees are cut for wood, ore mined for metals, groundwater extracted, rivers dammed, pollutants emitted, and so on.

Even if the resulting built environment may be difficult to visually differentiate from its "natural" equivalent, it is actually the result of deliberate modifications of the landscape by people in pursuit of popular leisure activities.

Skiing in Dubai

An example of a popular culture feature placed in an unsustainable landscape is Ski Dubai (Figure 4.11.5). Ski Dubai is located adjacent to a shopping mall in Dubai, the capital of the United Arab Emirates (UAE).

Temperatures in the UAE exceed 100°F much of the year, but people can snow ski year-round indoors. The UAE is one of the world's wealthiest countries, thanks to its supply of petroleum.

The spread of a cultural feature like skiing across larger spaces results in a loss of localized folk culture diversity and convergence of popular cultural preferences. This convergence is known as **cultural homogenization**. Some question the actions of Western countries, who have been the most successful in commodifying and exporting elements of their popular culture, because the resulting dominance in the world of American popular culture is regarded as a form of cultural imperialism.

Ultimately, the diffusion of popular culture can adversely impact environmental quality and cultural diversity in multiple ways:

- Pollution of the landscape.
- Depletion of scarce natural resources.
- Crowding out of locally owned businesses.

Golf

Popular culture can pollute the landscape by modifying it with little regard for local environmental conditions, such as climate and soil. To create a uniform landscape, hills may be flattened and valleys filled in. The same building and landscaping materials may be employed regardless of location.

Golf courses, because of their large size (80 hectares, or 200 acres), provide a prominent example of imposing popular culture on the environment. Features of popular culture such as golf courses consume large quantities of land and water; non-native grass species are planted, and fertilizers and pesticides are laid on the grass to ensure an appearance considered suitable for the game.

Golf courses can be designed sustainably partially in response to local physical conditions. Grass species may be selected to thrive in the local climate and still be suitable for the needs of greens, fairways, and roughs. Existing trees and native vegetation are retained if possible. For example, few fairways in Michigan are lined with palms as fairways might be in Florida and California.

As with other popular customs, golf courses remake the environment—creating or flattening hills, cutting grass or letting it grow tall, carting in or digging up sand for traps, and draining or expanding bodies of water to create hazards and maintain the course.

And what happens when a golf course is laid out in the desert, where grass doesn't grow at all? Even creating a "green" is a massive remaking of the landscape (Figure 4.11.6).

▲ 4.11.6 GOLF COURSE IN NAMIBIA

CHAPTER

4 Review, Analyze, & Apply

KEY ISSUE 1 How Are Leisure Activities Distributed?

- Folk culture and popular culture have distinctive patterns of origin, diffusion, and distribution.
- Electronic communications and social media facilitate the distribution of not only popular culture but folk culture as well.

THINKING GEOGRAPHICALLY

▲ 4.CR.1 GOOGLE STREET MAPPING, LONDON

Google Maps has extensive street view coverage in the developed regions of North America and Europe but remains spotty in the developing regions of Africa and Asia. What factors might account for these differences in coverage?

KEY ISSUE 2 How Does Culture Diffuse?

- Material folk culture, including food, clothing, and housing, has limited patterns of distribution and diffuses primarily through relocation diffusion.
- Popular culture is increasingly likely to diffuse widely through expansion diffusion, facilitated by social media.

THINKING GEOGRAPHICALLY

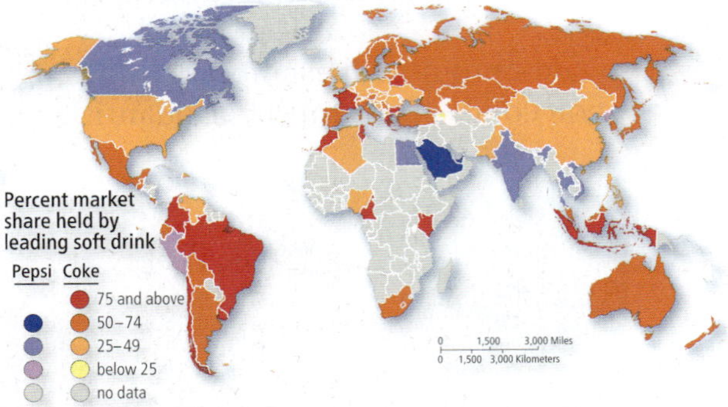

▲ 4.CR.2 COKE VS. PEPSI PREFERENCES

Coke and Pepsi predominate in different countries. What factors might contribute to this variation?

KEY ISSUE 3 Why Is Access to Culture Unequal?

- Various forms of social media from TV to streaming to the internet vastly and rapidly disperse popular culture.
- Not every country permits its citizens complete freedom to access social media.
- Social media facilitate the diffusion of cyberattacks and fake news.

THINKING GEOGRAPHICALLY

▲ 4.CR.3 PROTESTING FACEBOOK'S PRIVACY POLICY

Social media platforms and electronic devices have settings and apps that offer varying degrees of privacy from unwanted material and collection of personal information. Do you use privacy settings on any of your devices or apps? Why or why not?

KEY ISSUE 4 Why Do Cultures Face Sustainability Challenges?

- Distinctive folk cultures survive, and in many cases continue to respond sustainably to environments.
- Material popular culture may disperse from places where the environment sustains it to places where the environment cannot sustain it.

THINKING GEOGRAPHICALLY

▲ 4.CR.4 STUDENT VOLUNTEERS CLEAN UP, LAGOS, NIGERIA

Some young people volunteer to work in another country as a way to interact with people of other cultures. Is this form of cultural convergence more sustainable than other forms of international tourism? Why or why not?

Key Terms

Alternative fact (alt fact) (4.9.1) A statement that can be proved to be false.

Cultural homogenization (4.11.2) The process of reduction in cultural diversity through the diffusion of popular culture.

Culture (4.1.1) The body of material possessions, customary beliefs, and social norms that together constitute the distinct tradition of a group of people.

Custom (4.1.1) The frequent repetition of an act, to the extent that it becomes characteristic of the group of people performing the act.

Cyber espionage (4.9.2) The unauthorized and clandestine deployment of a computer virus to observe or destroy data in the computer systems of government agencies and large corporations.

Esports (4.6.2) Sports competitions using video games.

Fake news (4.9.1) A false report disseminated under the guise of an authentic news report, created to maliciously spread misinformation and mislead consumers of the content.

Folk culture (4.1.1) Culture traditionally practiced by a small, homogeneous, rural group living in relative isolation from other groups.

Franchise (4.2.1) An agreement between a corporation and businesspeople to market that corporation's products in a local area.

Habit (4.1.1) A repetitive act performed by a particular individual.

Malware (malicious software) (4.9.2) Hostile or intrusive computer software designed to cause intentional harm.

Popular culture (4.1.1) Culture found in a large, heterogeneous society that shares certain habits and seeks similar material goods despite differences in other personal characteristics.

Ransomware (cryptoviral extortion) (4.9.2) A form of malware that encrypts the victim's computer files, making them inaccessible, until a ransom is paid to decrypt them.

Taboo (4.2.2) A restriction on behavior imposed by social custom.

Terroir (4.10.2) The contribution of a location's distinctive environment to the way food produced there tastes.

Trolling (4.9.1) The practice of posting deliberately inflammatory, extraneous, or off-topic messages in social media in order to provoke quarrels or otherwise upset people.

Geospatial Analysis

Log in to the Mastering Geography Study Area to access MapMaster 2.0.

Connectivity in Sub-Saharan Africa

Sub-Saharan Africa has relatively poor mobile phone coverage and relatively low internet usage. But there are exceptions.

Add the *Internet Users* layer. ***Select*** the *Settings* icon from the *Legend*, and ***select*** *Show Political Labels*. ***Add*** the *Cellular Phones* layer and **select** *Split Map Window*. **Zoom** to Sub-Saharan Africa.

1. What are the only four countries in sub-Saharan Africa that have at least 400 internet users per 1,000 population?
2. How many of the four countries with relatively high internet users have relatively extensive cellular phone subscriptions (at least 800 subscriptions per 1,000 inhabitants)?
3. How might people stay connected in the countries with high internet usage but low cellular phone subscriptions?
4. How might people stay connected in the countries with high cellular phone subscriptions but low internet usage?

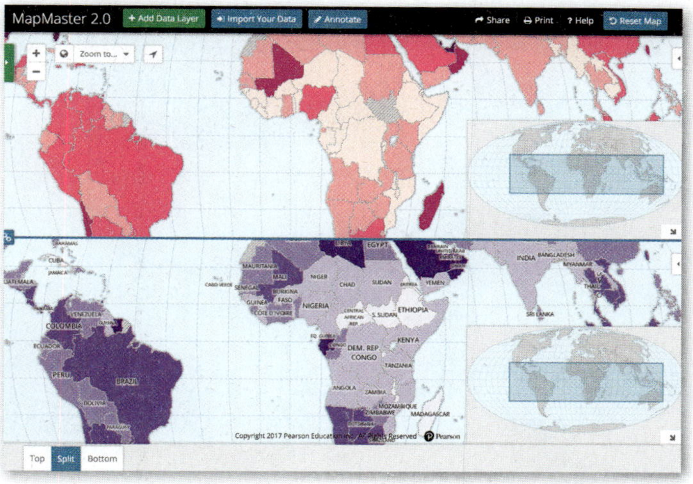

▲ **4.CR.5 INTERNET AND CELLULAR PHONE COVERAGE**

Explore

Ski Dubai

Pictures of skiing in the desert are shown in Figure 4.11.5 and on the right. Let's see what else is in the vicinity of the desert ski resort. **Search** for *Ski Dubai*.

1. What is housed in the structure immediately adjacent to Ski Dubai?
2. What places to eat are located in the Ski Dubai structure? Why might these restaurants wish to locate here?
3. Use Street View to enter inside Ski Dubai. Check out the skiing. How challenging does the course appear?
4. How are the women dressed, and what are they doing in Ski Dubai? How might this differ from a ski area in the United States?

▲ **4.CR.6 SKI DUBAI SNOW PARK**

CHAPTER 5

Languages

▲ Preschool children in Czechia learn English words.

Language is an important part of culture. It is the means through which other cultural values are communicated. People are trying to preserve local languages as one of the basic elements of cultural identity. At the same time, people around the world are learning English to participate in a global economy and culture. In Europe, for example, 96 percent of high school students learn English. In contrast, only 25 percent of students in the United States study another language.

KEY ISSUES

1 How Are Languages Distributed?

Languages can be classified as belonging to a family. Individual languages and families display distinctive distributions.

2 How Do Languages Diffuse?

Like other cultural features, the contemporary distribution of languages results from processes of origin and diffusion.

3 How Do Languages Vary Among Places?

Geographers observe variations within languages. In some countries, multiple languages are used.

4 Why Do Languages Survive or Perish?

The list of world languages is in constant flux. Some lesser-used languages are disappearing, whereas others are being preserved and invented.

LOCATIONS IN THIS CHAPTER

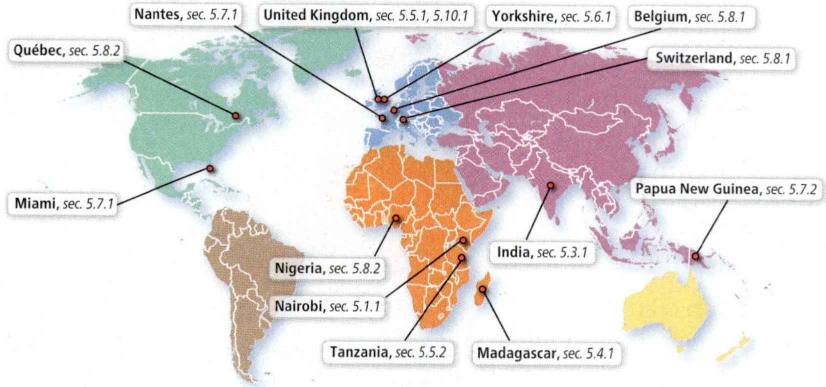

Classifying Languages

- Describe how languages are classified.

Earth's collection of languages is one of the most obvious examples of cultural diversity. The study of language follows logically from the study of migration and culture because the contemporary distribution of languages around the world is largely a result of past migrations. People living in different regions use the same language because of migration from one region to another.

▲ 5.1.1 MASS IN SIGN LANGUAGE
Nairobi, Kenya.

5.1.1 Introducing Languages

Language is a system of communication through speech, movement, sounds, or symbols that a group of people understands to have the same meaning. Most people communicate through speech, though some people use symbols or movements, such as sign language. An estimated 500,000 Americans and 72 million people worldwide with disabling hearing loss communicate through sign language (Figure 5.1.1).

Geographers observe that cultural elements such as language, religion, and ethnicity can be strong centripetal and centrifugal forces, two terms that are adapted from physical science. For geographers, a **centripetal force** is a force that tends to unify people, whereas a **centrifugal force** is a force that tends to pull people apart. Sharing a common language is a good example of a centripetal force. On the other hand, conflict between two groups of people living in the same place but using different languages would be an example of a centrifugal force.

Organizing Languages

Ethnologue, one of the most authoritative sources of languages (online at ethnologue.com), estimates that the world has 7,139 languages, including 90 used by at least 10 million people, 307 used by between 1 and 10 million people, and 6,702 used by fewer than 1 million people. The distribution of some of these languages is easy for geographers to document, whereas with others—especially in Africa and Asia—determining the distribution of languages is difficult, if not impossible.

Ethnologue classifies languages as institutional, stable, and endangered:

- An **institutional language** is used in education, work, mass media, and government. *Ethnologue* identifies 540 institutional languages, including English. An institutional language has a **literary tradition**, which means it is written as well as spoken. An **official language** is an institutional language used by the government to enact legislation, publish documents, and conduct other public business.
- A **stable language** is used for face-to-face communication within all generations, including children.
- An **endangered language** is threatened or dying because it is losing users and may not be retained by the younger generation.

The world's languages can be organized into families, branches, and groups:

- A **language family** is a collection of languages related through a common ancestral language that existed long before recorded history (Figure 5.1.2).
- A **language branch** is a collection of languages within a language family related through a common ancestral language that existed several thousand years ago. Differences in language branches are not as extensive or as old as between language families, and archaeological evidence can confirm that a language branch derived from a particular family.
- A **language group** is a collection of languages within a language branch that share a common origin in the relatively recent past and display many similarities in grammar and vocabulary.

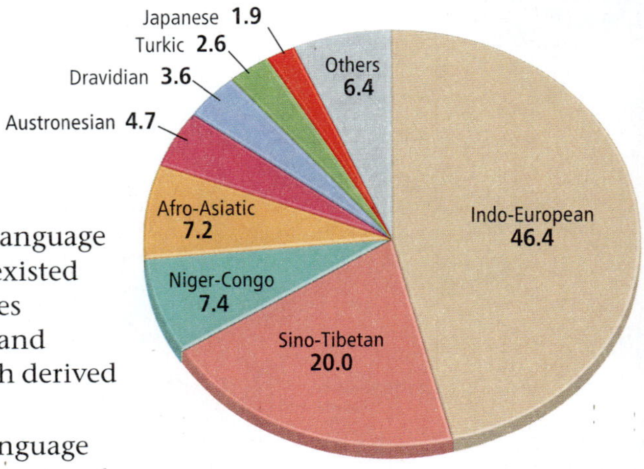

▲ 5.1.2 LANGUAGE FAMILIES BY SHARE OF WORLD POPULATION

5.1.2 Language Families

Figure 5.1.3 depicts the world's major languages and language families as a forest. Individual languages are displayed as leaves. The larger the leaf, the more users of that language. Language families are represented as trunks of the trees, and some trunks divide into branches, which logically represent language branches.

Ethnologue identifies 18 language families with at least 5 million users. The two largest—Indo-European and Sino-Tibetan—together account for around two-thirds of the world's speakers. The other 16 account for all but 2 percent of the remainder.

▼ 5.1.3 LANGUAGE FAMILY TREES

Language families with at least 5 million users, according to *Ethnologue*, are shown as trunks of trees. Individual languages that have more than 5 million users are shown as leaves. Some trunks divide into several branches, which represent language branches. Possible superfamilies are shown as roots below the surface because their existence is speculative and highly controversial.

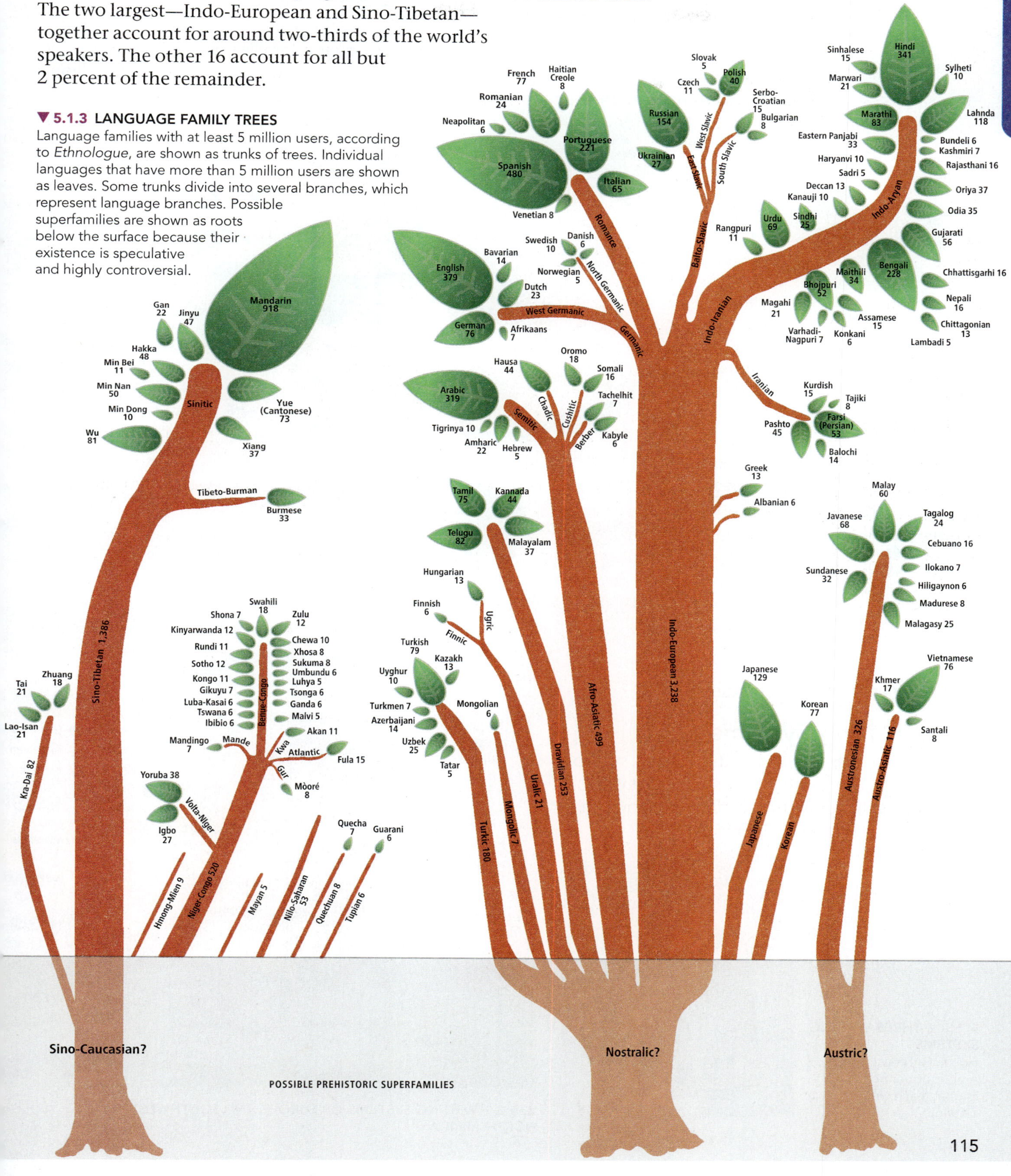

Distribution of Languages

- **Summarize the distribution of languages.**

Ethnologue identifies 141 language families. Language families with at least 8 million native users are shown in Figure 5.2.1. Individual languages with at least 50 million users are named on the map.

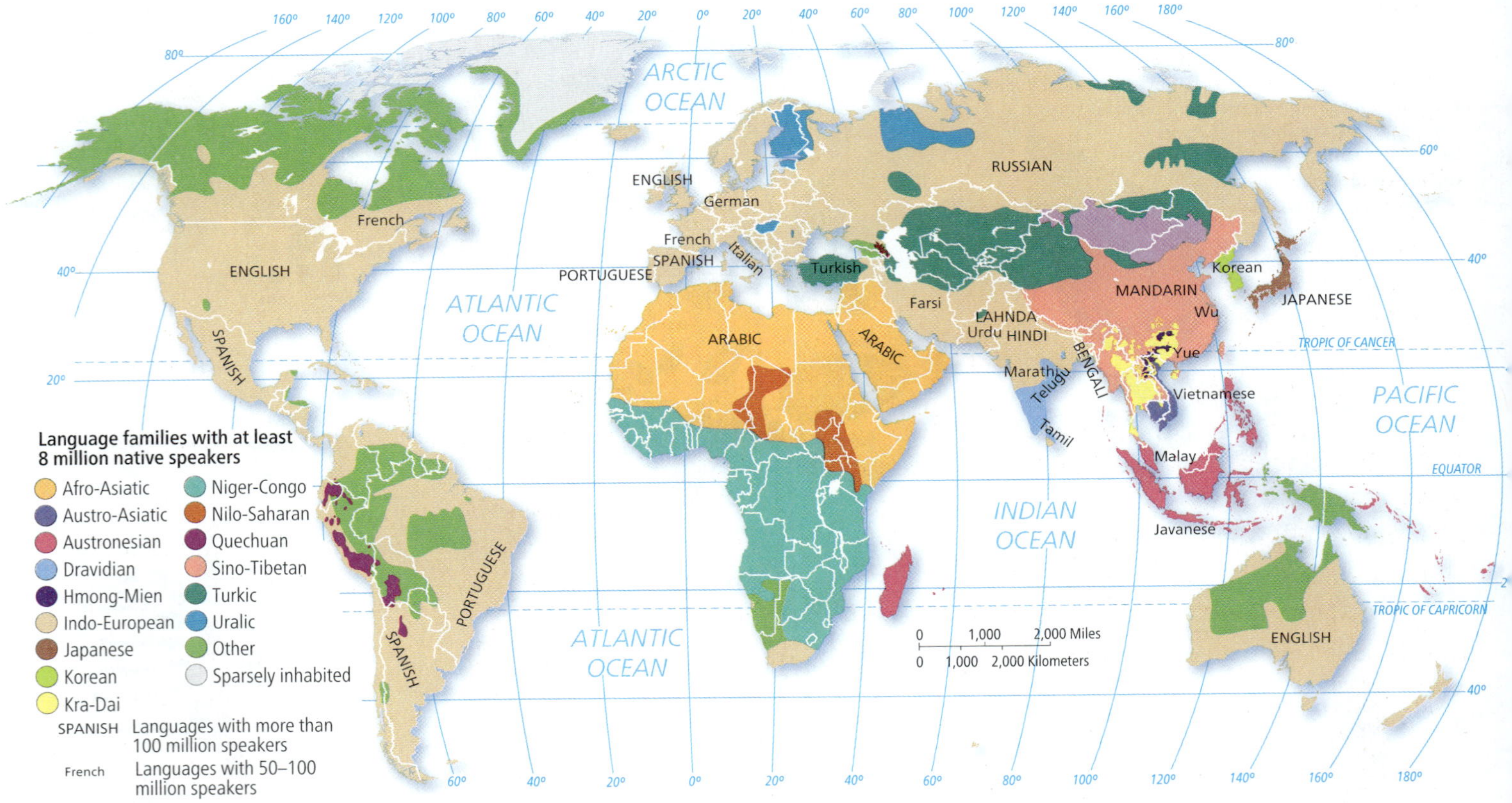

▲ **5.2.1** DISTRIBUTION OF LANGUAGE FAMILIES

5.2.1 Writing Systems

Ethnologue identifies 3,856 writing systems, symbols such as letters in an alphabet that represent words of a language. Three writing systems are shown in Figure 5.2.2 and seven in Figure 5.2.3.

▶ **5.2.2** THREE WRITING SYSTEMS
Sign in Hebrew (bottom), Arabic (upper right), and English (Latin writing system).

▲ **5.2.3** WRITING SYSTEMS OF EUROPE AND SOUTHWEST ASIA & NORTH AFRICA

5.2.2 Distribution of Language Families

The language forest (Figure 5.1.3) shows 12 language families with at least 50 million speakers. This page summarizes 11 of them. The 12th language family—Indo-European—is discussed in Section 5.3.1.

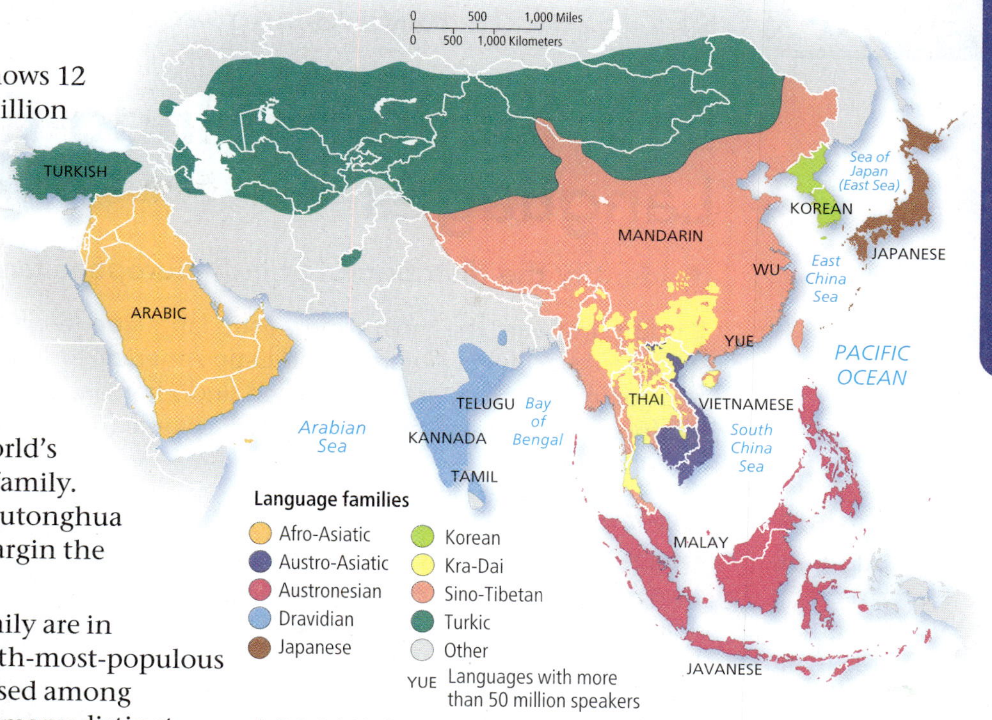

▲ **5.2.4 LANGUAGE FAMLIES OF SOUTHWEST, EAST, AND SOUTHEAST ASIA**

Asian Language Families

Eight language families with at least 50 million speakers are located primarily in Asia (Figure 5.2.4).

Sino-Tibetan. Sino-Tibetan is the world's second-most widely used language family. Mandarin, which the Chinese call Putonghua ("common speech"), is by a wide margin the world's most-used language.

Austronesian. Most users of this family are in Indonesia, which is the world's fourth-most-populous country. With its inhabitants dispersed among thousands of islands, Indonesia has many distinct languages and dialects.

Dravidian. Dravidian is the second-most widely used language family in South Asia, following Indo-European, and is the principal language family in southern India.

Turkic. The Turkic language family is thought to have originated in the steppes bordering the Qilian Shan and Altai mountains between Tibet and China. Present distribution covers an 8,000-kilometer band of Asia. The Turkic language that has by far the most users is Turkish. The Turkic language family was formerly called "Altaic."

Austro-Asiatic. Vietnamese is the most used language of the Austro-Asiatic language family.

Kra-Dai. The principal languages of the Kra-Dai language family are spoken in Thailand and neighboring portions of China.

Japanese and Korean. Japanese and Korean are two examples of **isolated languages**, which are languages that are unrelated to any other and therefore not attached to any language family.

African Language Families

The world's third- and fourth-most used language families are based in Africa (Figure 5.2.5).

Afro-Asiatic. Arabic is the major language of the Afro-Asiatic language family, an official language in two dozen countries of Southwest Asia & North Africa, and one of six official languages of the United Nations.

Niger-Congo. More than 95 percent of the people in sub-Saharan Africa use languages of the Niger-Congo language family. The three most widely spoken are Yoruba, Igbo, and Swahili.

Nilo-Saharan. Languages of the Nilo-Saharan language family are used by more than 50 million people in north-central Africa, sandwiched between the Afro-Asiatic and Niger-Congo language regions. The total number of speakers of each individual Nilo-Saharan language is extremely small.

▲ **5.2.5 LANGUAGE FAMILIES OF AFRICA**

Distribution of Indo-European Languages

- **Identify the distribution of Indo-European languages.**

Indo-European, the most widely used language family, is the predominant one in Europe, South Asia, North America, and Latin America. Indo-European has eight branches, including four that are widely used (Indo-Iranian, Germanic, Romance, and Balto-Slavic) and four that are used by relatively few people (Albanian, Armenian, Celtic, and Greek).

5.3.1 Asia's Indo-European Branch

The Indo-Iranian language branch is divided into the Iranian (or Western) group and the Indo-Aryan (or Eastern) group. The most widely used Indo-Aryan language is Hindi, which is the official language of India. India also recognizes 22 so-called scheduled languages (Figure 5.3.1). India's constitution obligates the government of India to encourage the use of these languages.

Bengali, the second most widely spoken Indo-Aryan language, is the official language of Bangladesh. Urdu, though very similar to Hindi, is classified as a distinct language and is Pakistan's official language. Lahnda, another Indo-Aryan language, is the most widely used language in Pakistan.

The distribution of Indo-Iranian differs from that of other Indo-European branches, which are centered in Europe. All eight branches are depicted in Figure 5.3.2.

▶ 5.3.1 LANGUAGES OF INDIA

▲ 5.3.2 INDO-EUROPEAN BRANCHES

5.3.2 Europe's Indo-European Branches

The three most widely used Indo-European language branches in Europe are Germanic, Romance, and Balto-Slavic.

Romance Branch

The most widely used contemporary Romance languages are Spanish, Portuguese, French, and Italian (Figure 5.3.3). Their language names derive from Latin, the principal language of the Roman Empire (the "Romans' language"). The regions in which these four languages are spoken correspond somewhat to the boundaries of the modern states of Spain, Portugal, France, and Italy. Rugged mountains serve as boundaries among these countries. The fifth most widely used Romance language, Romanian, is the principal language of Romania and Moldova, separated from the other Romance-speaking countries by Slavic-speaking peoples.

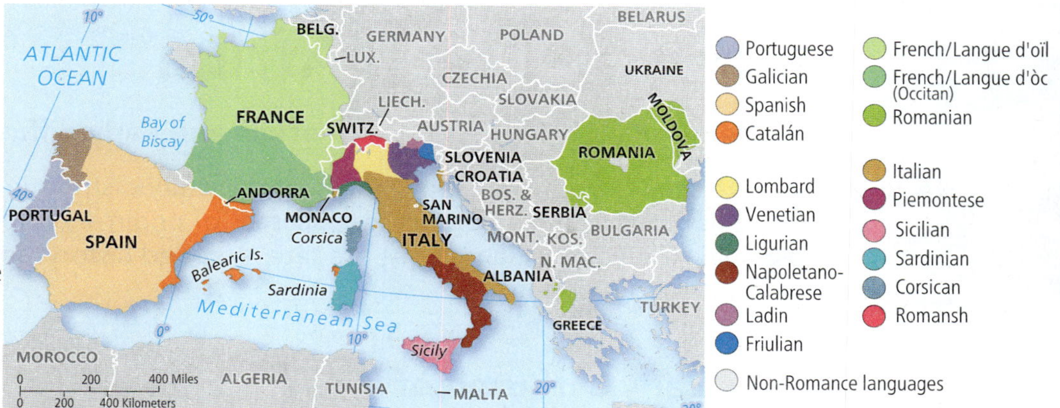

▲ 5.3.3 ROMANCE BRANCH OF INDO-EUROPEAN

Germanic Branch

English belongs to the West Germanic group of the Germanic language branch of the Indo-European family (Figure 5.3.4). German and Dutch are also West Germanic group languages. The Germanic language branch also includes languages in the North Germanic group, spoken in Scandinavia. Swedish, Danish, Norwegian, and Icelandic all derive from Old Norse, the principal language used throughout Scandinavia before 1000 C.E. The four distinct Scandinavian languages emerged because of migration and the political organization of the region into four independent and isolated countries.

▲ 5.3.4 GERMANIC AND BALTO-SLAVIC BRANCHES OF INDO-EUROPEAN

Balto-Slavic Branch

Balto-Slavic languages predominate in Eastern Europe. Slavic was once a single language, but differences developed in the seventh century C.E., when several groups of Slavs migrated from Asia to different areas of Eastern Europe and thereafter lived in isolation from one another. The Balto-Slavic language branch divided into East, West, and South Slavic groups and a Baltic group (refer to Figure 5.3.4).

The most widely used Slavic languages are the eastern ones, primarily Russian, which is spoken by more than 80 percent of the Russian people. The importance of Russian increased with the Soviet Union's rise to power. Soviet officials forced native speakers of other languages to learn Russian as a centripetal force to foster cultural unity among the country's diverse peoples. In Eastern European countries that were dominated by the Soviet Union, Russian was taught as the second language.

The most used West Slavic language is Polish, followed by Czech and Slovak. The latter two are quite similar, and speakers of one can understand the other. The most widely used South Slavic language is the one spoken in Bosnia & Herzegovina, Croatia, Montenegro, and Serbia. When they were all part of Yugoslavia, the language was called Serbo-Croatian. This name now offends Bosniaks and Croatians because it recalls when they were once in a country that was dominated by Serbs. Instead, the names Bosnian, Croatian, and Serbian are preferred by people in these countries to demonstrate that each language is unique, even though linguists consider them one. Bosniaks and Croatians write the language in the Latin alphabet, and Serbs use the Cyrillic alphabet.

▲ 5.4.1 MALAGASY, AN AUSTRONESIAN LANGUAGE SPOKEN IN MADAGASCAR

Into & Out of Africa

- Identify possible processes of origin and early diffusion of languages.

Like other cultural elements, the contemporary distribution of languages exists because of geographic processes of origin and diffusion. The origin and initial diffusion of languages predate recorded history, so we can only speculate about them.

5.4.1 Early Diffusion of a Language

Refer back to the distribution of the Austronesian language family on the world map of languages (refer to Figure 5.2.1). Most users of Austronesian languages live in Southeast Asia, especially in Indonesia. However, note that an Austronesian language is shown for Madagascar, the large island off the east coast of Africa (Figure 5.4.1). Malagasy, the principal language of Madagascar, can be seen on the Austronesian "tree" in Figure 5.1.3.

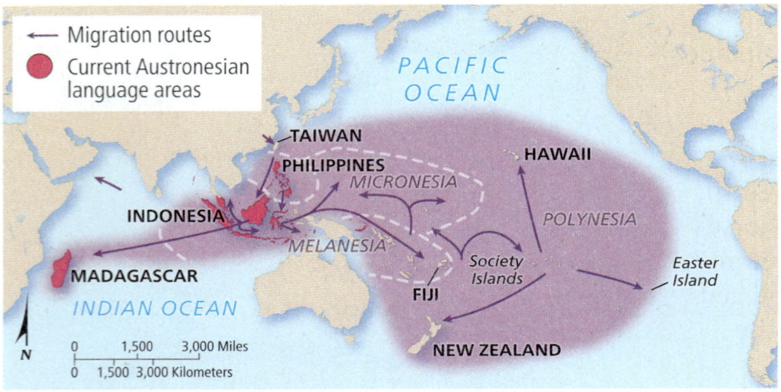

▲ 5.4.2 ANCIENT MIGRATION OF AUSTRONESIAN LANGUAGE FAMILY

That the people of Madagascar use a language belonging to the same family as the languages of most of Indonesia and the Philippines is strong evidence of migration a long time ago between these two places. Researchers have concluded that migrants sailed the 6,000 kilometers (4,000 miles) across the Indian Ocean from Indonesia to Madagascar approximately 3,000 to 1,200 years ago and that migration diffused the same language family east across the Pacific to Polynesia, eventually reaching Hawaii and New Zealand (Figure 5.4.2). We don't know what their boats were like or why they migrated.

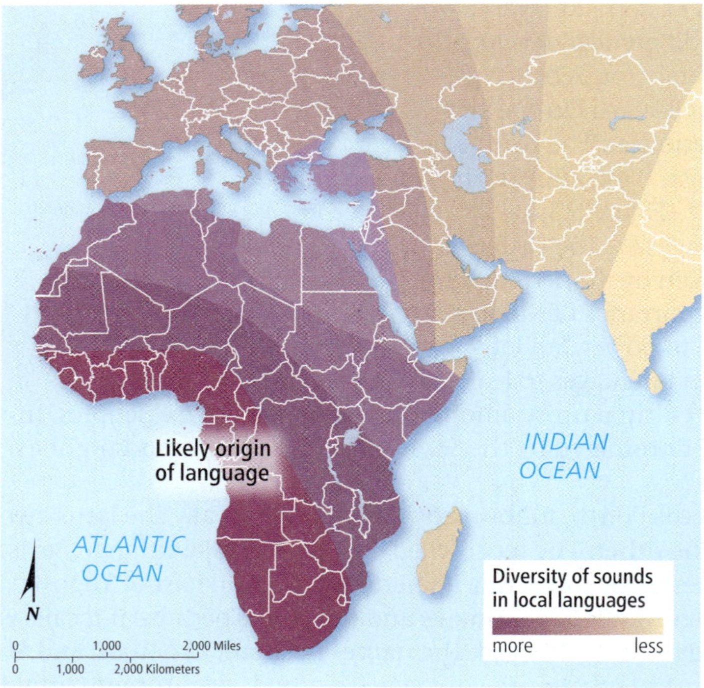

▲ 5.4.3 POSSIBLE ORIGIN AND EARLY DIFFUSION OF LANGUAGES

Possible Origin of Languages

The emergence of distinct language families predates recorded history, so we can only speculate about their origin and initial diffusion. A biologist in New Zealand, Quentin Atkinson, argues that language originated in Africa. He argues that languages first diffused eastward across Africa before crossing into Asia and Europe (Figure 5.4.3).

According to Atkinson, languages are most complex and diverse in Africa. Atkinson thinks humans outside Africa display less linguistic diversity because their languages have had a shorter time in which to evolve into new languages than have African languages.

5.4.2 Common Ancestry of Indo-European

If Germanic, Romance, Balto-Slavic, and Indo-Iranian languages are all part of the same Indo-European language family, then it is reasonable to assume they are descended from a single common ancestral language. The existence of a single ancestor language—which can be called Proto-Indo-European—cannot be proved with certainty because it would have existed thousands of years before the invention of writing or recorded history.

Since the origin of language families predates recorded history, the evidence that the Indo-European language originated with a single Proto-Indo-European language comes primarily from words related to the physical environment. For example, individual Indo-European languages share similar words for winter and snow but not for ocean. Therefore, linguists conclude that original Proto-Indo-European speakers probably lived in a cold climate or one that had a winter season but did not come in contact with oceans. However, linguists and anthropologists disagree on when and where a Proto-Indo-European language originated and the process and routes by which it diffused. Two theories are "nomadic warrior" and "sedentary farmer."

Nomadic Warrior Theory

The first Proto-Indo-European speakers were the Kurgan people, according to archaeologist Marija Gimbutas. The earliest archaeological evidence of the Kurgans dates to around 4300 B.C.E., near the border between present-day Russia and Kazakhstan.

Among the first people to domesticate horses and use chariots, the Kurgans migrated in search of grasslands for their animals. This took them westward through Europe, eastward to Siberia, and southeastward to Iran and South Asia. Between 3500 and 2500 B.C.E., Kurgan warriors, having established military superiority with their horse-drawn chariots, conquered much of Europe and South Asia (Figure 5.4.4).

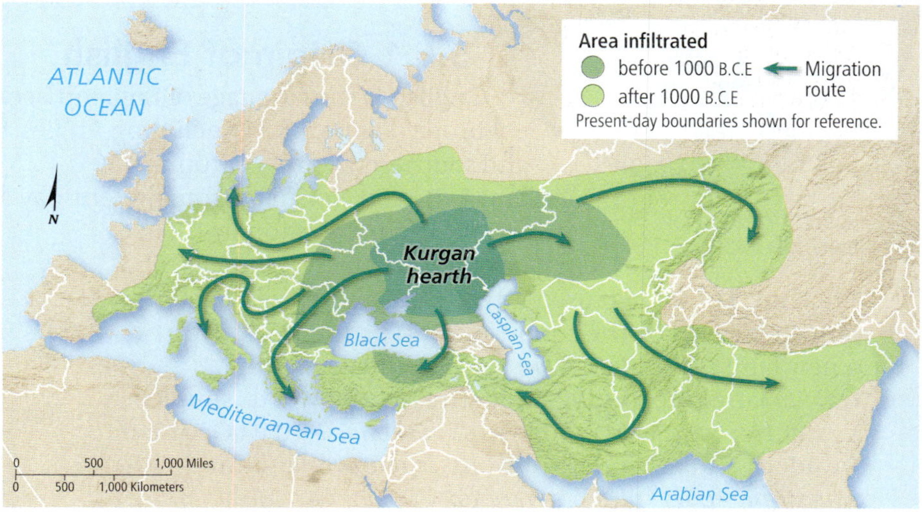

▲ 5.4.4 ORIGIN AND DIFFUSION OF INDO-EUROPEAN: NOMADIC WARRIOR THEORY

Sedentary Farmer Theory

Archaeologist Colin Renfrew argues that the first speakers of Proto-Indo-European lived 2,000 years before the Kurgans, in the eastern part of present-day Turkey (Figure 5.4.5). Supporting Renfrew, biologist Russell D. Gray dates the first speakers even earlier, at around 6700 B.C.E. This hypothesis argues that Indo-European diffused into Europe and South Asia along with agricultural practices rather than by military conquest. The language triumphed because its speakers became more numerous and prosperous by growing their own food instead of relying on hunting.

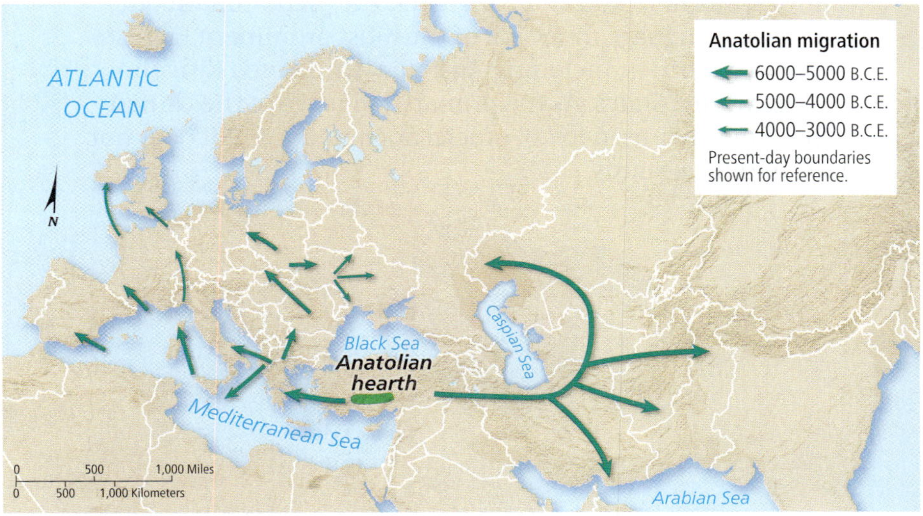

▲ 5.4.5 ORIGIN AND DIFFUSION OF INDO-EUROPEAN: SEDENTARY FARMER THEORY

Thus, the diffusion of Indo-European speaks to a fundamental question for humanity: Do cultural elements such as language diffuse primarily through warfare and conquest or primarily through peaceful sharing of resources? Regardless of how Indo-European diffused, communication was poor among different peoples, whether warriors or farmers. After many generations of complete isolation, individual groups evolved increasingly distinct languages.

Origin & Diffusion of English

- **Analyze processes underlying current distribution of English.**

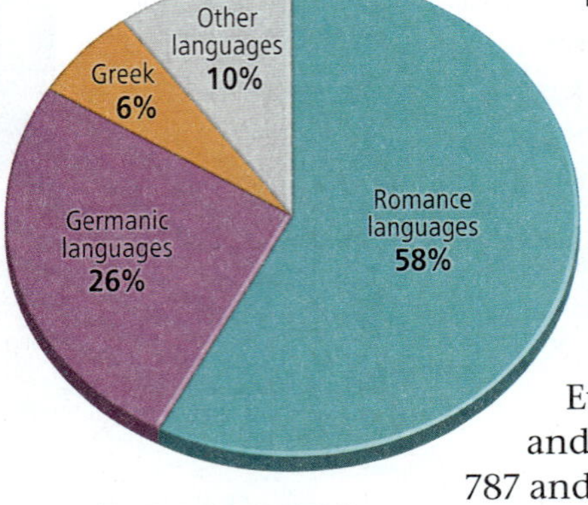

▲ **5.5.1 ORIGIN OF ENGLISH WORDS**

The British Isles have been inhabited for thousands of years, but we know nothing of their early languages until tribes called the *Celts* arrived around 2000 B.C.E., speaking languages we call *Celtic*. Around 450 C.E., tribes from mainland Europe invaded, pushing the Celts into the remote northern and western parts of Britain.

5.5.1 Origin of English

English is the language of England because of migration from various parts of Europe (Figure 5.5.1), including Angles, Saxons, and Jutes from northern Germany and southern Denmark around 450 C.E.; Vikings from present-day Norway between 787 and 1171; and Normans from present-day Normandy in France in 1066 (Figure 5.5.2).

After the Norman invasion in 1066, French became the official language of England. The French used by the leaders and the Germanic languages used by the common people mingled to form a new language, English, which became the country's official language in 1362. Modern English owes its simpler straightforward words, such as *sky* and *horse*, to its Germanic invaders, and more elegant words, such as *celestial* and *equestrian*, to its French invaders.

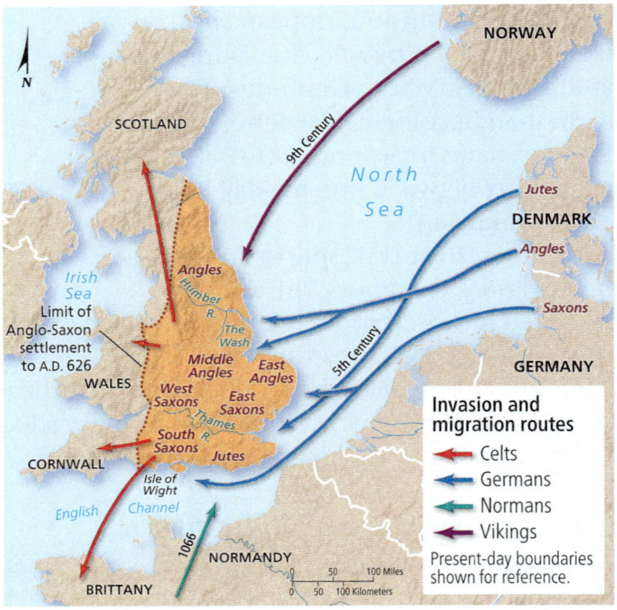

▲ **5.5.2 INVASIONS OF ENGLAND**

Contemporary Distribution of English

English is an official language in 56 countries, more than any other language (Figure 5.5.3). Two billion people live in a country where English is an official language, even if they cannot speak it. In addition, English is the predominant but not official language in several of the most prominent English-speaking countries, including Australia, the United Kingdom, and the United States. As the primary language of the United States, English is used for all official documents, but it does not have an official status.

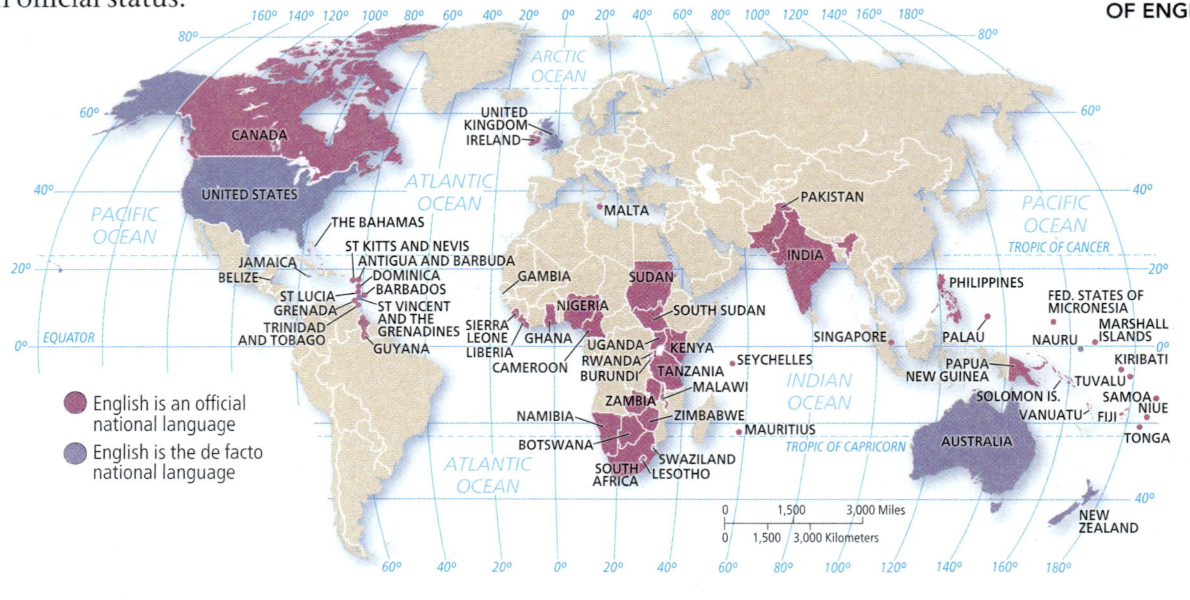

◀ **5.5.3 ENGLISH-SPEAKING COUNTRIES**

5.5.2 Sharing Languages

One of the most fundamental needs in a global society is a common language for communication. In the modern world, the most important language of international communication is English. A Polish airline pilot who flies over Spain speaks to the traffic controller on the ground in English. Swiss bankers speak German among themselves, but with German bankers they prefer to speak English. English is the language spoken at an aircraft factory in France and an appliance company in Italy.

▲ 5.5.4 LINGUA FRANCA, TANZANIA
(top to bottom) English, Arabic, Swahili, and Hindi utilize different writing systems.

Lingua Franca

A language of international communication, such as English, is known as a **lingua franca**. To facilitate trade, speakers of two different languages agree to use a language known to both as a lingua franca. The term, which means "language of the Franks," was originally applied by Arab traders during the Middle Ages to the language they used to communicate with Europeans, whom they called Franks.

Other than English, modern lingua franca languages include Swahili in East Africa (Figure 5.5.4), Hindi in South Asia, Indonesian in Southeast Asia, and Russian in the former Soviet Union. A number of African countries that became independent in the twentieth century adopted English or Swahili as an official language for government business, as well as for commerce, even if the majority of the people couldn't speak it.

A **pidgin language** is a simplified form of a lingua franca. To communicate with speakers of another language, two groups construct a pidgin language by learning a few elements of the lingua franca.

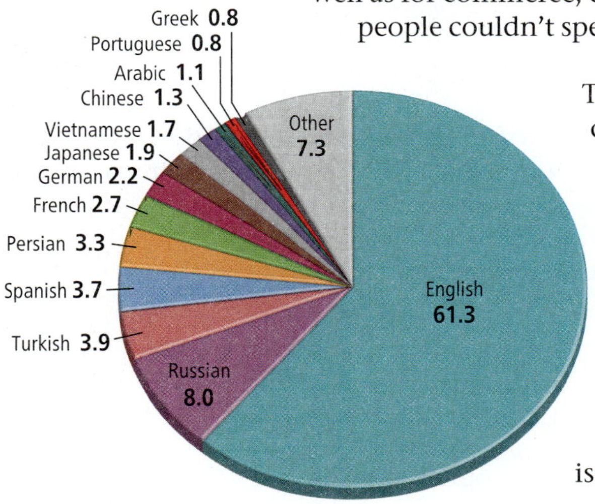

▲ 5.5.5 LANGUAGES OF WEBSITES

English on the Internet

The emergence of the internet as an important means of communication has further strengthened the dominance of English. Around 60 percent of all internet content is in English (Figure 5.5.5), even though only around 25 percent of internet users are English speakers (Figure 5.5.6). Given the large percentage of internet content in English, knowledge of English is essential for internet users around the world.

Chinese: The Next Lingua Franca?

English may be less dominant as the language of the internet in the years ahead. Future leadership in social media may pass to China, in part because of the large number of people worldwide who speak Chinese languages. The attraction of Chinese languages also comes from the way they are written. Rather than sounds (as in English), Chinese languages are written primarily with **logograms**, which are symbols that represent words or meaningful parts of words. Most logograms are compounds; words related to bodies of water, for example, include a symbol that represents a river plus additional strokes that alter the river in some way. Chinese is an attractive language to use in social media that restrict the number of characters, such as Twitter. An English message that uses the maximum 280 symbols permitted by Twitter could be written in Chinese in only around 140 characters.

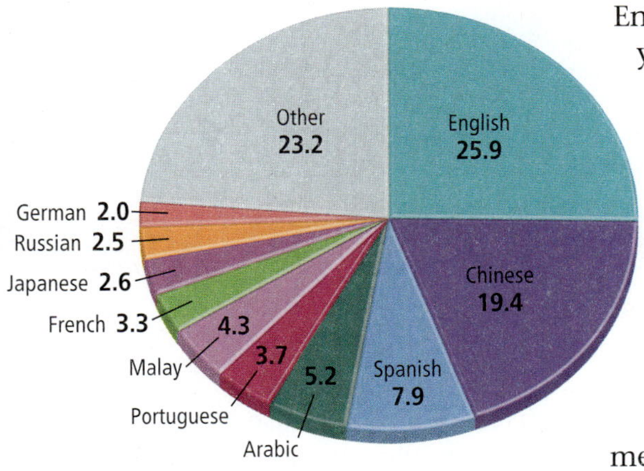

▲ 5.5.6 LANGUAGES OF INTERNET USERS

5.6 KEY ISSUE 3 How do languages vary among places?

▲ 5.6.1 ENGLISH SUBDIALECT
Yorkshire, England.

Dialects

- **Identify regional variations of languages.**

A **dialect** is a regional variation of a language distinguished by distinctive vocabulary, spelling, and pronunciation. A **subdialect** is a subdivision of a dialect. Because of its large number of speakers and widespread distribution, English has an especially large number of dialects and subdialects (Figure 5.6.1).

5.6.1 American & British English

When speakers of a language migrate to other locations, various dialects of that language may develop. This was the case with the migration of English speakers to North America several hundred years ago. North Americans are well aware that they speak English differently from the British, not to mention people living in India, Pakistan, Australia, and other English-speaking countries.

Vocabulary. The vocabulary of U.S. English differs from the English of England largely because settlers in America encountered many new objects and experiences. The new continent contained physical features, such as large forests and mountains, that had to be given new names. Indigenous Native Americans also enriched American English with names for objects such as canoe, moccasin, and squash, and animals such as moose, raccoon, and chipmunk. As new inventions appeared, they acquired different names on either side of the Atlantic. For example, the British call the hood of a car the bonnet and the trunk the boot.

Pronunciation. From the time of their arrival in North America, colonists began to pronounce words differently from the British. Such divergence is normal, for interaction between the two groups was largely confined to exchange of letters and other printed matter rather than direct speech.

Spelling. American spelling diverged from the British standard because of a strong feeling in the United States for an independent identity. Noah Webster, the creator of the first comprehensive American dictionary and grammar books, was determined to develop a uniquely American dialect of English. He either ignored or was unaware of recently created rules of grammar and spelling developed in England. Webster argued that spelling and grammar reforms would help establish a national language, reduce cultural dependence on England, and inspire national pride.

Dialects in the United Kingdom

English varies by regions within individual countries. In England, dialects can be traced back to the invading groups from Northern Europe who settled in different parts of the country (Figure 5.6.2). The boundaries between English dialects have been moving (Figure 5.6.2c). The changes reflect patterns of migration, especially of Londoners moving out.

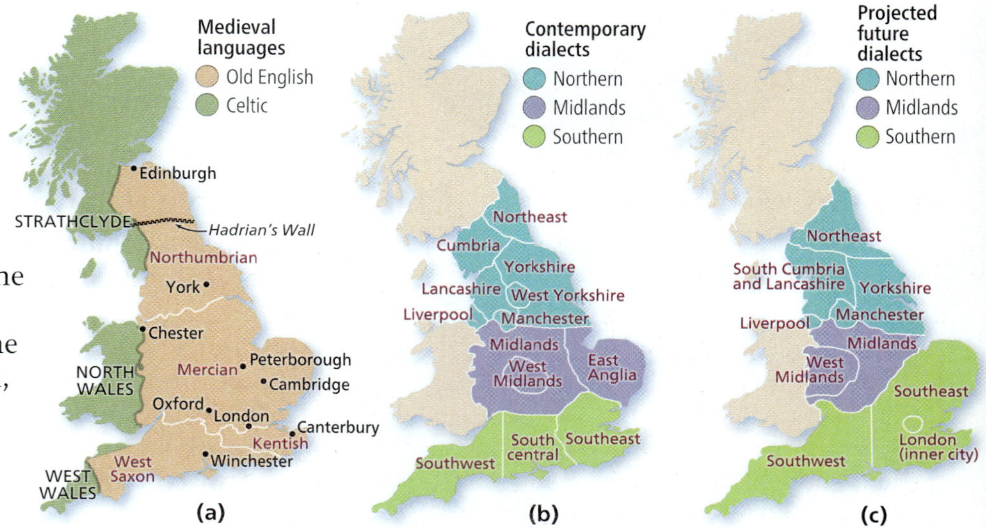

▲ 5.6.2 DIALECTS IN THE UNITED KINGDOM

In a language with multiple dialects, one dialect may be recognized as the **standard language**, which is a dialect widely recognized as the most acceptable for government, business, education, and mass communications like TV. In the case of England, the standard language is known as **Received Pronunciation (RP)**. It is well known around the world as the dialect commonly used by politicians, broadcasters, and actors.

5.6.2 U.S. Dialects

The United States has four major dialect regions: North, Midland, South, and West (Figure 5.6.3). The three eastern dialect regions can also be divided into several subdialects. As with other elements of language, the distribution of U.S. dialects is a function of migration.

The distribution of dialects is documented through the study of particular words. Every word that is not used nationally has some geographic extent within the country and therefore has boundaries. Such a word-usage boundary, known as an **isogloss**, can be constructed for each word. Although every word has a unique isogloss, boundary lines of different words coalesce in some locations to form regions.

The current distribution of U.S. dialects can be traced to differences in the origin of the English colonists along the East Coast. Three distinct dialect regions developed in the early colonies—North, South, and Midlands.

- **North.** Two-thirds of the New England colonists were Puritans from East Anglia in southeastern England, and only a few came from the north of England. The characteristic dropping of the /r/ sound at the end of words is shared with speakers from the south of England.
- **South.** About half of immigrants to the South came from southeastern England, although they represented a diversity of social-class backgrounds, including deported prisoners, indentured servants, and political and religious refugees.
- **Midland.** Immigrants to the Midlands were more diverse. The early settlers of Pennsylvania were predominantly Quakers from the north of England. Scots and Irish also went to Pennsylvania as well as to New Jersey and Delaware. The Middle Atlantic colonies also attracted many German, Dutch, and Swedish immigrants, who learned their English from the English-speaking settlers in the area.

The diffusion of particular English dialects across the United States is a result of the westward movement of colonists from the three East Coast dialect regions. The North and South accents sound unusual to the majority of Americans because the standard pronunciation throughout the American West comes from the Midland rather than the North and South regions. This pattern occurred because most western settlers came from the Midland.

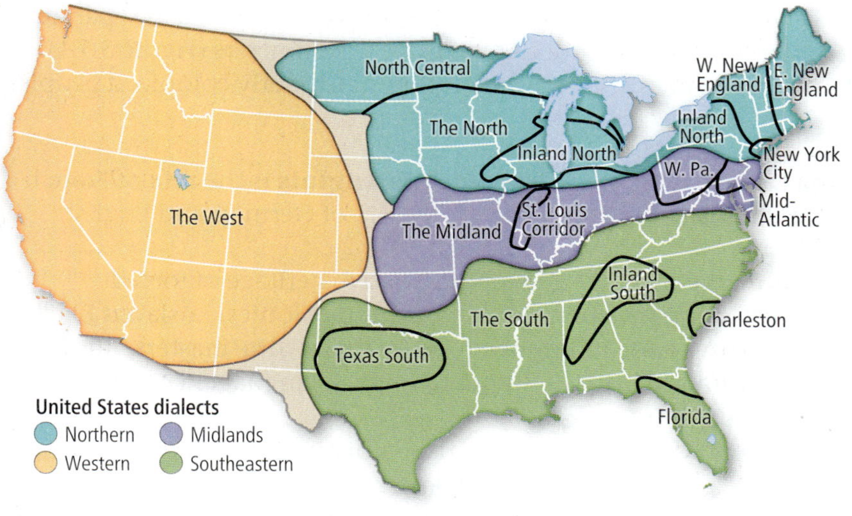

▲ 5.6.3 U.S. DIALECTS AND SUBDIALECTS

Changing Distribution of U.S. Dialects

Many words that were once regionally distinctive are now national in distribution. Mass media, especially TV, influence the adoption of the same words throughout the country. Nonetheless, regional dialect differences persist in the United States. For example, the word for soft drink varies. Most people in the Northeast and Southwest as well as the St. Louis area use soda to describe a soft drink. Most people in the Midwest, Great Plains, and Northwest prefer pop. Southerners refer to all soft drinks as Coke (Figure 5.6.4).

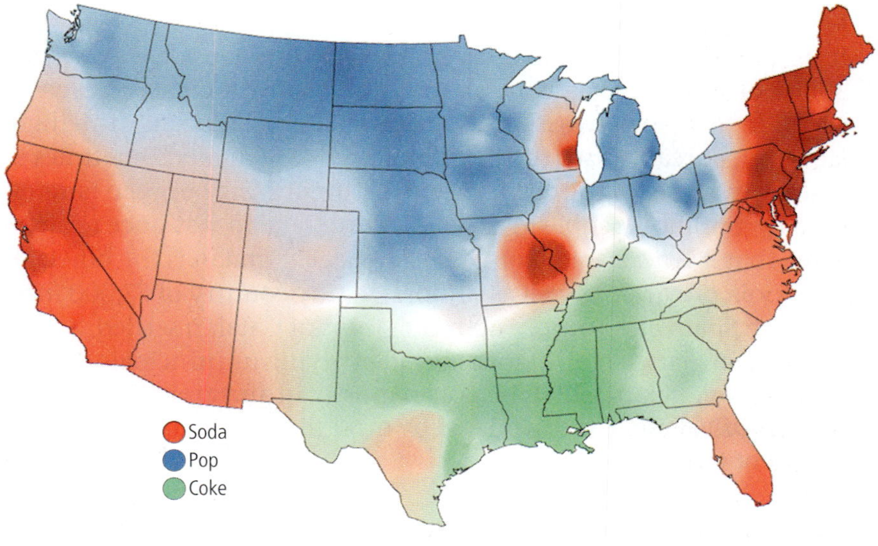

▲ 5.6.4 U.S. SOFT-DRINK DIALECTS

Dialect or Language?

- **Summarize challenges in recognizing and identifying languages and dialects.**

Distinguishing between dialects and distinct languages is a good example of global–local tensions. Migration, increased travel and business interactions worldwide, and other globalization processes have resulted in strengthening of standard languages and suppression of dialects. On the other hand, desire for more local cultural identity has resulted in the emergence of distinct languages that were once considered dialects.

▲ 5.7.1 SIGN IN ENGLISH, SPANISH, AND CREOLE
Miami, Florida.

5.7.1 Creole Languages

A **creole**, or **creolized language**, is a language that forms over time from the mixing of a colonizer's language with the indigenous language of the people being dominated. A creolized language forms when the colonized group adopts the language of the dominant group but makes some changes, such as simplifying the grammar and adding words from the former language. Examples include French Creole in Haiti, Papiamento (creolized Spanish) in Netherlands Antilles (West Indies), and Portuguese Creole in the Cabo Verde, an island country off the African coast. These creole languages spoken in former colonies are classified as separate languages because they differ substantially from the original introduced by European colonizers (Figure 5.7.1). The use of a creole language is a way to maintain cultural identity and unity amongst formerly colonized people.

Informal Languages

New languages are being created through mixing English with other languages. **Franglais** is the mix of French and English, **Spanglish** is the mix of Spanish and English, and **Denglish** is the mix of German and English (Figure 5.7.2).

Some Blacks speak a dialect of English heavily influenced by the group's distinctive heritage of forced migration from Africa during the eighteenth century for enslavement in the southern colonies. Enslaved Blacks preserved a distinctive dialect in part to communicate in a code not understood by their white masters. Black dialect words such as *gumbo* and *jazz* have long since diffused into the standard English language. That dialect has been termed **African American Vernacular English (AAVE)**.

Natives of Appalachian communities, such as in rural West Virginia, also have a distinctive dialect. An Appalachian dialect is a source of regional identity and pride, but it has long been regarded by other Americans as a sign of poor education and an obstacle to obtaining employment outside Appalachia.

Dialects Become Languages

It is sometimes difficult to distinguish between a language and a dialect. **Mutual intelligibility** refers to the ability of people speaking in two ways to readily understand each other without prior familiarity or special effort. In principle, two dialects of a language are mutually intelligible, whereas two different languages are not. However, in reality, two distinct languages are actually sometimes intelligible, whereas two dialects may not be.

One example of the challenge in distinguishing between languages and dialects is Catalán-Valencian-Balear, used by 4 million people in eastern Spain. Catalán was once regarded as a dialect of Spanish, but linguists now agree that it is a separate Romance language. Catalán is the official language of Spain's highly autonomous Catalonia province as well as Andorra, a tiny country of 77,000 inhabitants situated in the Pyrenees Mountains between Spain and France. With the status of Catalán settled as a separate language, linguists are identifying its principal dialects. Valencian, which is spoken mostly in and around the city of Valencia, is considered by most linguists to be a dialect of Catalán.

▶ 5.7.2 FRANGLAIS
The name of this café in Nantes, France, is a mix of English (*death*) and French (*porc*).

5.7.2 Sign Languages

Sign languages are full-fledged natural languages with their own grammar and lexicon (Figure 5.7.3). Sign languages are not universal, and they are not mutually intelligible with each other. *Ethnologue* lists 144 sign languages, but other sources have higher totals.

American Sign Language (ASL) is thought to have originated at the American School for the Deaf, founded in Hartford, Connecticut, in 1817 by Thomas Gallaudet. The sign language that Gallaudet adopted was based on methods that he had observed in France. As a result, ASL shares around 80 percent of its sign language movements with those used in France. Around half a million Americans are estimated to use ASL.

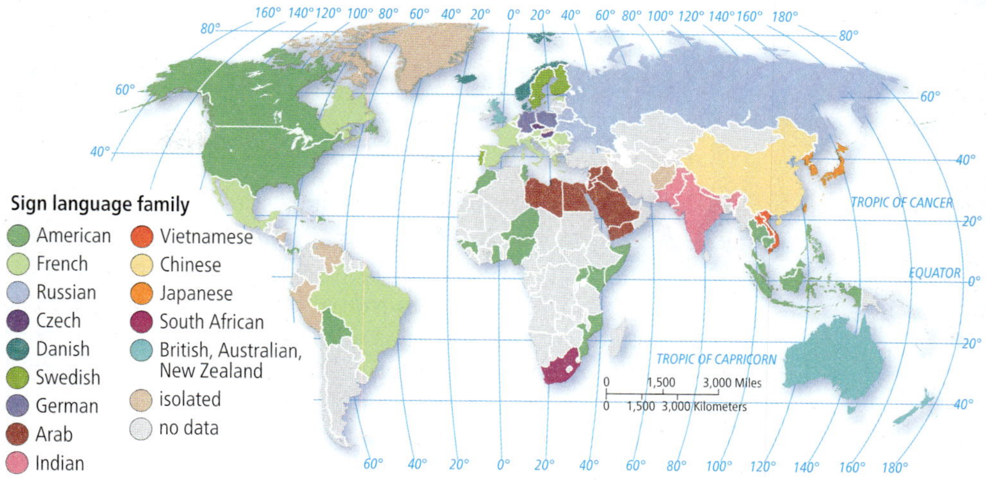

▲ **5.7.3 SIGN LANGUAGE FAMILIES**

Gender Differences in Languages

Languages vary in their ability to absorb and to reflect a cultural change, such as gender equality. Languages are divided about evenly between those that distinguish between masculine and feminine and those that do not. Of the 20 most widely used languages, 12 make gender distinctions, and 8 do not. Nouns and verbs may carry distinctive forms or endings depending on gender, such as adding *ova* to designate feminine names in Russian. Others use identifying articles, such as the French *la* in front of a feminine noun such as *la géographie* (geography) and *le* in front of a masculine one such as *le cartographe* (cartographer).

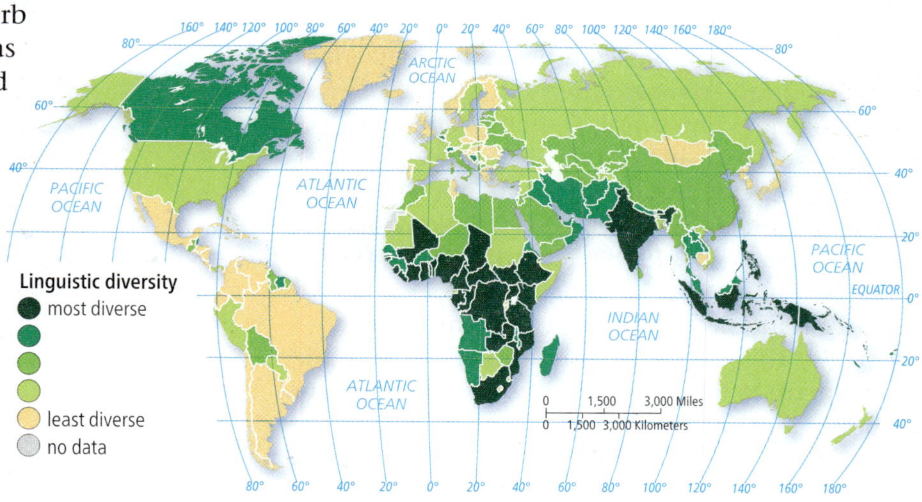

▲ **5.7.4 NUMBER OF LANGUAGES PER COUNTRY**

Language Diversity

Multiple languages coexist in some countries (Figure 5.7.4). Two countries in Southeast Asia—Papua New Guinea and Indonesia—together are home to users of more than 20 percent of all of the world's languages. In Papua New Guinea, an island country with only 9 million inhabitants, the median number of speakers of each language is only 1,315 people, according to *Ethnologue* (Figure 5.7.5).

▲ **5.7.5 LANGUAGE FAMILIES IN PAPUA NEW GUINEA**

Multilingual Countries

- **Examine examples of countries whose citizens use more than one language.**

Multiple languages coexist in some countries, with varying degrees of success. In some countries, distinct cultural groups speaking different languages may occupy separate regions. In other countries, speakers of various languages intermingle and are able to converse in more than one language. Some countries have devised strategies to promote language as a centripetal force through peaceful coexistence among speakers of different languages. Other countries face challenges among cultural groups speaking more than one language, a condition threatening to act as a centrifugal force.

5.8.1 Multilingual Countries in Europe

Figure 5.3.1 shows that the boundary between the Romance and Germanic branches runs through the middle of two small European countries, Belgium and Switzerland. Belgium has had more difficulty than Switzerland in reconciling the interests of the different language speakers.

Switzerland: Institutionalized diversity. Switzerland has four official languages: German (used by 63 percent of the population), French (23 percent), Italian (8 percent), and Romansh (1 percent). These four languages predominate in different parts of the country (Figure 5.8.1). Swiss voters made Romansh an official language in a 1938 referendum, even though only 1 percent use the language.

Switzerland peacefully exists with multiple languages. The key is a long tradition of decentralized government, in which local authorities hold most of the power and decisions are frequently made by voter referenda.

▲ 5.8.1 LANGUAGE DIVERSITY IN SWITZERLAND

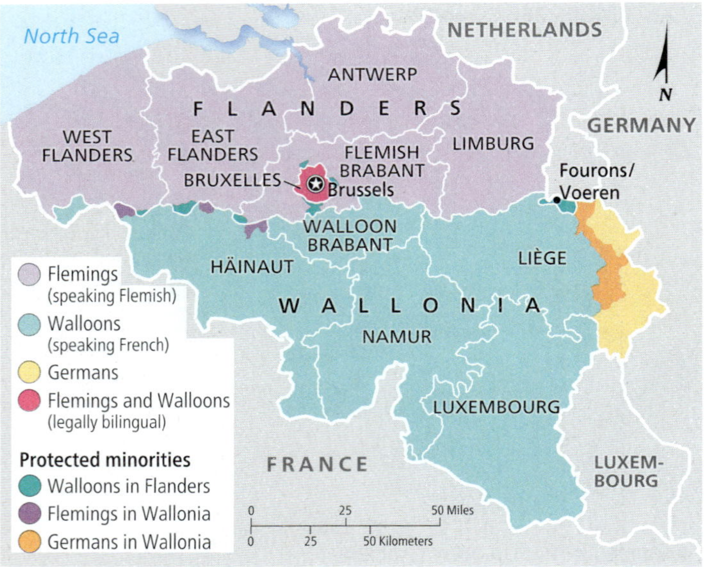

Belgium: Barely speaking. A language boundary sharply divides the small country of Belgium into two regions, Flanders and Wallonia. Southern Belgians (known as Walloons) speak French, whereas northern Belgians (known as Flemings) speak Flemish, a dialect of the Germanic language Dutch (Figure 5.8.2). Brussels, the capital city, is officially bilingual.

Antagonism between the Flemings and Walloons is aggravated by economic and political differences. Historically, the Walloons dominated Belgium's economy and politics, and French was the official state language. But in recent years Flanders has been much more prosperous than Wallonia, and the Flemish-speaking northerners do not wish to see their taxes spent in the poorer south.

◀ 5.8.2 LANGUAGE DIVERSITY IN BELGIUM

5.8.2 Other Multilingual Countries

Of the world's most multilingual countries, according to *Ethnologue*, Papua New Guinea is first (with 840 languages spoken there, see Figure 5.7.5). The United States, with 326 languages, is the fifth most multilingual country in the world.

Nigeria: Spatial compromise. Africa's most populous country, Nigeria, provides an example of the tensions that can arise from the presence of many speakers of many languages living near each other. Of Nigeria's 524 distinct languages, only three (Hausa, Igbo, and Yoruba) are used by more than 10 percent of the country's population (Figure 5.8.3). Religion further divides the country; the north is predominantly Muslim, and the south is predominantly Christian.

Groups living in different regions of Nigeria have often battled. To reduce these regional tensions, the government moved the capital from Lagos in the Yoruba-dominated southwest to Abuja in the center of the country, where none of the three major languages or two major religions predominates.

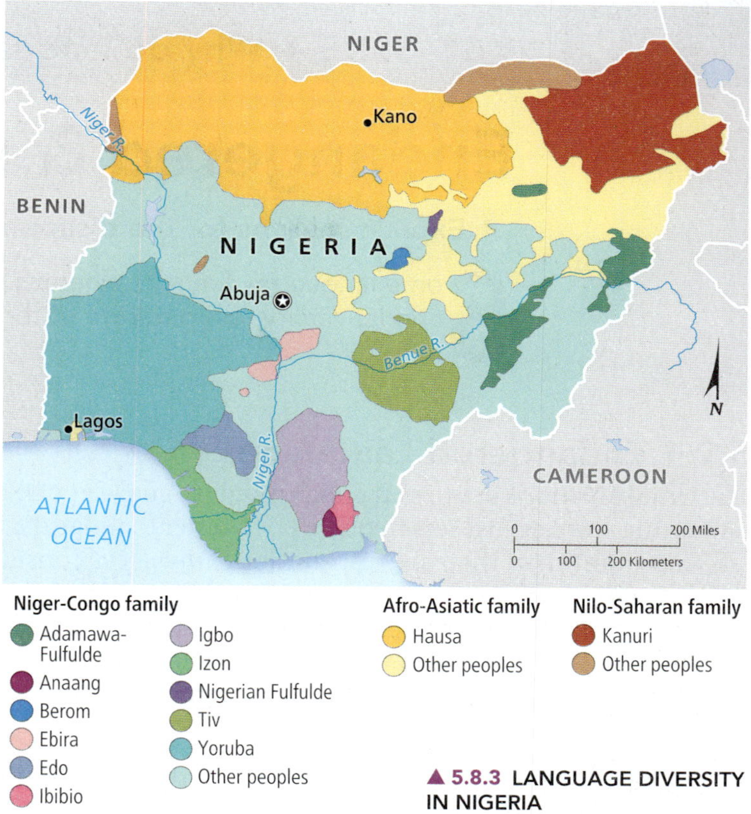

▲ **5.8.3 LANGUAGE DIVERSITY IN NIGERIA**

Canada: Bilingual autonomy. French is one of Canada's two official languages, along with English. French speakers comprise one-fifth of the country's population and are clustered in Québec, where they account for more than three-fourths of the province's speakers (Figure 5.8.4).

Colonized by the French in the seventeenth century, Québec was captured by the British in 1763, and in 1867, it became one of the provinces in the Confederation of Canada.

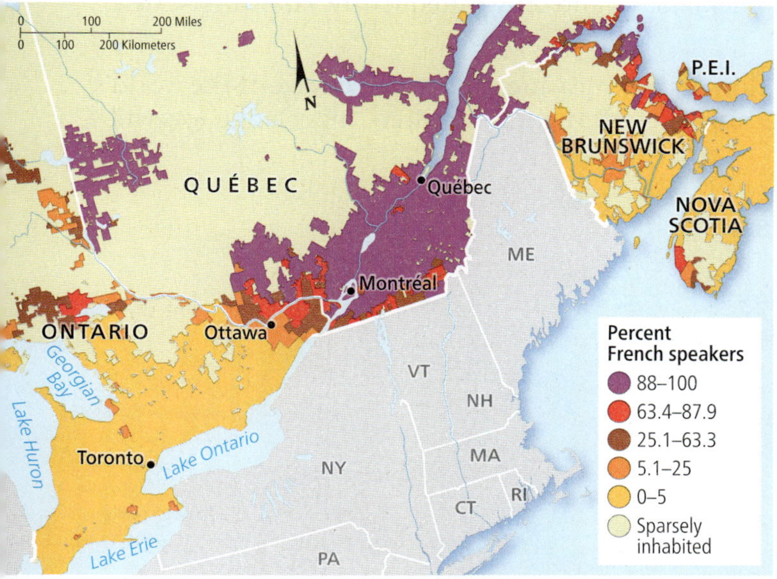

▲ **5.8.4 LANGUAGE DIVERSITY IN CANADA**

The Québec government has made the use of French mandatory in many daily activities (Figure 5.8.5). Québec's Commission de Toponymie renamed French towns, rivers, and mountains that had been given English-language names. French must be the predominant language on all commercial signs in Québec.

Until the late twentieth century, Québec was one of Canada's poorest and least-developed provinces. Its economic and political activities were dominated by an English-speaking minority and lack of French-speaking leaders, and the province suffered from cultural isolation.

To promote French-language cultural values, the Parti Québécois—one of the province's leading political parties—advocates sovereignty (effectively, independence from Canada), but voters have thus far not supported it.

Confrontation has been replaced in Québec by increased cooperation between French and English speakers. The neighborhoods of Montréal, Québec's largest city, have become more linguistically mixed, and one-third of Québec's native English speakers have married French speakers in recent years.

▲ **5.8.5 STREET SIGNS IN FRENCH, QUÉBEC**

Endangered & Extinct Languages

- **Explain reasons for the disappearance of languages.**

Some languages flourish, whereas others disappear. The distribution of a language, and its vitality, is a measure of the ability of a cultural group to retain a distinctive identity.

5.9.1 Endangered Languages

The United Nations defines an endangered language as one that children are no longer learning and whose remaining speakers use less frequently. In order to judge if a language is endangered, the number of speakers is less important than their age distribution. *Ethnologue* considers a language to be endangered when its users begin to teach and speak a more dominant language to their children. Due to their nature, endangered languages often have few speakers left. The U.N. considers 2,346 languages to face various levels of endangerment. *Ethnologue* has a slightly different total of endangered languages: 3,018.

Languages face various levels of endangerment. The U.N. and *Ethnologue* employ different terms to describe the levels of endangerment:

- Vulnerable languages are still used for face-to-face communication among generations, but they are losing users. *Ethnologue* calls these threatened languages.
- Definitely endangered languages are no longer taught to children as their principal language at home. *Ethnologue* calls these shifting languages.
- Severely endangered languages are spoken by older generations, but parents do not speak them with their children or among themselves. *Ethnologue* calls these moribund languages.
- Critically endangered languages are spoken only by older people, and only infrequently. *Ethnologue* calls these nearly extinct languages.

Regardless of the precise count and method of definition, by any measure, a large percentage of the world's languages are endangered. Because parents are no longer able to teach them to children, these languages will survive only through concerted community efforts.

Distribution of Endangered Languages

More than 50 percent of the world's endangered languages are located in just eight countries: India, Brazil, Mexico, Australia, Indonesia, Nigeria, Papua New Guinea, and Cameroon (Figure 5.9.1). In such countries and around them are the areas that are the most linguistically diverse in the world (refer to Figure 5.7.4).

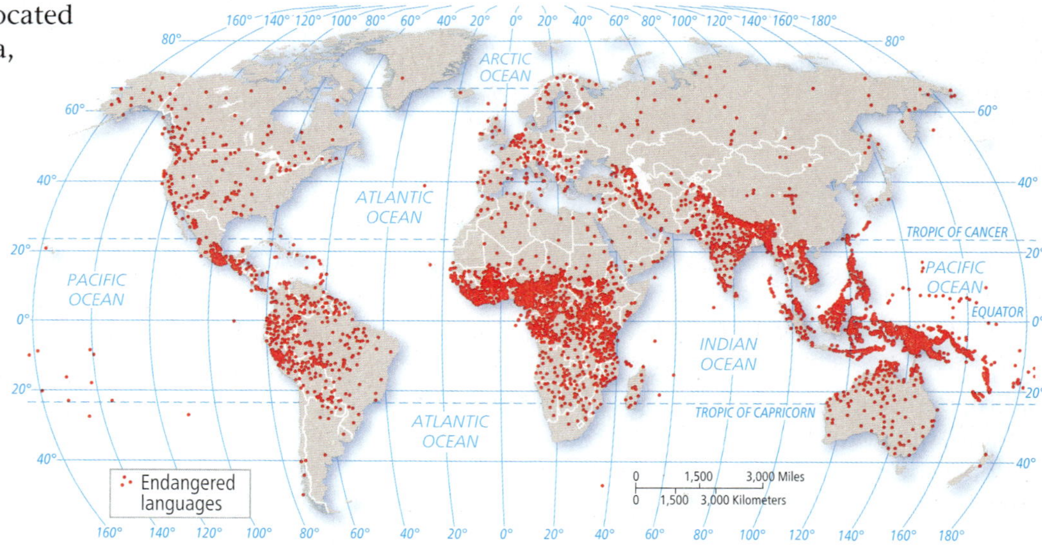

▲ **5.9.1** ENDANGERED LANGUAGES

5.9.2 Extinct Languages

An **extinct language** is a language that was once used by people in daily activities but is no longer in use. The extinction may follow this process: (1) A language becomes vulnerable to extinction when its use is restricted to certain domains, like only the home, but not in public places, like shops and offices. A vulnerable language is spoken by children, but only in some places. (2) A language becomes endangered when children are no longer learning it. (3) A language becomes moribund when it is spoken only by older people. (4) A language becomes extinct when it is no longer spoken.

For example, when Spanish missionaries reached the eastern Amazon region of Peru in the sixteenth century, they found more than 500 languages. Only 92 survive today, according to *Ethnologue*, and 14 of them face extinction in the very near future because fewer than 100 speakers remain. Of Peru's 92 surviving indigenous languages, only Cusco, a Quechuan language, is currently used by more than 1 million people. *Ethnologue* estimates that 367 languages have become extinct since 1950, a rate of 6 per year. The U.N. identifies 228 recently extinct languages (Figure 5.9.2).

The loss of many languages is a reflection of globalization. To be part of a global economy and culture, people choose to use a widely used language, leaving their traditional or indigenous language to disappear.

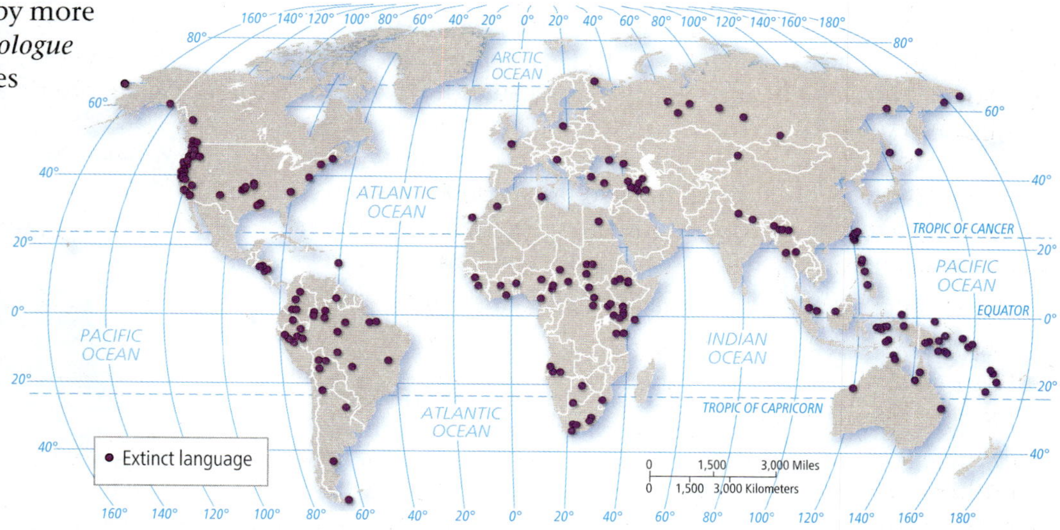

▲ 5.9.2 LANGUAGES EXTINCT SINCE 1950

Extinct Languages in the United States

Ethnologue lists 74 languages based in the United States that are now extinct, and the U.N. lists 54. These are languages once spoken by groups of Native Americans, especially in the West (Figure 5.9.3).

One example is Klallam (also known as S'Klallam), a language once spoken in Washington's Olympic Peninsula and Vancouver Island, Canada. Klallam became extinct on February 4, 2014, when its last-known speaker, Hazel Sampson, died at age 103. The language is taught at Port Angelos High School in Washington, and six youths are reported to be able to speak some words of Klallam as their second language. A dictionary of Klallam was published in 2012. If someday a child is taught Klallam as his or her first language, perhaps by one of the six who know some words, then the language might be reclassified as reawakening.

The COVID-19 pandemic has taken its toll on languages. Dakota, a Sioux language, is critically endangered, with only 290 speakers, according to *Ethnologue*. The average age of the remaining speakers is around 70. The language was being taught at local schools and colleges, but the programs were halted during the pandemic. Mortality rates among both the elders and the teachers have been especially high on the reservation where most of the remaining Dakota speakers live. Tuscarora, an Iroquoian language of the Tuscarora people, became extinct when its last fluent speaker died in 2020.

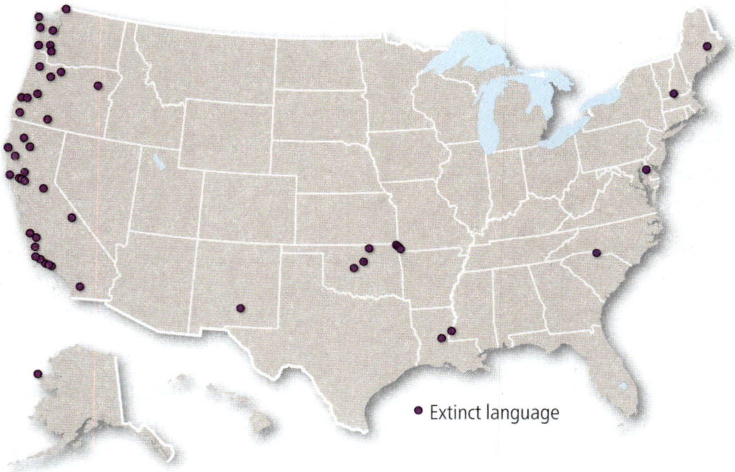

▶ 5.9.3 LANGUAGES EXTINCT SINCE 1950 IN THE UNITED STATES

Preserving & Reviving Languages

- **Describe how some lesser-used languages are being revived and protected.**

Although the number of languages in the world is declining, some are being preserved and protected. Others are being revived. These efforts reflect the importance that people place on language as an element of local culture.

5.10.1 Preserving Languages

Recent efforts have bolstered languages belonging to the Celtic branch of the Indo-European language family. Preserving Celtic languages is of particular interest to English speakers because these languages offer insights into the cultural heritage of places that now speak English.

Celtic languages were spoken in much of present-day Germany, France, northern Italy, and the British Isles 2,000 years ago. Today, Celtic languages survive only in remote parts of Scotland, Wales, and Ireland and on the Brittany peninsula of France (Figure 5.10.1). Celtic is divided into Brythonic and Goidelic groups.

▲ 5.10.1 CELTIC LANGUAGES

Brythonic Celtic

Speakers of Brythonic (also called Cymric or Britannic) fled westward during the Germanic invasions to Wales, southwestward to Cornwall, or southward across the English Channel to the Brittany peninsula of France.

Welsh is the principal Brythonic Celtic language. Cornish and Breton are less commonly used.

Welsh. Welsh was the dominant language in Wales until the nineteenth century, when many English speakers migrated there to work in coal mines and factories. Welsh language training has been compulsory in Welsh schools since 1988, and Welsh was made the official language of Wales in 2011. Welsh-language road signs have been posted throughout Wales, and the British Broadcasting Corporation (BBC) produces Welsh-language television and radio programs.

Cornish. The last-known speaker of Cornish died in 1777, but the language was revived by a handful of people in the twentieth century. It is taught in some schools and used in some church services.

Breton. Brittany, like Cornwall, is an isolated peninsula that juts out into the Atlantic Ocean. As part of France, Brittany has a language containing many French words.

Goidelic Celtic

The Goidelic (also called Gaelic) group originated in Ireland. Two Goidelic languages survive: Irish Gaelic and Scottish Gaelic.

Irish. Irish is one of two official languages in the Republic of Ireland, along with English. As with Welsh, the percentage of people speaking Irish is higher in the more remote areas of the country. An Irish-language TV station began broadcasting in 1996. English road signs were banned from portions of western Ireland in 2005.

Gaelic. In Scotland 1 percent of the people speak Scottish Gaelic. Scottish Gaelic should not be confused with Scots, which *Ethnologue* considers to be an Indo-European Germanic branch language separate from English. Scots is said to be the first language for 1 percent of the people in Scotland. Scots is more prevalent in Scotland's southern lowlands nearest to England, whereas Scottish Gaelic is more likely to be found in more remote northern highlands and islands.

5.10.2 Reviving Languages

Examples can be found of languages that were rarely used or even considered extinct that have been revived and are again in daily use.

Hebrew. Hebrew is an example of a language that was once rarely used but is more commonly used now. Most of the Jewish Bible and Christian Old Testament were written in Hebrew. A language of daily activity in biblical times, Hebrew diminished in use in the fourth century B.C.E. and was thereafter retained primarily for Jewish religious services. At the time of Jesus, most people in present-day Israel spoke Aramaic, which in turn was replaced by Arabic.

When Israel was established as an independent country in 1948, Hebrew became one of the new country's two official languages, along with Arabic (Figure 5.10.2). Since 2018, Hebrew has been the country's only official language. Hebrew was chosen because the Jewish population of the State of Israel consisted of refugees and migrants from many countries who spoke many languages. Because Hebrew was still used in Jewish prayers, no other language could so symbolically unify the disparate cultural groups in the new country.

▲ **5.10.2 HEBREW**
Sign at sea level in Hebrew, French, and English on the road between Jerusalem (754 meters or 2,474 feet above sea level) and the Dead Sea (431 meters or 1,411 feet below sea level).

Myaamia. The Miami Native American Tribe traditionally spoke the Myaamia language (listed in *Ethnologue* as Miami), bringing it with them until after their forced migration in the nineteenth century from their ancestral homelands in the Lower Great Lakes to northeastern Oklahoma. The last native speakers died in the 1960s.

Daryl Baldwin, a citizen of the Miami Tribe of Oklahoma and Director of the Myaamia Center at Miami University in Ohio, has acquired second language proficiency in Myaamia. Baldwin's children learned Myaamia as their first language in the home—the first in a half-century to be first-language speakers of the language (Figure 5.10.3). As part of the revival of the Myaamia language, an online dictionary has been created, accessible at myaamiadictionary.org.

Based on the language skills of the Baldwin family, as well as other Miami people, *Ethnologue* has reclassified Myaamia from an extinct language to a reawakening language. Before reclassifying Myaamia, *Ethnologue* first undertook an in-depth study of the Baldwin family's language skills, conducted by Dr. Wesley Leonard, a Native American language expert at the University of California, Riverside.

▼ **5.10.3 REVIVING THE MYAAMIA LANGUAGE**

Inventing Languages

Isolated languages continue to be identified and documented, and entirely new languages are found. Here are three recent examples:

Koro Aka. A research team from Oregon's Living Tongues Institute for Endangered Languages was in India to study rarely spoken languages. They heard people in the area speaking a language that was not listed in *Ethnologue*. The researchers concluded that what they were hearing was a distinct language belonging to the Sino-Tibetan family. The language, named "Koro Aka," is now listed in *Ethnologue* as a language of northeastern India with 1,500 speakers.

Warlpiri Rampaku. A language called Warlpiri Rampaku, or Light Warlpiri, is spoken by about half of the 700 residents of the village of Lajamanu in Australia's Northern Territory. No one outside the village knew about the language until a University of Michigan researcher, Carmel O'Shannessy, heard it in 2013. Young people had invented the new language. It started with parents speaking to their babies in a combination of three languages: Strong Warlpiri, English, and a Creole language that combines English and Strong Warlpiri. As they got older, the children combined the three languages into an entirely new one through actions such as inventing new verb endings and tenses.

Jedek. A language spoken by 280 people in a village in Malaysia was identified in 2018 by linguists at Lund University. The language is classified in the Aslian branch of the Austro-Asiatic language family. Despite the small number of speakers, Jedek is being used by all generations, including children.

CHAPTER 5 Review, Analyze, & Apply

KEY ISSUE 1 How Are Languages Distributed?

- Languages are classified as institutional, stable, and endangered.
- Languages are organized into families and branches.
- Eighteen language families are used by at least 5 million people worldwide.
- The Indo-European language family has four widely used branches.

THINKING GEOGRAPHICALLY

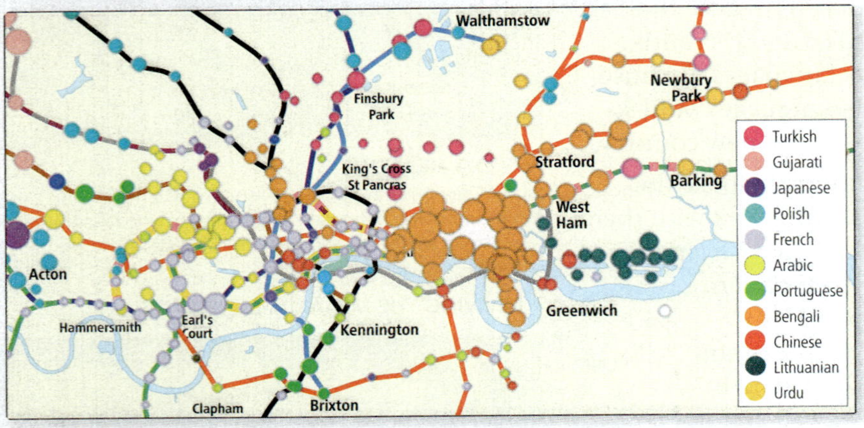

▲ 5.CR.1 MOST USED SECOND LANGUAGE AROUND EACH SUBWAY STOP, LONDON

Although English is the principal language of the United Kingdom, its capital London is home to many people who speak another language. This map shows the most used second language around each subway stop. Why might a city like London exhibit language diversity? What might account for differences among different parts of the city?

KEY ISSUE 2 How Do Languages Diffuse?

- The origin and early diffusion of languages occurred before recorded history.
- Indo-European languages originated in either present-day Turkey or present-day Kazakhstan and spread either by warriors or farmers.
- The origin of English can be traced to the invasion of England by groups from Germany, Denmark, Norway, and France.

THINKING GEOGRAPHICALLY

▲ 5.CR.2 CHINESE LOGOGRAMS

Chinese languages are written with logograms. For example, Figure 5.CR.2 shows several logograms related to bodies of water. What similarities and differences can you detect in these various water-related logograms? In what ways might you make use of logograms when you communicate through social media?

KEY ISSUE 3 How Do Languages Vary Among Places?

- A dialect is a regional variation of a language.
- The United States and the United Kingdom have several major dialects that have distinctive regional distributions.
- Some countries more peacefully embrace multiple languages than do others.

THINKING GEOGRAPHICALLY

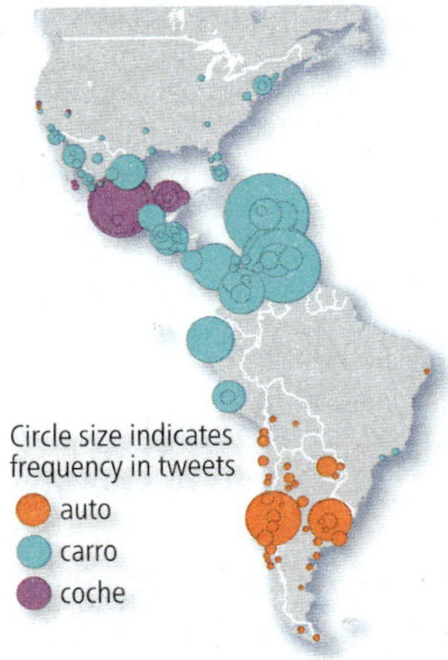

▲ 5.CR.3 THE WORD FOR "CAR" IN SPANISH DIALECTS

Dialects and subdialects exist in many languages, not just English. For example, the word for "car" varies among Spanish dialects. What patterns do you see in the distribution of different words for "car" through the Western Hemisphere? What might account for those patterns?

KEY ISSUE 4 Why Do Languages Survive or Perish?

- Many languages have become extinct, and others are in danger of extinction.
- Some endangered languages are being preserved and protected.
- Other languages are being invented, in some cases through combination with English.

THINKING GEOGRAPHICALLY

▲ 5.CR.4 CHILDREN IN OAXACA, MEXICO, WHERE A NEW SIGN LANGUAGE WAS RECENTLY DOCUMENTED

New languages are constantly being discovered. Most frequently discovered are new sign languages. What might account for the relative frequency of new sign languages compared with spoken ones?

Key Terms

African American Vernacular English (AAVE) (5.7.1) A dialect used by some Blacks.

Centrifugal force (5.1.1) A cultural value that tends to pull people apart.

Centripetal force (5.1.1) A cultural value that tends to unify people.

Creole (or **creolized**) **language** (5.7.1) A language that results from the mixing of a colonizer's language with the indigenous language of the people being dominated.

Denglish (5.7.1) A combination of Deutsch (the German word for German) and English.

Dialect (5.6.1) A regional variety of a language distinguished by vocabulary, spelling, and pronunciation.

Endangered language (5.1.1) A language that is endangered or dying because it is losing users and may not be retained by the younger generation.

Extinct language (5.9.2) A language that was once used by people in daily activities but is no longer used.

Franglais (5.7.1) A combination of français and anglais (the French words for French and English, respectively).

Institutional language (5.1.1) A language used in education, work, mass media, and government.

Isogloss (5.6.2) A boundary that separates regions in which different language usages predominate.

Isolated language (5.2.2) A language that is unrelated to any other languages and therefore not attached to any language family.

Language (5.1.1) A system of communication through speech, movement, sounds, or symbols that a group of people understands to have the same meaning.

Language branch (5.1.1) A collection of languages related through a common ancestor that can be confirmed through archaeological evidence.

Language family (5.1.1) A collection of languages related to each other through a common ancestor long before recorded history.

Language group (5.1.1) A collection of languages within a branch that share a common origin in the relatively recent past and display relatively few differences in grammar and vocabulary.

Lingua franca (5.5.2) A language mutually understood and commonly used in trade by people who have different native languages.

Literary tradition (5.1.1) A language that is written as well as spoken.

Logogram (5.5.2) A symbol that represents a word rather than a sound.

Mutual intelligibility (5.7.2) The ability of people communicating in two ways to readily understand each other without prior familiarity or special effort.

Official language (5.1.1) The language adopted for use by a government for the conduct of business and publication of documents.

Pidgin language (5.5.2) A form of language that adopts a simplified grammar and limited vocabulary of a lingua franca; used for communications among speakers of two different languages.

Received Pronunciation (RP) (5.6.1) The dialect of English commonly used by politicians, broadcasters, and actors in the United Kingdom.

Spanglish (5.7.1) A combination of Spanish and English spoken by Hispanic Americans.

Stable language (5.1.1) A language that is used for face-to-face communication within all generations, including children.

Standard language (5.6.1) The form of a language used for official government, business, education, and mass communication.

Subdialect (5.6.1) A subdivision of a dialect.

Geospatial Analysis

Log in to the Mastering Geography Study Area to access MapMaster 2.0.

South America's Indigenous Languages

On a world map, most of South America appears in the Indo-European language family. Some diversity exists within the region, as shown in Figure 5.2.4.

1. ***Add*** the *Spanish as Official or De Facto Language* layer and ***zoom*** to Latin America.
   1. What is the one large country in South America where Spanish is not the official language? If you do not know the answer, in the *Legend*, ***select*** the *Settings* icon, and ***select*** *Show Political Labels.*
   2. ***Add*** the *Indigenous Languages (South America)* layer. ***Reset*** the map. ***Add*** the *Population Density* layer and select *Split Map Window*. Are indigenous languages more likely to be spoken in areas of high population density or low population density?
   3. ***Add*** *World Elevations*, and ***replace*** the *Population Density* layer. ***Hover*** over the area of highest elevation on the *World Elevations* map. Describe the spatial association between high elevation and indigenous languages.

▲ **5.CR.5 INDIGENOUS LANGUAGES AND WORLD ELEVATIONS**

Explore

Multilingual Belgium

Brussels, the capital of Belgium, is officially bilingual, but many languages comingle there.

In Google Earth, ***search*** for *54 rue de Flandre, Brussels*.

1. What two languages are displayed in the window of the store at 54 rue de Flandre? What does the store sell?
2. What language is displayed on the sign by the entrance to #47, directly across the street from #54? Why might a language other than the two official ones be used at the entrance to this shop?
3. ***Scan up and down*** rue de Flandre. Which of Brussels' two official languages is more common? Why might that be the case? Can you find examples along the street where both official languages are displayed? Why might both official languages be used in these cases?

▼ **5.CR.6 RUE DE FLANDRE, BRUSSELS**

CHAPTER

6 Religions

Interfaith prayer service in remembrance of George Floyd.

Religion can be a source of pride and an expression of identity for members of a particular culture. Geographers recognize that religions are essential to understanding the spatial patterns underlying how humans occupy Earth. An important element of culture, religion gives many people a sense of who they are, what they value, and how they relate to other people, the natural world, and questions of ultimate meaning.

KEY ISSUES

1 Where Are Religions Distributed?

Some religions are highly clustered in one or two of Earth's regions, whereas others are distributed throughout the world. Individual religions also have distinctive distributions within regions.

2 Why Do Religions Have Distinctive Distributions?

Some religions have well-documented places of origin and have diffused extensively. Others have unknown origins and have experienced limited diffusion.

3 How Do Religions Organize Space?

Religious groups derive distinctive meaning from the physical landscape and construct places for worship and other religious practices.

4 Why Do Territorial Conflicts Arise Among Religions?

Adherents of various religions have disputes over territory. The attempt by adherents of one religion to control a region can create conflicts with governments or with other religious groups.

LOCATIONS IN THIS CHAPTER

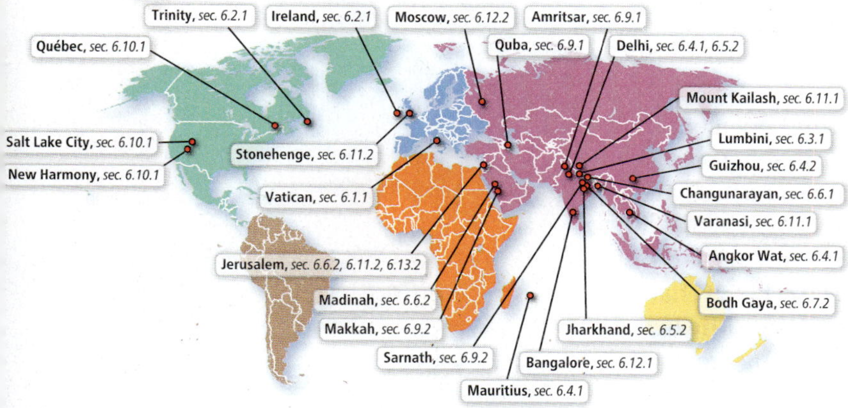

Distribution of Religions

- **Identify the world's major religions and classify them by type.**

Geographers document the places where various religions are located in the world and offer explanations for why some religions have widespread distributions and others are highly clustered in particular places.

▲ 6.1.1 **CHRISTIANITY: ST. PETER'S**
The Papal Basilica of St. Peter, Vatican City, the world's largest church.

6.1.1 Religions & Geography

Identifying the major religions and their distributions in the world is a challenging endeavor because governments are often reluctant to get involved in religions and because religions have difficulty in counting their members. Reliable statistics on the number of adherents to any given religion are therefore not readily available. No official count of religious membership is taken in the United States or in many other countries. Statistics on the number of followers of religions can be controversial; adherents may feel that their religion has been undercounted and therefore accorded less prominence than deserved in world and regional data.

Most international statistics in this chapter come from Adherents.com, the Association of Statisticians of American Religious Bodies, the Pew Research Center, and World Religion Database. None of these organizations is affiliated with a particular religion. The statistics are built on self-identification—that is, whatever people themselves view as their religion. Statisticians do not measure how actively an individual practices a religion.

Classifying Religions

Only a few religions claim the adherence of large numbers of people. The world's religions can be grouped as follows:

- **The three largest universalizing religions.** A **universalizing religion** attempts to be global—to appeal to all people, wherever they may live in the world, not just those of one culture or location. The three universalizing religions with the largest number of adherents in the world are Christianity (Figure 6.1.1), Islam, and Buddhism. Together, they claim the adherence of around 64 percent of the world's people (Figure 6.1.2).
- **The three largest ethnic religions.** An **ethnic religion** appeals primarily to one ethnic or cultural group, or the people of a specific region. The three with the largest number of adherents are Hinduism, Chinese Folk, and Ethno-religions. Together, they claim the adherence of around 23 percent of the world's people.
- **Other religions.** Two percent of the world's people adhere to a religion other than the three largest universalizing religions and three largest ethnic religions. The most numerous of the other religions are New Religions, Sikhism, Spiritism, and Judaism.
- **Nonreligious.** The remaining 11 percent of the world's people express no religious interest or preference and do not participate in any organized religious activity. The World Religion Database identifies two beliefs among non-religionists. **Atheism** is the belief that God does not exist. **Agnosticism** is the belief that the existence of God cannot be proven or disproven empirically.

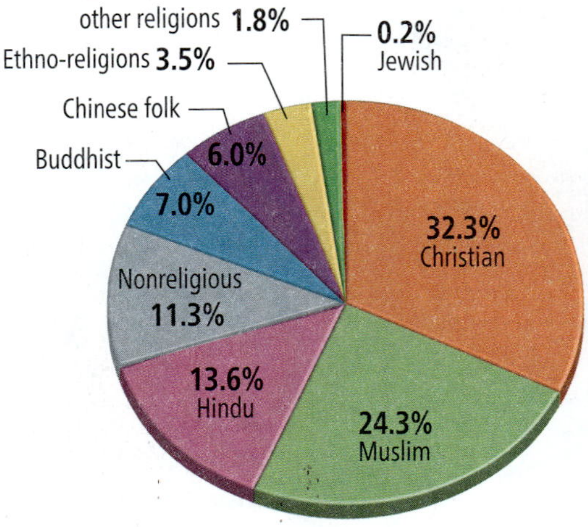

▲ 6.1.2 **PERCENTAGE ADHERING TO VARIOUS RELIGIONS**

6.1.2 Regional Diversity of Religions

Some world regions have a large majority of adherents of one religion, whereas other regions have more of a mix. For example, around 90 percent of Latin Americans and more than 70 percent of Europeans and North Americans identify themselves as adhering to Christianity. Muslims comprise more than 90 percent of the population of Central Asia and of Southwest Asia & North Africa (Figure 6.1.3).

These four regions have the most diverse religious composition:

- In East Asia, more than one-third of the people are nonreligious. Around one-fourth of the people adhere to Chinese Folk religions, and around one-fifth are Buddhists.
- In South Asia, around one-half of the people are Hindus and one-third Muslims. However, adherents are sharply divided by country; India is around 80 percent Hindu, whereas Bangladesh and Pakistan are more than 90 percent Muslim.
- In Southeast Asia, around one-third are Muslims, one-fourth are Buddhists, and one-fourth are Christians. Indonesia (the region's most populous country) is 87 percent Muslim, Cambodia and Thailand are more than 90 percent Buddhist, and the Philippines is more than 90 percent Christian.
- In sub-Saharan Africa, around two-thirds are Christian and one-fifth are Muslim. Christians are more numerous in the south, and Muslims are more numerous in the north.

▶ 6.1.3 MOST NUMEROUS RELIGIONS BY REGION
Sources: Pew Research Center and World Religion Database.

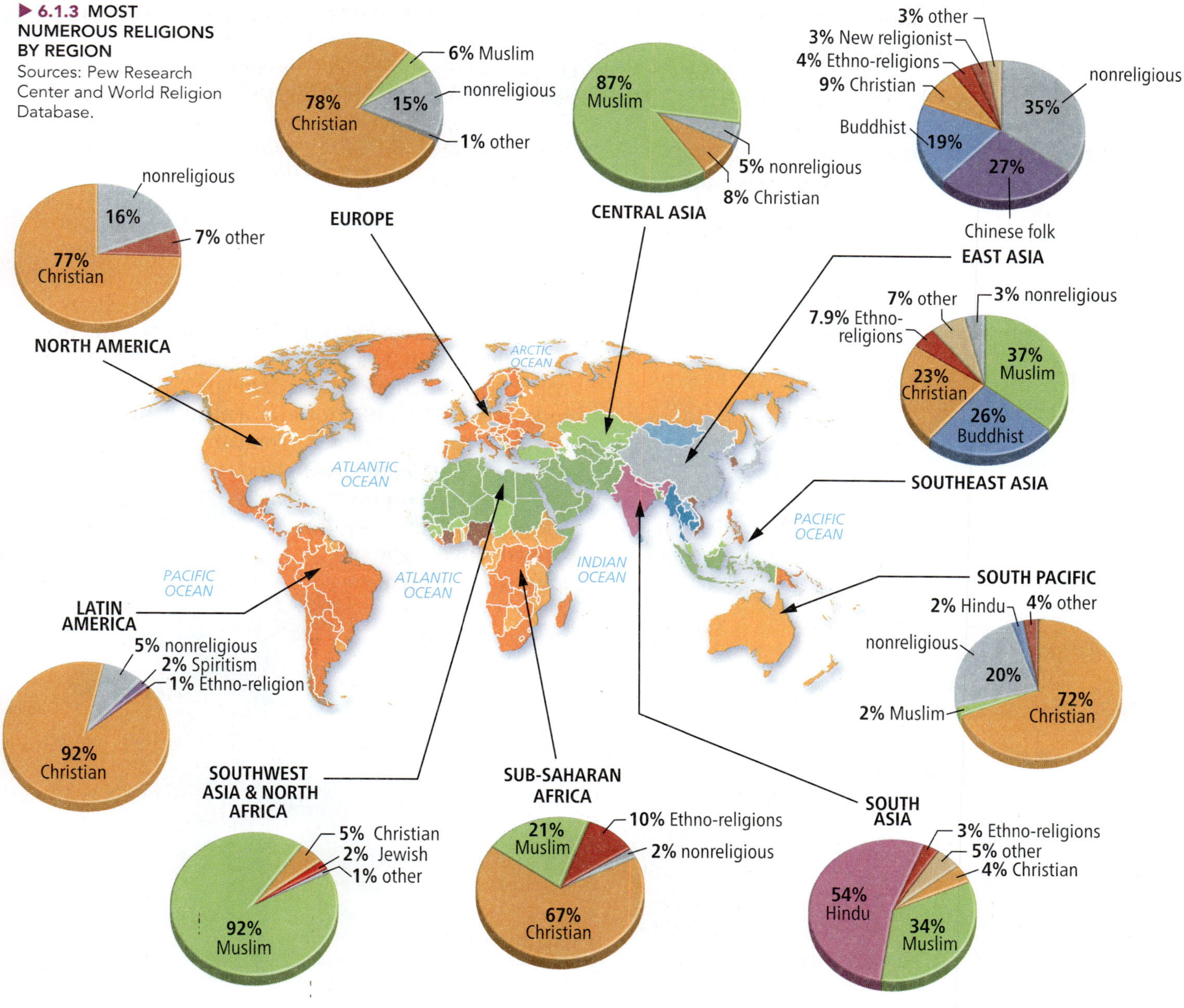

Distribution of Christians

- **Describe the distribution of Christianity's principal branches.**

Many religions, including the three most widely practiced universalizing religions, are divided into congregations, denominations, and branches. A **congregation** is a local assembly of persons brought together for common religious worship. A **denomination** unites a number of local congregations in a single legal and administrative body. A **branch** is a large and fundamental division within a religion.

▲ **6.2.1 ST. PAUL'S ANGLICAN CHURCH** Trinity, Newfoundland, Canada.

6.2.1 Branches of Christianity

Christianity has three major branches: Roman Catholic, Protestant (Figure 6.2.1), and Orthodox. In addition, many Christians belong to churches that do not consider themselves to be within any of these three branches. Roman Catholics comprise approximately 50 percent of the world's Christians and Orthodox around 12 percent. The other 38 percent of the world's Christians are divided between Protestants and others, but sources do not agree on the magnitude of each. The *Encyclopaedia Britannica* classifies 24 percent of the world's Christians as Protestant and 14 percent as other, whereas Pew classifies 37 percent as Protestant and only 1 percent as other.

Distribution of Branches in Europe

Overall in Europe, 47 percent of Christians are Roman Catholic, 18 percent are Protestant, and 35 percent are Orthodox. Roman Catholicism is the most widely practiced branch of Christianity in the southwest and east of Europe, Protestantism in the northwest of Europe (Figure 6.2.2), and Orthodoxy in the east and southeast of Europe.

The regions of Roman Catholic and Protestant majorities frequently have sharp boundaries, even when they run through the middle of countries. For example, Germany, the Netherlands, and Switzerland have approximately equal percentages of Roman Catholics and Protestants, but the Roman Catholic populations are concentrated in the south of these countries and the Protestant populations in the north.

The island of Eire (Ireland) was an English colony for many centuries and was made part of the United Kingdom in 1801. Following a succession of bloody confrontations, the Republic of Ireland achieved independence in 1937. However, a majority in the northern part of Eire voted to remain in the United Kingdom. The Republic of Ireland is 78 percent Roman Catholic, whereas the portion included in the United Kingdom, now known as Northern Ireland, is 44 percent Protestant and 41 percent Roman Catholic (Figure 6.2.3). Adherents of Northern Ireland's two religious branches have fought a series of wars for control of the territory.

▲ **6.2.2 MOST NUMEROUS BRANCHES OF CHRISTIANITY IN EUROPE**

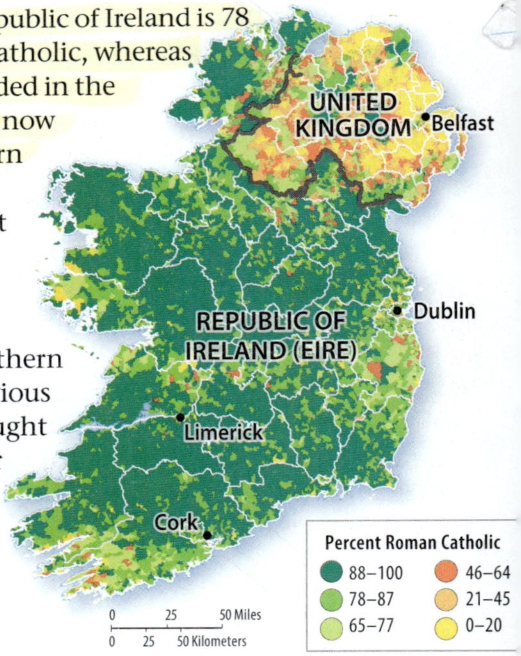

▶ **6.2.3 DISTRIBUTION OF ROMAN CATHOLICS IN IRELAND**

6.2.2 Distribution of Christian Branches in the Western Hemisphere

Christianity is by far the most widely practiced religion in the Western Hemisphere. Christians comprise 87 percent of the population of the Western Hemisphere. This includes 92 percent of Latin Americans and 77 percent of North Americans. People unaffiliated with any religion comprise 5 percent of Latin Americans and 16 percent of North Americans.

At the regional scale within the Western Hemisphere, a sharp boundary exists between North America and Latin America in the predominant branches of Christianity. Roman Catholics comprise 81 percent of Christians in Latin America and 32 percent in North America, whereas Protestants comprise 18 percent of Christians in Latin America and 63 percent in North America.

Religions of the United States

The diversity of faiths in the United States is displayed in Figure 6.2.4. Roman Catholics comprise 21 percent of the U.S. population, Evangelical Protestants 25 percent, Mainline Protestants 15 percent, historically Black churches 7 percent, Latter-day Saints and other Christians 2 percent each, and other faiths 6 percent.

The Southern Baptist Convention is the most numerous Evangelical Protestant church, and Methodist is the most numerous Mainline Protestant church. Roman Catholics are more numerous in the Northeast and Southwest, whereas Evangelical Protestants are most numerous in the Southeast. Around 23 percent of Americans are unaffiliated, and most of them respond that they are "nothing in particular."

Most of the historically Black congregations and churches formed before 1800 were founded by free Blacks. After slavery was abolished, segregationist attitudes among Whites discouraged and even prevented Blacks from worshiping in the same churches as Whites. Blacks most often established congregations and church facilities separate from their White neighbors. These new churches created communities and worship practices that were culturally distinct from other churches, including unique and empowering forms of Christianity that combined elements of Christian and African spiritual traditions.

Other less numerous faiths have distinctive distributions within the United States. Members of The Church of Jesus Christ of Latter-day Saints (Mormons) regard their church as separate from the three branches of Christianity. About 2 percent of Americans are members of the Latter-day Saints, and a large percentage is clustered in Utah and surrounding states.

Christian			70.6
Roman Catholic		20.8	
Evangelical Protestant		25.4	
Southern Baptist Convention	5.3		
Independent Baptist in the Evangelical Tradition	2.5		
Other Baptist in the Evangelical Tradition	1.4		
Lutheran in the Evangelical Tradition	1.5		
Assemblies of God	1.4		
Other Pentecostal in the Evangelical Tradition	2.2		
Restorationalist in the Evangelical Tradition (primarily Church of Christ)	1.6		
Nondenominational in the Evangelical Tradition	4.9		
Other Evangelical Protestant Traditions	5.4		
Mainline Protestant Churches		14.7	
Baptist in the Mainline Tradition	2.1		
Methodist in the Mainline Tradition	3.9		
Lutheran in the Mainline Tradition	2.1		
Presbyterian in the Mainline Tradition	1.4		
Anglican/Episcopal in the Mainline Tradition	1.2		
Other Mainline Protestant Traditions	4.0		
Historical Black Churches		6.5	
National Baptist Convention	1.4		
Other Baptist in the Historically Black Tradition	2.6		
Other Historically Black Churches	2.5		
Latter-day Saints		1.6	
Other Christian		1.6	
Jewish			1.9
Buddhist			0.7
Muslim			0.9
Hindu			0.7
Other religions or don't know			2.4
Unaffiliated			22.8

▶ **6.2.4** RELIGIONS OF THE UNITED STATES **Source:** Pew Research Center

Distribution of Buddhists & Muslims

- **Describe the distribution of the principal branches of Buddhism and Islam.**

Buddhism is clustered primarily in East Asia and Southeast Asia. Islam is the predominant religion of Central Asia and of Southwest Asia & North Africa. Like Christianity, Buddhism and Islam are divided into major branches with distinctive geographic distributions.

▲ 6.3.1 STATUE OF THE BUDDHA, LUMBINI, NEPAL

6.3.1 Foundation of Buddhism

The foundation of Buddhism is represented by the Four Noble Truths. These are traditionally identified as the first teaching given by the founder of Buddhism, Siddhartha Gautama, known as the Buddha (Figures 6.3.1 and 6.3.2):

1. All living beings must endure suffering.
2. Suffering, which is caused by a desire to live, leads to reincarnation (repeated rebirth in new bodies or forms of life).
3. The goal of all existence is to escape suffering and the endless cycle of reincarnation into Nirvana (a state of complete redemption), which is achieved through mental and moral self-purification.
4. Nirvana is attained through an Eightfold Path: rightness of belief, resolve, speech, action, livelihood, effort, thought, and meditation.

An accurate count of Buddhists is especially difficult because only a few people participate in Buddhist institutions. Religious functions are performed primarily by monks rather than by the general public. The number of Buddhists is also difficult to count because Buddhism, although a universalizing religion, differs in significant respects from the Western concept of a formal religious system. Someone can be both a Buddhist and a believer in other Eastern religions, whereas Christianity and Islam both require exclusive adherence. Most Buddhists in China and Japan, in particular, believe at the same time in an ethnic religion.

▲ 6.3.2 WORLD PEACE PAGODA IN LUMBINI, NEPAL, BIRTHPLACE OF THE BUDDHA

Branches of Buddhism

Buddhism split into more than one branch, as followers disagreed on interpreting statements by the founder, Siddhartha Gautama. The three main branches of Buddhism are Mahayana, Theravada, and Vajrayana (Figure 6.3.3). Mahayanists account for about 56 percent of Buddhists, primarily in China, Japan, and Korea. Theravadists comprise about 38 percent of Buddhists, especially in Cambodia, Laos, Myanmar, Sri Lanka, and Thailand. Vajrayanists, also known as Lamaists and Tantrayanists, comprise about 6 percent and are found primarily in Tibet and Mongolia.

▼ 6.3.3 DISTRIBUTION OF BRANCHES OF BUDDHISM

6.3.2 Distribution of Muslims

The word *Islam* in Arabic means "submission to the will of God," and it has a similar root to the Arabic word for "peace." An adherent of the Islamic religion is known as a Muslim, which in Arabic means "one who surrenders to God."

On a standard world map, such as Figure 6.1.3, Islam predominates in Central Asia and in Southwest Asia & North Africa. However, on a cartogram, most of the world's Muslims live further east, in South and Southeast Asia (Figure 6.3.4). The countries with the most Muslims are Indonesia, Pakistan, India, and Bangladesh. These four countries together are home to more than 40 percent of the world's Muslims.

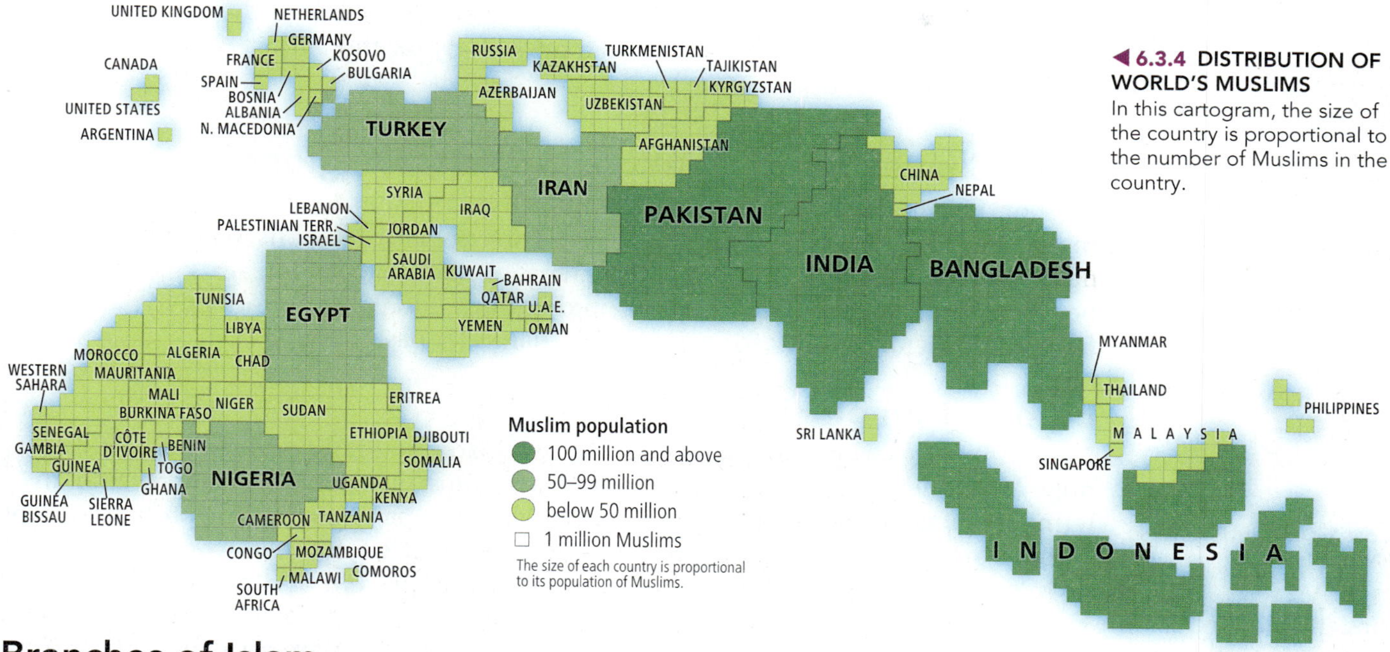

◀ **6.3.4 DISTRIBUTION OF WORLD'S MUSLIMS**
In this cartogram, the size of the country is proportional to the number of Muslims in the country.

Branches of Islam

Islam is divided into two principal branches: Sunni and Shiite. The word Sunni comes from the Arabic for "people following the tradition of Muhammad." The word Shiite (sometimes spelled Shia) comes from the Arabic word for "party" or "support group." Sunnis comprise 88 percent of Muslims and are the most numerous branch in most Muslim countries in Southwest Asia & North Africa, as well as in Southeast Asia.

Sunnis follow various schools of thought and religious law, which have distinctive regional distributions (Figure 6.3.5). The Hanafi, Hanbali, Maliki, and Shafi'i schools of thought and religious law are named for their founders.

Shiites are the largest branch in Azerbaijan, Bahrain, Iran, Iraq, Lebanon, and Yemen. Nearly 40 percent of all Shiites live in Iran, 15 percent in Pakistan, 12 percent in India, and 10 percent in Iraq. Shiite Islam is divided into three principal schools of thought, based in part on disputes over leadership succession after the Prophet Muhammad. The largest, known as Ithna Ashari (or Jafari), is the most widely followed tradition in Azerbaijan, Bahrain, Iran, Iraq, and Lebanon. Other traditions include the Ismaili and Zaidi. Ismailis are clustered in Pakistan and Zaidiyyahs in Yemen. A third branch of Islam, Ibadi, is the predominant form of Islam adhered to in Oman.

▼ **6.3.5 DISTRIBUTION OF BRANCHES OF ISLAM**

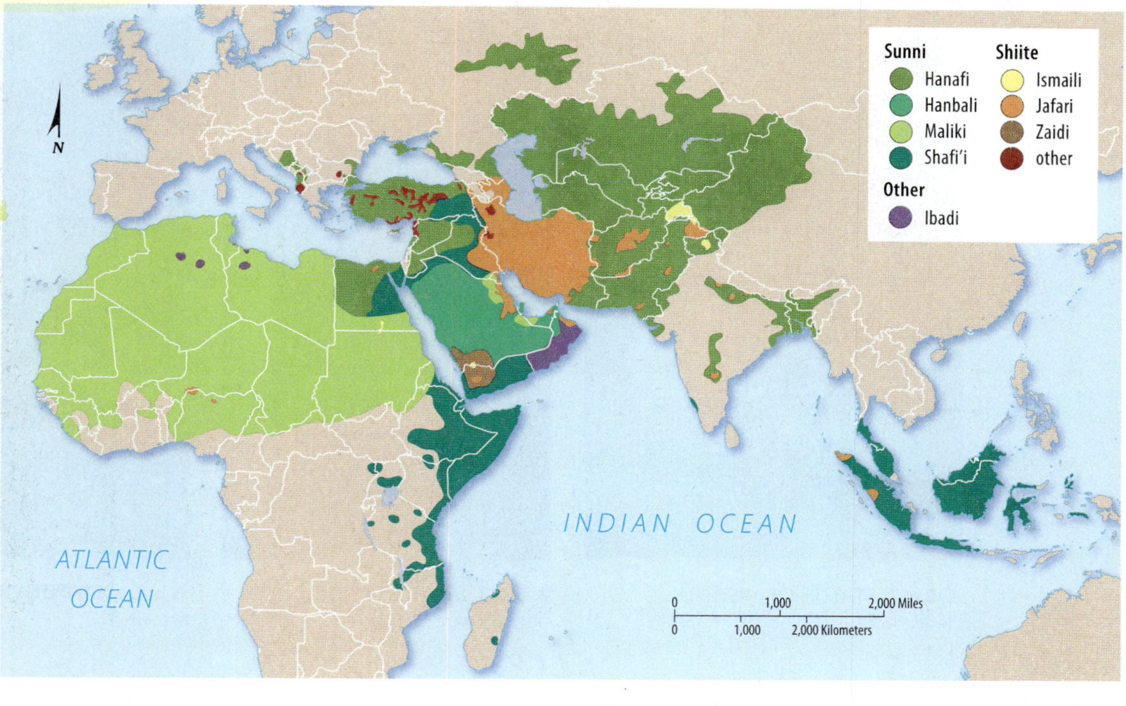

6.4 KEY ISSUE 1 Where are religions distributed?

Distribution of Ethnic Religions

- **Describe the distribution of Hinduism and other ethnic religions.**

Ethnic religions typically have much more clustered distributions than do universalizing religions. The ethnic religion with by far the largest number of followers is Hinduism.

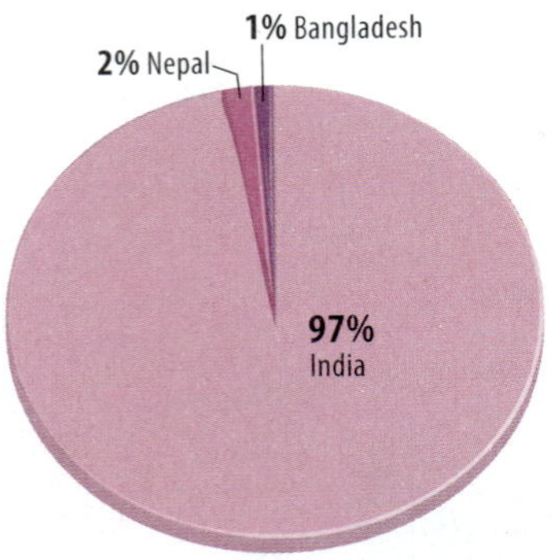

▲ 6.4.1 DISTRIBUTION OF HINDUS

6.4.1 Clustered Distribution of Ethnic Religions

Most of the adherents of the world's leading ethnic religions have remained embedded in the culture where they originated, unlike universalizing religions, which typically diffuse from one culture to another. Adherents of an ethnic religion are typically a people with a shared history, language, and destiny. Rituals are performed to pray for favorable agricultural conditions or to give thanks for past success.

Distribution of Hindus

Hinduism is the world's third-largest religion, with around 1 billion adherents. With no original founder or current leader, Hinduism is the most diverse of the major religions in terms of practices. The average Hindu has allegiance to a particular god or spiritual concept within a broad range of possibilities. Some revere one god, while others worship many.

In contrast to the large universalizing religions, 97 percent of Hindus are concentrated in just one country (India), 2 percent are in Nepal, 1 percent are in Bangladesh, and small numbers are elsewhere (Figure 6.4.1). Hindus comprise more than 80 percent of the population of India and Nepal and 49 percent of the population of Mauritius, a small island country in the Indian Ocean with only 1.3 million total inhabitants where Hindus from India have migrated for work.

Most Popular Hindu Gods

Many Hindus worship through a preferred god, who is a manifestation sent to help people communicate with the ultimate reality, known as Brahman. An estimated 80 percent follow Vaishnavism, which worships the god Vishnu, a loving god incarnated as Krishna (Figure 6.4.2). The second-largest is Shaivism, dedicated to Shiva, a god who can be both protective and destructive (Figure 6.4.3). The goddess Shakti is a creative source of energy (Figure 6.4.4).

▲ 6.4.2 STATUE OF VISHNU
Delhi, India.

▲ 6.4.3 STATUE OF SHIVA
Angkor Wat, Cambodia.

▲ 6.4.4 STATUE OF SHAKTI
Mauritius.

6.4.2 Chinese Folk Religions

Religions based in East Asia show the difficulty of classifying ethnic religions and counting adherents. Chinese folk religions comprise a diverse collection of traditional indigenous practices with roots in ancient Chinese culture. Core beliefs include a regard for heaven as the source of moral behavior and veneration of ancestors. Rituals serve as unifying activities, such as observing Chinese Lunar New Year (Figure 6.4.5).

Chinese folk religionists are **syncretic**, which means they combine ideas from several traditions. Many Chinese people believe and practice folk religions in addition to or alongside the more formal religions.

Adherence to Chinese folk religion is typically greater among the rural populations, whereas urban residents may adhere to Daoism or Confucianism (see Section 6.5.2). Practices may vary from one village to another and may be intermingled in a harmonious balance with the activities of the community's daily life, social relations, and politics.

▲ 6.4.5 CHINESE FOLK RELIGION
Celebration of Chinese New Year, Guizhou Province, China.

Ethno-religions

Approximately 10 percent of Africans are estimated to follow ethno-religions. African ethno-religions are sometimes called **animism**, which is belief that inanimate objects, such as plants and stones, or natural events, such as thunderstorms and earthquakes, are "animated," or have discrete spirits and conscious life.

Today, Africa is around 55 percent Christian—split about evenly among Roman Catholic, Protestant, and other—and around 35 percent Muslim. This distribution is in sharp contrast with the past. In 1900, more than 70 percent of Africans followed ethno-religions (Figure 6.4.6a). As recently as 1980, half of Africans—around 200 million people—were still classified as folk religionists. The growth in the two universalizing religions at the expense of ethnic religions reflects fundamental geographic differences between the two types of religions. Remaining ethno-religionists in Africa are clustered primarily in a belt that separates predominantly Muslim North Africa from what has become predominantly Christian sub-Saharan Africa (Figure 6.4.6b).

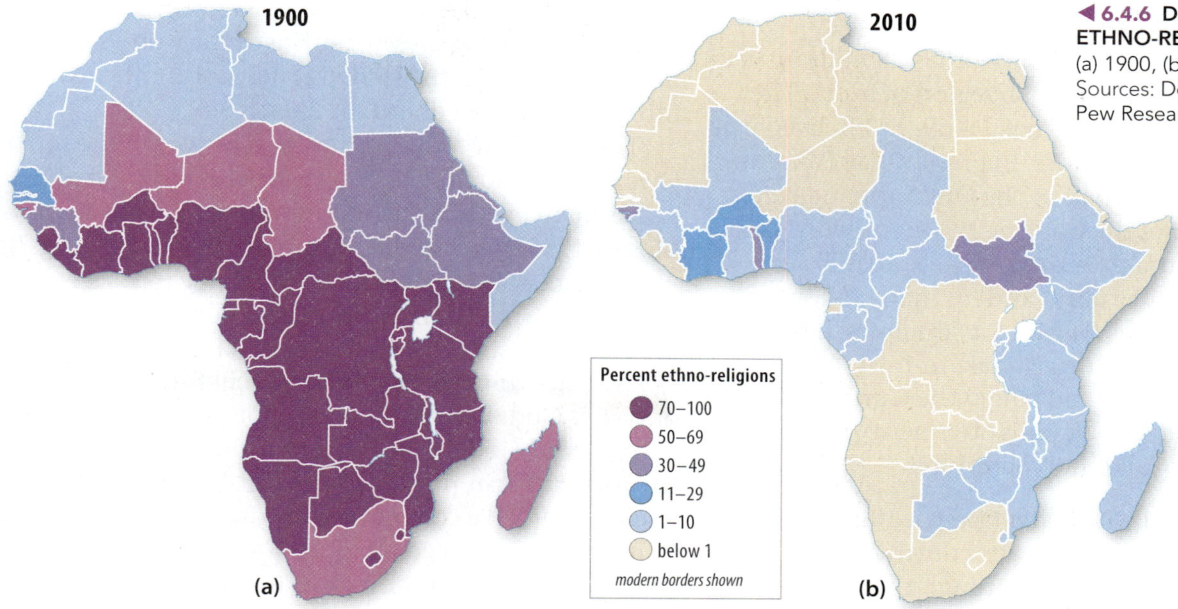

◀ 6.4.6 DISTRIBUTION OF ETHNO-RELIGIONS IN AFRICA
(a) 1900, (b) 2010
Sources: Decolonial Atlas (1900), Pew Research Center (2010).

Distribution of Other Religions

- **Describe the distribution of other established religions.**

Nine religions are briefly described in this section in order of the estimated numbers of adherents. The adherents of eight of these nine religions are highly clustered in one or two countries (Figure 6.5.1). The exception is Bahá'í.

6.5.1 Religions with at Least 15 million Adherents

Four religions, in addition to the universalizing and ethnic religions, have at least 15 million adherents: New Religion, Sikhism, Spiritism, and Judaism.

New Religions. Several religions created in the nineteenth and twentieth centuries are grouped into an umbrella designation of New Religions. A New Religion is an example of a syncretic religion, which means it combines several traditions.

Around 66 million people follow a New Religion, nearly all in East and Southeast Asia. Two examples of New Religions are Tenrikyo and Cao Dai.

Originally regarded as a branch of Shinto, Tenrikyo was organized as a separate religion in 1854 by a woman named Nakayama Miki (1798–1887). Cao Dai was founded in Vietnam during the 1920s. The number of adherents increased after Vietnam's Communist government granted Cao Dai legal status in 1997.

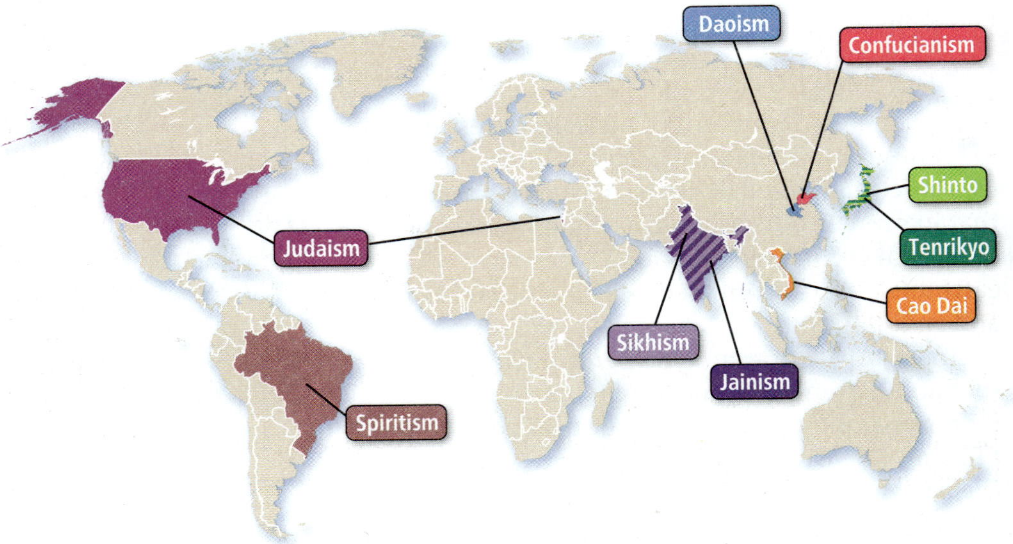

▲ 6.5.1 CLUSTERED RELIGIONS WITH AT LEAST 3 MILLION ADHERENTS

Sikhism. All but 3 million of the world's 28 million Sikhs are clustered in the Punjab region of India. The founder of Sikhism, Guru Nanak (1469–1538), lived in a village near the city of Lahore, in present-day Pakistan. God was revealed to Nanak as The One Supreme Being, or Creator, who rules the universe by divine will. Nanak traveled widely through South Asia around 500 years ago, preaching his new faith, and his many followers became known as Sikhs (Hindu for "disciples"). Nine other gurus succeeded Guru Nanak. In 1604, Arjan, the fifth guru, compiled and edited the Guru Granth Sa-hib (the Holy Granth of Enlightenment), which became the book of Sikh holy scriptures.

Spiritism. Spiritism is the belief that the human personality continues to exist after death and can communicate with the living through the agency of a medium or psychic. Most of the world's 15 million Spiritists reside in Brazil.

Judaism. The name Judaism derives from Judah, one of the patriarch Jacob's 12 sons; Israel is another biblical name for Jacob. Roughly two-fifths of the world's 15 million Jews live in the United States and another two-fifths in Israel (refer ahead to Figure 6.8.6).

The Tanakh recounts the ancient history of the Jewish people and the laws of the Jewish faith. Tanakh is an acronym for Torah (also known as the Five Books of Moses), Nevi'im ("Prophets"), and Ketuvim ("Writings").

Judaism plays a more substantial role in Western civilization than its number of adherents would suggest. Judaism is the first recorded religion to espouse **monotheism**, which is belief that there is only one God. Fundamental to Judaism is belief in one all-powerful God. Judaism offered a sharp contrast to the **polytheism** practiced by neighboring people, who worshiped a collection of gods. The world's two most widely practiced religions—Christianity and Islam—find some of their roots in Judaism.

6.5.2 Religions with between 3 & 9 Million Adherents

Five religions have an estimated 3 to 9 million adherents. Daoism, Confucianism, and Bahá'í each have approximately 9 million adherents, Jainism has around 6 million, and Shintoism has around 3 million.

Daoism. Laozi (born 571 B.C.E., date of death unknown, also spelled Lao-tzu) developed Daoism (also spelled Taoism). Although a government administrator by profession, Laozi's writings emphasized the mystical and magical aspects of life rather than the importance of public service, which Confucius had emphasized. Tao, which means "the way" or "the path," cannot be comprehended by reason and knowledge because not everything is knowable. It emphasizes the importance of studying nature to find one's place in the world instead of striving to change the world.

▲ **6.5.2 BAHÁ'Í HOUSE OF WORSHIP**
Delhi, India.

Confucianism. Confucius (551–479 B.C.E.) was a philosopher and teacher in the Chinese province of Lu. His sayings, which were recorded by his students, emphasized the importance of the ancient Chinese tradition of *li*, which can be translated roughly as "propriety" or "correct behavior." Confucianism prescribed a series of ethical principles for the orderly conduct of daily life in China, such as following traditions, fulfilling obligations, and treating others with sympathy and respect. These rules applied to China's rulers as well as to their subjects.

Bahá'í. Roughly equal numbers of Bahá'ís are found in India, other Asian countries, Africa, and the Western Hemisphere (Figure 6.5.2). A universalizing religion, Bahá'í was established in Shíráz, Iran, in 1844. It grew out of the Bábi faith, under the leadership of Siyyid 'Ali Muhammad, known as the Báb (Persian for "gateway"). Bahá'í provoked strong opposition from Shiite Muslims, and the Báb was executed in 1850, as were 20,000 of his followers. Bahá'ís believe that one of the Báb's disciples, Husayn 'Ali Nuri, known as Bahá'u'lláh (Arabic for "Glory of God"), was the prophet and messenger of God. Bahá'u'lláh's function was to overcome the disunity of religions and establish a universal faith through abolition of racial, class, and religious prejudices. Bahá'u'lláh was arrested and then exiled. In 1863, his claim that he was the messenger of God anticipated by the Báb was accepted by other followers. Before he died in 1892, Bahá'u'lláh appointed his eldest son, 'Abdu'l-Bahá (1844–1921), to be the leader of the Bahá'í community and the authorized interpreter of his teachings.

Jainism. Jainism originated in South Asia around 2,500 years ago. Its membership declined with the spread of Buddhism and Hinduism followers in the region, especially since the eighth century C.E. Jains believe that nonviolence and self-control are the means to achieve liberation (Figure 6.5.3). India is the home to 95 percent of the world's 4 million Jains, although Jain centers are located in 25 of the 50 U.S. states.

▼ **6.5.3 JAINISM**
Shikharji Temple, Jharkhand, India.

Shinto. Shinto, Japan's ethnic religion, is strongly rooted in the cultural history of the country. Japanese government statistics report around 100 million Shintos, or 78 percent of the country's population. However, in opinion polls only 4 million Japanese, or 3 percent of the population, identify themselves as Shinto. The large discrepancy stems in part from the fact that a seventeenth-century law in Japan assigns Shinto organizations with the task of maintaining records on Japanese citizens, so many people who do not self-identify as Shinto are recorded as Shinto. The discrepancy also stems from the perception by some Japanese people that Shinto is a cultural feature rather than a religion.

Geographic Origin of Religions

- **Recount the place of origin for four important religions.**

The current global and regional distributions of religions and their branches result from geographic processes of origin and diffusion. These processes may be well defined or contested.

▲ 6.6.1 **OLDEST HINDU TEMPLE IN NEPAL**
Parts of Changu Narayan Temple in Changunarayan, Nepal, date from 464 C.E.

6.6.1 Origin of Hinduism & Buddhism in South Asia

Geographic elements of the origin for the four most widely followed religions are discussed next in chronological order from the oldest to the most recent.

Unknown Date of Origin of Hinduism

As an ethnic religion based in India, Hinduism has unknown origins. The word *Hinduism* originated in the sixth century B.C.E. to refer to people living in what is now India, but religious practices existed prior to recorded history. Elements of Hinduism may have originated in the Indus Valley civilization, which flourished between approximately 2500 B.C.E. and 1500 B.C.E. along the Indus River in present-day eastern Pakistan. Archaeological evidence of possible rituals from that era includes bathing rituals, animal sacrifices, and sculptures that may depict Hindu gods.

Aryan tribes from Central Asia invaded South Asia about 1400 B.C.E. and brought with them Indo-European languages, as discussed in Chapter 5. In addition to their language, the Aryans brought their religion. The Aryans first settled in the area now called the Punjab in northwestern India and later migrated east to the Ganges River valley, as far as Bengal. Centuries of intermingling with the Dravidians already living in the area modified their religious beliefs.

The earliest surviving Hindu texts, known as Vedas, were written around 1100 B.C.E. Some rituals from this era survive in contemporary Hinduism, including worship of various gods representing Earth, atmosphere, and sky. By around 2,000 years ago, key texts were composed and rituals were developed that remain central to contemporary Hinduism. The earliest surviving Hindu temples were constructed between 1,500 and 2,000 years ago (Figure 6.6.1).

Origin of Buddhism in Present-Day Nepal

The founder of Buddhism, Siddhartha Gautama, was born about 563 B.C.E. in Lumbini in present-day Nepal, near the border with India (Figure 6.6.2). The son of a lord, he led a privileged existence, sheltered from life's hardships. Gautama had a beautiful wife, palaces, and servants.

According to Buddhist tradition, Gautama's life changed after a series of four trips. He encountered a decrepit old man on the first trip, a disease-ridden man on the second trip, and a corpse on the third trip. After witnessing these scenes of pain and suffering, Gautama began to feel he could no longer enjoy his life of comfort. On a fourth trip, Gautama saw a monk, who taught him about withdrawal from the world.

At age 29, Gautama left his palace one night and lived in a forest for the next 6 years, thinking and experimenting with forms of meditation. Gautama emerged as the Buddha, the "awakened or enlightened one," and spent 45 years preaching his views across India. In the process, he trained monks, established orders, and preached to the public.

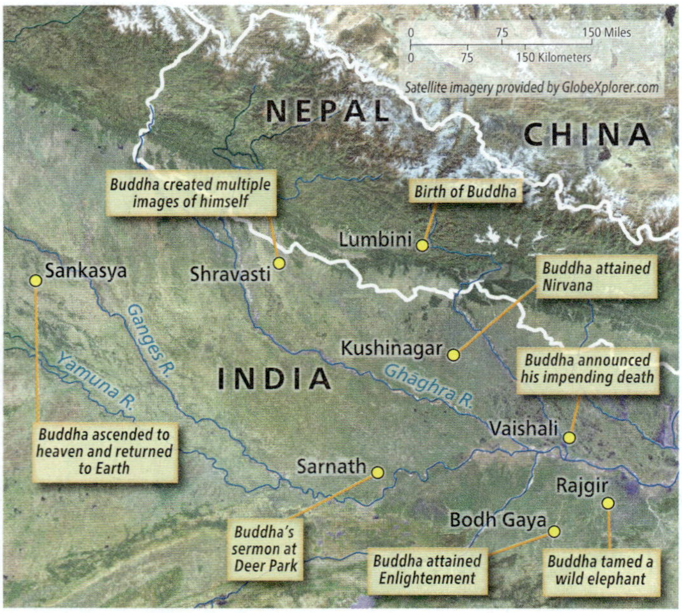

◀ 6.6.2 **HOLY PLACES IN BUDDHISM**
Buddha and his disciples would have walked from place to place amid the river valleys and plains of northern India, then a region of farming villages divided among several kingdoms.

6.6.2 Origin of Christianity & Islam in Southwest Asia

The two religions with the most adherents—Christianity and Islam—both originated in Southwest Asia. It is typical of universalizing religions such as Christianity and Islam that their places of origin are known and derived from events in the life of a specific person.

Origin of Christianity

Christianity was founded upon the teachings of Jesus, who was born in Bethlehem between 8 and 4 B.C.E. and died on a cross in Jerusalem about 30 C.E. Raised as a Jew, Jesus gathered a small group of disciples as well as a large group of followers and preached the coming of the Kingdom of God. The four Gospels of the Christian Bible—Matthew, Mark, Luke, and John—document miracles and extraordinary deeds that Jesus is believed to have performed. He was referred to as Christ, from the Greek word for the Hebrew word *messiah*, which means "anointed."

According to the Gospels, in the third year of his mission, Jesus was betrayed to the authorities by Judas Iscariot, one of the twelve Disciples. After sharing the Last Supper (possibly the Jewish Passover Seder) with his disciples in Jerusalem, Jesus was arrested by the Romans and put to death as an agitator. On the third day after his death, his tomb was found empty (Figure 6.6.3). Christians believe that Jesus died to atone for human sins, that he was raised from the dead by God, and that his resurrection from the dead provides people with hope for salvation.

▲ **6.6.3 ORIGIN OF CHRISTIANITY**
The tomb, known as the Edicule, inside the Church of the Holy Sepulchre in Jerusalem was erected at the place where Christians accept that Jesus was buried and resurrected.

Origin of Islam

Like other universalizing religions, Islam arose from the teachings of a historical founder, Muhammad. Muhammad was born in Makkah about 570. At age 40, while engaged in a meditative retreat, Muhammad is believed by Muslims to have received his first revelation from God through the angel Gabriel. The Quran, the holiest book in Islam, is accepted by Muslims to be a record of God's words, as revealed to the Prophet Muhammad through Gabriel. Arabic is a lingua franca, or language of communication, within the Islamic world because it is the language in which the Quran is written.

Islam teaches that as he began to preach the truth that God had revealed to him, Muhammad and his followers suffered persecution, and in 622, he was commanded by God to emigrate. His migration from Makkah to the city of Yathrib—an event known as the Hijra (from the Arabic word for "migration," sometimes spelled *hegira*)—marks the beginning of the Islamic calendar. Yathrib was subsequently renamed Madinah, Arabic for "the City" (Figure 6.6.4). After several years, Muhammad and his followers returned to Makkah and established Islam as the city's religion. By Muhammad's death, in 632 at about age 63, Islam had spread through most of present-day Saudi Arabia.

According to the Islamic narrative, Muhammad is a descendant of Abraham. According to the Jewish Torah and Christian Old Testament, Abraham married Sarah, who did not bear children. As polygamy (having more than one wife) was a custom of the culture, Abraham then married Hagar, who bore a son, Ishmael. Sarah was later able to conceive, and she bore a son, Isaac. Jews and Christians trace their story through Abraham's wife Sarah and her son Isaac. Muslims trace their story through his wife, Hagar, and her son Ishmael. The Islamic narrative tells that Abraham brought Hagar and Ishmael to Makkah, in present-day Saudi Arabia.

▲ **6.6.4 ORIGIN OF ISLAM**
Muhammad is buried in Al-Masjid an-Nabawī (Prophet's Mosque) in Madinah, Saudi Arabia.

Historical Diffusion of Religions

- **Analyze the process of diffusion of universalizing religions.**

The three widely practiced universalizing religions have diffused from specific hearths, or places of origin, to other regions of the world. In contrast, most ethnic religions have typically remained clustered.

6.7.1 Early Diffusion of the Three Largest Universalizing Religions

The hearths where each of the three largest universalizing religions originated are based on the events in the lives of the three key individuals. Followers transmitted the messages preached in the hearths to people elsewhere, diffusing them across Earth's surface along distinctive paths, as shown in Figure 6.7.1.

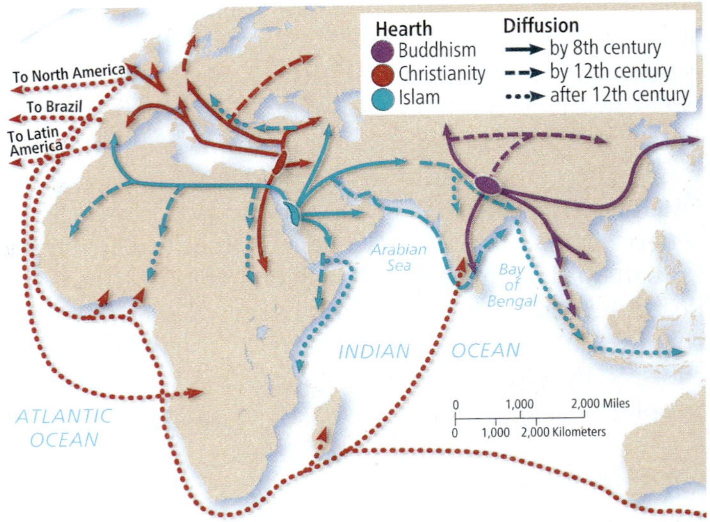

▲ 6.7.1 DIFFUSION OF UNIVERSALIZING RELIGIONS

Diffusion of Christianity

Christianity's diffusion has been rather clearly recorded since the time of Jesus in the Roman province known at the time as Judea.

In Chapter 1, two processes of diffusion were identified: relocation diffusion and expansion diffusion. Within expansion diffusion, we distinguished between hierarchical diffusion (diffusion through key leaders) and contagious diffusion (widespread diffusion). Christianity diffused through a combination of these forms of diffusion.

Christianity first spread from its hearth in Judea through relocation diffusion. **Missionaries**—individuals who help to transmit a universalizing religion through relocation diffusion—carried the teachings of Jesus along the Roman Empire's protected sea routes and excellent road network to people in other locations (Figure 6.7.2).

▼ 6.7.2 DIFFUSION OF CHRISTIANITY
Christianity began to diffuse throughout Europe during the time of the Roman Empire and continued after the empire's collapse. Paul of Tarsus, a disciple of Jesus, traveled through the eastern Roman Empire as a missionary.

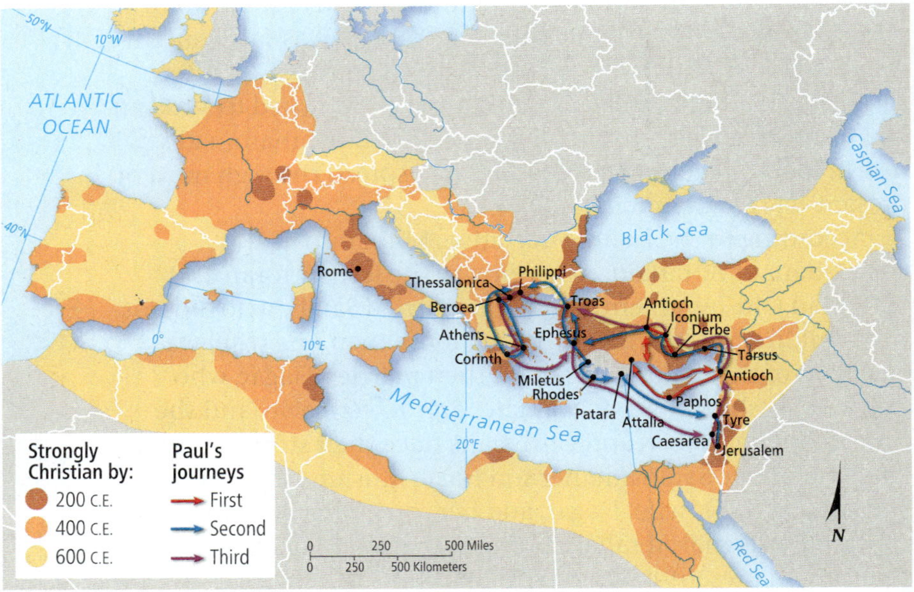

The dominance of Christianity throughout the Roman Empire was assured during the fourth century through the two types of expansion diffusion. Christianity spread widely within the Roman Empire through contagious diffusion—daily contact between believers in the towns and nonbelievers in the surrounding countryside. The spread of Christianity was further encouraged through hierarchical diffusion, especially once the Roman Empire's key elite figure, Emperor Constantine (ca. 272–337 C.E.), embraced the religion in 313 C.E., and another elite figure, Emperor Theodosius, (347–395 C.E.) proclaimed it the empire's official religion in 380 C.E.

6.7.2 Historical Diffusion of Buddhism

Buddhism diffused from its point of origin in South Asia to East Asia (Figure 6.7.3). Most responsible for the early spread of Buddhism was Ashoka, emperor of the Maurya Empire from about 268 to 232 B.C.E. The Maurya Empire formed the nucleus of several powerful kingdoms in South Asia between the sixth century B.C.E. and the eighth century C.E.

In the first century C.E., merchants along the trading routes from northeastern India introduced Buddhism to China (Figure 6.7.4). Many Chinese were receptive to the ideas brought by Buddhist missionaries, and Buddhist texts were translated into Chinese languages. Chinese rulers allowed their people to become Buddhist monks during the fourth century, and Buddhism spread from Korea to Japan two centuries later. During the same era, Buddhism lost its original base of support in India although Sri Lanka remains partially Buddhist.

▼ 6.7.3 ORIGIN OF BUDDHISM
Mahabodhi (Great Awakening) Temple is in Bodh Gaya, the site in present-day India where Siddhartha Gautama, known as the Buddha, attained enlightenment.

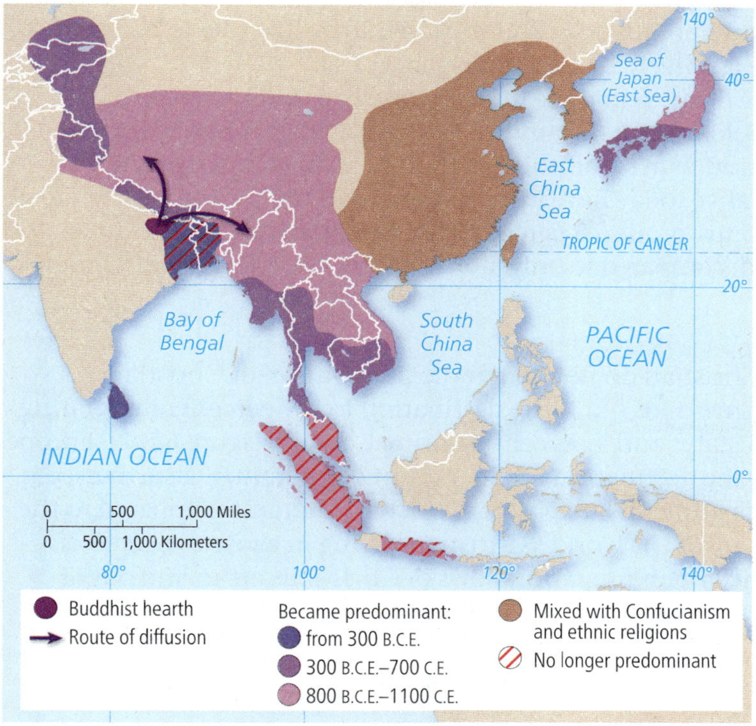

◀ 6.7.4 DIFFUSION OF BUDDHISM

Historical Diffusion of Islam

Muhammad's successors organized followers who extended Islamic control over an extensive area of Africa, Asia, and Europe through military conquest (Figure 6.7.5). Within a century of Muhammad's death, Muslims controlled Palestine, the Persian Empire, and much of India, resulting in the conversion of many non-Arabs to Islam, often through intermarriage.

To the west, Muslims diffused across North Africa, crossed the Strait of Gibraltar, and retained much of present-day Spain, until 1492. During the same century in which the Christians regained all of Western Europe, Muslims took control of much of southeastern Europe and Turkey.

◀ 6.7.5 DIFFUSION OF ISLAM
Islam diffused rapidly from its point of origin in present-day Saudi Arabia. Within 200 years, Muslims controlled much of Southwest Asia & North Africa as well as southwestern Europe.

Recent Diffusion of Religions

- **Describe contemporary migration patterns of several religious groups.**

The distribution of religions continues to be altered in modern times by migration. Although most people do not migrate primarily for religious reasons, when they do migrate, they take their religion with them.

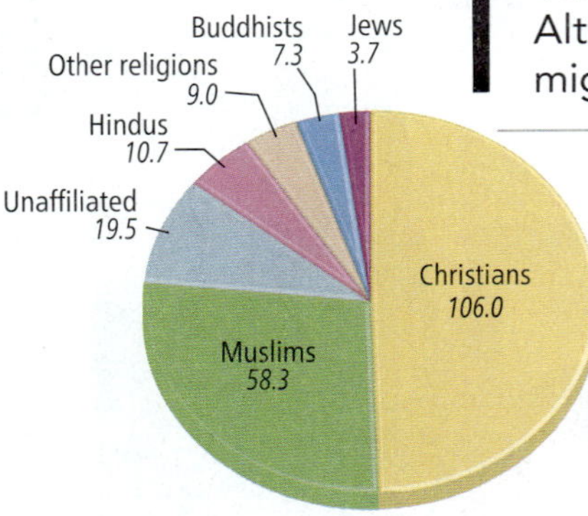

▲ **6.8.1 RELIGION OF INTERNATIONAL MIGRANTS**
Numbers are in millions of people. Compare the pie chart to Figure 6.1.2. Christians comprise less than one-third of the world's population but around half of international migrants.

6.8.1 Religious Composition of Migrants

The religious composition of international migrants does not match the overall share of adherents of various religions. Christians comprise only one-third of the world's population but account for half of the world's international migrants. Muslims and Jews also comprise higher percentages of the world's migrants than their shares of the world's population. Buddhists, Hindus, folk religionists, and unaffiliated people are less likely to migrate (Figure 6.8.1).

Jews are much more likely to migrate internationally than their share of the world's population would suggest. Around 25 percent of all Jews have migrated from one country to another at some point in their lives. In comparison, only 3 percent of all people alive today have migrated internationally. Among the more numerous religious groups, around 5 percent of all Christians and 4 percent of all Muslims have migrated, compared to only 1 or 2 percent of other religious groups.

Migration Patterns of Christians

The destinations of international migrants who are Christian do not match the distribution of Christians. North America is home to 12 percent of the world's Christians but is the destination for 34 percent of migrating Christians, including 30 percent entering the United States and 4 percent entering Canada (Figure 6.8.2). Europe is home to 26 percent of the world's Christians but is the destination for 38 percent of migrating Christians. On the other hand, Latin America and sub-Saharan Africa attract relatively few Christian migrants compared to the share of the world's Christians living in those two regions. The largest migration flows of Christians are in and out of Russia and into the United States (Figure 6.8.3).

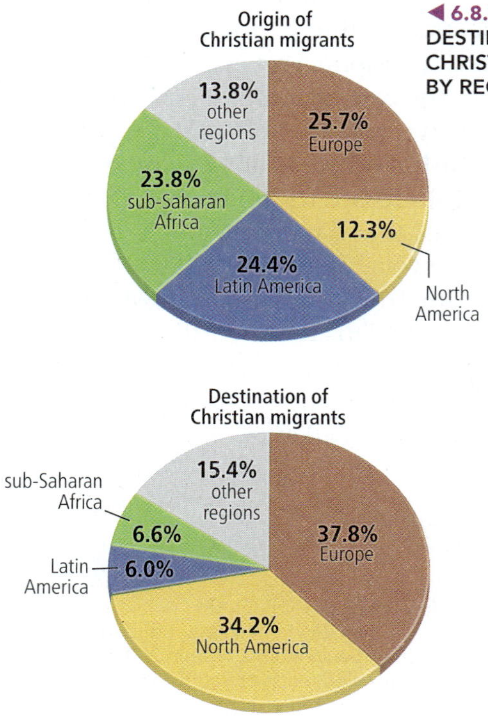

◀ **6.8.2 ORIGIN AND DESTINATION OF CHRISTIAN MIGRANTS BY REGION**

▼ **6.8.3 MIGRATION OF CHRISTIANS BY LARGEST FLOWS**

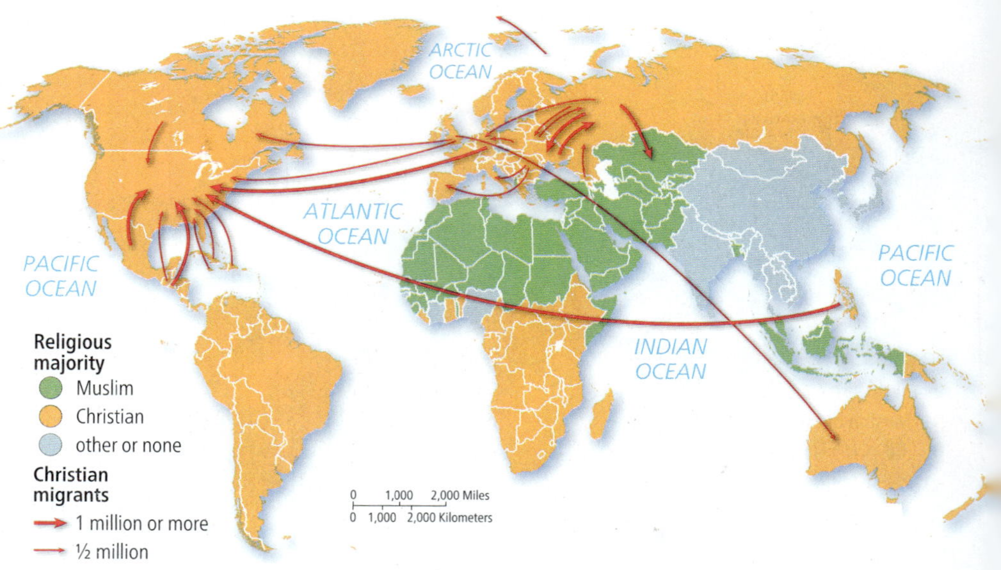

6.8.2 Migration Patterns of Muslims

Southwest Asia & North Africa is home to only 20 percent of the world's Muslims but attracts 34 percent of the Islamic migrants. Saudi Arabia is the country that attracts the largest number of Islamic migrants; Egypt sends the largest number of Islamic migrants to Saudi Arabia (Figure 6.8.4).

Europe, with only 3 percent of the world's Muslims, attracts 34 percent of Muslims who migrate internationally (Figure 6.8.5). Russia, Germany, and France are all world leaders in attracting Muslims. The largest numbers of Muslims who have migrated to Europe have come from Turkey to Germany and from Algeria to France. Morocco has been the source of a large number of Islamic migrants to France and Spain. North America receives only around 1 percent of Muslim migrants.

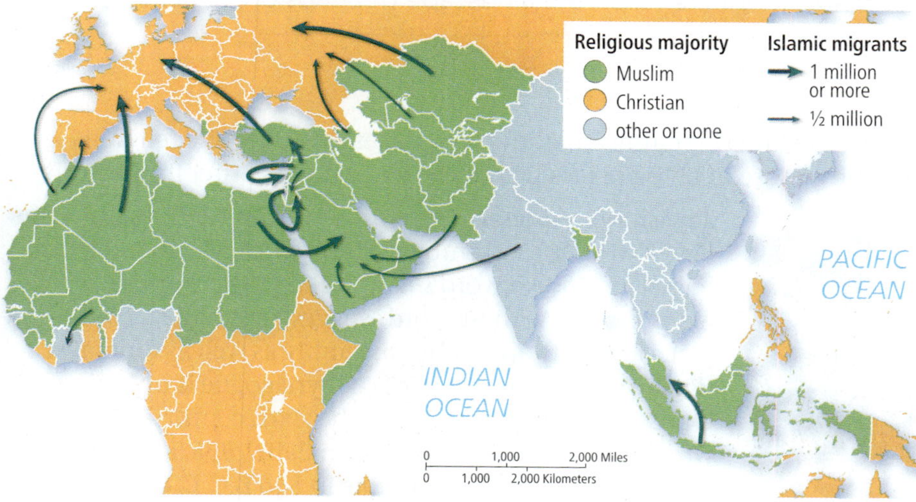

▲ **6.8.4 LARGEST FLOWS OF ISLAMIC MIGRANTS**

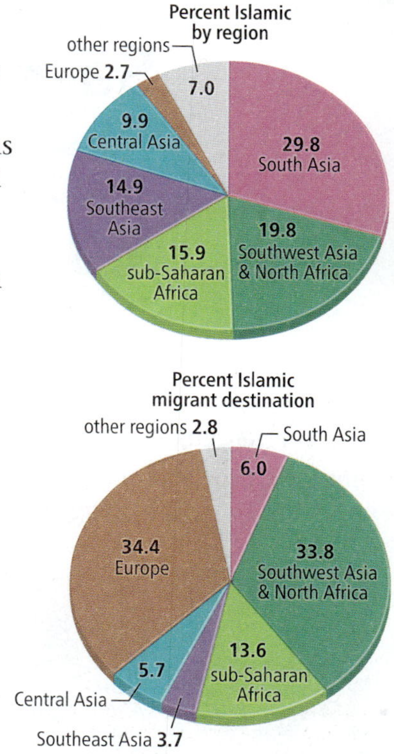

▲ **6.8.5 PERCENT ISLAMIC BY REGION AND DESTINATION OF ISLAMIC MIGRANTS BY REGION**

Migration Patterns of Jews

Israel is the destination for 73 percent of Jews who migrate internationally. As the world's only state with a Jewish majority, Israel offers an especially strong pull factor for Jewish migrants.

Only in the modern period, since the creation of the State of Israel in 1948, has a significant percentage of the world's Jews lived in their ancestral homeland (Figure 6.8.6). Jews trace their origins as a people to the land of Israel and King David's establishment of the Kingdom of Israel around 3,000 years ago. Most Jews have not lived there since 70 C.E., when the Romans forced them to disperse throughout the world, an action known as the diaspora, from the Greek word for "dispersion."

Many Jews migrated from the eastern Mediterranean to Europe. Other nationalities often persecuted the Jews living in their midst. Historically, the Jews of many European countries were forced to live permanently in **ghettos**, defined as city neighborhoods set up by law to be inhabited only by Jews. The term *ghetto* originated during the sixteenth century in Venice, Italy, as a reference to the city's former copper foundry or metal-casting district. Persecution and rising nationalist movements in Europe sparked the growth of Jewish nationalism, known as Zionism. This led to increased Jewish migration to the land of Israel (then under Ottoman rule) in the late nineteenth and early twentieth centuries.

During the Holocaust (1939–1945), the Nazis systematically murdered 6 million Jews, about two-thirds of the Jews of Europe. After its establishment in 1948, the State of Israel absorbed several hundred thousand Jewish refugees from Europe as well as from Southwest Asia & North Africa. Today, less than 10 percent of the world's 15 million Jews live in Europe, compared to 90 percent a century ago.

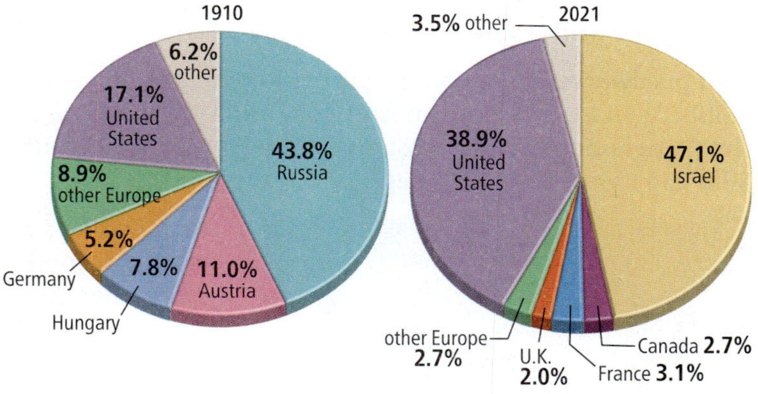

▲ **6.8.6 DISTRIBUTION OF JEWS, 1910 AND 2021**
In 1910, most Jews lived in Europe (including Russia). Now most live in Israel or in the United States.

Sacred Spaces

- **Describe places of worship and of pilgrimage for several religions.**

Geographers study the ways in which religious beliefs and practices can have a significant impact on the landscape through the organization of space and people. The impact of religion is clearly seen in the arrangement of human activities on the landscape at several scales, from relatively small parcels of land to entire regions. How each religion distributes its elements on the landscape depends on its beliefs.

6.9.1 Places of Worship

Church, *basilica*, *mosque*, *temple*, and *synagogue* are familiar names that identify places for collective religious expression in various religions. They share the feature of being places where people come together for prayer.

▲ **6.9.1 JEWISH SYNAGOGUE**
Six Dome Synagogue, Quba, Azerbaijan.

Jewish synagogues. The word *synagogue* derives from the Greek word for "assembly." The building is often referred to by the Yiddish word *shul*, similar to the German word for "school." As these words suggest, a synagogue is a place for study and public assembly as well as for prayer (Figure 6.9.1). The origin of the synagogue is unknown; it possibly came to be during the sixth century B.C.E., when Jews were living in exile in Babylonia, after the destruction of the First Temple in 586 B.C.E. (refer to Section 6.13.2). Synagogues took on more importance as the place for communal prayer after the destruction of the Second Temple in 70 C.E.

Christian churches. The Christian landscape is dominated by a high density of churches. The church is an expression of religious principles, an environment in the image of God. The word *church* derives from a Greek term meaning "lord," "master," and "power." Church also refers to a gathering of believers as well as the building at which the gathering occurs. Since Christianity split into many denominations, no single style of church construction has dominated. Churches reflect both the cultural values of the denomination and the region's architectural heritage. Availability of building materials also influences church construction.

Islamic mosques. The word *mosque* derives from the Arabic for "place of worship." Muslims consider a mosque to be a location for the community to gather together for worship. Mosques are found primarily in larger cities of the Islamic world; simple structures may serve as places of prayer in rural villages.

Sikh gurdwaras. Sikhs come together for worship at a gurdwara. The most important gurdwara is the Harmandir Sahib, or Golden Temple, in Amritsar, India (Figure 6.9.2). The Golden Temple houses Sikhism's holiest book, the Guru Granth Sahib. Most gurdwaras imitate the layout of the Golden Temple. A gurdwara is identified by the Sikh flag flying from a tall flagpole.

Bahá'í houses of worship. Bahá'ís have built a House of Worship in every continent to dramatize that Bahá'í is a universalizing religion with adherents all over the world. A Bahá'í House of Worship has been built in Illinois, Australia, Uganda, Lagenhaim (Germany), Panama, Samoa, and India. Every Bahá'í House of Worship is required to be built in the shape of a nonagon (nine-sided).

◀ **6.9.2 SIKH GURDWARA**
The Harmandir Sahib, known informally as the Golden Temple, Amritsar, India.

6.9.2 Pilgrimages to Religious Shrines

A **pilgrimage** is a journey for religious purposes to a place considered sacred, and pilgrimages are rituals in some religions. Buddhism and Islam are the universalizing religions that place the most emphasis on identifying shrines, or holy places. Shrines are holy and destinations of pilgrimages because they are the locations of important events in the life of Buddha or Muhammad.

Pilgrimages in Islam. Muslims are especially encouraged to make pilgrimages to visit holy places in accordance with recommended itineraries. Every healthy Muslim who has adequate financial resources is expected to undertake a pilgrimage, called a *hajj*, to Makkah (Mecca), which is the holiest city for Muslims because it is the birthplace of Muhammad. The word *mecca* now has a general meaning in the English language as a goal sought or a center of activity.

Now a city of 2 million inhabitants, Makkah contains the holiest object in the Islamic landscape, namely al-Ka'ba, a cube-like structure encased in silk, which stands at the center of the Great Mosque, Masjid al-Haram, Islam's largest mosque (Figure 6.9.3). The Ka'ba, thought by Muslims to have been built by Abraham and Ishmael, contains a black stone that Muslims believe was given to Abraham by Gabriel as a sign of a covenant with Ishmael and the Muslim people. According to Islamic tradition, the Masjid al-Haram mosque also contains the well of Zamzam, considered to have the same source as the water given to Hagar by the angel Gabriel to quench the thirst of her infant, Ishmael.

Regardless of nationality and economic background, most pilgrims dress alike, in plain white robes, to emphasize common loyalty to Islam and the equality of people in the eyes of Allah. A precise set of rituals is practiced, culminating in a visit to the Ka'ba. The hajj attracts millions of Muslims annually to Makkah. Hajj visas are issued by the government of Saudi Arabia according to a formula of 1 per 1,000 Muslims in a country. Roughly 80 percent come from Southwest Asia & North Africa and 20 percent from elsewhere in Asia. Although Indonesia is the country with the most Muslims, it has not sent the largest number of pilgrims to Makkah because of the relatively long travel distance.

▲ **6.9.3 MASJID AL-HARAM, MAKKAH, SAUDI ARABIA**
Muslims who undertook a hajj surround al-Ka'ba.

Pilgrimages in Buddhism. Eight places in a small area of northeastern India and southern Nepal are holy to Buddhists because of their association with events in the Buddha's life. Perhaps the most holy is Lumbini in Nepal, where Buddha was born around 563 B.C.E. Deer Park in Sarnath, India, is where the Buddha gave his first sermon (Figure 6.9.4).

Stupas and pagodas contain relics that Buddhists believe to be a portion of the Buddha's body or clothing. After the Buddha's death, his followers scrambled to obtain these relics. As part of the process of diffusing the religion, Buddhists carried these relics to other countries and built pagodas for them. Pagodas typically include tall, many-sided towers arranged in a series of tiers, balconies, and slanting roofs. Pagodas are not designed for congregational worship. Individual prayer or meditation is more likely to be undertaken at an adjacent temple, at a remote monastery, or in a home.

▶ **6.9.4 BUDDHIST MONKS AT DHAMEK STUPA, DEER PARK, SARNATH, INDIA**
Built to mark the spot where the Buddha gave his first sermon.

6.10 KEY ISSUE 3 How do religions organize space?

Religious Settlements & Administration

- **Give examples of religious settlements and of religious toponyms.**

Buildings for worship and sites for burial are individual expressions of religious belief systems on the landscape, but there are also other examples of religious faith on the landscape such as religious settlements. Most human settlements serve an economic purpose (see Chapter 12), but some are established primarily for religious reasons.

▲ **6.10.1 UTOPIAN SETTLEMENT**
New Harmony was founded in 1814 by Harmonists, people who migrated to the United States to escape persecution in Germany. The Harmonists left Indiana and returned to their first U.S. settlement in Pennsylvania in 1824. They sold New Harmony to a second group of utopians, who were social reformers rather than associated with a religious group. The nondenominational Roofless Church was dedicated in 1960.

6.10.1 Utopian Settlements

A **utopian settlement** is a community built to reflect the ideals of a particular religious or social group. Buildings are sited and economic activities organized to integrate the group's values into all aspects of daily life. By 1858, some 130 utopian settlements had been established in the United States, in conformance with a group's distinctive religious beliefs. Examples include Oneida, New York; Ephrata, Pennsylvania; Nauvoo, Illinois; and New Harmony, Indiana (Figure 6.10.1).

The culmination of the utopian movement in the United States was the founding of Salt Lake City by the Latter-day Saints (Mormons), beginning in 1848. The layout of Salt Lake City is based on a plan of the city of Zion given to the church elders in 1833 by the Mormon prophet Joseph Smith. The city has a regular grid pattern, unusually broad boulevards, and church-related buildings situated at strategic points (Figure 6.10.2).

Most utopian communities declined in importance or disappeared altogether. Some died out because the inhabitants were celibate. Others dissolved because they failed to attract immigrants.

Religious Place Names

Roman Catholic immigrants have frequently given religious place names, or toponyms, to their settlements in the New World, particularly in Québec and the U.S. Southwest. Québec's boundaries with Ontario and the United States clearly illustrate the difference between toponyms selected by Roman Catholic and Protestant settlers. Religious place names are common in Québec but rare in its two neighbors (Figure 6.10.3).

▲ **6.10.2 SALT LAKE CITY**
The Salt Lake Temple of the Church of Jesus Christ of Latter-day Saints is positioned at the center of the city.

▶ **6.10.3 RELIGIOUS TOPONYMS**

6.10.2 Hierarchical Religions

A **hierarchical religion** has a well-defined geographic structure and organizes territory into local administrative units. The Church of Jesus Christ of Latter-day Saints and Roman Catholicism are examples of hierarchical religions.

Latter-day Saints. Latter-day Saints exercise strong organization of the landscape. The territory occupied by Latter-day Saints, primarily Utah and portions of surrounding states, is organized into wards, with populations of several hundred each. Several wards are combined into a stake of approximately 5,000 people. The highest authority in the Church—the board and president—frequently redraws ward and stake boundaries in rapidly growing areas to reflect the ideal population standards.

Roman Catholics. The Roman Catholic Church has organized much of Earth's inhabited land into an administrative structure ultimately accountable to the pope in Rome. The basic unit of geographic organization in the Roman Catholic Church is the diocese (Figure 6.10.4). A diocese is administered by a bishop, who has a headquarters, called a "see," which is typically the largest city in the diocese. A diocese is spatially divided into parishes, each headed by a priest who reports to the bishop.

A province consists of several dioceses. A province is headed by an archbishop, who is also the bishop of one diocese within the province. Bishops report to an archbishop. Some distinguished archbishops are elevated to the rank of cardinal. Archbishops report to the pope, who is also the bishop of the Diocese of Rome.

The area and population of parishes and dioceses vary according to historical factors and the distribution of Roman Catholics across Earth's surface. In parts of Europe, where the population density is high and most people are Roman Catholics, a typical parish may encompass only a few square kilometers and fewer than 1,000 people.

On the other hand, Latin American parishes may encompass several hundred square kilometers and 5,000 people. The more dispersed Latin American distribution is attributable partly to a lower population density than in Europe. The less intensive network of Roman Catholic institutions also results in part from colonial traditions, as both Portuguese and Spanish rulers discouraged parish development in Latin America.

The Roman Catholic population is growing rapidly in the U.S. Southwest and in suburbs of some large North American and European cities. Some of these areas have a low density of parishes and dioceses compared to the population, so the Church must adjust its territorial organization. Conversely, the Roman Catholic population is declining in inner cities and rural areas. Maintaining services in these areas is expensive, and the process of combining parishes and closing Roman Catholic schools is very difficult.

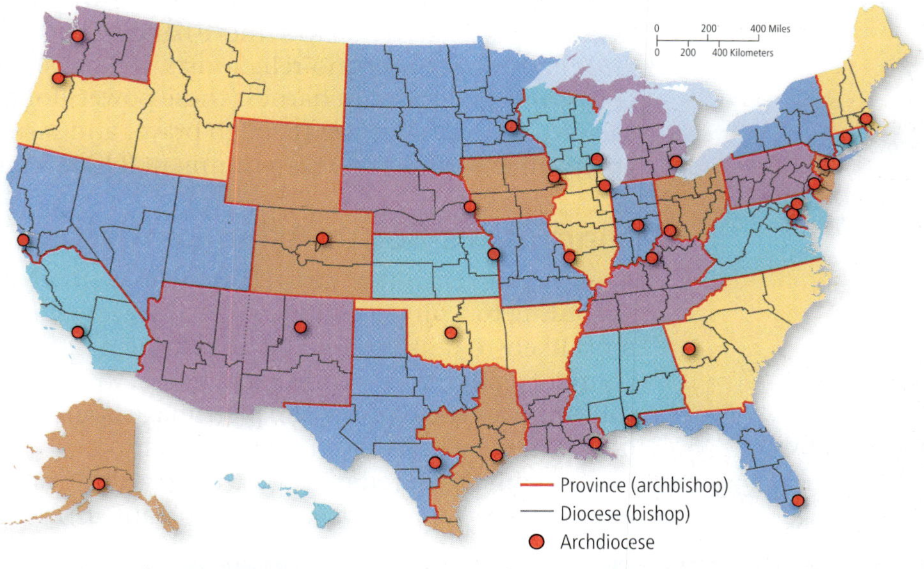

▲ 6.10.4 ROMAN CATHOLIC HIERARCHY IN THE UNITED STATES

Locally Autonomous Religions

An **autonomous religion** is one that does not have a central authority but shares ideas informally. Islam and some Protestant denominations are good examples.

Local autonomy in Islam. Among the three large universalizing religions, Islam is the most locally autonomous. Islam has neither a religious hierarchy nor a formal territorial organization. In the absence of a hierarchy, formal organization of territory in Islam is through the coincidence of religious territory with secular states. Governments in some predominantly Islamic countries include in their administration people who oversee Islamic institutions. These administrators interpret Islamic law and run social services programs.

Protestant denominations. Protestant Christian denominations vary in geographic structure from extremely autonomous to somewhat hierarchical. Extremely autonomous denominations include Baptists and United Church of Christ. Each congregation is self-governing, establishes the precise form of worship, and selects the leadership.

Religions & the Physical Environment

- **Discuss the importance of the physical environment for some religions.**

Some religions are closely tied to the physical geography of a particular place. Pilgrimages may be undertaken to view these physical features. Religions also adapt features of the environment to the annual calendar.

▲ **6.11.1 BATHING IN THE GANGES** Hindus bathe in the Ganges River in Varanasi, India, to wash away their sins.

6.11.1 Cosmogony & Spirits

Cosmogony is a set of religious beliefs concerning the origin of the universe. The universalizing religions Christianity and Islam consider that God (or Allah, for Islam) created the universe, including Earth's physical environment and human beings. Religious people in these traditions believe that they can best serve the Creator by cultivating the land, draining wetlands, clearing forests, building new settlements, and otherwise making productive use of natural features that the Creator made.

The cosmogony underlying some Chinese folk religions is that the universe is made up of two forces, yin and yang, which exist in everything. The force of yin (earth, darkness, female, cold, depth, passivity, and death) interacts with the force of yang (heaven, light, male, heat, height, activity, and life) to achieve balance and harmony. An imbalance results in disorder and chaos. To some ethno-religionists, the powers of the universe are mystical, and only a few people on Earth can harness these powers for medical or other purposes. Spirits or gods can be placated, however, through prayer and sacrifice. Rather than attempt to transform the environment, they accept environmental hazards as normal and unavoidable.

Hinduism's South Asian Landscape

As an ethnic religion of India, Hinduism is closely tied to the physical geography of India. According to a survey conducted by the geographer Surinder Bhardwaj, the natural features most likely to rank among the holiest shrines in India are riverbanks and coastlines. Hindus consider a pilgrimage, known as a tirtha, to be an act of purification. Although not a substitute for meditation, the pilgrimage is an important act in achieving redemption. Hindu holy places are organized into a hierarchy.

Particularly sacred places attract Hindus from all over India, despite the relatively remote locations of some; less important shrines attract primarily local pilgrims. Because Hinduism has no central authority, the relative importance of shrines is established by tradition, not by doctrine. For example, some Indians come from all over the country to bathe in the Ganges (Figure 6.11.1). Some Hindus make long-distance pilgrimages to Mt. Kailash, located at the source of the Ganges in the Himalayas, which is holy because Shiva lives there (Figure 6.11.2). Other mountains may attract only local pilgrims: Local residents may consider a nearby mountain to be holy if Shiva is thought to have visited it at one time.

▼ **6.11.2 MOUNT KAILASH** Holy to Hindus because Shiva lives here.

6.11.2 The Calendar

Religions adopt different types of annual calendars. A **solar calendar** has months that correspond to the season or the apparent position of the sun in relation to the stars. The solar calendar, such as followed by Christians, has 12 months, each containing 30 or 31 days, taking up the astronomical slack with 28 or 29 days in February.

A **lunar calendar** has months that correspond to cycles of moon phases. Islam uses a lunar calendar, which has around 354 days.

A **lunisolar calendar** has lunar months that are brought into alignment with the solar year through periodic adjustment. The Jewish calendar is lunisolar, with most years comprising around 354 days but adding a thirteenth month every two or three years to line up with agricultural and solar calendars.

▲ **6.11.3 HOLIDAYS IN JUDAISM**
Jews at the Western Wall in Jerusalem bless the *etrog* (citron) and *lulav* (date palm branches) during the holiday of Sukkot.

Religious Holidays

Regardless of the type that is used, the annual calendar of a religion typically includes holidays with theological themes. In universalizing religions, major holidays often relate to events in the life of the founder. A prominent feature of ethnic religions is the idea that adherents have a shared history, language, and destiny, and the calendar and holidays often refer to this shared history. An ethnic religion may also refer to distinctive physical geography such as celebration of the seasons—the calendar's annual cycle of variation in climatic conditions.

Universalizing religions. The principal purpose of the holidays in universalizing religions is to commemorate events in the founder's life rather than to the physical environment. Because Islam uses a lunar calendar, its holidays arrive in different seasons from generation to generation. For example, Ramadan, during which Muslims fast from sunup to sundown, occurred in the Northern Hemisphere in the summer of 2010 and the spring of 2020 and will occur in the winter of 2030. Christians in the Northern Hemisphere associate the birth of Jesus on December 25 with snow and cold, whereas Christians in the Southern Hemisphere observe December 25 with warm days and abundant sunlight.

The calendar in Judaism. Major holidays in Judaism are based on events in the Hebrew Bible and connected to the agricultural calendar of the religion's homeland in present-day Israel. These holidays are known as pilgrimage festivals because the Torah called for the Israelites to come to the Temple in Jerusalem. Following the destruction of the Second Temple, observance of these festivals shifted, though Jews in Israel may attend prayer services at the Western Wall (the remaining accessible part of the Second Temple).

Pesach (Passover) celebrates the liberation of the Jews from slavery in Egypt and the time for harvesting barley in the land of Israel. Shavuot (Feast of Weeks), which falls exactly seven weeks after Pesach, comes at the time for harvesting wheat. It is also considered the date when Moses received the Ten Commandments from God. Sukkot recalls the period of 40 years when the Israelites were in the desert and relied on God for food and shelter. The term derives from the Hebrew word for the "booths," or "temporary shelters," which the Israelites used in the desert. Sukkot also celebrates the final harvest of the year. A prayer for rain is offered to bring success in the upcoming agricultural year (Figure 6.11.3).

The Solstice

The **solstice** has special significance in some ethnic religions. A major holiday in some pagan religions is the winter solstice, December 21 or 22 in the Northern Hemisphere. The winter solstice is the shortest day and longest night of the year, when the Sun appears lowest in the sky and the apparent movement of the Sun's path north or south comes to a stop before reversing direction (solstice comes from the Latin for "stand still"). Stonehenge, a collection of stones erected in southwestern England some 4,500 years ago (Figure 6.11.4), is a prominent remnant of a pagan structure apparently aligned so that the Sun rises directly between two stones on the summer and winter solstices.

▼ **6.11.4 STONEHENGE**
The 30 enormous stones that comprise Stonehenge were erected around 4,500 years ago.

6.12 KEY ISSUE 4 Why do territorial conflicts arise among religions?

Religions & Social Change

- **State reasons for geographic conflicts between religious and secular cultural groups.**

Religion is an element of cultural diversity that has led to disputes in some places. The attempt by intense adherents of one religion to organize Earth's surface can conflict with the spatial expression of other faiths or of nonreligious ideas. Conflicts arise in all continents, but disputes involving religious groups have been especially intense across Asia.

▲ 6.12.1 HINDU CASTE SYSTEM
Women demonstrate for Dalit rights, Bangalore, India.

6.12.1 Hinduism & Social Equality

Changing cultural, political, and economic customs can sometimes conflict with traditional religious values. In India, Hinduism has been strongly challenged since the 1800s, when British colonial administrators introduced their social and moral concepts. The most vulnerable aspect of the Hindu religion was its rigid **caste system**, which indicated the class or distinct hereditary order into which a Hindu was born, according to religious law.

The caste system apparently originated around 1500 B.C.E., when Aryans invaded India from the west. The Aryans divided themselves into four castes that developed strong differences in social and economic position: Brahmans (priests and top administrators), Kshatriyas (warriors), Vaishyas (merchants), and Shudras (agricultural workers and artisans). The Shudras occupied a distinctly lower status than the other three castes.

Below the four castes were the Dalits, which is Sanskrit and Hindi for "broken" or "scattered" (Figure 6.12.1). The Dalits—also known as outcasts or untouchables—did work considered too dirty for other castes. In theory, the Dalits were descended from the Indigenous people who lived in India prior to the Aryan conquest. They comprise around 16 percent of India's total population.

Over the centuries, the original castes split into thousands of subcastes. Until recently, social relations among the castes were limited, and the rights of non-Brahmans, especially Dalits, were restricted. British administrators and Christian missionaries pointed out the shortcomings of the caste system, such as neglect of the untouchables' health and economic problems.

The rigid caste system has been considerably relaxed in recent years. But consciousness of caste persists: A government plan to devise a quota system designed to give untouchables more places in the country's universities generated strong opposition. People looking for a marriage partner advertise their caste and the castes they are willing to consider for a spouse.

Fundamentalism

Contributing to more intense religious conflict has been a resurgence of religious fundamentalism. As the name implies, **fundamentalism** is a literal interpretation and a strict and intense adherence to what the fundamentalists define as the basic principles of a religion (or a religious branch, denomination, or sect).

Fundamentalism started in the nineteenth century as a term for some fundamental principles held by some Protestants in the United States. More recently, fundamentalism has instead been used to refer to groups who mix politics with aspects of their religion and make claims that their views are the only correct views. These groups sometimes advocate and practice violence to enforce their claims.

6.12.2 Religion & Communism

Organized religion was challenged in the twentieth century by the rise of communist regimes in Europe and Asia that discouraged religious belief and practice. Following the 1917 Bolshevik Revolution, the communist government of the Soviet Union pursued especially strong antireligious programs. Karl Marx had called religion "the opium of the people," a view shared by V. I. Lenin and other early communist leaders. Marxism became the official doctrine of the Soviet Union, so religious doctrine was a potential threat to the success of the revolution.

In 1721, Czar Peter the Great had made the Russian Orthodox Church a part of the Russian government (Figure 6.12.2). The Soviet government in 1918 eliminated the official church–state connection. All church buildings and property were nationalized and could be used only with government permission.

▲ 6.12.2 CATHEDRAL OF VASILY THE BLESSED (ST. BASIL'S CATHEDRAL), MOSCOW, RUSSIA

The Orthodox religion retained adherents, especially among the elderly, but younger people generally had little contact with the church beyond attending a service perhaps once a year. With religious organizations prevented from conducting social and cultural work, the role of religion in Soviet life was sharply reduced.

The end of communist rule in the late twentieth century brought a religious revival in Eastern Europe. Church property confiscated by the communist governments reverted to church ownership, and church attendance increased.

In countries in Central Asia that were once part of the Soviet Union—Kazakhstan, Kyrgyzstan, Tajikistan, Turkmenistan, and Uzbekistan—most people are Muslims. These newly independent countries are struggling to determine the extent to which laws should be rewritten to conform to Islamic custom rather than to the secular tradition inherited from the Soviet Union.

China, Buddhism, & the Dalai Lama

Conflict between the doctrine of communism and the authority of Buddhist religious leaders is especially acute in China. The territory known as Tibet in most of the world but called Xizang by the government of China is the traditional home of the spiritual leader of Tibetan Buddhists, the Dalai Lama. Before the most recent communist takeover in 1950, daily life in Tibet was traditionally dominated by Buddhist rites.

China, which had ruled Tibet from 1720 until its independence in 1911, invaded the isolated western country, renamed it Xizang Province in 1951, and later installed a communist government. The Chinese sought to reduce the domination of Buddhist traditions by attacking the economic and cultural pillars of daily life. Monasteries and temples were destroyed, and people were required to work on agricultural communes rather than pursue their nomadic way of life as livestock herdsmen.

Tibetan Buddhists believe that when the Dalai Lama dies, his spirit enters the body of a child so that the leader is reincarnated. In this tradition, in 1937, a group of priests located a 2-year-old boy named Tenzin Gyatso and recognized him as the fourteenth Dalai Lama (Figure 6.12.3). The child was brought to Lhasa in Tibet when he was 4, enthroned a year later, and trained by priests to assume leadership.

An unsuccessful Tibetan rebellion against Chinese rule in 1959 resulted in the execution or imprisonment of tens of thousands and forced another 100,000 Tibetans, including the Dalai Lama, to flee. Buddhist temples were closed and demolished, and religious artifacts and scriptures were destroyed. Since then, the Dalai Lama and other Tibetan Buddhist leaders have been based in Dharamshala, India.

When the fourteenth Dalai Lama passes away, Buddhist leaders assert that they will find the next Dalai Lama, wherever he is living in the world. However, Chinese government leaders counter that they will be responsible for finding the next Dalai Lama somewhere in Xizang, and they accuse Buddhist leaders of plotting to restore Tibet as an independent country.

▼ 6.12.3 THE 14TH DALAI LAMA

Geographic Perspectives in Israel

- **Understand geographic elements underlying conflicts in Southwest Asia.**

Religious conflict in Southwest Asia is among the world's longest standing and most intractable. Jews, Christians, and Muslims have fought for more than 1,000 years to control the same small strip of land, most of which is now part of the State of Israel.

6.13.1 Religious Importance

Jews make a special claim to the territory they call the Promised Land. The major events in the development of Judaism took place there, and the religion's customs and rituals acquired meaning from the agricultural life of the ancient Israelite tribes. Jerusalem is especially holy to Jews as the location of the Temple, their center of worship in ancient times. Christians consider Palestine the Holy Land and Jerusalem the Holy City because the major events in Jesus's life, death, and resurrection were concentrated there. Muslims regard Jerusalem as their third-holiest city, after Makkah and Madinah, because it is the place from which Muhammad is thought by Muslims to have ascended to heaven. Islam became the region's most widely practiced religion beginning in the seventh century C.E.

Competing Israeli & Palestinian Perspectives

The Holy Land was incorporated into a succession of empires, culminating with the British after World War I. The United Nations voted in 1947 to partition the British Mandate for Palestine into two independent states, one Jewish and one Arab. Jerusalem was to be an international city, open to all religions, and run by the U.N. (Figure 6.13.1a).

When the British withdrew in 1948, Jews declared an independent State of Israel within the boundaries prescribed by the U.N. resolution (Figure 6.13.1b). Over the next quarter-century, Israel was forced to fight for its survival in four wars with its neighbors in 1948–1949, 1956, 1967, and 1973. During the 1967 Six-Day War, Israel captured territory from its neighbors. Israel has returned some of the captured territory and retains control of some of it (Figure 6.13.1c).

After the 1973 War, the Palestinians emerged as Israel's principal opponent. Five groups of people consider themselves Palestinians: people living in the territories captured by Israel in 1967; some citizens of Israel who are Arabs; people who fled from Israel to other countries after the 1948–1949 war; people who fled from the West Bank or Gaza to other countries after the 1967 Six-Day War; and some citizens of other countries, especially Jordan, Lebanon, Syria, Kuwait, and Saudi Arabia.

Israel has permitted the organization of a limited form of government in much of the West Bank and Gaza, called the Palestinian Authority, but Palestinians are not satisfied with either the territory or the power they have received thus far. Some Palestinians are willing to recognize the State of Israel with its Jewish majority in exchange for return of all territory taken by Israel in the 1967 Six-Day War. Other Palestinians do not recognize the right of Israel to exist and want to continue fighting for control of the entire territory between the Jordan River and the Mediterranean Sea. The United States, European countries, and Israel consider the Palestinian organization Hamas to be a terrorist organization.

Israeli Jews have been divided between those who wished to retain some of the West Bank and those who wished to make compromises with the Palestinians in return for formal recognition and a stable peace. In recent years, a large majority of Israelis have supported construction of a barrier to deter Palestinian attacks.

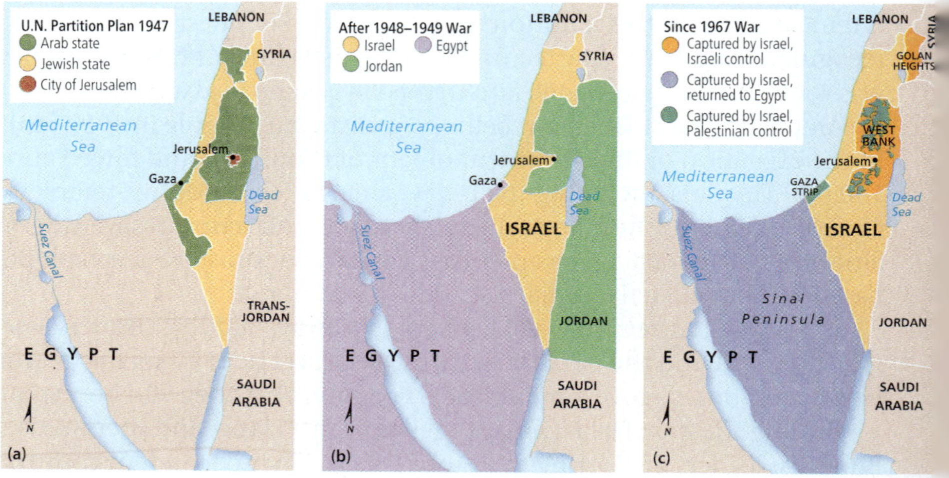

▶ 6.13.1 TERRITORIAL CHANGES IN ISRAEL AND ITS NEIGHBORS
(a) The 1947 U.N. Partition Plan. (b) After the 1948–1949 war. (c) After the 1967 Six-Day War.

6.13.2 Jerusalem's Challenging Geography

At the heart of the conflicts among religious groups in Southwest Asia lies the city of Jerusalem, an important place for several religions (Figure 6.13.2). Geography makes it difficult to settle long-standing disputes over access to Jerusalem by Jews and Muslims. The challenge is that the most sacred space in Jerusalem for Muslims was literally built on top of the most sacred space for Jews.

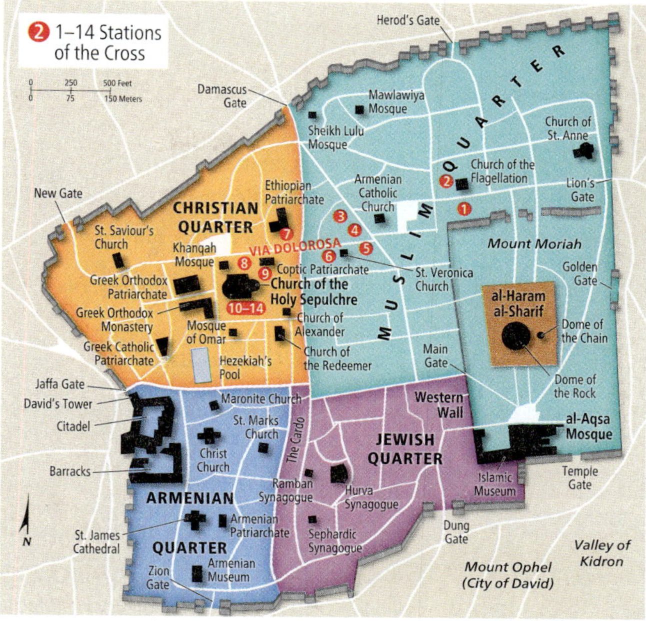

▲ **6.13.2 OLD CITY OF JERUSALEM**
The Old City of Jerusalem, less than 1 square kilometer (0.4 square miles), is divided into four quarters. The Stations of the Cross mark the route that Christians believe Jesus was forced to walk to his crucifixion.

Judaism's Jerusalem

Jerusalem is Judaism's holiest city, established as the capital of the Kingdom of Israel approximately 3,000 years ago, and home to the Temple, their center of life and worship for hundreds of years. The First Temple, built by King Solomon in approximately 957 B.C.E., was destroyed by the Babylonians in 586 B.C.E. After the Persian Empire, led by Cyrus the Great, gained control of Jerusalem, Jews were allowed to build a Second Temple, which was completed in 515 B.C.E. The Romans destroyed the Jewish Second Temple in 70 C.E. The Western Wall of the Temple and an elevated plaza above the wall known as the Temple Mount survive (Figure 6.13.3).

Christians and Muslims call the Western Wall the Wailing Wall because for many centuries Jews were allowed to visit the surviving Western Wall only once a year to lament the Temple's destruction. After Israel captured the entire city of Jerusalem during the 1967 Six-Day War, officials removed the barriers that had prevented Jews from visiting and living in the Old City of Jerusalem, including the Western Wall. The Western Wall soon became a site for daily prayers by observant Jews.

Islam's Jerusalem

The most important Muslim structure in Jerusalem is the Dome of the Rock, which was completed in 691 C.E. Muslims believe that the large rock beneath the building's dome is the place from which Muhammad ascended to heaven as well as the altar on which Abraham prepared to sacrifice his son Isaac (according to Jews and Christians) or his son Ishmael (according to Muslims). Immediately south of the Dome of the Rock is the al-Aqsa Mosque. The challenge facing Jews and Muslims is that al-Aqsa Mosque was built on the Temple Mount, which is holy to Jews and Muslims.

▼ **6.13.3 WESTERN WALL AND DOME OF THE ROCK, JERUSALEM**
Jews are praying at the Western Wall (foreground), situated immediately below the Mount containing Islam's Dome of the Rock (top left) and al-Aqsa Mosque (top right). The ramp to the right provides Muslims access to the Mount without requiring them to walk in front of the Western Wall.

CHAPTER

6 Review, Analyze, & Apply

KEY ISSUE 1 Where Are Religions Distributed?

- Religions can be classified into two major groups: universalizing and ethnic.
- The three largest universalizing religions are Christianity, Islam, and Buddhism.
- Christians are most numerous in Europe, North America, and Latin America; Muslims in Southeast Asia, Central Asia, and Southwest Asia & North Africa; and Buddhists in East Asia.
- The largest ethnic religion is Hinduism, which is found primarily in South Asia.

THINKING GEOGRAPHICALLY

▲ 6.CR.1 LARGE MUSLIM FAMILY, ISTANBUL, TURKEY

The percentage of the world's people adhering to Islam is forecast by the Pew Research Forum to increase from 23 percent in 2010 to 30 percent in 2050. What demographic factors presented in Chapter 2 (Population & Health) would underlie this increase? Conversely, why might the percentage of people adhering to Christianity not increase?

KEY ISSUE 2 Why Do Religions Have Distinctive Distributions?

- Universalizing religions have well-documented places of origin, based on events in the lives of the founders.
- Ethnic religions typically have unknown or unclear origins.
- Universalizing religions typically diffuse widely from their place of origin, whereas ethnic religions typically do not.

THINKING GEOGRAPHICALLY

▲ 6.CR.2 EDICULE DURING RENOVATION WORK

Most Christians accept that Jesus was buried and resurrected from a tomb that is now housed in the Edicule inside the Church of the Holy Sepulchre in Jerusalem. A few years ago during renovations, researchers unsealed the Edicule. What did they discover? In your internet browser, search for "Jesus tomb uncovered."

KEY ISSUE 3 How Do Religions Organize Space?

- Some religions have sacred spaces and sacred locations in the physical environment that serve as places of worship and veneration.
- Making pilgrimages to sacred spaces and environments is important for some religions.
- For universalizing religions, the most sacred places are typically related to events in the founder's life.

THINKING GEOGRAPHICALLY

▲ 6.CR.3 ORIGINS OF PILGRIMS TO MAKKAH, 2019

Based on your understanding of reasons for migration discussed in Chapter 3 (Migration), as well as the distribution of Muslims by country, what factors might account for the distinctive origin of pilgrims? How might the COVID-19 pandemic impact the distribution of pilgrims?

KEY ISSUE 4 Why Do Territorial Conflicts Arise Among Religions?

- Changing cultural, political, and economic customs can sometimes conflict with traditional religious values.
- Religious groups have been known to oppose government policies, especially those of communist governments.
- An especially long-standing conflict among religious groups has been centered in the area of the Eastern Mediterranean considered holy by Jews, Christians, and Muslims; the dispute is especially intractable in Jerusalem.

THINKING GEOGRAPHICALLY

▲ 6.CR.4 POTALA PALACE, LHASA, THE DALAI LAMA'S RESIDENCE UNTIL 1959

Three-fourths of the Chinese people adhere to a Chinese folk religion or are unaffiliated with any religion. Given the relatively small percentage of Chinese who adhere to the principal universalizing religions, why might the government of China be engaged in conflicts with Tibetan Buddhists and Muslims living in the western portion of the country?

Key Terms

Agnosticism (6.1.1) The belief that the existence of God cannot be proven or disproven empirically.

Animism (6.4.2) The belief that objects, such as plants and stones, or natural events, like thunderstorms and earthquakes, have discrete spirits and conscious life.

Atheism (6.1.1) The belief that God does not exist.

Autonomous religion (6.10.2) A religion that does not have a central authority but shares ideas and cooperates informally.

Branch (6.2.1) A large and fundamental division within a religion.

Caste system (6.12.1) The class or distinct hereditary order into which a Hindu is assigned, according to religious law.

Congregation (6.2.1) A local assembly of persons brought together for common religious worship.

Cosmogony (6.11.1) A set of beliefs concerning the origin of the universe.

Denomination (6.2.1) A division of a branch that unites a number of local congregations into a single legal and administrative body.

Ethnic religion (6.1.1) A religion with a relatively concentrated spatial distribution whose principles are likely to be based on the physical characteristics of the particular location in which its adherents are concentrated.

Fundamentalism (6.12.1) Literal interpretation and strict adherence to basic principles of a religion (or a religious branch, denomination, or congregation).

Ghetto (6.8.2) During the Middle Ages, a neighborhood in a city set up by law to be inhabited only by Jews; now used to denote a section of a city in which members of any minority group live because of social, legal, or economic discrimination.

Hierarchical religion (6.10.2) A religion in which a central authority exercises a high degree of control.

Lunar calendar (6.11.2) A calendar with months that correspond to cycles of moon phases.

Lunisolar calendar (6.11.2) A calendar with lunar months that are brought into alignment with the solar year through some process.

Missionary (6.7.1) An individual who helps to diffuse a universalizing religion.

Monotheism (6.5.1) The doctrine of or belief in the existence of only one God.

Pilgrimage (6.9.2) A journey to a place considered sacred for religious purposes.

Polytheism (6.5.1) Belief in or worship of more than one god.

Solar calendar (6.11.2) A calendar that relates to the season or the apparent position of the sun in relation to the stars.

Solstice (6.11.2) An astronomical event that happens twice each year, when the tilt of Earth's axis is most inclined toward or away from the Sun, causing the Sun's apparent position in the sky to reach its northernmost or southernmost extreme, and resulting in the shortest and longest days of the year.

Syncretic (6.4.2) Combining several religious traditions.

Universalizing religion (6.1.1) A religion that attempts to appeal to all people, not just those living in a particular location.

Utopian settlement (6.10.1) A community built around an ideal way of life, often based on a religion.

Geospatial Analysis

Log in to the Mastering Geography Study Area to access MapMaster 2.0.

South Asia's Religions

South Asia has several religious boundaries that appear to be sharp on a world map.

Add ***Religions*** and zoom to South Asia. In the *Legend*, select the *Settings* icon, and select *Show Political Labels*.

Add **The British Empire, *Early 1920s*** layer, select *Add Group Data Layers to Map*, select *Split Map Window*.

1. What present-day countries in *South Asia* were part of the British Empire in the early 1920s?
2. What religions were contained within the British Empire? Why might this have been a challenge for the Empire rulers? Why might this have been a challenge when the Empire was broken into independent countries?

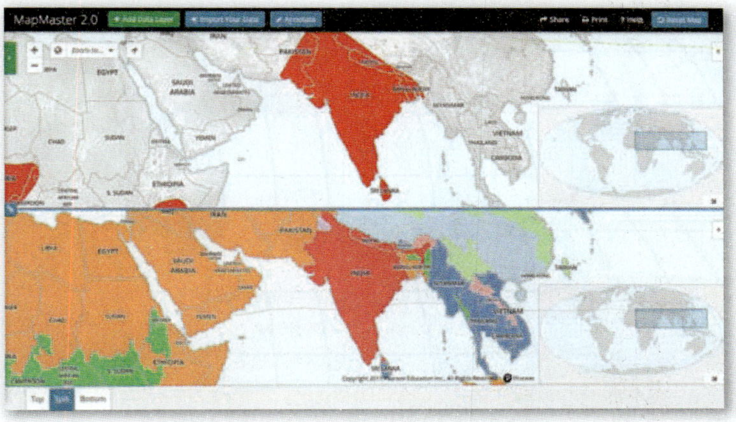

▲ **6.CR.5** SOUTH ASIA RELIGIONS

Explore

Masjid's Surroundings

Use Google Earth to explore the surrounding of Masjid al-Haram, Islam's most important mosque, in Makkah, Saudi Arabia. ***Search*** for **Masjid al-Haram, Makkah**.

1. What is the name of the tall building complex immediately to the south of the Masjid? Click on the name. What is in the building complex?
2. Why might this building complex be located immediately next to Islam's holiest place (refer to Section 6.9.2)?

▶ **6.CR.6** ABRAJ AL-BAIT

CHAPTER

7 Ethnicity & Race

▲ Oxford Street, London.

Our ethnicity is an important element of our cultural identity. It is a link to the experiences of our ancestors and to our cultural traditions. Geographers are especially interested in ethnicity because the concept is place-based. Ethnic identity is rooted in the geography of a particular place on Earth but can remain important even when people live far from the place where their ethnicity originated.

KEY ISSUES

1 Where Are Ethnicities & Races Distributed?

Ethnicities and races draw identity and meaning from the physical and cultural features of a particular place where they currently live or where they once did.

2 Why Do Ethnicities Have Distinctive Distributions?

Blacks have a distinctive ethnic history shaped by forced migration from Africa and enforced segregation within the United States and elsewhere.

3 How Do Ethnicities & Nationalities Differ?

The territory inhabited by an ethnic group does not always match the boundaries of a country.

4 Why Do Ethnic Cleansing & Genocide Occur?

Conflicts among ethnic groups have sometimes led to mass killing and forced relocation.

LOCATIONS IN THIS CHAPTER

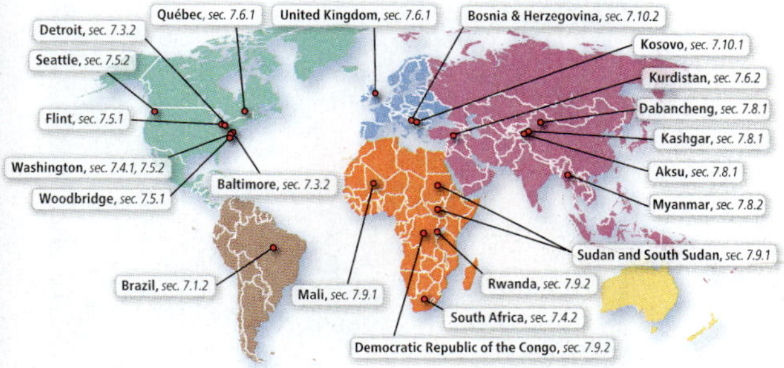

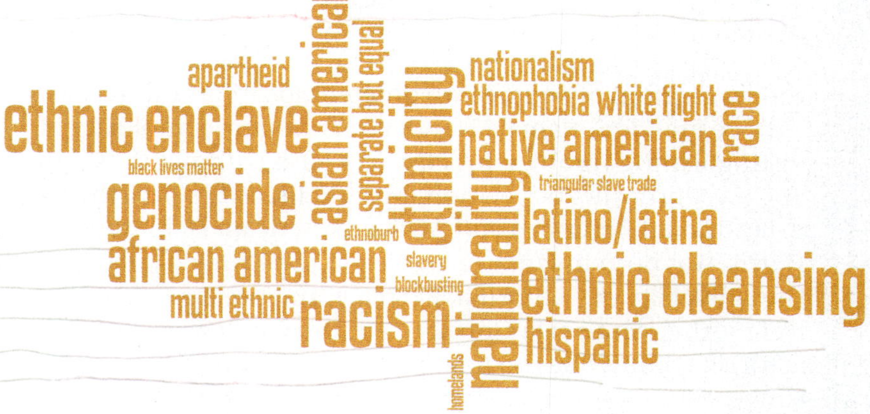

Ethnicity, Race, & Nationality

- **Distinguish among ethnicity, race, and nationality.**

Ethnicity is an important cultural element of local diversity because our ethnic identity is immutable. We can suppress our ethnicity, but we cannot change it in the same way we can speak a new language or practice a different religion.

7.1.1 Ethnicities & Geography

Geographers are interested in where ethnicities, like other elements of culture, are distributed. Ethnicity is especially important to geographers because in the face of globalization trends in culture and economy, ethnicity stands as a strong bulwark for the preservation of local diversity. An ethnic group is tied to a particular place because members of the group—or their ancestors—were born and raised there. The cultural traits displayed by an ethnicity derive from particular conditions and practices in the group's homeland.

The meaning of ethnicity is frequently confused with definitions of race and nationality:

- **Ethnicity** is identity with a group of people who share the cultural traditions of a particular homeland or hearth. An ethnicity could refer to a group occupying a very small area, such as the Tutsis of Central Africa, or it could refer to a large heterogeneous group, such as Asian Americans.
- **Race** is identity with a group of people who are perceived to share a physical trait, such as skin color.
- **Nationality** is identity with a group of people who share legal attachment to a particular country. The term is introduced in this chapter and discussed in more detail in Chapter 8.

Distinguishing Among Ethnicity, Race, & Nationality

Ethnicity and nationality are of importance to geographers because they are both place-based. Though not place-based, race is an important concept because it is often used as a synonym for ethnicity. Features of race, such as skin color, hair type and color, blood traits, and shape of body, head, and facial features, were once thought to be scientifically classifiable. Contemporary geographers, along with other scientists, reject the discredited biological basis of classifying humans into a handful of races and recognize that these features are not rooted in specific places.

However, one feature of race is especially important for geographers: the color of skin. The distribution of persons by skin color is studied by geographers because it is the fundamental basis by which people in many societies sort out where they reside, attend school, spend their leisure time, and perform many other activities of daily life.

The attempt to classify people according to race is regarded by geographers as a **social construction**, which is an idea or a meaning that is widely accepted as natural by a society but may not represent a reality shared by those outside the society.

At worst, classification by race is the basis for **racism**, which is the belief that race is the primary determinant of human traits and capacities and that racial differences produce an inherent superiority of a particular race. A **racist** is a person who displays discrimination or feels prejudice against people of particular races. One result of racism is the segregation of people into different locations on the basis of race.

Vice President Kamala Harris illustrates the complexity of distinguishing among ethnicity, race, and nationality (Figure 7.1.1). She was born with U.S. nationality and identifies her ethnicity as both Asian-Indian American and African American. Her father, Douglas Harris, was born in Jamaica, moved to the United States to attend the University of California, Berkeley, and became a U.S. citizen. Her mother, Shyamala Gopalan, was born in India when it was still a British colony. She too migrated to the United States to attend the University of California, Berkeley, and became a U.S. citizen.

▼ 7.1.1 KAMALA HARRIS WITH HER MOTHER, SHYAMALA GOPALAN

Harris is receiving the oath of office as San Francisco District Attorney in 2004. Her mother died in 2009.

7.1.2 Ethnically Complex Brazil

Brazil struggles with defining its population by race or ethnicity. Like the United States, Brazil is composed of people whose ancestors emigrated from many places. Portugal and West Africa have been the leading places of origin, but the ancestors of large numbers of Brazilians came from other European countries, Japan, Southwest Asia, and elsewhere. In addition, a large number of Indigenous people inhabited Brazil prior to the immigration of people from other continents.

Genetic studies show that roughly 70 percent of Brazilians have predominantly European ancestry, 20 percent predominantly African, and 10 percent predominantly Indigenous. However, through many generations of marriages and births, most Brazilians have a mix of ethnicities.

Rather than ethnicity, though, Brazil's census classifies people according to skin color. The race to which a person chooses to be classified is an important element of cultural identity in Brazil.

The Brazilian Institute of Geography and Statistics, a government agency that conducts the official census, asks Brazilians to identify themselves as belonging to one of five so-called races: branco (white), pardo (brown), preto (black), amarelo (yellow), and Indigenous (Figure 7.1.2). Brancos and pardos each comprise more than 40 percent of Brazil's population and together account for more than 90 percent.

Further complicating Brazil's racial classifications, Brazilian Institute of Geography and Statistics researcher José Luiz Petrucelli found that Brazilians don't care for the census choices. When asked an open-ended question about their race, Brazilians responded with 143 different answers. Most significant were the large numbers who identified themselves as moreno (brunette or olive) or moreno claro (light brown), races not even included in the census (Figure 7.1.3).

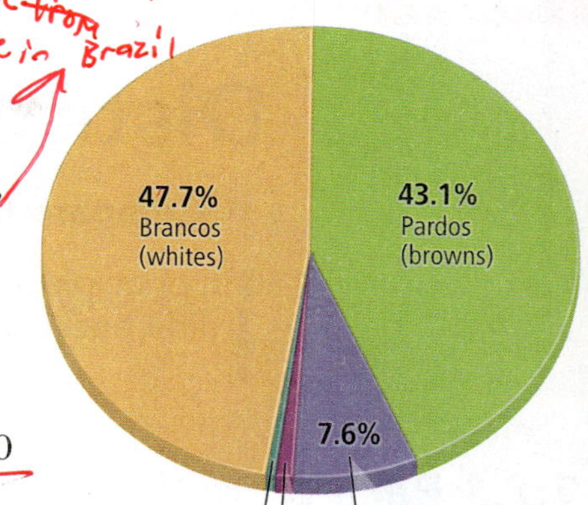

▲ 7.1.2 RACES IN BRAZIL BASED ON CENSUS
Official results from 2010 census.

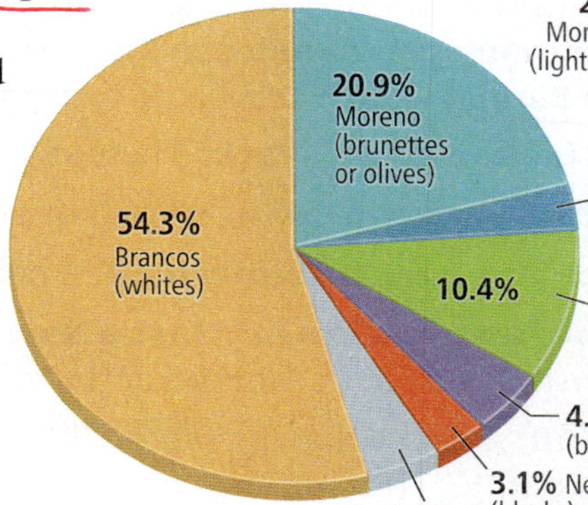

▲ 7.1.3 RACES IN BRAZIL BASED ON SELF-IDENTIFICATION
Results from a self-identification study by the Brazilian Institute of Geography and Statistics.

Distribution of Races in Brazil

Brazil also displays sharp regional variations in the distribution of races (Figure 7.1.4). Brancos are clustered in the south and pardos in the north (Figure 7.1.5).

▲ 7.1.4 DISTRIBUTION OF RACES IN BRAZIL

▲ 7.1.5 SCHOOLCHILDREN, BRASÍLIA, BRAZIL

Distribution of U.S. Ethnicities

- **Compare the distribution of U.S. ethnicities.**

In principle, ethnicity, nationality, and race are distinct concepts. In actuality, differences among the three concepts are not always clear-cut.

7.2.1 Race & the U.S. Census

Information about race and ethnicity in the United States is derived from the Census of Population that is administered every 10 years. The U.S. census shows the difficulty in distinguishing between ethnicity and race.

The census asks two questions concerning race and ethnicity:

- **Is the person of Hispanic, Latino, or Spanish origin?** Approximately 17 percent of Americans say they are Hispanic. If the answer is yes, the census offers several options. Nearly two-thirds of the people who responded yes came from Mexico. This correlates with the immigration patterns noted in Chapter 3. The second largest group is Puerto Rican (Figure 7.2.1).

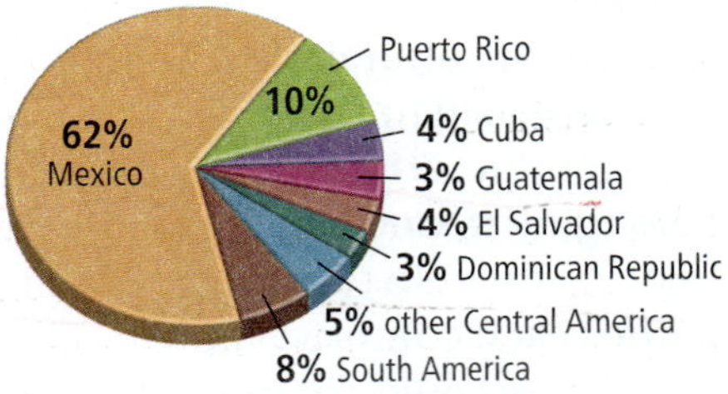

▲ 7.2.1 HISPANICS BY PLACE OF ORIGIN

- **What is the person's race?** These are the choices offered by the census:
  - **White:** A person having origins in any of the original peoples of Europe, the Middle East, or North Africa.
  - **Black or African American:** A person having origins in any of the Black racial groups of Africa.
  - **American Indian or Alaska Native:** A person having origins in any of the original peoples of North, Central, and South America and who maintains tribal affiliation or community attachment.
  - **Asian:** A person having origins in any of the original peoples of the Far East, Southeast Asia, or the Indian subcontinent (Figure 7.2.2).
  - **Native Hawaiian or other Pacific Islander:** A person having origins in any of the original peoples of Hawaii, Guam, Samoa, or other Pacific Islands.

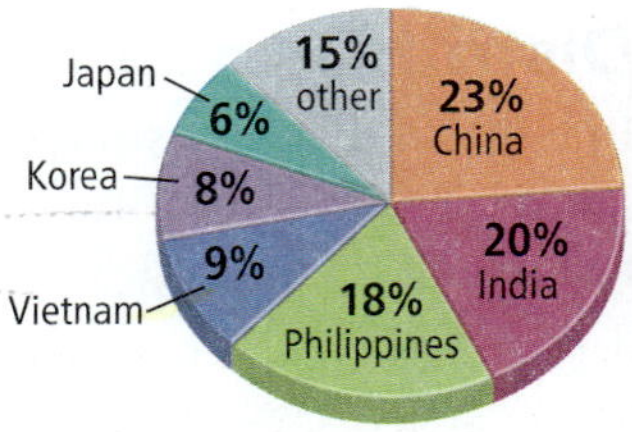

▲ 7.2.2 ASIAN AMERICANS BY PLACE OF ORIGIN

Distribution of Hispanic Americans

In the United States, Hispanics are clustered in the Southwest. They exceed one-third of the population of Arizona, New Mexico, and Texas and one-quarter of California (Figure 7.2.3). California is home to one-third of all Hispanics in the United States, Texas one-fifth, and Florida and New York one-sixth each.

The terms *Latino* (for males) and *Latina* (for females) are generally used interchangeably with the term *Hispanic*. The U.S. government adopted the term *Hispanic* in 1973 because it was considered an inoffensive label that could be applied to all people from Spanish-speaking countries while avoiding the gender-specific limitation of the terms *Latino* and *Latina*. *Latinx* and *Latino(a)* are used as well as *Hispanic* to avoid gender-specific terms. However, according to a survey by the Pew Research Center, most Americans of Central and South American heritage prefer to identify with a more specific ethnicity or national origin, such as Mexican, rather than terms like *Hispanic* that encompass several dozen countries.

▼ 7.2.3 DISTRIBUTION OF HISPANICS IN THE UNITED STATES

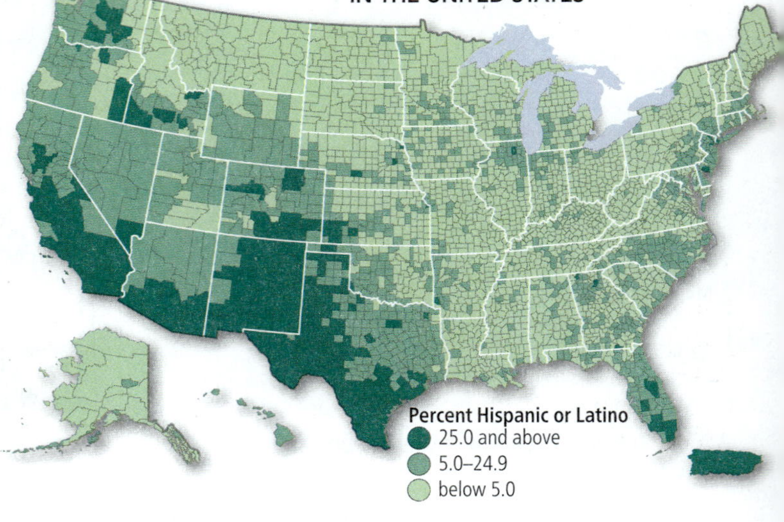

7.2.2 Distribution of Blacks in the United States

Blacks are clustered in the Southeast, where they comprise at least one-fourth of the population in Alabama, Georgia, Louisiana, Maryland, and South Carolina and more than one-third in Mississippi (Figure 7.2.4). Concentrations are even higher in selected counties. At the other extreme, nine states in upper New England and the West have less than 1 percent African Americans.

It is difficult to pinpoint precisely the ethnic origins of Blacks in the United States. The ancestors of most Blacks arrived around 300 years ago from places in Africa that were not yet organized into independent countries. Records were not kept of the ethnic origin of Blacks who were enslaved on arrival.

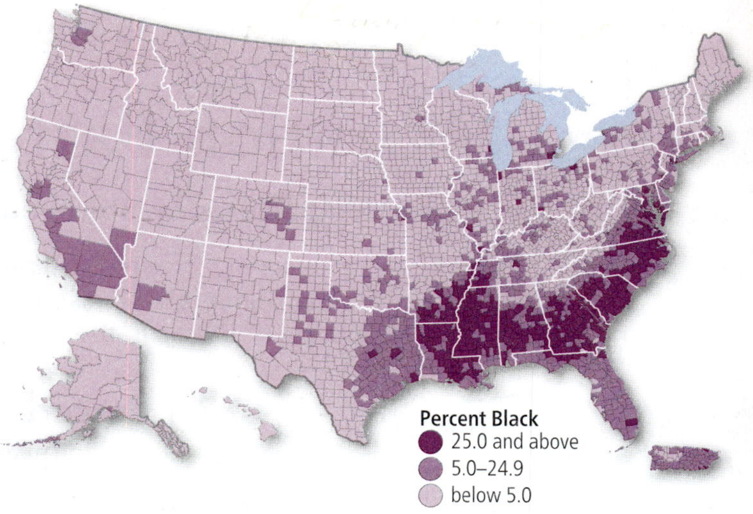

▲ **7.2.4** DISTRIBUTION OF BLACKS IN THE UNITED STATES

Distribution of Asian Americans

Asian Americans are clustered in the West, comprising more than 40 percent of the population of Hawaii. Half of all Asian Americans in the United States live in California, where they comprise 12 percent of the population (Figure 7.2.5). Most Asian Americans identify their ethnicity as the country of origin of themselves or their ancestors.

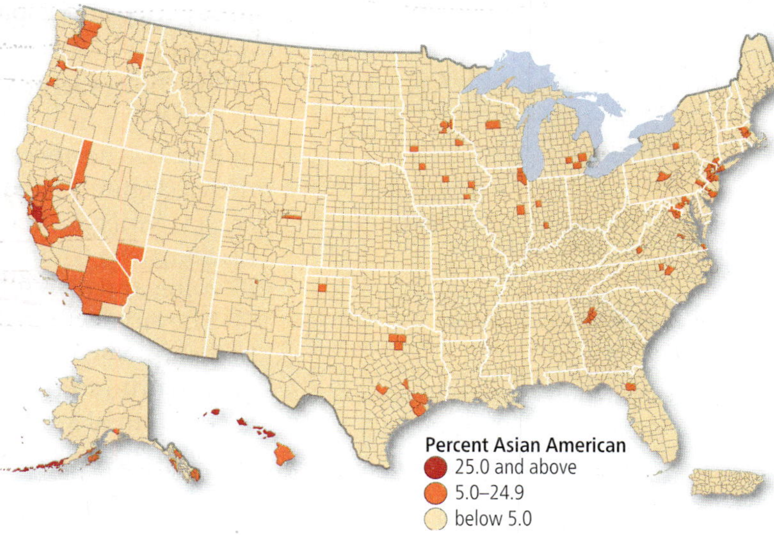

▲ **7.2.5** DISTRIBUTION OF ASIAN AMERICANS IN THE UNITED STATES

Distribution of Indigenous Peoples

The descendants of Indigenous people—who lived in North America prior to the arrival of Europeans—comprise numerous ethnicities. They are grouped in the United States into three principal ethnic identities: Native American, Alaska Native, and Native Hawaiian. The most numerous are Native Hawaiians, Cherokee, Navajo, Chippewa, Sioux, and Choctaw. The largest numbers of Native Americans, however, do not specify a group. Native Americans are clustered in the southwest and north-central regions of the United States as well as Alaska. Native Americans exceed 10 percent of the population of Alaska, Oklahoma, New Mexico, and South Dakota (Figure 7.2.6).

Indigenous people occupied much of the land for thousands of years before the first European explorers arrived. Forced migration during the nineteenth century then played a major role in the current distribution of Native Americans (refer to Section 3.7.1).

▼ **7.2.6** DISTRIBUTION OF AMERICAN INDIAN, ALASKA NATIVE, AND NATIVE HAWAIIAN IN THE UNITED STATES

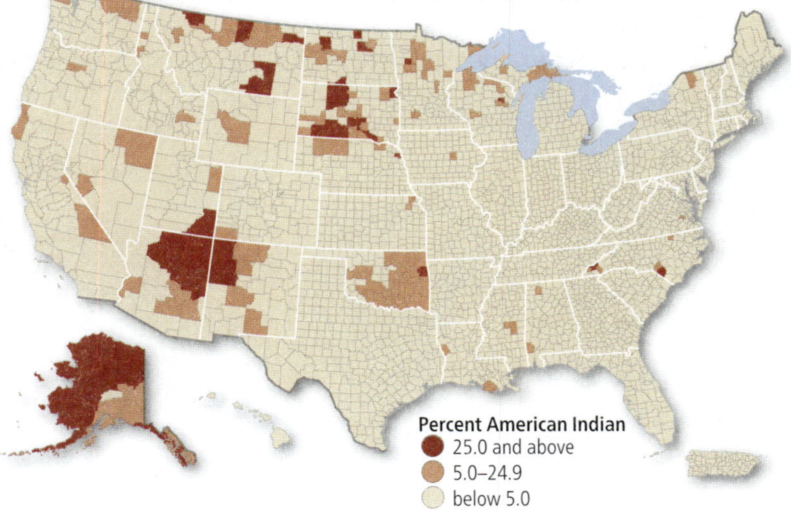

Migration Patterns of U.S. Blacks

- Describe patterns of international and internal migration of U.S. Blacks.

The clustering of ethnicities within the United States is partly a function of migration. In Chapter 3, migration was divided into international and internal scales. The distribution of Blacks in the United States provides examples of both processes.

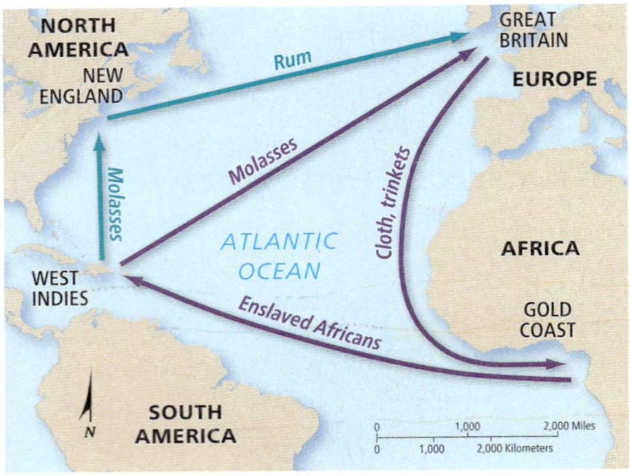

▲ 7.3.1 TRIANGULAR SLAVE TRADE
Ships left Europe for Africa with cloth and other goods used to buy people. The same ships transported these enslaved people across the Atlantic. Completing the triangle, the ships returned to Europe with sugar and molasses. Some ships carried molasses to North America and rum from North America to Europe.

7.3.1 Forced International Migration of Blacks

Some Blacks in the United States are descended from Africans forced to migrate to the Western Hemisphere to be enslaved during the eighteenth century. Slavery is a system whereby one person owns another person as a piece of property and can force the enslaved to work for the owner's benefit. Europeans were responsible for diffusing the practice to the Western Hemisphere. At the height of the eighteenth-century slave demand, a number of European countries adopted a trading pattern called the **triangular slave trade** (Figure 7.3.1).

DNA testing is now helping narrow the range of the ethnic heritage of Blacks. The ancestors of most Blacks came from three areas in West Africa (Figure 7.3.2). The modern-day countries of these three areas are: (1) Senegal, Mali, The Gambia, Guinea, Sierra Leone, and Liberia; (2) Southern Ghana, Togo, Benin, Nigeria, and southeastern Côte d'Ivoire; and (3) Western Democratic Republic of the Congo and Angola.

The large-scale forced migration of Africans caused them unimaginable hardship, separating families and destroying villages. The Africans were packed onto ships at extremely high density, kept in chains, and provided with minimal food and sanitary facilities (Figure 7.3.3). Approximately one-fourth died crossing the Atlantic.

▼ 7.3.2 ORIGINS AND DESTINATIONS OF ENSLAVED AFRICANS

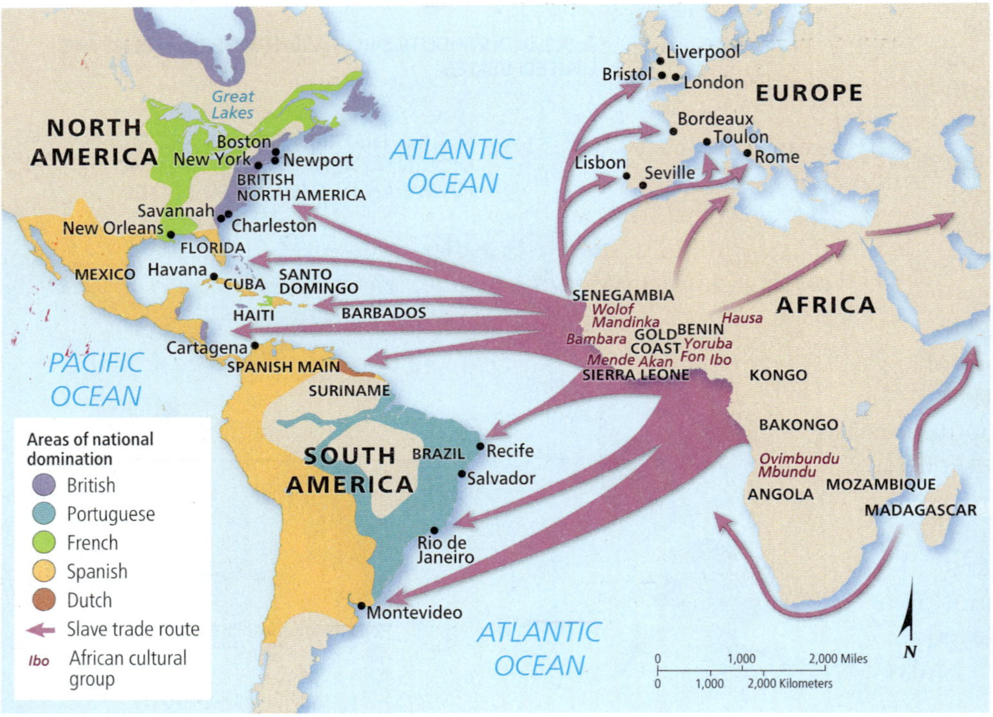

▼ 7.3.3 SHIP PACKED WITH AFRICANS TO BE ENSLAVED IN THE UNITED STATES, 1860
Drawing published in Harper's Weekly.

7.3.2 Internal Migration of Blacks in the United States

The spatial distribution of Blacks in the twenty-first century reflects the legacy of two internal migration patterns during the twentieth century:

- Interregional migration from the U.S. South to northern cities during the first half of the twentieth century.
- Intraregional migration from the inner city to outer city and inner suburban neighborhoods during the second half of the twentieth century.

Interregional Migration of Blacks

At the close of the Civil War, freed from enslavement, most Blacks remained in the rural South during the late nineteenth century, working as sharecroppers. A **sharecropper** works fields rented from a landowner and pays the rent by turning over to the landowner a share of the crops. Sharecropping became less common into the twentieth century, as farm machinery reduced labor demand and industrial jobs pulled migrants to the North.

Today, as a result of interregional migration, many Blacks live in northern and western cities. Southern Blacks migrated north and west in two main waves, the first in the 1910s and 1920s before and after World War I and the second in the 1940s and 1950s before and after World War II. Blacks migrated out of the South along several clearly defined channels (Figure 7.3.4).

▼ 7.3.4 INTERREGIONAL MIGRATION OF BLACKS

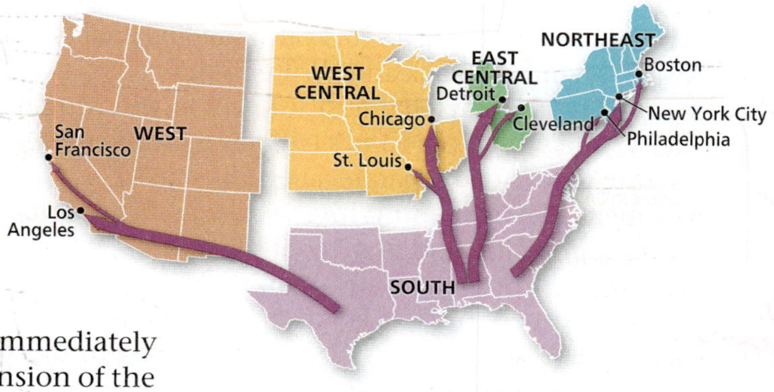

Intraregional Migration of Blacks

When they reached cities, Black immigrants clustered in the one or two neighborhoods where the small numbers who had arrived in the nineteenth century were already living. These areas became known as ghettos, after the term for where Jews were forced to live in the Middle Ages (see Section 6.8.2).

Blacks moved from highly clustered communities into immediately adjacent neighborhoods during the 1950s and 1960s. Expansion of the predominantly Black areas typically followed major avenues that radiated out from the center of the city (Figure 7.3.5).

The expansion of predominantly Black areas in American cities was made possible by "white flight," the emigration of Whites from an area in anticipation of Blacks immigrating into the area. Rather than integrate, Whites fled (Figure 7.3.6).

White flight was encouraged by unscrupulous real estate practices, especially blockbusting and redlining. Under **blockbusting**, real estate agents convinced White homeowners living near a Black area to sell their houses at low prices, preying on their fears that Black families would soon move into the neighborhood and cause property values to decline. The agents then sold the houses at much higher prices to Black families desperate to escape overcrowded conditions.

Redlining is a process by which financial institutions draw red-colored lines on a map and refuse to lend money for people to purchase or improve property within the lines. Through redlining, Blacks were prevented from getting mortgages to buy houses in the neighborhoods where Whites also lived.

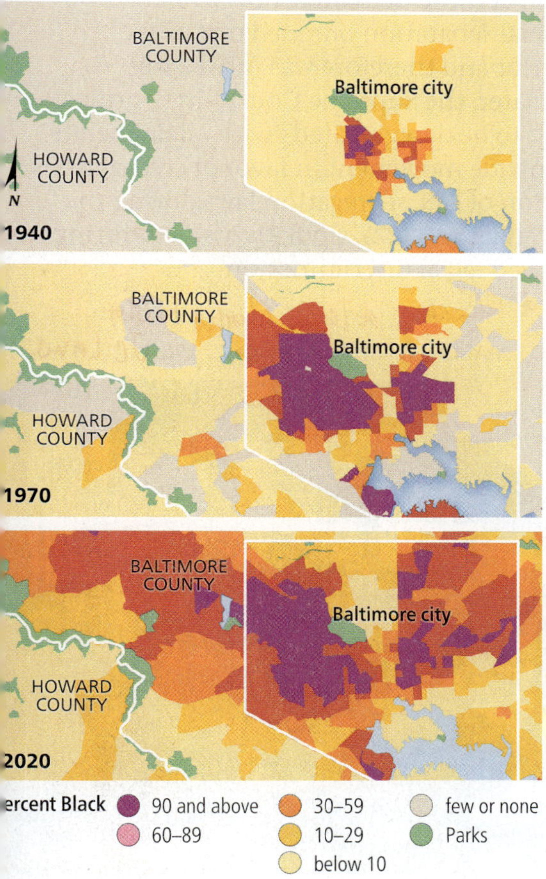

◄ 7.3.5 EXPANSION OF BLACK NEIGHBORHOODS IN BALTIMORE

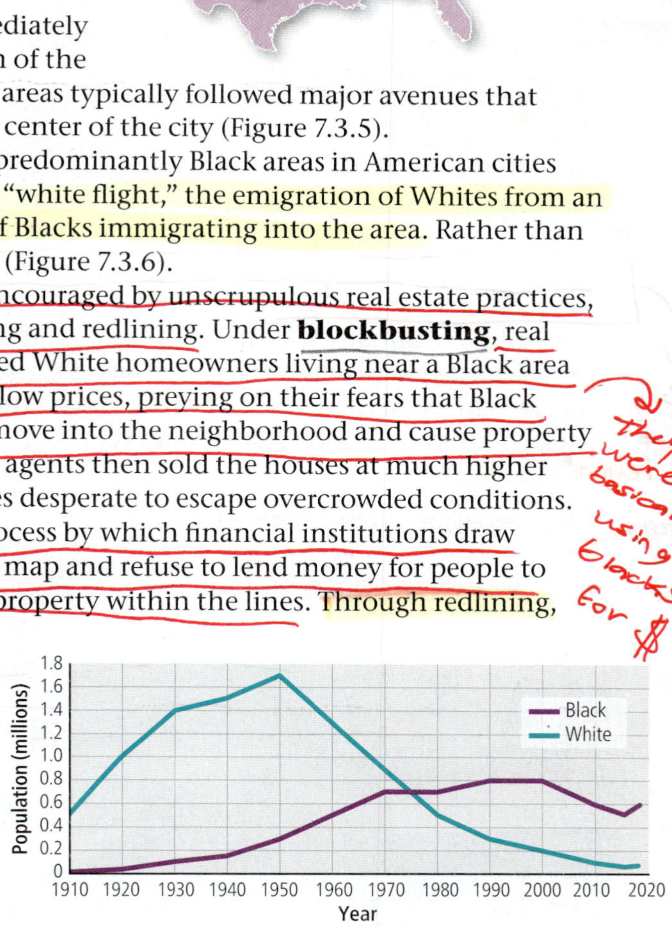

▲ 7.3.6 BLACK AND WHITE POPULATION CHANGE IN DETROIT

Ethnic & Racial Segregation

- **Describe laws that once segregated races.**

In explaining spatial regularities, geographers look for patterns of spatial interaction. A distinctive feature of ethnic and race relations in the United States and South Africa has been the strong discouragement of spatial interaction in the past through legal means. Although segregation laws are no longer in effect, their legacy remains fundamental to understanding the geography of ethnicity and race in both countries.

7.4.1 "Separate but Equal" in the United States

The concept of "separate but equal" was upheld by the U.S. Supreme Court in 1896. Louisiana had enacted a law that required Black and White passengers to ride in separate train cars. In *Plessy v. Ferguson*, the Supreme Court stated that Louisiana's law was constitutional because it provided separate, allegedly equal, treatment of Blacks and Whites, and equality did not mean that Whites had to mix with Blacks.

Once the Supreme Court permitted "separate but equal" treatment of races, southern states enacted a comprehensive set of laws to segregate Blacks from Whites as much as possible. These laws were called "Jim Crow" laws, named for a nineteenth-century song-and-dance act that depicted Blacks offensively. Blacks had to sit in the backs of buses and use different restrooms and drinking fountains (Figure 7.4.1). Shops, restaurants, and hotels could choose to serve only Whites. Separate schools were established for Blacks and Whites. The "Jim Crow" separate but equal laws were not discriminatory, White southerners argued, because buses got Blacks to a destination at the same time as Whites, Blacks could patronize shops, and Black children were allowed to attend school.

Throughout the country, not just in the South, house deeds contained restrictive covenants that prevented the owners from selling to Blacks, as well as to Roman Catholics or Jews in some places. Restrictive covenants also kept Blacks from moving into all-White neighborhoods. And because schools, especially at the elementary level, were located to serve individual neighborhoods, most were segregated by discriminatory housing practices, even if not by legal mandate.

U.S. Civil Rights Movement

The landmark Supreme Court decision *Brown v. Board of Education of Topeka*, in 1954, found that having separate schools for Blacks and Whites was unconstitutional because no matter how equivalent the facilities, racial separation branded minority children as inferior and therefore was inherently unequal. A year later, the Supreme Court further ruled that schools had to be desegregated "with all deliberate speed." A nationwide movement in favor of civil rights forced elimination of U.S. segregation laws during the 1950s and 1960s (Figure 7.4.2). Civil Rights Acts during the 1960s outlawed racial discrimination.

▼ 7.4.1 SEGREGATION IN THE UNITED STATES

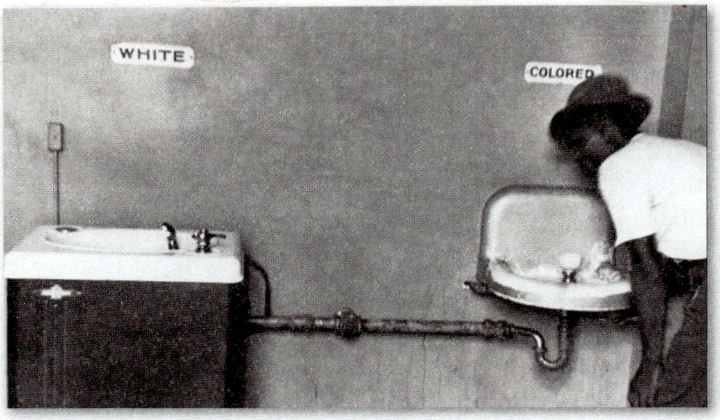

▼ 7.4.2 CIVIL RIGHTS MARCH, WASHINGTON, D.C., 1969

7.4.2 Apartheid in South Africa

Discrimination by race reached its peak in the late twentieth century in South Africa. The cornerstone of the South African policy was the creation of **apartheid**, which was the legal separation of races into different geographic areas. Under apartheid, a newborn baby was classified as being one of four government-designated races: Black, White, Colored (mixed race), or Asian (mostly Indian). According to the 2011 census, 79 percent of South Africans were Black, 9 percent each were Colored and White, and 3 percent were Asian.

The apartheid laws determined where different races could live, attend school, work, shop, travel, and own land (Figure 7.4.3). Blacks were restricted to certain occupations and were paid far lower wages than were Whites for similar work. Blacks could not vote or run for political office in national elections.

▲ 7.4.3 TRAIN CAR IN SOUTH AFRICA FOR WHITES ONLY, 1984

The apartheid system was created by descendants of Whites who arrived in South Africa from the Netherlands in 1652 and settled in Cape Town, at the southern tip of the territory. They were known either as Boers, from the Dutch word for "farmer," or Afrikaners, from the word "Afrikaans," the name of their language, which is a dialect of Dutch. The British seized the Dutch colony in 1795 and controlled South Africa's government until 1948, when the Afrikaner-dominated Nationalist Party won elections. The Afrikaners vowed to resist pressures to turn over South Africa's government to Blacks, and the Nationalist Party created the apartheid laws in the next few years to perpetuate White dominance of the country (Figure 7.4.4).

Legacy of Apartheid

To ensure geographic isolation of races, in 1970 the South African government designated 10 so-called homelands for Blacks (Figure 7.4.5). The White minority government expected every Black to become a citizen of one of the homelands and to move there, unless employed in a White area. In 1970, more than 99 percent of the population in the 10 homelands was Black.

The White-minority government of South Africa repealed the apartheid laws in 1991. The principal anti-apartheid organization, the African National Congress, was legalized, and its leader, Nelson Mandela, who had been convicted of sabotage against the government, was released from jail after more than 27 years of imprisonment. When all South Africans were permitted to vote in national elections for the first time, in 1994, Mandela was overwhelmingly elected the country's first Black president.

Though South Africa's apartheid laws have been repealed, the legacy of apartheid will linger for many years through continued segregation. South Africa's Blacks have achieved political equality, but they are economically disadvantaged compared with White South Africans. For example, South Africa is a major producer of wine, but only a small number of wineries are owned by Blacks.

▼ 7.4.4 BATHROOM IN SOUTH AFRICA FOR BLACKS, COLOREDS, AND ASIANS, 1977

▼ 7.4.5 SOUTH AFRICA'S APARTHEID HOMELANDS

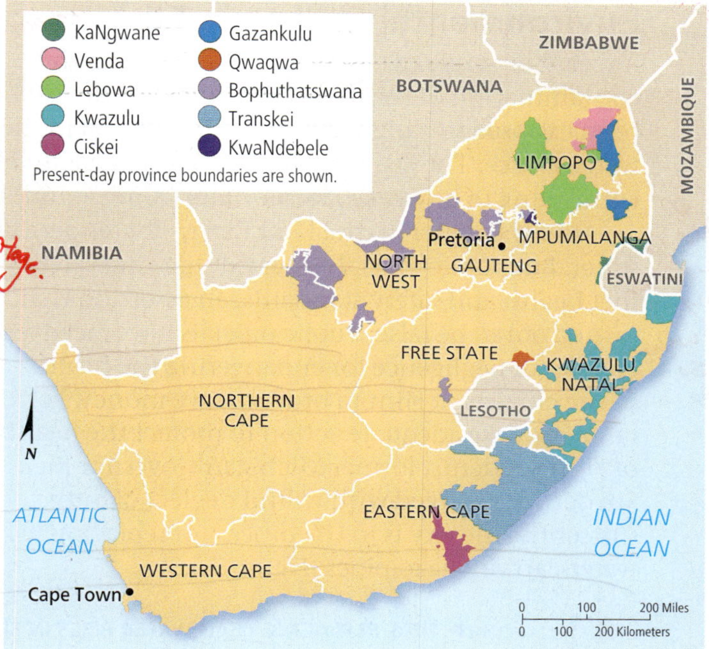

▲ 7.5.1 JEFFERSON DAVIS HIGHWAY
Woodbridge, Virginia.

Black Lives Matter

- Describe laws that once segregated races.

The legacy of racial discrimination and segregation is still felt in the United States. Many Blacks, recent immigrants, and those of other ethnicities remain clustered in urban neighborhoods because of economic and cultural factors. In urban schools, Black and other students of color are now often the majority as White residents have moved out.

7.5.1 Racism on the Landscape

In February 1861, seven states announced that they were seceding from the United States and forming an independent Confederate States of America (CSA). When the Civil War started two months later, four more states joined the CSA. CSA Vice President Alexander Stephens declared at the time that the CSA was formed "upon the great truth that the negro is not equal to the white man; that slavery, subordination to the superior race, is his natural and normal condition."

In the decades after the Civil War, more than 1,500 statues were erected to honor the CSA. Schools, parks, and other elements of the landscape were named for CSA leaders. Most of the CSA memorials were added around 1900, at the time of the creation of the Jim Crow laws (see Section 7.4.1).

For example, the Jefferson Davis Highway was planned in the 1910s and 1920s as a transcontinental route between the East and West Coasts, traversing most of the Confederate states. The road was named for the President of the Confederate States. Portions of the road exist today (Figure 7.5.1).

In recognition of the racist attitudes of the CSA leaders, more than 70 of the monuments honoring the CSA were removed between 2016 and 2020 (Figure 7.5.2). But that represents only a small percentage of those erected in the years following the Civil War.

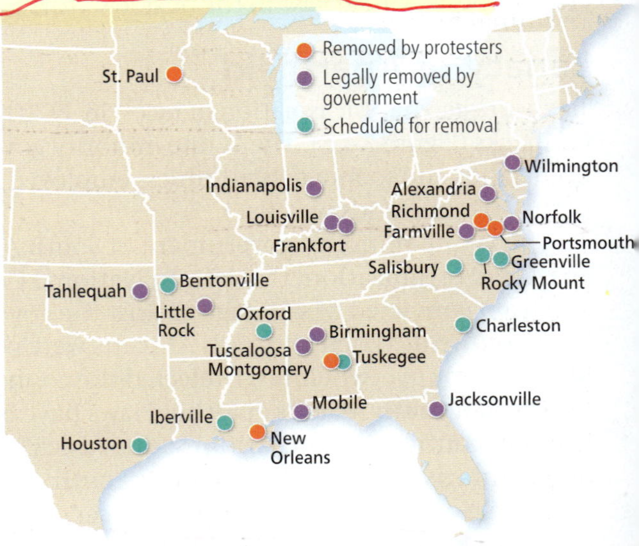

▲ 7.5.2 REMOVED MONUMENTS TO THE CONFEDERACY

Environmental Racism in Flint

A recent example of environmental racism is the contamination of the drinking water in Flint, Michigan. Flint's residents, who are predominantly Black, have been exposed to a high concentration of lead in the drinking water because of insufficient treatment of the water source. The crisis started in 2014 when Flint changed its source of water during a budget crisis. Instead of continuing to buy Lake Huron water from Detroit, the city switched to Flint River water.

After the switch to the Flint River, city residents began complaining about the color, taste, and odor of their water. But public officials asserted that the water met health and safety standards, an assertion that turned out to be false. Public officials have faced legal charges of negligence for endangering the health of the citizens of Flint in order to save money. For failing to take adequate action to protect the health of Flint residents, former Michigan Governor Rick Snyder was charged with willful neglect of duty. Currently, the city is in the process of replacing its water-carrying lead pipes (Figure 7.5.3).

▶ 7.5.3 REPLACING LEAD WATER PIPES IN FLINT

7.5.2 Black Lives Matter

Black Lives Matter (BLM) is a decentralized political and social movement protesting incidents of police brutality and other racially motivated violence against Black people. The BLM movement campaigns against perceived racism toward Black people and educates others about the challenges that Blacks continue to face in the United States.

The hashtag #BlackLivesMatter first appeared in social media in 2013 after the acquittal of a White man, George Zimmerman, in the shooting death of a Black teen, Trayvon Martin, in Sanford, Florida. Widespread street protests started the following year after a Black man, Michael Brown, was fatally shot by a White police officer, Darren Wilson, in Ferguson, Missouri, and another Black man, Eric Garner, died in a prohibited chokehold by a White police officer, Daniel Pantalero, in New York. The movement has since protested the deaths of numerous other Blacks through police actions.

The movement gained strength and widespread support nationally in 2020 following the death of a Black man, George Floyd, at the hands of a White police officer, Derek Chauvin, in Minneapolis. Protests demanding criminal justice reform erupted in several hundred U.S. cities, involving between 15 million and 26 million people, making it one of the largest protest movements in U.S. history (see Section 1.9.1). Washington, D.C., has renamed part of the street outside the White House *Black Lives Matter Plaza* (Figure 7.5.4). For his use of excessive force, failing to comply with Minneapolis Police procedures during the arrest of Floyd, and consciously risking and ultimately taking Floyd's life, Chauvin was convicted of second- and third-degree murder and second-degree manslaughter.

▲ 7.5.4 WASHINGTON, D.C.

▲ 7.5.5 T-MOBILE PARK, SEATTLE

Diffusion of BLM

BLM has diffused from the United States to other countries, in part to show solidarity with demonstrators in the United States and in part to recognize that issues of racism and discrimination are present in many countries. For example, BLM originated in the United Kingdom in 2016 with demonstrations in several cities protesting police brutality and marking the fifth anniversary of the death of Mark Duggan, a Black man, who was killed by police in London under circumstances that were not made clear by the police nor through legal action. Numerous protests were held across the world in 2020 in response to the killing of George Floyd.

BLM has received support from professional sports around the world, including Major League Baseball (Figure 7.5.5), soccer players in England's Premier League (Figure 7.5.6), and players in the National Basketball Association (Figure 7.5.7).

▲ 7.5.6 TYLER ROBERTS AND PATRICK BAMFORD OF LEEDS UNITED FOOTBALL CLUB

▼ 7.5.7 KEVIN GARNETT OF THE BROOKLYN NETS

Ethnicities & Nationalities

- **Explain differences between ethnicities and nationalities.**

In principle, the cultural values shared with others of the same ethnicity derive from religion, language, and material culture, whereas those shared with others of the same nationality derive from voting, obtaining a passport, and performing civic duties. However, this distinction is not always clear.

▲ 7.6.1 RALLY IN SUPPORT OF QUÉBEC INDEPENDENCE, 1995

7.6.1 Ethnicities & Nationalities in North America

Nationality is generally kept reasonably distinct from ethnicity and race in common usage in the United States. Nationality identifies citizens of the United States of America, including those born in the country and those who immigrated and became citizens. Ethnicity identifies groups with distinct ancestry and cultural traditions, such as Mexican Americans, Chinese Americans, or Polish Americans.

In Canada, the distinction between ethnicity and nationality is less clear and more controversial. Québécois are clearly distinct from other Canadians in cultural traditions, especially language. But do the Québécois form a distinct ethnicity within the Canadian nationality or a second French-speaking nationality separate altogether from English-speaking Canadians? The distinction is critical because, if Québécois is recognized as a separate nationality from English-speaking Canadians, the Québec government would have a much stronger justification for breaking away from Canada to form an independent country (Figure 7.6.1).

Ethnicities & Nationalities in the British Isles

Sorting out the inhabitants of the British Isles into ethnicities and nationalities is increasingly controversial. The British Isles are divided between two countries—the Republic of Ireland and the United Kingdom (known officially as the United Kingdom of Great Britain and Northern Ireland). The Republic of Ireland comprises the southern 84 percent of the island of Ireland (called Eire in Irish), and the United Kingdom comprises Great Britain and the northern 16 percent of Eire. The United Kingdom is divided into four main parts: England, Northern Ireland, Scotland, and Wales (Figure 7.6.2).

The nationality of the citizens of the Republic of Ireland is clearly Irish. However, the nationality of the citizens of the United Kingdom is disputed. Does the United Kingdom contain one nationality (called British)? Or does it contain four nationalities—English, Scottish, Welsh, and the Irish of Northern Ireland?

- **English.** The English are descendants of Germanic tribes who crossed the North Sea and invaded the country in the fifth century. These invasions were summarized in Chapter 5 (Languages) in the discussion of the origin of the English language (see Section 5.5.1).
- **Welsh.** The Welsh were Celtic people conquered by England in 1282 and formally united with England through the Act of Union of 1536. Welsh laws were abolished, and Wales became a local government unit.
- **Scots.** The Scots were Celtic people who had an independent country for more than 700 years, until 1603, when Scotland's King James VI also became King James I of England, thereby uniting England and Scotland. The Act of Union in 1707 formally merged the two governments, although Scotland was allowed to retain its own systems of education and local laws.
- **Irish.** The Irish were Celtic people who were ruled by England until the twentieth century, when most of the island became the independent country of Ireland. The northern portion remained part of the United Kingdom.

▶ 7.6.2 COMPONENTS OF THE BRITISH ISLES

7.6.2 Nationalism

Nationalism is loyalty and devotion to a nationality. Nationalism typically promotes a sense of national consciousness that exalts one nation above all others and emphasizes its culture and interests as opposed to those of other nationalities. People display nationalism by supporting a country that preserves and enhances the culture and attitudes of their nationality. States foster nationalism by promoting symbols of the country, such as flags and songs.

Nationalism is an important example of a centripetal force, which was defined in Section 5.1.1 as a cultural value that tends to unify people. (The word *centripetal* means "directed toward the center"; it is the opposite of *centrifugal*, which means "to spread out from the center.") Most countries find that the best way to achieve citizen support is to emphasize shared attitudes that unify the people in support of the country's cultural values.

The Kurds: Ethnicity Without Nationality

A prominent example of an ethnicity without its own nationality is the Kurds, who live in an area comprising parts of eastern Turkey, western Iran, and northern Syria and Iraq (Figure 7.6.3). The Kurds are Sunni Muslims who speak a language in the Iranian group of the Indo-Iranian branch of Indo-European and have distinctive literature, dress, and other cultural traditions.

When the victorious European allies carved up the Ottoman Empire after World War I, they created an independent state of Kurdistan to the south and west of Van Gölü (Lake Van) under the 1920 Treaty of Sèvres. Before the treaty was ratified, however, the Turks, under the leadership of Mustafa Kemal (later known as Kemal Ataturk), fought successfully to expand the territory under their control beyond the small area the allies had allocated to them. The Treaty of Lausanne in 1923 established the modern state of Turkey, and three years later the League of Nations determined that much of Kurdistan would become part of Turkey.

Today around 35 million Kurds are split among several countries; 15 million live in eastern Turkey, 9 million in western Iran, 7 million in northern Iraq, 2 million in Syria, and 2 million in other countries (primarily Germany). Kurds comprise 18 percent of the population in Turkey, 17 percent in Iraq, 11 percent in Syria, and 10 percent in Iran.

▲ 7.6.3 KURDISTAN
The area on the map is the territory that Kurds wish to consolidate into an independent Kurdistan.

Semi-Independent Kurdistan

To foster the development of Turkish nationalism, the Turks have tried repeatedly to suppress Kurdish culture. Use of the Kurdish language was illegal in Turkey until 1991, and laws banning its use in broadcasts and classrooms remain in force. Kurdish nationalists, for their part, have waged a guerrilla war since 1984 against the Turkish army.

Iraq's Kurds have made several unsuccessful attempts to gain independence, including in the 1930s, 1940s, and 1970s. A few days after Iraq was defeated in the 1991 Gulf War, the country's Kurds launched another unsuccessful rebellion. The United States and its allies decided not to intervene in the fight against Iraq on behalf of the Kurdish rebels, but after the revolt was crushed, they sent troops to protect the Kurds from further attacks by the Iraqi army. Iraqi Kurds achieved even more autonomy after the United States attacked Iraq and deposed its president Saddam Hussein on suspicions of possessing weapons of mass destruction and ties to the terrorist organization al-Qaeda. The Kurdistan Regional Parliament governs northern Iraq, with little control exercised by the Iraqi national government (Figure 7.6.4).

▼ 7.6.4 POPE FRANCIS AND MASOUD BARZANI, FORMER PRESIDENT OF THE KURDISTAN REGION OF IRAQ

Ethnic Diversity in Asia

- Identify principal ethnic conflicts in Western & West-Central Asia.

The lack of correspondence between the territories occupied by ethnicities and by nationalities is especially severe in Western Asia and in West-Central Asia.

7.7.1 Ethnic Diversity in Western Asia

Western Asia has been the center of long-standing conflicts among ethnicities. Ethnic diversity is intermixed with religious diversity (refer to Chapter 6). The region has also experienced military intervention by states outside the region and terrorist conflict. Figure 7.7.1 depicts the diversity of ethnicities in Western Asia. The most numerous ethnicities in the region are Turks, Iraqi Arabs, Azerbaijanis, Kurds, and Syrian Arabs.

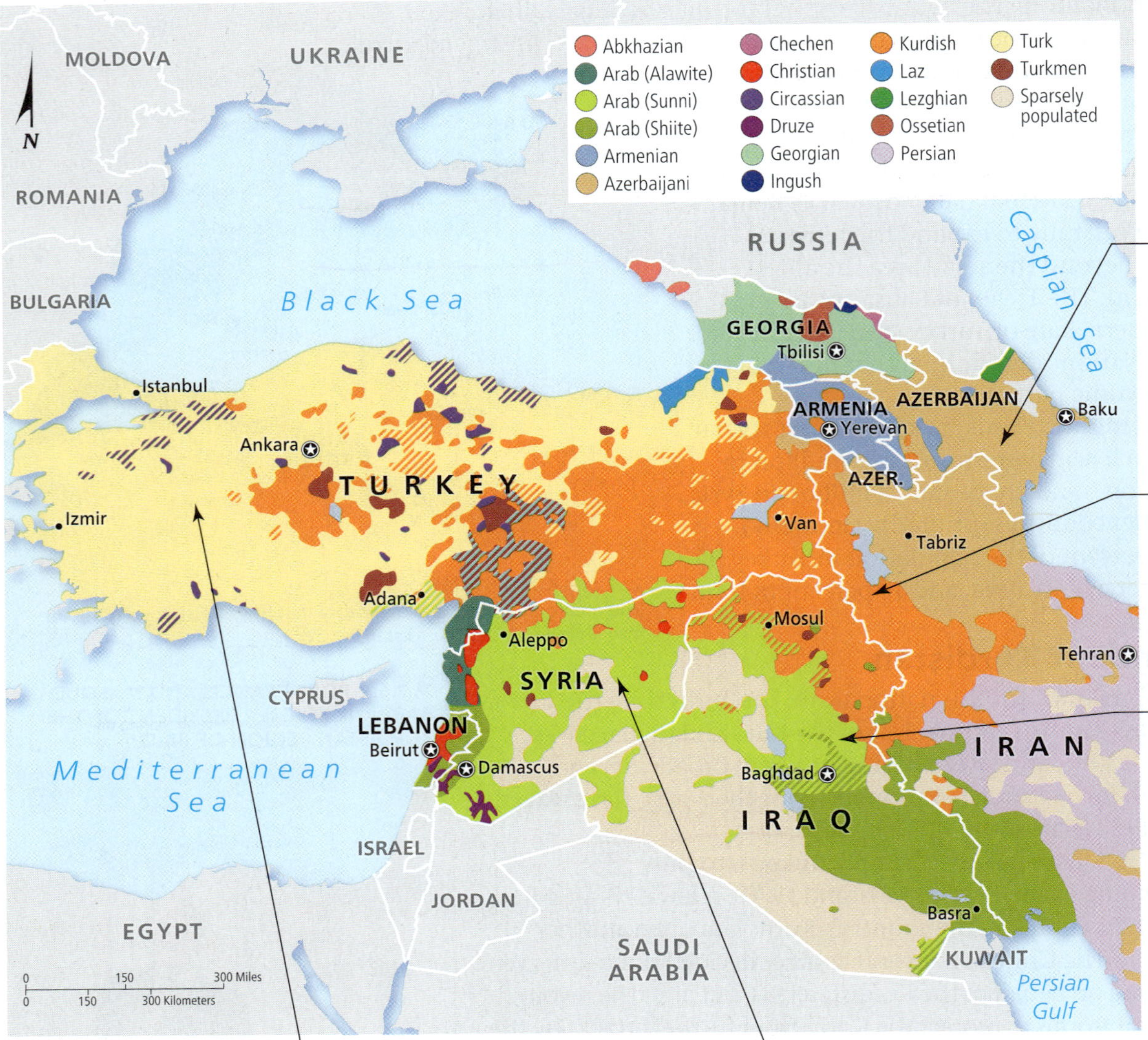

7.7.1 ETHNICITIES IN WESTERN ASIA

AZERBAIJANIS
Azerbaijanis trace their roots to Turkish invaders who migrated from Central Asia in the eighth and ninth centuries and merged with the existing Persian population. Azerbaijanis are predominantly Shiite Muslims and speak Azerbaijani, a Turkic language.

KURDS
The Kurds are predominantly Sunni Muslims who speak Kurdish, a language in the Indo-Iranian branch of Indo-European. The Kurds have inhabited Western Asia since ancient times (refer to section 7.6.2).

IRAQI ARABS
Iraqi Arabs are descended from Arabs who emigrated in ancient times from Arabia and mingled with peoples already living in the area known then as Mesopotamia. Iraqi Arabs are predominantly Muslim but are divided between Shiite and Sunni.

TURKS
Ethnic Turks are descended from migrants to present-day Turkey around 1,000 years ago, most likely from Siberia. Turks are predominantly Sunni Muslims and speak Turkish, a Turkic language.

SYRIANS
Arabs comprise 90 percent of Syria's population. The ethnic Arabs are divided by religion: 64 percent are Sunni Muslim, 11 percent Alawi Muslim, 10 percent Christian, 3 percent Druze, and 2 percent other branches of Islam. The most numerous Christian denominations are Greek Orthodox and Greek Catholic.

7.7.2 Ethnic Diversity in West-Central Asia

Dozens of ethnicities inhabit a 4-million-square-kilometer region of West-Central Asia, allocated among eight nationalities (Figure 7.7.2). The five most numerous ethnicities are Punjabis, Persians, Pashtuns, Uzbeks, and Azerbaijanis.

Ethnicities in Afghanistan

Ethnicities in Afghanistan include 42 percent Pashtun, 27 percent Tajik, and 9 percent each Hazara and Uzbek. Afghanistan's ethnic diversity can be attributed to its position as a key trading route between Central, Southern, and Western Asia. The current unrest among Afghanistan's ethnicities dates from 1979.

After many years of infighting among ethnicities, a Pashtun faction called the Taliban gained control over most of the country in 1996. The Taliban imposed strict laws on Afghanistan, according to Islamic values as the Taliban interpreted them. The United States invaded Afghanistan in 2001 and overthrew the Taliban-led government because it was harboring terrorists (see Section 8.10.2). Removal of the Taliban unleashed a new struggle for control of Afghanistan among the country's many ethnicities. The Taliban regained control of Afghanistan in 2021 and reimposed harsh laws. As a result of the fighting, more than 4 million people have become refugees or internally displaced persons.

◀ **7.7.2 ETHNICITIES IN WEST-CENTRAL ASIA**

UZBEKS

Turkic people migrated here from China around 1,500 years ago. They were among the first people of Central Asia to convert to Islam around 900 years ago. They speak a Turkish language.

PUNJABIS

Punjabi is a term applied 200 years ago to people of varying cultural heritages inhabiting the Punjab region of South Asia. A distinct ethnic identity was forged initially through sharing the Punjabi language, which is classified in the Indo-Iranian branch of Indo-European. The partition of South Asia in 1947 divided the Punjab between the newly independent countries of India and Pakistan.

PERSIANS

Persians are believed to be descendants of Indo-European peoples who migrated from Central Asia into what is now Iran several thousand years ago. Persians are predominantly Shiite Muslims and speak Persian, a language in the Indo-Iranian branch of Indo-European.

PASHTUNS

The Pashtuns have inhabited this region since prehistoric times. Cultural unity has come primarily from speaking various dialects of Pashto, a language in the Indo-Iranian branch of Indo-European. Most Pashtuns have practiced Islam since the arrival of that religion in the area in the seventh century.

▲ 7.8.1 UYGHUR STREET VENDORS, KASHGAR, CHINA

Ethnic Cleansing in Asia

- Identify examples of ethnic cleansing in East & Southeast Asia.

Throughout history, ethnic groups have been forced to flee from other ethnic groups' more powerful armies. In the most extreme cases, hatred of other ethnicities can result in ethnic cleansing and genocide.

7.8.1 Ethnic Cleansing & Genocide

Fear of people of a particular ethnicity is **ethnophobia**. It can overlap with **xenophobia**, which is fear of people who are from other countries. These fears of others sometimes result in the forced migration of ethnic groups. At its most extreme, an ethnic group can be subject to ethnic cleansing and genocide.

The United Nations defines **ethnic cleansing** as a purposeful policy designed by one ethnicity or religious group to remove by violent and terror-inspiring means the civilian population of another ethnic or religious group from certain geographic areas. **Genocide** is the mass killing of a group of people in an attempt to eliminate the entire group from existence. In recent years, instances of ethnic cleansing and genocide have occurred in Asia, Africa, and Europe.

Ethnic cleansing and genocide are especially important for cultural geography because they change the spatial distribution of ethnicities, and they do so through force and criminal violence. These violent and calculated processes are undertaken to rid an area of an entire ethnicity so that the surviving ethnic group can be the sole inhabitants of an ethnically homogeneous region. The point is not simply to defeat an enemy or to subjugate them, as was the case in traditional wars. Rather than a clash between armies of male soldiers, ethnic cleansing and genocide involve the removal of every member of the less powerful ethnicity—women as well as men, children as well as adults, the frail elderly as well as the strong youth.

Ethnic Cleansing of Uyghurs

In Asia, the most extensive example of ethnic cleansing is the forced removal of the Uyghurs in China. Uyghurs are ethnic Turks clustered primarily in Xinjiang Uyghur Autonomous Region, in Western China (Figure 7.8.1). Up to 2 million of the 13 million Uyghurs in Xinjiang are believed to have been placed in a network of detention centers across the region, according to the U.S. State Department.

Former detainees allege they were subjected to indoctrination, sexually abused, and in some cases forcibly sterilized. In other cases, they were bused away from their villages and families to be used as cheap labor in factories.

China denies allegations of human rights abuses. China's government calls the centers where Uyghurs have been placed Vocational Education and Training Centers (Figures 7.8.2 and 7.8.3). The government says the centers are necessary to prevent religious extremism and terrorism.

▼ 7.8.2 A FACILITY SAID BY INTERNATIONAL OBSERVERS TO BE A DETENTION CENTER FOR UYGHURS, AKSU, CHINA

▼ 7.8.3 AIR PHOTO OF URUMQI NUMBER 3 DETENTION CENTER, DABANCHENG, CHINA

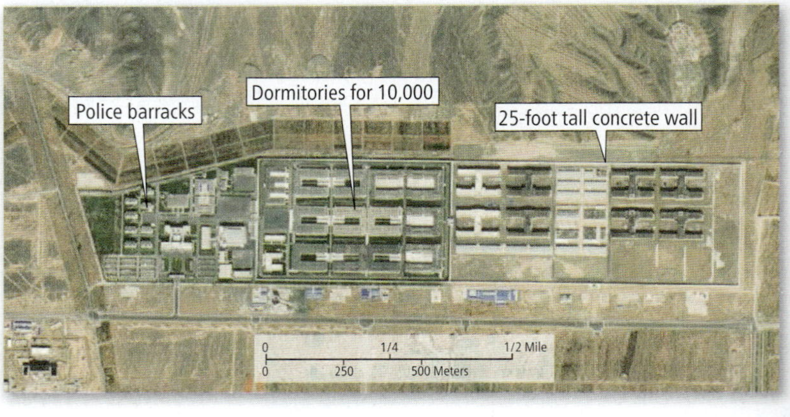

7.8.2 Ethnic Cleansing & Genocide in Myanmar

Asia's most extreme example of ethnic cleansing and genocide in recent years has been in Myanmar, a country of 54 million inhabitants. The attacks have been directed against the Rohingya, an ethnic group living in Myanmar's far western Rakhine State (Figure 7.8.4). Rakhine is the least economically developed state in Myanmar, which in turn is one of the world's least economically developed countries.

▲ **7.8.4 ETHNIC CLEANSING OF ROHINGYA**
These ethnically cleansed Rohingya have been forced to migrate to Bangladesh.

The Rohingya are predominantly Sunni Muslims who speak an Indo-European language. In contrast, the largest number of people in Myanmar are Theravada Buddhists who speak Sino-Tibetan languages.

Most Rohingya migrated in the nineteenth century from present-day Bangladesh to present-day Myanmar, when both were British colonies. The Myanmar government claims that because the Rohingya immigrated during the British colonial era, they are living in Myanmar illegally. In 1982, the Myanmar government enacted laws that took away the Rohingya's citizenship, land, and rights to attend school and hold jobs.

In 2016, Rohingya attacked some police and military posts. In retaliation, the Myanmar military launched a massive ethnic cleansing operation against the Rohingya. The pretext was to eliminate the group that had attacked the police and military, but the United Nations concluded that more than 1 million Rohingya have been ethnically cleansed. Their villages have been destroyed, and they have been forced to move from Rakhine to other countries, primarily Bangladesh.

According to the United Nations, the attacks against the Rohingya led by the Myanmar military have included massacres, mass killings, looting and then burning of villages, and sexual violence including gang rapes. Throughout the region, Rohingya also fall victim to human trafficking and smugglers in their attempts to reach safety.

The government of Myanmar issued directives in 2020 ordering the end to the genocide activities against the Rohingya. But the roughly 1 million ethnically cleansed Rohingya are living in tents and other temporary structures (Figure 7.8.5).

Conditions in the camps continue to be dire, especially after the onset of the COVID-19 pandemic in 2020. Nearly 1 million Rohingya remain completely dependent on humanitarian aid, and malnutrition and other health conditions persist in what is the largest refugee settlement in the world.

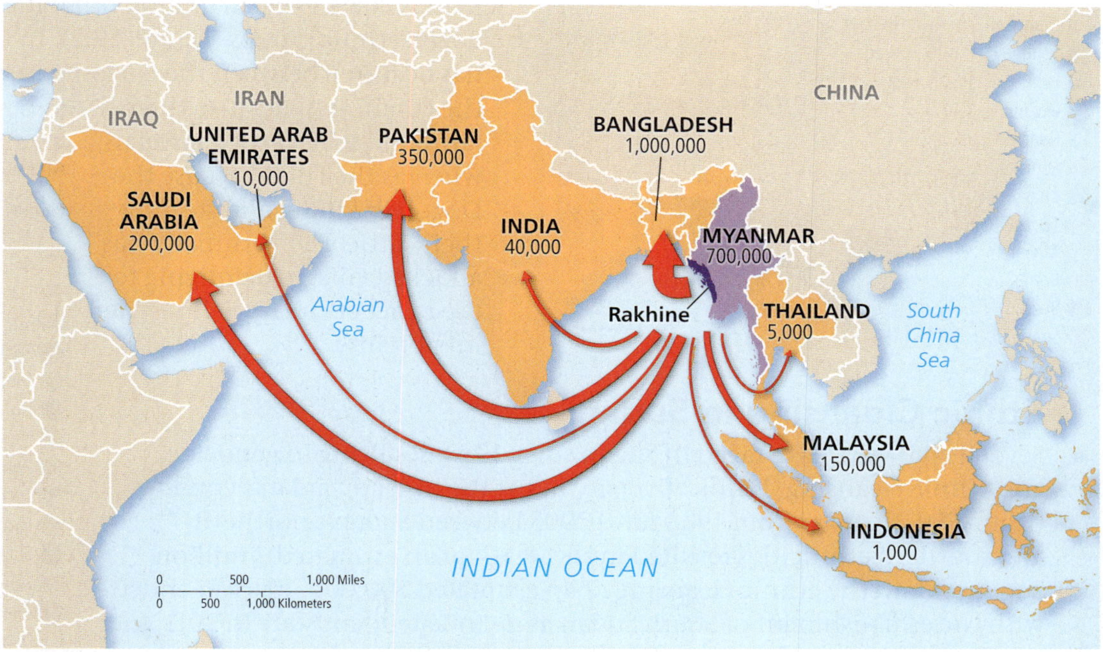

▲ **7.8.5 DESTINATIONS OF ETHNICALLY CLEANSED ROHINGYA**

Ethnic Cleansing in Africa

- **Identify principal ethnic conflicts and examples of genocide in Africa.**

Africa contains several thousand ethnicities with distinct languages, religions, and material culture. For most Africans, ethnicity has traditionally been a more important element of cultural identity than nationality. Competition among ethnicities has led to ethnic cleansing and in some cases genocide.

7.9.1 Ethnicities & Nationalities in Africa

The precise number of ethnicities in Africa is hard to determine because it is often unclear whether a particular group forms a distinct ethnicity or is part of an ethnicity that encompasses several similar groups. During the late nineteenth and early twentieth centuries, European countries carved up Africa into a collection of colonies, with little regard for the distribution of these ethnicities.

When European colonies in Africa became independent states, especially during the 1950s and 1960s, the areas of the new states typically matched the colonial administrative units imposed by the Europeans rather than the historical distribution of ethnicities. As a result, each new state contained more than one ethnicity, and some ethnicities were divided among more than one state (Figure 7.9.1). Conflict among ethnicities is widespread in Africa, largely because the historical distribution of ethnicities bears little relationship to present-day nationalities.

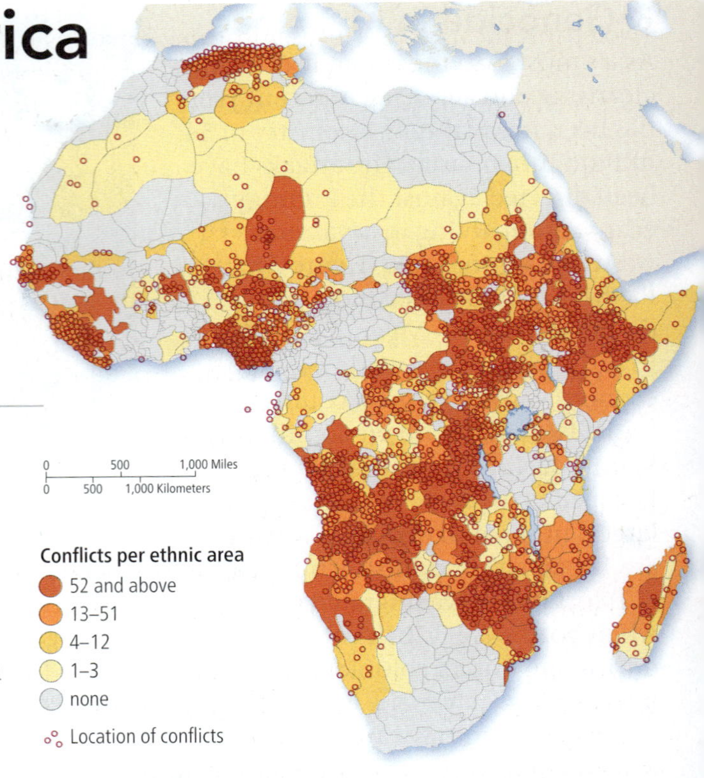

▲ 7.9.1 CONFLICTS IN AFRICA BY ETHNIC AREA

▼ 7.9.2 ETHNIC DIVERSITY IN MALI

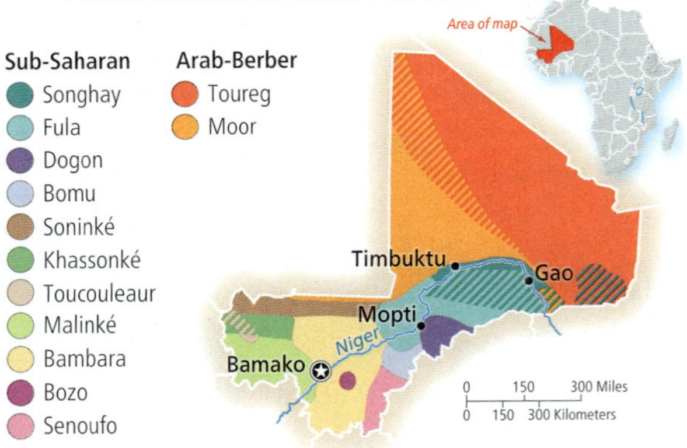

Ethnic Cleansing in Mali

Mali, a country of 20 million inhabitants, is home to more than a dozen ethnic groups (Figure 7.9.2). In recent years, ethnic cleansing has occurred between the Dogon and Fulani. The Dogon claim that the Fulani trespass into their farmland to feed their animals, whereas the Fulani accuse the Dogon of killing and stealing their cattle. Each ethnicity has attacked villages belonging to the other ethnicity, killing the inhabitants.

▼ 7.9.3 ETHNIC DIVERSITY IN SUDAN AND SOUTH SUDAN

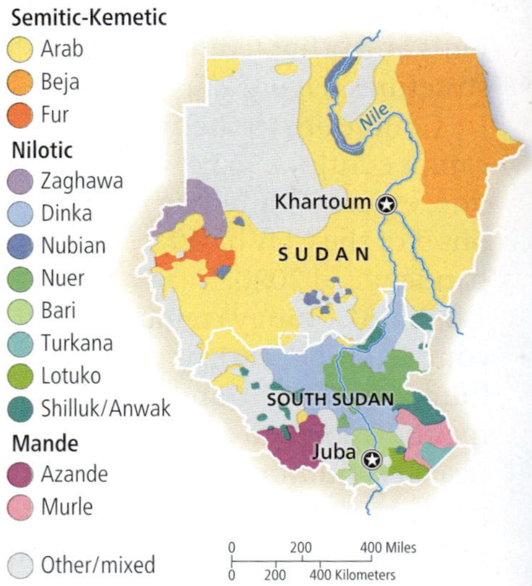

Ethnic Cleansing in Sudan

Several civil wars have raged in Sudan since 1983, resulting in genocide and ethnic cleansing. Ethnic diversity lies at the heart of Sudan's conflicts (Figure 7.9.3). A war from 1983 until 2005 between Sudan's northern and southern ethnicities resulted in the death of an estimated 2 million Sudanese and the ethnic cleansing of an estimated 700,000. The war ended with the establishment of South Sudan as an independent state in 2011. In the Darfur region of Sudan, 3 million people have been victims of ethnic cleansing and 300,000 have been victims of genocide.

7.9.2 Ethnic Cleansing & Genocide in Central Africa

Rwanda and Burundi, tiny countries in central Africa, have suffered long-standing conflicts between the two countries' two ethnic groups, the Hutus and Tutsis. The two ethnicities speak the same language, hold similar beliefs and practice similar social customs, and intermarriage has lessened the physical differences between the two ethnic groups. Yet Hutus and Tutsis have engaged in large-scale ethnic cleansing and genocide.

Hutus constituted a majority of the population of Rwanda and Burundi historically, but Tutsis controlled kingdoms there for several hundred years and turned the Hutus into their serfs. Rwanda and Burundi became colonies of Germany during the 1880s and then of Belgium between 1924 and 1962. During the colonial period, the Tutsis retained leadership positions in both Rwanda and Burundi.

When Rwanda became an independent country in 1962, Hutus gained power and undertook ethnic cleansing and genocide against the Tutsis (Figure 7.9.4). Descendants of the ethnically cleansed Tutsis invaded Rwanda in 1990, launching a three-year civil war. Meanwhile in Burundi, where the Tutsis remained in power, a civil war resulted in genocide committed by and against both Hutus and Tutsis.

An agreement to share power in Rwanda was signed in 1993, but genocide resumed after an airplane carrying the presidents of Rwanda and neighboring Burundi—both Hutus—was shot down by a surface- to-air missile in 1994. International intelligence groups and independent researchers have never been able to determine whether the attacker was a Hutu or a Tutsi presumably trying to scuttle the peace agreement. This attack followed an assassination a few months earlier of the previous president of Burundi, the first Hutu to be elected president of that country.

After the assassination of the two presidents in 1994, Hutus launched a genocide campaign, killing an estimated 800,000 Tutsis in Rwanda and 300,000 in Burundi. However, the Tutsis prevailed in both countries, and reprisals by Tutsis added to the total fatalities. Rwanda continues to be governed by Tutsis, but Burundi has been led since 2005 by democratically elected Hutus.

▲ **7.9.4 ETHNIC CLEANSING, RWANDA**
Kigali Genocide Memorial Centre contains mass graves of victims of the 1994 genocide.

Conflict in Democratic Republic of the Congo

The conflict between Hutus and Tutsis spilled over into the Democratic Republic of Congo (DRC), central Africa's largest and most populous country, with considerable mineral wealth. The DRC is estimated to be home to more than 200 distinct ethnicities. Most Congolese are classified as ethnic Bantus, but this encompasses a large number of specific ethnic groups (Figure 7.9.5). The DRC is considered to have suffered from the world's deadliest wars in the past 70 years. More than 5 million have died in the DRC's ongoing civil wars, mostly from malaria, diarrhea, pneumonia, and malnutrition, aggravated by displacement and unsanitary and overcrowded living conditions.

Tutsis were instrumental in the successful overthrow of the DRC's longtime president Joseph Mobutu in 1997. After succeeding Mobutu as president, Laurent Kabila relied heavily on Tutsis and permitted them to kill some of the Hutus who had been responsible for atrocities against Tutsis a few years earlier.

Kabila soon split with the Tutsis and turned for support to Hutus and other ethnic groups that also hated Tutsis. Kabila was assassinated in 2001 and succeeded by his son, who negotiated an accord with many of the ethnic groups. Despite the accord, conflict among the country's many ethnicities has continued for the last two decades.

◀ **7.9.5 ETHNICITIES IN THE DEMOCRATIC REPUBLIC OF THE CONGO**

Ethnic Cleansing in Europe

- **Describe the fate of ethnicities after the breakup of Yugoslavia.**

Ethnic cleansing in the former Yugoslavia is part of a complex pattern of ethnic diversity in the region of southeastern Europe known as the Balkans, named for the Balkan Mountains. The Balkans has long been a hotbed of ethnic unrest.

7.10.1 Ethnic Diversity in Former Yugoslavia

After World War I, the Allies created a new country, Yugoslavia, to unite several Balkan ethnicities that spoke similar South Slavic languages. The prefix "Yugo" in the country's name derived from the Slavic word for "south." Longtime leader Josip Broz Tito (prime minister 1945–1953 and president 1953–1980) was instrumental in forging a Yugoslav nationality. Central to Tito's vision of a Yugoslav nationality was acceptance of ethnic diversity in language and religion. Individuals from the five most numerous ethnicities—Croat, Macedonian, Montenegrin, Serb, and Slovene—were allowed to exercise considerable control over the areas they inhabited within Yugoslavia.

▲ 7.10.1 ETHNICITIES IN FORMER YUGOSLAVIA

Rivalries among ethnicities resurfaced in Yugoslavia during the 1980s after Tito's death, leading ultimately to its breakup into several small countries (Figure 7.10.1). Because the boundaries of the new countries did not match the distribution of ethnicities, the breakup of Yugoslavia did not happen peacefully. Several episodes of ethnic cleansing ensued.

Ethnic Cleansing in Kosovo

At the time of the breakup of Yugoslavia, 82 percent of the population of Kosovo were ethnic Albanians and 10 percent were Serbs. Nonetheless, Kosovo was controlled by Serbia. Serbia had a historical claim to Kosovo, having controlled it between the twelfth and fourteenth centuries. Serbs fought an important—though losing—battle in Kosovo against the Ottoman Empire in 1389. In recognition of its role in forming the Serb ethnicity, Serbia was given control of Kosovo when Yugoslavia was created in the early twentieth century.

With the breakup of Yugoslavia, Serbia took direct control of Kosovo and launched a campaign of ethnic cleansing of the Albanian majority. At its peak in 1999, Serb ethnic cleansing forced more than 800,000 of Kosovo's 2 million ethnic Albanian residents from their homes, mostly to camps in Albania (Figure 7.10.2). Outraged by the ethnic cleansing, the United States and Western European countries, operating through the North Atlantic Treaty Organization (NATO), launched an air attack against Serbia. The bombing campaign ended when Serbia agreed to withdraw all of its soldiers and police from Kosovo.

Kosovo declared its independence from Serbia in 2008. Around 115 countries, including the United States, recognize Kosovo as an independent country, but Serbia, Russia, China, and their allies oppose it. However, Kosovo's declaration of independence induced nearly 90 percent of the country's Serbs to leave.

▶ 7.10.2 DAMAGE IN PEJA, KOSOVO, AFTER ETHNIC CLEANSING

7.10.2 Ethnic Cleansing in Bosnia & Herzegovina

The largest scale of ethnic cleansing in the former Yugoslavia occurred in the newly formed country Federation of Bosnia & Herzegovina. An estimated 2.7 million people were ethnically cleansed in Bosnia & Herzegovina in 1991 and 1992, in a country of around 3.3 million people.

Bosnia & Herzegovina was the most ethnically diverse republic of the former Yugoslavia. At the time of the breakup of Yugoslavia, the population of Bosnia & Herzegovina was 44 percent Bosniaks, 31 percent Serbs, and 17 percent Croats (Figure 7.10.3a). Bosniaks are frequently called Bosnian Muslims, in recognition of their predominant religion. Rather than live in an independent multiethnic state with a Muslim plurality, Bosnia & Herzegovina's Serbs and Croats fought to unite the portions of the republic that they inhabited with Serbia and Croatia, respectively.

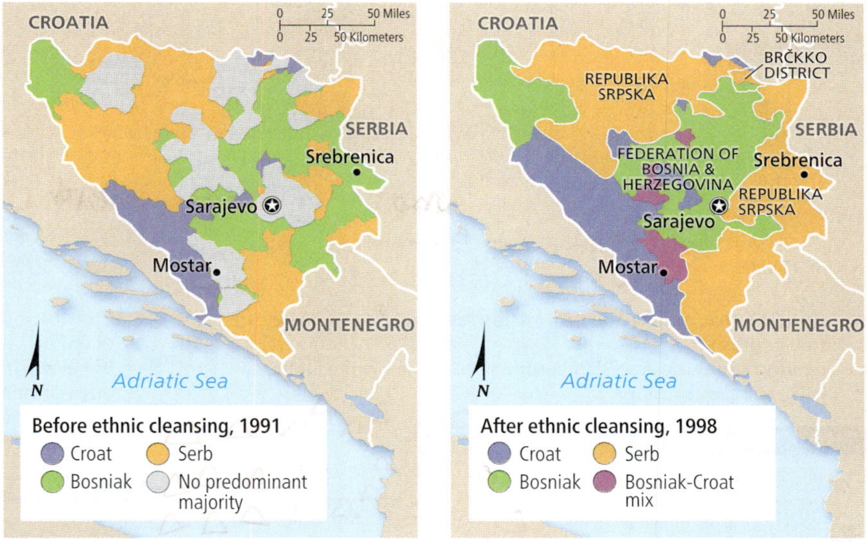

▲ **7.10.3 ETHNICITIES IN BOSNIA & HERZEGOVINA**
(a) Before the war during the 1990s. (b) After ethnic cleansing.

To strengthen their cases for breaking away from Bosnia & Herzegovina, Serbs and Croats engaged in ethnic cleansing of Bosniaks. Ethnic cleansing ensured that areas did not merely have majorities of Bosnian Serbs and Bosnian Croats but were ethnically homogeneous and therefore better candidates for union with Serbia and Croatia. Ethnic cleansing by Serbs against Bosniaks was especially severe because the territory inhabited by Serbs in Bosnia comprised several discontinuous areas, many of which were separated from Serbia by areas with Bosniak majorities. By ethnically cleansing Bosniaks from intervening areas, Bosnian Serbs created one continuous area of Serb domination rather than several discontinuous ones (Figure 7.10.3b).

Bosnia & Herzegovina After Ethnic Cleansing

Accords reached in Dayton, Ohio, in 1996 by leaders of the various ethnicities divided Bosnia & Herzegovina into three regions, one each dominated, respectively, by Bosniaks, Croats, and Serbs. The Bosniak and Croat regions were combined into a federation, with some cooperation between the two groups, but the Serb region has operated with almost complete independence in all but name from the others.

The International Criminal Tribunal for the former Yugoslavia, established by the United Nations, convicted Croat and Serb leaders of war crimes for the ethnic cleansing of Bosniaks. Yet ethnic cleansing was ultimately successful. The Dayton agreements gave Serbs control of around 50 percent of the land even though they comprised only 40 percent of the population of Bosnia & Herzegovina. Croats got around 20 percent of the land but comprised only 15 percent of the population. Bosniaks, with 45 percent of the population, got only 30 percent of the land. Nonetheless, Bosnia & Herzegovina is once again a relatively peaceful place (Figure 7.10.4).

(a)

(b)

(c)

◀ **7.10.4 A SYMBOL OF ETHNIC CLEANSING IN BOSNIA & HERZEGOVINA**
(a) The Stari Most (old bridge), in the city of Mostar, was built in 1566. (b) The bridge was blown up by Croats in 1993 as part of ethnic cleansing against the Bosniaks. (c) With the end of the war, the bridge was rebuilt in 2004.

CHAPTER 7

Review, Analyze, & Apply

KEY ISSUE 1 Where Are Ethnicities & Races Distributed?

- Ethnicity, race, and nationality are distinct concepts, but they are often confused.
- The most numerous races and ethnicities in the United States have distinctive distributions.

THINKING GEOGRAPHICALLY

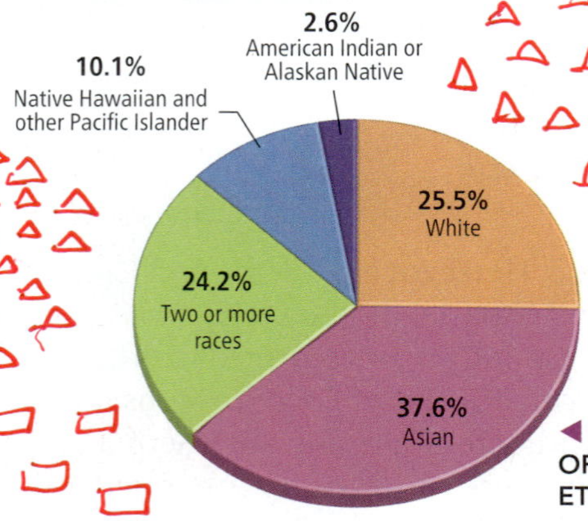

◀ 7.CR.1 DISTRIBUTION OF RACES AND ETHNICITIES IN HAWAII

Hawaii demonstrates a racial and ethnic diversity distinct from the other 49 states. What geographic factors contribute to Hawaii's distinctive ethnic and racial distribution?

KEY ISSUE 2 Why Do Ethnicities Have Distinctive Distributions?

- Ancestors of some Blacks were forcibly enslaved and transported to the South of the United States.
- Many Blacks migrated from the South during the early twentieth century.
- Segregation of races was legal in the United States and South Africa for much of the twentieth century.
- Black Lives Matter is a decentralized movement protesting continued racism and police brutality against Blacks.

THINKING GEOGRAPHICALLY

◀ 7.CR.2 BLACK TOWNSHIP, SOUTH AFRICA

Since the repeal of apartheid laws, races in South Africa continue to be highly segregated. Why might spatial separation persist despite the end of legal segregation?

KEY ISSUE 3 How Do Ethnicities & Nationalities Differ?

- The distribution of ethnicities does not always match the boundaries of countries.
- Some ethnicities, such as the Kurds, are divided among more than one nationality.
- Lack of correspondence between the boundaries of ethnicities and nationalities is especially severe in western areas of Asia.

THINKING GEOGRAPHICALLY

▲ 7.CR.3 KURDS CELEBRATE NEWROZ (KURDISH NEW YEAR)

The Kurds claim to be the world's largest ethnic group not in control of a country. Based on Figures 7.7.1 and 7.7.2 , what other ethnic groups might have a strong claim to reorganize territorial boundaries so that they could become the majority? What are some of the obstacles that might be faced in sharing power with other ethnicities?

KEY ISSUE 4 Why Do Ethnic Cleansing & Genocide Occur?

- Ethnic cleansing is a process in which a more powerful ethnic group forcibly removes a less powerful one in order to create an ethnically homogeneous region.
- Genocide is the mass killing of a group of people in an attempt to eliminate the entire group from a region.
- Examples of ethnic cleansing and genocide can be found in Asia, Africa, and Europe.

THINKING GEOGRAPHICALLY

▲ 7.CR.4 STARI MOST BRIDGE DIVE, MOSTAR

Professional divers hold competitions to dive off the Stari Most bridge in Mostar (refer to Figure 7.10.4). Why might the bridge be an important place to hold professional diving competitions? The United Nations Educational, Scientific and Cultural Organization (UNESCO) has placed the bridge on its list of World Heritage sites. Why would the bridge be considered a World Heritage site? Consult unesco.org and search for World Heritage site #946.

Key Terms

Apartheid (7.4.2) Laws (no longer in effect) in South Africa that physically separated different races into different geographic areas.

Black Lives Matter (BLM) (7.5.2) Movement that campaigns against violence and perceived racism toward Blacks and educates others about the challenges that Black Americans continue to face in the United States.

Blockbusting (7.3.2) A deceptive business tactic whereby real estate agents convince White property owners to sell their houses at low prices by warning that persons of color will soon move into the neighborhood and cause a significant fall in housing prices.

Ethnic cleansing (7.8.1) A purposeful policy designed by one ethnic or religious group to remove by violent and terror-inspiring means the civilian population of another ethnic or religious group from certain geographic areas.

Ethnicity (7.1.1) Identity with a group of people who share the cultural traditions of a particular homeland or hearth.

Ethnophobia (7.8.1) Fear of people of a particular ethnicity.

Genocide (7.8.1) The mass killing of a group of people in an attempt to eliminate the entire group from existence.

Nationalism (7.6.2) Loyalty and devotion to a particular nationality.

Nationality (7.1.1) Identity with a group of people who share legal attachment to a particular country.

Race (7.1.1) Identity with a group of people who are perceived to share a physical trait, such as skin color.

Racism (7.1.1) The belief that race is the primary determinant of human traits and capacities and that racial differences produce an inherent superiority of a particular race.

Racist (7.1.1) A person who subscribes to the beliefs of racism.

Redlining (7.3.2) A process by which financial institutions draw red-colored lines on a map and refuse to lend money for people to purchase or improve property within the lines.

Sharecropper (7.3.2) A person who works fields rented from a landowner and pays the rent and repays the loans by turning over to the landowner a share of the crops.

Social construction (7.1.1) An idea or a meaning that is widely accepted as natural by a society but may not represent a reality shared by those outside the society.

Triangular slave trade (7.3.1) A practice, primarily during the eighteenth century, in which European ships transported enslaved people from Africa to Caribbean islands, molasses from the Caribbean to Europe, and trade goods from Europe to Africa.

Xenophobia (7.8.1) Fear of people who are from other countries.

Geospatial Analysis

Log in to the Mastering Geography Study Area to access MapMaster 2.0.

Ethnic Diversity in Sub-Saharan Africa

Sub-Saharan Africa has the world's most ethnically diverse countries.

Add the *Ethnic Diversity* layer. ***Select*** the *Settings* icon from the *Legend*, and ***select*** *Show Political Labels*. ***Zoom*** to sub-Saharan Africa.

1. ***Select*** one country that is most ethnically diverse and one country that is least or less ethnically diverse. Place the cursor over these two countries. How do the data show that one country is more ethnically diverse than the other?
2. ***Add*** the *Refugees, by Country of Origin* layer, ***select*** *Split map window*. Does the spatial association between the two maps appear to be strong or weak? Why might that be the case?

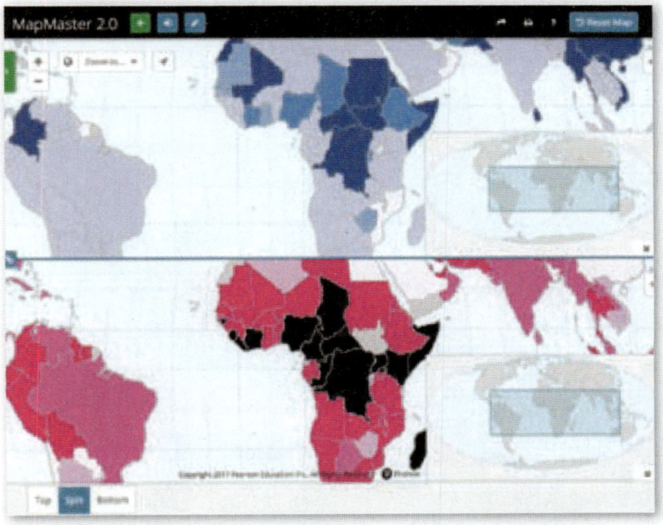

▶ 7.CR.5 ETHNIC DIVERSITY AND REFUGEES IN AFRICA

Explore

Lingering Apartheid in South Africa

Use Google Earth to explore a township where only Blacks lived during South Africa's apartheid period. ***Search*** for *500 Barracuda Road, Lawley, Gauteng, South Africa.*

1. Continue east along Barracuda Road until you reach the end. What nonresidential activities did you see along the road?
2. Describe the condition of the houses. Large or small? Modern or old? In good shape or poor?
3. Describe the landscaping around the houses, especially when you get to the end of the road.
4. Why is it important for people to erect fences or walls around their houses?

▶ 7.CR.6 LAWLEY TOWNSHIP, SOUTH AFRICA

CHAPTER

8 Political Geography

▲ Lining up to vote, Zimbabwe.

How many states of the world can you name? Old-style geography sometimes required memorization of countries and their capitals. Human geographers now emphasize a thematic approach that studies the locations of activities in the world, the reasons underlying the distribution of these activities, and the significance of the arrangements. Despite this change in emphasis, you still need to know the locations of the states of the world. Without such knowledge, you lack a basic frame of reference—knowing where things are.

KEY ISSUES

1 Where Are States Distributed?

Earth's land area is divided into states, although what constitutes a state is not always clear-cut.

2 How Are States Created?

The organization of Earth's surface into a collection of sovereign states is recent. In principle, states are designed to match the territory occupied by cultural groups. But in practice, the creation of nation-states is sometimes difficult.

3 Why Do States Have Distinctive Boundaries?

Boundaries divide Earth's surface into a collection of states that assume various shapes. Individual states assume varying forms of governance.

4 What Threats Do States Face?

States have come together in international organizations, but many people do not feel connected to some of these organizations. Some recent conflicts have been initiated by terrorist organizations.

LOCATIONS IN THIS CHAPTER

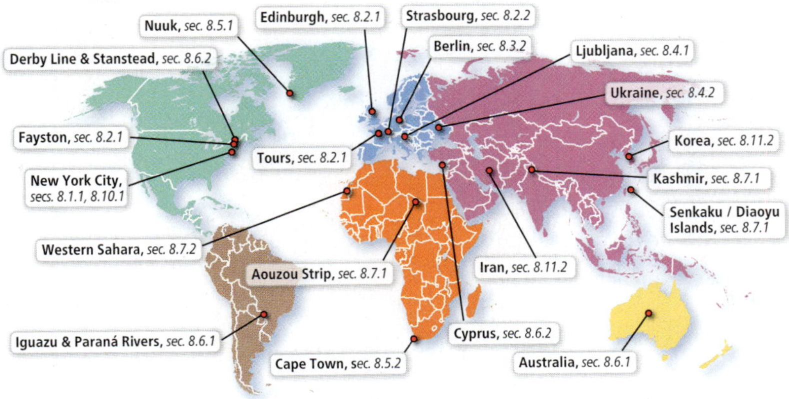

A World of States

- **Identify the states of the world.**

One of Earth's most fundamental characteristics is the division of our planet's surface into a collection of individual countries. From the air, we see landmasses and water features, but when we are traveling across Earth's surface, its organization into countries is highly visible and critically important.

▲ 8.1.1 UNITED NATIONS HEADQUARTERS, NEW YORK

8.1.1 What Is a State?

A **state** is an area organized into a political unit and ruled by an established government that has control over its internal and foreign affairs. It occupies a defined territory on Earth's surface and contains a permanent population. *Country* is a synonym for *state*. The term *state*, as used in political geography, does not refer to the 50 regional governments inside the United States. The 50 states of the United States are local government units within a single state—the United States of America.

The land area of states varies considerably. The largest state is Russia, which encompasses 17.1 million square kilometers (6.6 million square miles). At the other extreme are about two dozen **microstates**, which are states with very small land areas. The Vatican is the world's smallest microstate at 0.44 square kilometers (0.17 square miles).

The United Nations

Nearly every state of the world is a member of the United Nations (U.N.). Headquarters of the U.N. are in New York (Figure 8.1.1). When it was organized in 1945, the U.N. had 51 members, including 49 sovereign states plus Byelorussia (now Belarus) and Ukraine, then part of the Soviet Union. The number of U.N. members reached 193 in 2011 (Figure 8.1.2). U.N. members in the Western Hemisphere are shown in Figure 8.1.3 and those in the Eastern Hemisphere in Figure 8.1.4.

▼ 8.1.2 INCREASE IN U.N. MEMBERSHIP

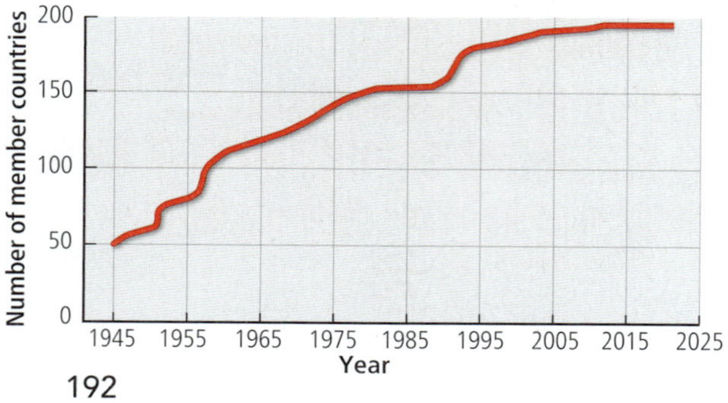

▼ 8.1.3 U.N. MEMBERS, WESTERN HEMISPHERE

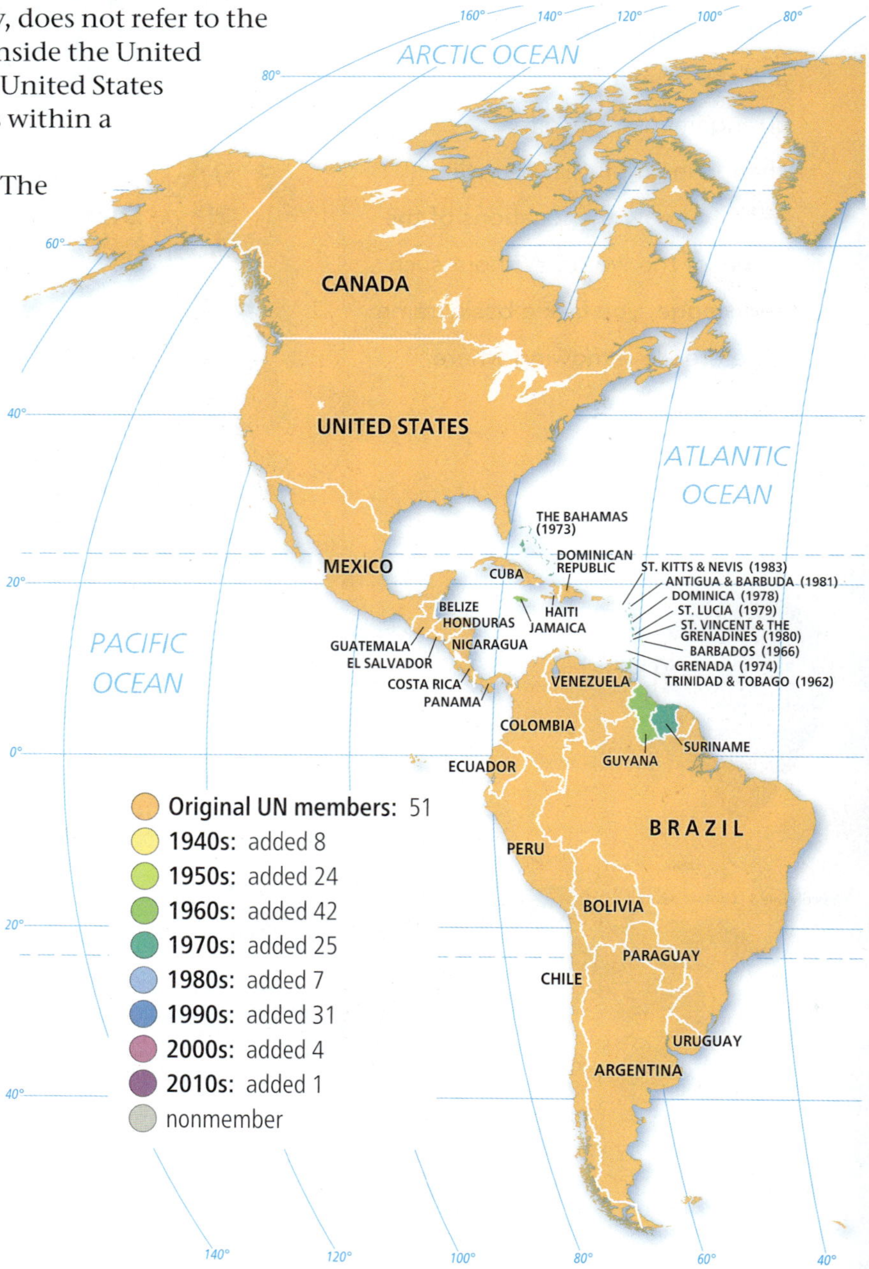

8.1.2 Growth of U.N.

U.N. membership increased rapidly on three occasions (refer to Figure 8.1.2):

- **1955.** Sixteen countries joined in 1955, mostly European countries that had been liberated from Nazi Germany during World War II.
- **1960.** Seventeen new members were added in 1960, all but one a former African colony of the United Kingdom or France. Only four African states were original members of the United Nations—Egypt, Ethiopia, Liberia, and South Africa—and only six more joined during the 1950s.
- **1990–1993.** Twenty-six countries were added between 1990 and 1993, primarily due to the breakup of the Soviet Union and Yugoslavia.

The only populous state not in the U.N. is Taiwan (officially the Republic of China). Taiwan's current status arises from a civil war in China during the late 1940s between the Nationalists and the Communists. After losing in 1949, Nationalist leaders fled to Taiwan, 200 kilometers (125 miles) off the Chinese coast, and proclaimed that they were still the legitimate rulers of the entire country of China. In 1971, the U.N. voted to transfer China's seat from the Nationalist government of Taiwan to the Communist government of China (note Taiwan is in gray in Figure 8.1.4).

▲ **8.1.4 UNITED NATIONS MEMBERS, EASTERN HEMISPHERE**

Sovereignty of States

- **Describe various ways by which states are organized.**

States are organized in a variety of ways, both internally within individual states and externally through formal relationships with other states.

▲ 8.2.1 UNITARY STATE
Voting for president, Tours, France.

8.2.1 Governing States

A state has **sovereignty**, which means independence from control of its internal affairs by other states. The internal governments of states are organized according to one of two approaches: unitary and federal.

Unitary state. In a **unitary state**, most power is placed in the hands of central government officials. In principle, a unitary government works best in a state characterized by few internal cultural differences and a strong sense of national unity. Unitary states are especially common in Europe. France, for example, has a long tradition of unitary government in which a very strong national government dominates local government decisions (Figure 8.2.1).

Federal state. Strong power is allocated to units of local government in a **federal state**. In principle, the federal system is more suitable for very large states because the national capital may be too remote to provide effective control over isolated regions. Most of the world's largest states are federal, including Russia, Canada, Brazil, India, and the United States (Figure 8.2.2). For example, the federal state principle is embedded in the Tenth Amendment to the United States Constitution, which states, "The powers not delegated to the United States by the Constitution, nor prohibited by it to the States, are reserved to the States respectively, or to the people."

▼ 8.2.2 FEDERAL STATE
Town hall meeting, Fayston, Vermont.

The size of the state is not always an accurate predictor of the form of government: Tiny Belgium is a federal state to accommodate the two main cultural groups, the Flemish and the Walloons (refer to Section 5.8.1 and Figure 5.8.2). On the other hand, the world's most populous state, China, is a unitary state (to promote communist values).

Trending Away from Unitary States

In recent years, there has been a strong global trend away from unitary government in favor of federal government. The transfer or delegation of power to a lower level, especially by a national government to local or regional administration is a process known as **devolution**.

In the face of increasing demands by ethnicities for more self-determination, states have restructured their governments to devolve some authority from the national government to local government units. An ethnicity that is not sufficiently numerous to gain control of the national government may be content with control of its territory through a regional or local unit of government.

The United Kingdom has been a strong example of a unitary state for more than a thousand years. First the monarchy and then Parliament exercised strong rule over the entire country. In recent years, though, the country has turned over considerable autonomy to Northern Ireland, Scotland, and Wales (refer to Section 7.6.1 and Figure 7.6.2). However, desire for even more separation from decisions made at the national level has pushed the Scots to seek complete independence from the United Kingdom (Figure 8.2.3). And many in Northern Ireland would like to leave the United Kingdom and join the Republic of Ireland.

▶ 8.2.3 TRENDING AWAY FROM A UNITARY STATE: DEMONSTRATION IN SUPPORT OF SCOTLAND LEAVING THE UNITED KINGDOM
Edinburgh, Scotland.

8.2.2 Cooperation Among European States

States cooperate with each other for economic and military reasons. An economic alliance enlarges markets for one state's goods, services, and labor. A military alliance offers protection to one state through the threat of retaliation by the combined force of its allies. European states have been especially active in creating economic and military organizations.

Military Cooperation in Europe

Most European states are members of the North Atlantic Treaty Organization (NATO), a military alliance that also includes Canada and the United States (Figure 8.2.4). NATO was founded in 1949 to protect European states from an attack by the Soviet Union. Around the same time, the Soviet Union and its allies in Eastern Europe formed their own military alliance, the Warsaw Pact. NATO and the Warsaw Pact were designed to maintain a **balance of power**, which is a condition of roughly equal strength between opposing alliances.

After the breakup of the Soviet Union in 1991, in a Europe no longer dominated by military confrontation between two blocs, the Warsaw Pact was disbanded, and NATO expanded its membership to include most of the former Warsaw Pact countries.

▲ **8.2.4 NATO MEMBERS IN EUROPE**
Canada and the United States are also members.

Economic Cooperation in Europe

An economic alliance enlarges markets for one state's goods, services, and labor. The most important economic alliance in Europe is the European Union (EU), which had 27 member states as of 2022 (Figure 8.2.5). The main task of the European Union is to promote economic and political cooperation. With a few exceptions, a citizen of one EU state is permitted to work and live in the other states. Most goods move across borders of EU member states in trucks and trains without stopping. A bank or retailer can open branches in any EU member country with supervision only by the corporation's home country. A European Parliament is elected by the people in each of the member states simultaneously (Figure 8.2.6).

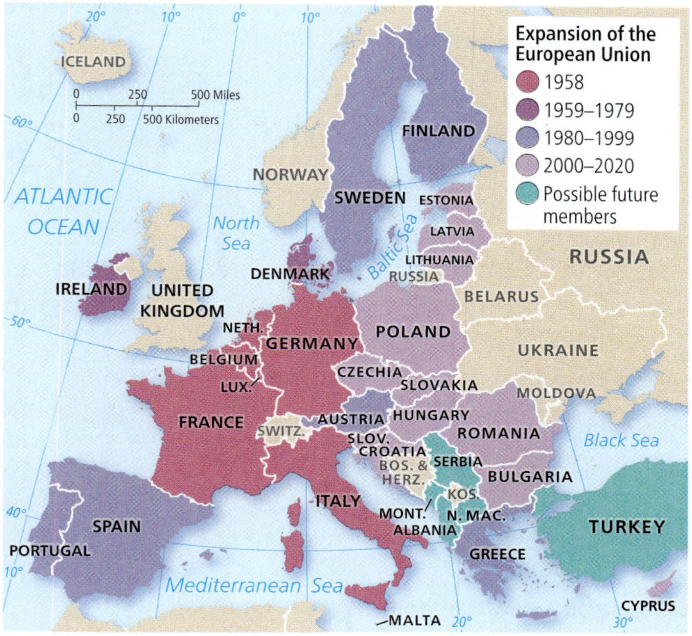

▲ **8.2.5 EUROPEAN UNION**

The most dramatic step taken toward integrating Europe's states into a "superstate" was the creation of a common currency, the euro. France's franc, Germany's mark, and Italy's lira—powerful symbols of sovereign nation-states—disappeared, replaced by the single currency. Twenty-five countries use the euro, including 19 EU members.

However, many Europeans do not feel connected to the institutions that govern the EU or the officials working in EU offices. Opposition to EU policies, especially free movement of citizens among EU member countries, induced the United Kingdom to withdraw from the EU, a move known as "Brexit."

◀ **8.2.6 TOURIST BOAT IN FRONT OF EUROPEAN PARLIAMENT BUILDINGS**
Strasbourg, France.

Evolution of States

- **Summarize how states developed historically.**

The concept of dividing the world into a collection of independent states is recent. Prior to the 1800s, Earth's surface was organized in other ways, such as into city-states, empires, kingdoms, and small land areas controlled by a hereditary class of nobles.

8.3.1 Ancient States

The development of states in ancient times can be traced to a region of Southwest Asia known as the Fertile Crescent (Figure 8.3.1). Situated at the crossroads of Europe, Asia, and Africa, the Fertile Crescent was a center for land and sea communications in ancient times.

The first states to evolve in Mesopotamia were known as city-states. A **city-state** is a sovereign state composed of a town and the surrounding countryside. Singapore is an example of a present-day city-state. City-states were sometimes organized into a succession of empires by the Sumerians, Assyrians, Babylonians, and Persians.

▲ 8.3.1 FERTILE CRESCENT

▲ 8.3.2 ROMAN EMPIRE, 100 C.E.

Roman Empire to Medieval States

Political unity in the ancient world reached its height with the establishment of the Roman Empire, which controlled most of Europe, North Africa, and Southwest Asia, from modern-day Spain to Iran and from Egypt to England (Figure 8.3.2). The Roman Empire collapsed in the fifth century, after a series of attacks by people living on its frontiers and because of internal disputes. Europe was fragmented into a large number of estates owned by competing kings, dukes, barons, and other nobles (Figure 8.3.3). The consolidation of estates under the unified control of a king formed the basis for the development of such modern European states as France and Spain (Figure 8.3.4).

▶ 8.3.3 EUROPE, 1300

◀ 8.3.4 EUROPE, 1800

8.3.2 Creating Nation-States in Europe

A state was defined in Section 8.1.1 as an area organized into a political unit. A **nation** is a large group of people who are united by common cultural characteristics, such as language and ethnicity, or by shared history. A **nation-state** is a state whose territory corresponds to that occupied by a particular nation.

The concept that nations have the right to govern themselves is known as **self-determination**. To preserve and enhance distinctive cultural characteristics, nations seek to govern themselves without interference. Nations have pushed to create nation-states because desire for self-rule is a very important shared attitude for many of them.

In modern times, the concept of nation-states developed first in Europe. After World War I, which engulfed nearly all of Europe, leaders of the victorious countries met at the Versailles Peace Conference to redraw the map of Europe. One of the chief advisers to President Woodrow Wilson, the geographer Isaiah Bowman, played a major role in the decisions.

The goal of the Allied leaders was to divide Europe into a collection of nation-states, using language as the principal criterion for identifying ethnic groups. New states were created and the boundaries of existing states were adjusted to conform as closely as possible to the territory occupied by speakers of different languages (Figure 8.3.5). This undertaking created some clear-cut examples of nation-states, but many of the states created a century ago in Europe have not survived as nation-states.

▲ 8.3.5 EUROPE, 1924

▲ 8.3.6 EUROPE, 1980

Europe After World War II

The attempt after World War I to divide Europe into nation-states was not a recipe for peace. During the 1930s, Germany's National Socialists (Nazis) claimed that all German-speaking parts of Europe constituted one nationality and should be unified into one state. After many years of appeasing the Nazis' expansion in Central Europe, the United Kingdom and France finally declared war when the Nazis invaded Poland, clearly not a German-speaking state.

Following its defeat in World War II, Germany was divided into two states: the German Democratic Republic (East Germany) and the German Federal Republic (West Germany). Two Germanys existed from 1949 until 1990 (Figure 8.3.6). A wall divided Berlin between the two states (Figure 8.3.7). A massive forced migration of people in Europe after World War II relocated many ethnic groups into the newly demarcated territory of the region's various nation-states.

With the end of communism, the German Democratic Republic ceased to exist, its territory became part of the German Federal Republic, and the Berlin Wall was dismantled. The present-day state of Germany bears little resemblance to the territory occupied by German-speaking people prior to the upheavals of the twentieth century.

▶ 8.3.7 REMNANTS OF THE BERLIN WALL AFTER MOST OF IT WAS DEMOLISHED
Berlin, Germany.

New Nation-States

- **Identify recently created nation-states in Eastern Europe.**

During the late twentieth century and into the twenty-first century, new nation-states were created in Eastern Europe through the breakup of the Soviet Union, Czechoslovakia, and Yugoslavia.

8.4.1 The Nation-State of Slovenia

The breakup of Yugoslavia during the early 1990s resulted in the creation of a good example of a nation-state in Slovenia (Figures 8.4.1 and 8.4.2). The state contains 2 million inhabitants, including 83 percent ethnic Slovenes. The country has one official language (Slovene), spoken by 92 percent of Slovenes. More than 80 percent of Slovenes who adhere to a religion are Roman Catholics.

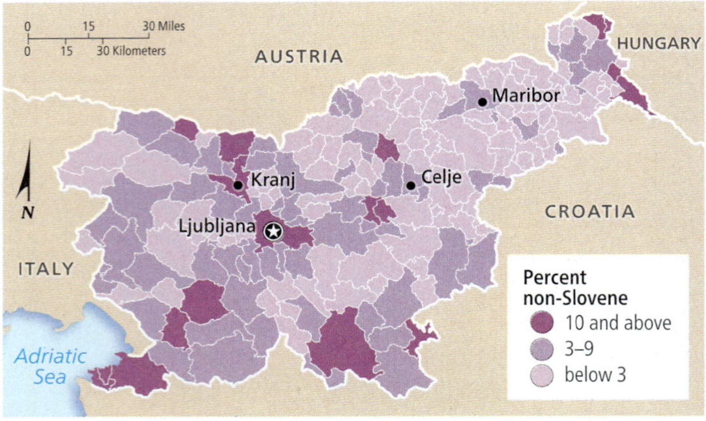

▲ 8.4.1 SLOVENIA

▲ 8.4.2 LJUBLJANA, CAPITAL OF SLOVENIA

Nation-States in the Former Soviet Union

During its existence between 1922 and 1991, the Union of Soviet Socialist Republics (U.S.S.R.) consisted of 15 republics, based on its 15 largest ethnicities. The breakup of the U.S.S.R. in 1991 resulted in the conversion of the 15 republics into 15 independent states (Figure 8.4.3). Some of these new states are good examples of nation-states, and some are not.

▼ 8.4.3 FORMER SOVIET UNION

Nation-states in the Baltics. Estonia, Latvia, and Lithuania are known as the Baltic states for their location on the Baltic Sea. They were independent countries between the end of World War I in 1918 and 1940, when the former Soviet Union annexed them under an agreement with Nazi Germany.

These three small neighboring Baltic countries have clear cultural differences and distinct historical traditions. Most Lithuanians speak Lithuanian, a language of the Baltic group within the Balto-Slavic branch of the Indo-European language family. Most Latvians speak Latvian, another Baltic group language. Most Estonians speak a Uralic language related to Finnish. Christianity is the principal religion among these three Baltic states. However, most Lithuanians are Roman Catholic; most Latvians are Lutheran, Orthodox, or Roman Catholic; and most Estonians are Orthodox or Lutheran.

8.4.2 Multinational States in the Former Soviet Union

A **multinational state** is a state that contains more than one nation. In some states, nations all contribute cultural features to the formation of a single nationality, which was defined in Chapter 7 as a group of people who share legal attachment to a particular country. Cultural groups can coexist peacefully, remaining culturally distinct while recognizing and respecting the distinctive traditions of other groups. However, in some states one cultural group may try to dominate others, sometimes by force.

The new states in the former Soviet Union include several multinational states. Prominent among them are Russia and Ukraine. Russia classifies ethnicities primarily by language family. Russia's census identifies 185 ethnic groups, including 81 percent who identify as Russian. Non-Russian ethnic groups are clustered in three principal locations (Figure 8.4.4).

▼ 8.4.4 ETHNICITIES IN RUSSIA

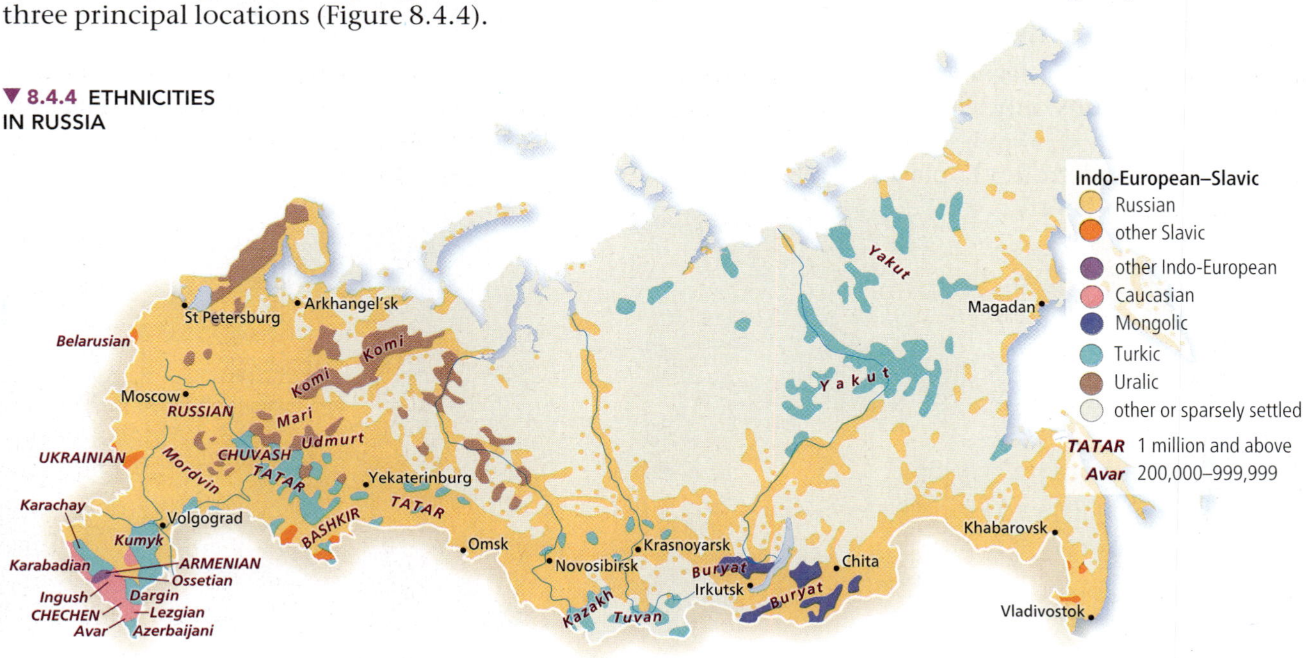

Ukraine

After the breakup of the Soviet Union, prospects for a stable nation-state were favorable in independent Ukraine because it possessed economic assets such as coal deposits, a steel industry, and proximity to the more economically developed countries of Western Europe. However, Ukraine's minority Russian population started an uprising in the eastern region of the country, where they were clustered (Figure 8.4.5). Claiming that the Russian ethnic minority in Ukraine was endangered, Russia invaded eastern Ukraine and seized Crimea.

Crimea, a 27,000-square-kilometer (10,000-square-mile) peninsula, has long been an area of conflict. Crimea's population is approximately 60 percent Russian, 24 percent Ukrainian, 10 percent Tatar, and 6 percent other ethnicities.

Russia took control of Crimea in 1783, and in 1921, it became an autonomous republic within the Russian Soviet Federative Socialist Republic, which in turn was a republic within the Soviet Union. In 1954, the Soviet government transferred responsibility for Crimea to the Ukrainian Soviet Socialist Republic, which was then also part of the Soviet Union.

When the Soviet Union broke up in 1991, Crimea became an autonomous republic in the newly independent Ukraine. In 2014, Russia invaded Crimea and annexed it, claiming that the majority of the Crimean people, who are ethnic Russians, supported the action. Nearly every other country in the world continues to recognize Ukraine's sovereignty over Crimea. However, the international community has not found a way to remove the Russians and restore Crimea to Ukraine.

▲ 8.4.5 ETHNICITIES IN UKRAINE

Colonies

- **Explain the concept of colonies, and describe their current distribution.**

Nearly all of Earth's land area has been allocated to a collection of around 200 sovereign states, but some territories remain that have not achieved self-determination and statehood.

8.5.1 Current Colonies

A **colony** is a territory that is legally tied to a sovereign state rather than being completely independent. In some cases, a sovereign state runs only the colony's military and foreign policy. In others, it also controls the colony's internal affairs.

The United Nations identifies 17 places in the world that it calls "non–self-governing territories" (Figure 8.5.1). All but Western Sahara are islands.

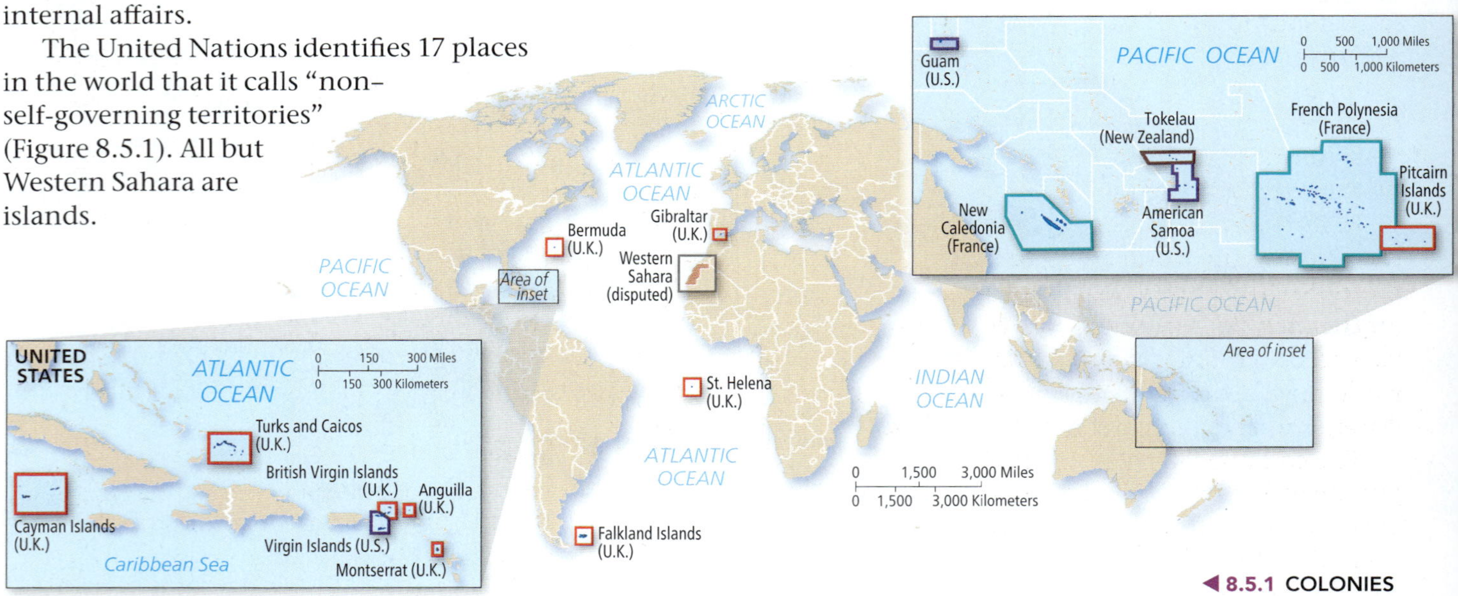

8.5.1 COLONIES

Are These Colonies?

The U.N. list does not include territories that are uninhabited, such as Baker and Midway islands, which are controlled by the United States. The U.N. also does not list inhabited territories that it considers to have considerable autonomy in self-governing.

These are examples of territories that the U.N. does not classify as colonies:

- **Puerto Rico.** As Puerto Rico is a commonwealth of the United States, Puerto Ricans do not participate in U.S. presidential elections or have a voting member of Congress.
- **Greenland.** As an autonomous unit within the Kingdom of Denmark, Greenland runs its internal affairs, but Denmark controls foreign affairs and defense (Figure 8.5.2).
- **Hong Kong & Macao.** Hong Kong and Macao are attached to the mainland of China as special administrative regions within the People's Republic of China. Hong Kong was a colony of the United Kingdom until it reverted to China in 1997, and a year later, Portugal returned its colony of Macao to China. Both Hong Kong and Macao have some autonomy in economic matters, but China controls their foreign affairs and defense.

▼ 8.5.2 NUUK, CAPITAL OF GREENLAND

8.5.2 Colonialism

At one time, colonies were widespread over Earth's surface. European states came to control much of the world through **colonialism**, which is an effort by one country to establish settlements in a territory and to impose its political, economic, and cultural principles on that territory (Figure 8.5.3).

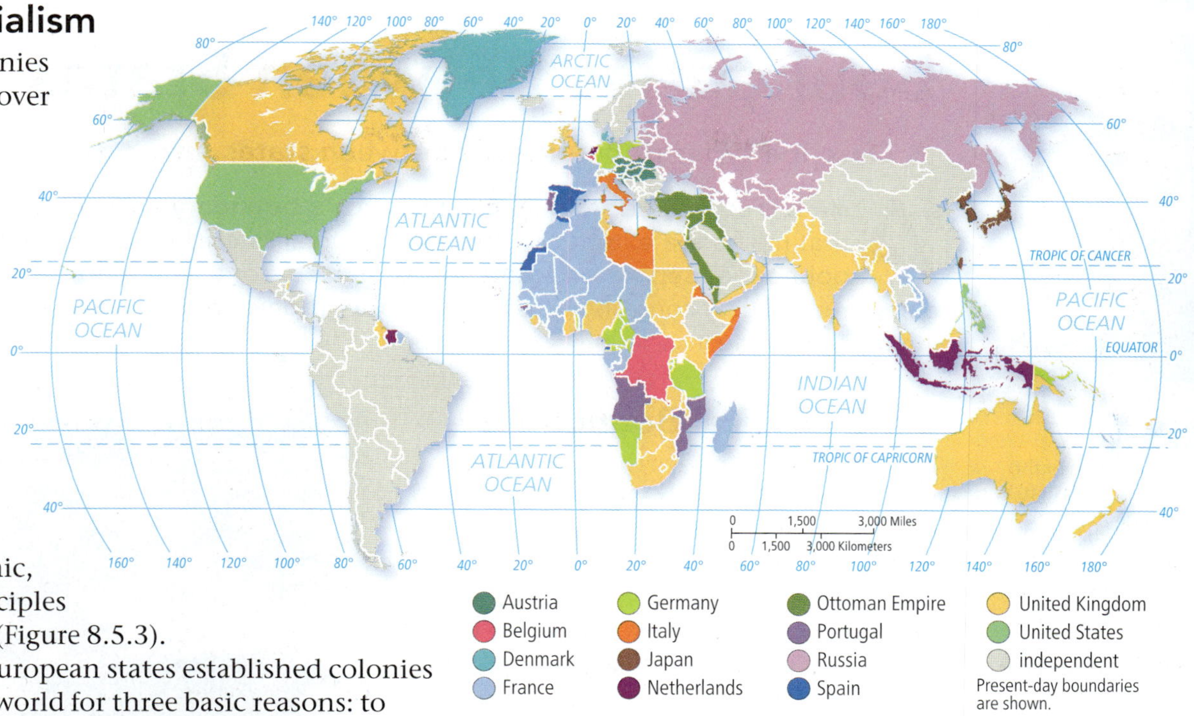

▲ 8.5.3 COLONIES IN 1914

Historically, European states established colonies elsewhere in the world for three basic reasons: to promote Christianity, to extract useful resources for the manufacture of products that were then sold back to the colonies at a profit, and to enhance their world power through the control of territory and domination of colonized people.

The colonial era began in the 1400s, when European explorers sailed westward for Asia but encountered and settled in the Western Hemisphere instead. Eventually, the European states lost most of their Western Hemisphere colonies. Independence was declared by the United States in 1776 and by most Latin American states between 1800 and 1824. European states then turned their attention to Africa and Asia. At the Berlin Conference of 1884–1885, European states decided that they could convert most of Africa into their colonies, and they defined areas within Africa that each would control.

The United Kingdom planted colonies on every continent, including much of eastern and southern Africa, South Asia, Southwest Asia, Australia, and Canada. With by far the largest colonial empire, the British proclaimed that the "Sun never set" on their empire.

France had the second-largest overseas territory, primarily in West Africa and Southeast Asia. France attempted to assimilate its colonies into French culture and educate an elite group to provide local administrative leadership. After the territories became independent, most of their leaders retained close ties with France.

Japan was the principal colonial power in Asia during the first half of the twentieth century. It controlled Korea, Taiwan, several smaller Pacific Ocean islands, and a portion of China. Colonial control ended in 1945 with Japan's defeat in World War II.

End of Colonialism

Most African and Asian colonies became independent after World War II. Only 15 African and Asian states were members of the United Nations when it was established in 1945, compared to 106 since 2011. The boundaries of the new states frequently coincide with former colonial provinces but not with the distribution of ethnicities (refer to Section 7.9.1).

However successful the decolonization of Africa, the legacy of colonialism has been hard to shake. Cecil Rhodes, a British imperialist, played a major role in extending British control through much of southern Africa (Figure 8.5.4). His treatment of indigenous people and attitudes of White racial superiority have been severely criticized in recent years (Figure 8.5.5).

▼ 8.5.4 CECIL RHODES

▼ 8.5.5 STATUE OF CECIL RHODES DISMANTLED
University of Cape Town, South Africa.

Boundaries Between States

- **Describe types of boundaries between states.**

A state is separated from its neighbors by a **boundary**, which is an invisible line that marks the extent of a state's territory. Historically, frontiers rather than boundaries separated states. A **frontier** is a zone where no state exercises complete political control. It is a tangible geographic area, whereas a boundary is an infinitely thin line.

▲ 8.6.1 WATER BOUNDARY
Confluence of the Iguazu River (left) and Paraná River is the boundary between Brazil (foreground), Argentina (left), and Paraguay (rear).

8.6.1 Physical Boundaries

Physical features on Earth's surface can make good boundaries because they are easily seen, both on a map and on the ground. A physical feature that serves as a boundary is known as an **antecedent boundary** because it existed before the present settlement. Three types of physical features typically serve as boundaries between states:

Water boundaries. Rivers and oceans are the physical features most commonly used as boundaries (Figures 8.6.1 and 8.6.2).

Desert boundaries. A boundary drawn in a desert can effectively divide two states because deserts are hard to cross and sparsely inhabited. Desert boundaries are common in Africa and Asia.

Mountain boundaries. Mountains can be effective boundaries if they are difficult to cross.

▲ 8.6.2 TRIPLE FRONTIER
Confluence of the Iguazu River (rear) and Paraná River is the boundary between Paraguay (foreground), Brazil (left), and Argentina (right).

The Law of the Sea

A complex set of boundaries surround countries that border the oceans. The U.N. Convention on the Law of the Sea defines the rights and responsibilities of countries with respect to the use of the world's oceans and the management of marine resources. It has been signed by 164 states (not including the United States). The Convention identifies three types of water boundaries (Figure 8.6.3):

Territorial waters. Up to 12 nautical miles from shore (about 22 kilometers or 14 land miles), a state may set laws regulating passage by ships registered in other states.

Contiguous zone. Between 12 and 24 nautical miles from shore, a state may enforce laws concerning pollution, taxation, customs, and immigration.

Exclusive economic zone. Between 24 and 200 nautical miles, a state has the sole right to the fish and other marine life. Disputes can be taken to a tribunal for the Law of the Sea or to the International Court of Justice.

Through enforcement of the exclusive economic zone, states bordering an ocean are able to claim vast areas of the ocean for control of valuable resources. This has become especially important in the delineation of boundaries in the polar regions.

▲ 8.6.3 THE LAW OF THE SEA APPLIED TO AUSTRALIA

8.6.2 Cultural Boundaries

Boundaries between countries are often placed to follow the distribution of cultural features. Geographers identify five types of cultural boundaries:

- A **superimposed boundary** is drawn on an area by a conquering or colonizing power that ignores existing cultural patterns. The division of Africa into independent states as the European powers dissolved their colonial empires resulted in numerous superimposed boundaries that ignored ethnic and other cultural borders (refer to Figure 7.9.1).
- A **subsequent boundary** is established after the settlement in an area and is changed to accommodate developments such as war. An example of a subsequent boundary is the partition of the former Yugoslavia into several independent states (refer to Figure 7.10.1).
- A **relic boundary** is one that no longer functions but can still be detected on the cultural landscape. An example of a relic boundary is the Berlin Wall that once served as a boundary between East and West Germany but was dismantled when Germany was reunified in 1990 (refer to Figure 8.3.7).
- A **consequent boundary** is drawn in order to separate groups based on ethnic, linguistic, religious, or economic differences. A consequent boundary is drawn between Greek and Turkish areas in Cyprus.
- A **geometric boundary** is based on a human construct such as a straight line or a parallel of latitude. The boundary between Canada and the United is an example of a geometric boundary devised by a parallel of latitude.

Consequent Boundary: Cyprus

Cyprus is divided between approximately 70 percent Greeks clustered in the south and 25 percent Turks clustered in the north. When Cyprus gained independence from the United Kingdom in 1960, the Turkish minority was guaranteed considerable autonomy, but after Greek Cypriot military officers, hoping to unify Cyprus with Greece, seized control of the government in 1974, Turkey invaded Cyprus. An elected government for the entire island was restored a few months later, but the north declared itself the independent Turkish Republic of Northern Cyprus in 1983 (Figure 8.6.4). A wall and a frontier patrolled by the United Nations were delineated across the country (Figure 8.6.5).

▲ **8.6.4 CONSEQUENT BOUNDARY: CYPRUS**
Cyprus is divided into Greek and Turkish areas, separated by a U.N. buffer zone.

▲ **8.6.5 CYPRUS GREEN LINE**
The city of Nicosia is divided between the Turkish side (foreground) and Greek side (background).

Geometric Boundary: North America

Part of the northern U.S. boundary with Canada is a 2,100-kilometer (1,300-mile) line along 49° north latitude, running from Lake of the Woods between Minnesota and Manitoba to the Strait of Georgia between Washington State and British Columbia (Figure 8.6.6). Québec's boundary with New York and Vermont is also geometric, along 45° north latitude. The two countries share an additional 1,100- kilometer (700-mile) geometric boundary between Alaska and the Yukon Territory along the north–south arc of 141° west longitude. The U.S.–Canada boundary was established through a series of treaties between 1783 and 1903 between the United States and the United Kingdom, which then still controlled Canada.

▶ **8.6.6 GEOMETRIC BOUNDARY: U.S.–CANADA BORDER**
The border separates Derby Line, Vermont (left), and Stanstead, Québec (right).

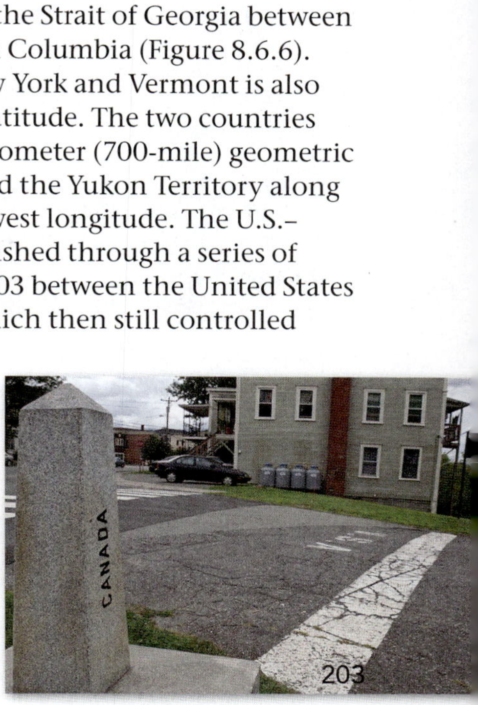

▲ 8.7.1 SENKAKU/ DIAOYU ISLANDS

Boundary Disputes

- **Offer examples of problems defining boundaries.**

States do not always agree on the correct location of the boundaries between them. Disagreements concerning the correct location can arise with both physical and cultural boundaries.

8.7.1 Physical Boundary Dispute

The People's Republic of China, Taiwan, and Japan all claim sovereignty over five uninhabited islands in the East China Sea. The islands are known as Diaoyu in China, Diaoyutai in Taiwan, and Senkaku in Japan. The largest island is only 4.32 square kilometers (1.67 square miles) (Figure 8.7.1).

China and Taiwan claim that the islands historically belonged to China until the Japanese government illegally seized them in 1895. Japan's position is that China did not have sovereignty over the uninhabited islands back in 1895, when Japan claimed them. To bolster their claims, China and Japan have both established air defense zones in the East China Sea with conflicting boundaries (Figure 8.7.2).

▲ 8.7.2 DISPUTED TERRITORY AMONG CHINA, JAPAN, AND TAIWAN

Cultural Boundary Disputes

Two examples of cultural boundary disputes are that of Kashmir between India and Pakistan and the Aouzou Strip between Chad and Libya.

Superimposed boundary dispute. When a boundary was superimposed on South Asia to delineate Pakistan and India, these two countries never agreed on the boundary location in the northern region of Kashmir. Since 1972, the two countries have maintained a "line of control" through the region, with Pakistan administering the northwestern portion and India the southeastern portion (Figure 8.7.3). Muslims, who comprise a majority in both portions, have fought to secure reunification of Kashmir, either as part of Pakistan or as an independent country.

Geometric boundary dispute. The 1,000-kilometer (600-mile) boundary between Chad and Libya is a straight line drawn across the desert in 1899 by the French and British to set the northern limit of French colonies in Africa. As an independent country, Libya claimed that the straight line should be 100 kilometers (60 miles) to the south, to include territory known as the Aouzou Strip (Figure 8.7.4). Citing an agreement in 1935 between France and Italy, which then controlled much of Libya, Libya seized the Aouzou Strip in 1973. Chad regained control of the strip in 1987, and Libya withdrew its troops after the International Court of Justice ruled in favor of Chad's claim in 1994.

▲ 8.7.3 KASHMIR

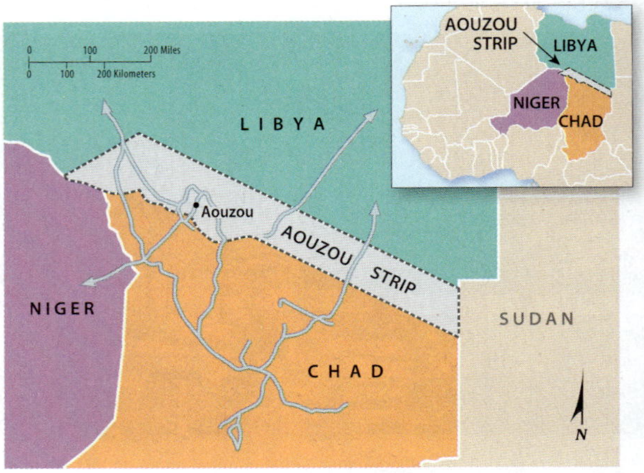

▲ 8.7.4 AOUZOU STRIP
Disputed between Chad and Libya.

8.7.2 Boundary Disputes in Polar Regions & Deserts

Some boundary disputes exist in sparsely inhabited areas. Examples include the polar regions and the Sahara Desert.

Polar Regions

The South Pole region contains the only large landmass on Earth's surface that is not part of a state. Several states, including Argentina, Australia, Chile, France, New Zealand, Norway, and the United Kingdom, claim portions of Antarctica (Figure 8.7.5). Argentina, Chile, and the United Kingdom have made conflicting, overlapping claims. The United States, Russia, and a number of other states do not recognize the claims of any country to Antarctica.

The Antarctic Treaty, signed in 1959 and entered into force two years later, provides a legal framework for managing Antarctica. States may establish research stations there for scientific investigations, but no military activities are permitted.

As for the Arctic Ocean, the 1982 U.N. Convention on the Law of the Sea permitted countries to submit claims inside the Arctic Circle by 2009 (Figure 8.7.6). The Arctic Ocean's seafloor is thought to be rich in energy resources. Canada, Denmark, Iceland, Norway, Russia, and the United States make conflicting claims within the Arctic Circle. The heart of the dispute is the Lomonosov Ridge, which runs 1,800 kilometers (1,100 miles) across the polar region. Russia and Denmark (which controls Greenland) both claim the Lomonosov Ridge as an extension of their respective landmasses.

▲ 8.7.5 NATIONAL CLAIMS TO ANTARCTICA

▲ 8.7.6 NATIONAL CLAIMS TO THE ARCTIC OCEAN

Morocco & Sahrawi Arab Republic

The Sahrawi Arab Democratic Republic, also known as Western Sahara, is considered by most African countries as a sovereign state. Morocco, however, claims the territory and to prove it has built a 2,700-kilometer (1,700-mile) wall around the territory to keep out rebels (Figure 8.7.7). Spain controlled the territory on the continent's west coast between Morocco and Mauritania until withdrawing in 1976. An independent Sahrawi Arab Republic was declared by the Polisario Front and recognized by most African countries, but Morocco and Mauritania annexed the northern and southern portions, respectively. Three years later Mauritania withdrew, and Morocco claimed the entire territory. Only the U.S. government recognizes Morocco's claim to the territory. Morocco controls most of the populated area, but the Polisario Front operates in the vast, sparsely inhabited deserts, especially the one-fifth of the territory that lies east of Morocco's wall.

▶ 8.7.7 WESTERN SAHARA SAND WALL
The Berm of Western Sahara is a 2,700-kilometer wall made of sand built between the territory occupied by Morocco and the territory occupied by the Polisario Front.

Shapes of States

- **Describe the five shapes of states.**

States have one of five basic shapes—compact, elongated, prorupted, fragmented, or perforated—and examples of each can be seen in sub-Saharan Africa. Sub-Saharan Africa also has several examples of states that lack access to the oceans.

▲ **8.8.1 LANDLOCKED STATES IN AFRICA**
Landlocked states must import and export goods by land-based transportation, primarily rail lines, to reach ocean ports in cooperating neighbor states.

8.8.1 Landlocked States

A **landlocked state** lacks a direct outlet to a sea because it is completely surrounded by other countries. Direct access to an ocean is critical to states because it facilitates international trade. Bulky goods, such as petroleum, grain, ore, and vehicles, are normally transported long distances by ship. This means that a country needs a seaport where goods can be transferred between land and sea. To send and receive goods by sea, a landlocked state must arrange to use another country's seaport.

Landlocked states are common in Africa, where 15 of the continent's 55 states have no direct ocean access (Figure 8.8.1). The prevalence of landlocked states in Africa is a remnant of the colonial era, when Britain and France controlled extensive regions that were eventually divided into independent states without regard for whether they were landlocked. The European powers built railroads in their colonial empires to connect the interior of Africa with the sea. Railroads moved minerals from interior mines to seaports, and in the opposite direction, rail lines carried mining equipment and supplies from seaports to the interior.

Compact States: Efficient

In a **compact state**, the distance from the center to any boundary does not vary significantly. An ideal theoretical compact state would be shaped like a circle, with the capital at the center and with the shortest possible boundaries to defend. Examples of compact states in sub-Saharan Africa include Burundi, Kenya, Rwanda, and Uganda (Figure 8.8.2).

Compactness can be a strong centripetal force in smaller states because good communications can be more easily established with all regions, especially if the capital is located near the center. However, compactness does not necessarily mean peacefulness, as compact states are just as likely as others to experience civil wars and ethnic rivalries.

▲ **8.8.2 COMPACT STATES**
Burundi, Kenya, Rwanda, and Uganda.

Perforated State: South Africa

A state that completely surrounds another one is a **perforated state**. The surrounded state may face problems of dependence on, or interference from, the surrounding state. A clear example of a perforated state is South Africa, which completely surrounds Lesotho, which is a compact state (Figure 8.8.3). Lesotho must depend almost entirely on South Africa for the import and export of goods. Dependency on South Africa was especially difficult for Lesotho when South Africa had a government controlled by Whites who discriminated against the Black majority population.

▼ **8.8.3 PERFORATED STATE**
South Africa.

8.8.2 Prorupted States: Access or Disruption

An otherwise-compact state with a large projecting extension is a **prorupted state**. Proruptions are created for two principal reasons, and examples of both are found in sub-Saharan Africa.

- **To provide a state with access to a resource.** The Democratic Republic of the Congo has a 500-kilometer (300-mile) proruption to the west along the Congo River. The Belgians created the proruption to give their colony access to the Atlantic.
- **To separate two states that otherwise would share a boundary.** Namibia has a 500-kilometer (300-mile) proruption to the east, the Caprivi Strip. When Namibia was a colony of Germany, the Caprivi Strip proruption disrupted communications among the British colonies of southern Africa. It also provided the Germans with access to the Zambezi, one of Africa's most important rivers (Figure 8.8.4).

▲ **8.8.4 PRORUPTED STATE**
Democratic Republic of the Congo and Namibia.

Elongated States: Potential Isolation

A handful of **elongated states** have a long and narrow shape. Examples in sub-Saharan Africa include The Gambia and Malawi.

The Gambia extends along the banks of the Gambia River about 500 kilometers (300 miles) east–west but is only about 25 kilometers (15 miles) north–south (Figure 8.8.5). The shape was determined in the late nineteenth century, when the British controlled the mouth of the river and the French most of the surrounding area.

Malawi measures about 850 kilometers (530 miles) north–south but only 100 kilometers (60 miles) east–west. In 1891, the British declared the territory a Protectorate in order to deter Portuguese on the west side of Lake Malawi from staking their own claim.

Elongated states may suffer from poor internal communications. A region located at an extreme end of the elongation might be isolated from the capital, which is usually placed near the center.

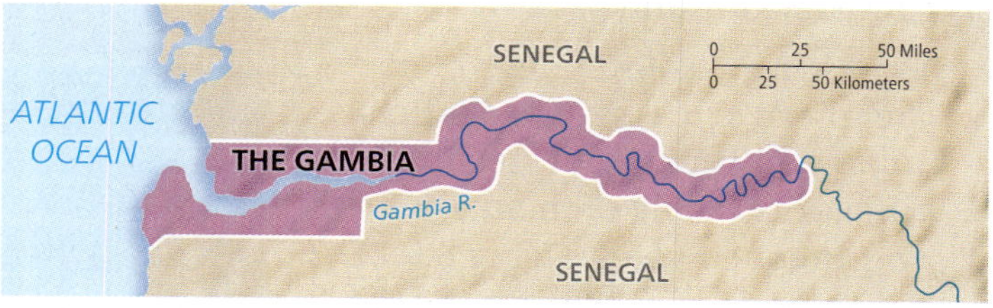

▲ **8.8.5 ELONGATED STATE**
The Gambia.

Fragmented States: Problematic

A **fragmented state** includes several discontinuous pieces of territory. Technically, all states that have offshore islands as part of their territory are fragmented. Those fragmented states as well as fragmented states on mainlands are particularly vulnerable to problems and costs associated with communications and maintaining national unity.

Angola is an example of a state that is divided into two fragments by another state, Democratic Republic of the Congo (Figure 8.8.6). An independence movement is trying to detach one fragment, Cabinda, as a separate state from the rest of Angola, with the justification that its population belongs to distinct ethnic groups.

Tanzania is an example of a state that is fragmented by water. The state was created in 1964 as a union of the island of Zanzibar with the mainland territory of Tanganyika. Although home to different ethnic groups, the two entities agreed to join together because they shared common development goals and political priorities.

▲ **8.8.6 FRAGMENTED STATE**
Angola.

▲ 8.9.1 THE ORIGINAL GERRYMANDERING CARTOON

Gerrymandering Boundaries

- **Explain the practice of gerrymandering and ways in which it is done.**

In democracies, politics must follow legally prescribed rules. But all parties to the political process can find ways of bending those rules to their advantage. A case in point is the drawing of legislative district boundaries.

8.9.1 Gerrymandering

The boundaries separating legislative districts within the United States and other countries are redrawn periodically to ensure that each district has approximately the same population. Boundaries must be redrawn because migration inevitably results in some districts gaining population and others losing population. The 435 districts of the U.S. House of Representatives are redrawn every 10 years, following the Census Bureau's release of official population figures.

Redrawing legislative boundaries to benefit the party in power is called **gerrymandering**. The term *gerrymandering* was named for Elbridge Gerry (1744–1814), governor of Massachusetts (1810–1812) and vice president of the United States (1813–1814). As governor, Gerry signed a bill that redistricted the state to benefit his party. An opponent observed that an oddly shaped new district looked like a "salamander," whereupon another opponent responded that it was a "gerrymander." A newspaper subsequently printed a cartoon of a monster named "gerrymander" with a body shaped like the district (Figure 8.9.1).

Two Ways to Gerrymander

Gerrymandering takes one of two forms—cracking and packing.

- **Cracking.** Like-minded voters are spread across several districts to prevent them from reaching a majority in any of them, thus wasting their votes (Figure 8.9.2). An example of cracking is in Utah, where Salt Lake City is split among multiple Congressional districts (Figure 8.9.3).
- **Packing.** Like-minded voters are stacked in one district to prevent them from affecting elections in other districts (Figure 8.9.4). An example of packing is Florida's elongated-shaped Fifth Congressional District, which packs a large percentage of the state's Black population (Figure 8.9.5).

▲ 8.9.2 GERRYMANDERING CRACKING
Like-minded voters are spread across many districts as a minority. Assume there are 50 voters, 20 supporters of the Red Party and 30 supporters of the Blue Party. If the Blue Party controls the redistricting process, it could create five districts all with slender majorities of Blue Party voters. As a result, the votes of all Red Party supporters are wasted.

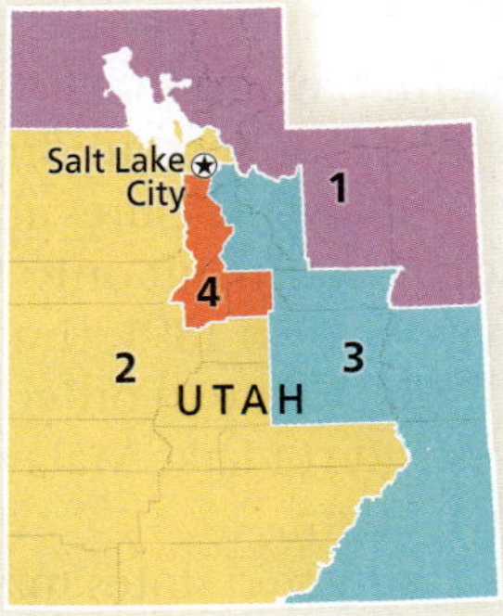

▲ 8.9.3 GERRYMANDERING CRACKING EXAMPLE: UTAH
Salt Lake City, which is home to relatively progressive voters in an otherwise conservative-leaning state, is cracked among several Congressional Districts.

◀ 8.9.4 GERRYMANDERING PACKING
Like-minded voters are stacked into a few districts. Assume there are 50 voters, 20 supporters of the Red Party and 30 supporters of the Blue Party. If the Blue Party controls the redistricting process, it could pack most of the Red Party voters into one district with an overwhelming majority, leaving four districts with solid Blue Party majorities.

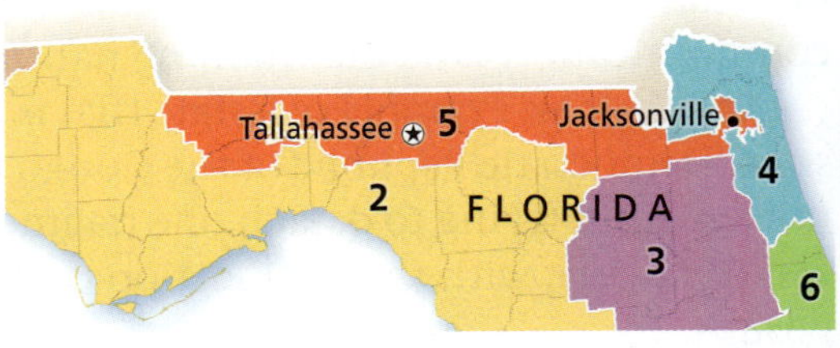

▲ 8.9.5 GERRYMANDERING PACKING EXAMPLE: FLORIDA
Florida's Fifth Congressional District, which stretches nearly 500 kilometers (300 miles), packs a large percentage of northern Florida's Black population.

8.9.2 State Variations in Gerrymandering

In most U.S. states, the job of redrawing boundaries is entrusted to the state legislature. The political party in control of the state legislature naturally attempts to redraw boundaries to improve the chances of its candidates to win. GIS has been an especially useful tool for creating gerrymandered districts. Only about one-tenth of congressional seats across the United States are competitive, making a shift of more than a few seats unlikely from one election to another in the United States, except in unusual circumstances.

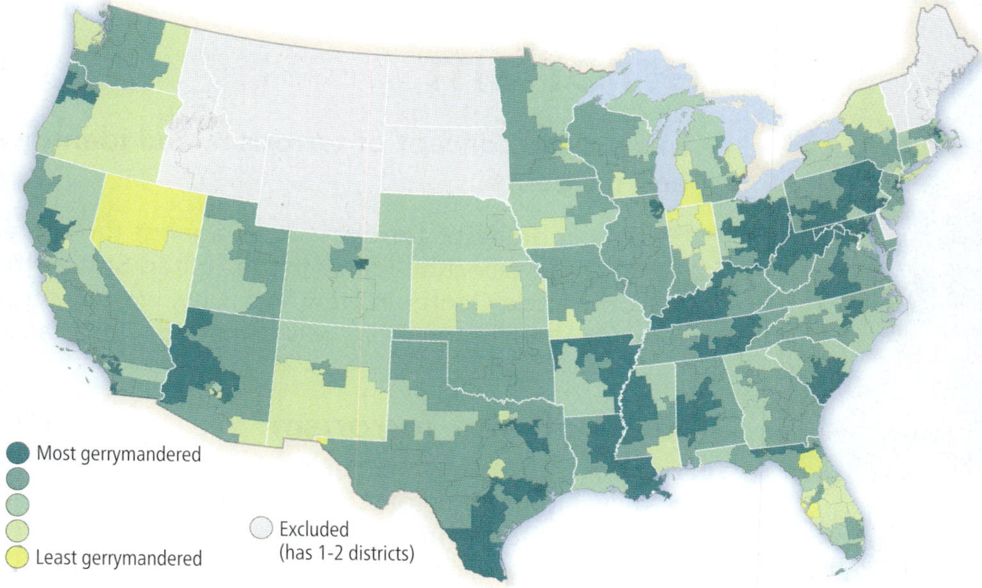

▲ 8.9.6 EXTENT OF GERRYMANDERING IN EACH CONGRESSIONAL DISTRICT

The job of redrawing boundaries in most European countries is entrusted to independent commissions. Commissions typically try to create compact homogeneous districts without regard for voting preferences or incumbents. In contrast, only a handful of U.S. states utilize a nonpartisan commission to draw the boundaries.

Figure 8.9.6 measures the extent of each U.S. congressional district's gerrymandering based on the compactness of the district's shape. States with only one or two congressional districts are excluded.

The worst gerrymandered state in 2020 was Maryland, followed by West Virginia, Louisiana, Pennsylvania, Virginia, Ohio, Kentucky, Illinois, Alabama, and New Jersey. The least gerrymandered state in 2020 was Nevada, followed by Indiana, Kansas, Florida, Iowa, Nebraska, New Mexico, New York, Minnesota, and Arizona.

▼ 8.9.7 SUBSTANTIAL GERRYMANDERING IN LOUISIANA

Louisiana: Substantial gerrymandering Louisiana is one of the most gerrymandered states. A large share of the state's Black population is packed into the Second Congressional District, a long narrow shape that stretches more than 160 kilometers (100 miles) from north of Baton Rouge to east of New Orleans (Figure 8.9.7). To connect the predominantly Black areas of Baton Rouge and New Orleans, the district is only one street wide in places.

Kansas: Minimal gerrymandering Kansas is one of the least gerrymandered states. Its five congressional districts follow county boundaries for the most part (Figure 8.9.8).

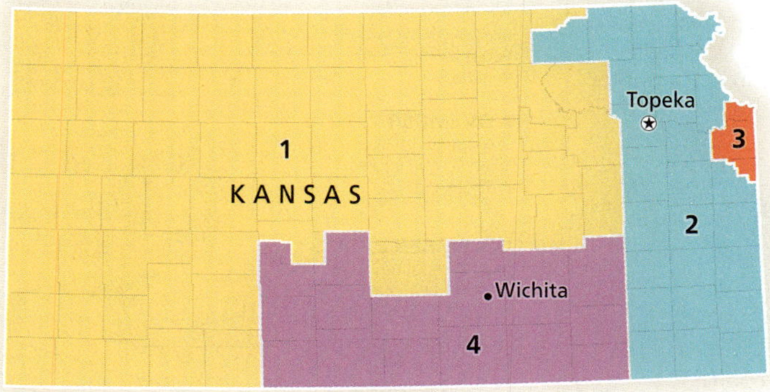

▲ 8.9.8 MINIMAL GERRYMANDERING IN KANSAS

▲ 8.10.1 TERRORIST ATTACK ON THE WORLD TRADE CENTER, 2001

Terrorism

- **Explain the concept of terrorism, and identify major terrorist organizations and incidents.**

The most dramatic terrorist attack against the United States came on September 11, 2001. The tallest buildings in the United States, the 110-story twin towers of the World Trade Center in New York City, were destroyed (Figure 8.10.1), and the Pentagon in Washington, D.C., was damaged. The attacks resulted in nearly 3,000 fatalities.

8.10.1 Defining Terrorism

Terrorism is the threatened or actual use of illegal force and violence by a non-state actor to attain a political, economic, religious, or social goal through fear, coercion, or intimidation. Because terrorism can be difficult to identify, international organizations concerned with terrorism, including the Institute for Economics & Peace and the National Consortium for the Study of Terrorism and Responses to Terrorism (START), have developed several criteria that must be met for an incident to be considered terrorism:

1. The incident must be intentional, the result of a conscious calculation on the part of a perpetrator.
2. The incident must entail some level of violence or threat of violence, including property damage as well as violence against people.
3. Acts of terrorism conducted by one state against another state are not included.

In addition, two of these three criteria must also be met in order to be considered an act of terrorism:

1. The violent act was aimed at attaining a political, economic, religious, or social goal.
2. The violent act included evidence of an intention to coerce, intimidate, or convey some other message to a larger audience other than to the immediate victims.
3. The violent act was outside the precepts of international humanitarian law.

Figure 8.10.2 displays attacks in the United States since 1990 that meet the definition of terrorism.

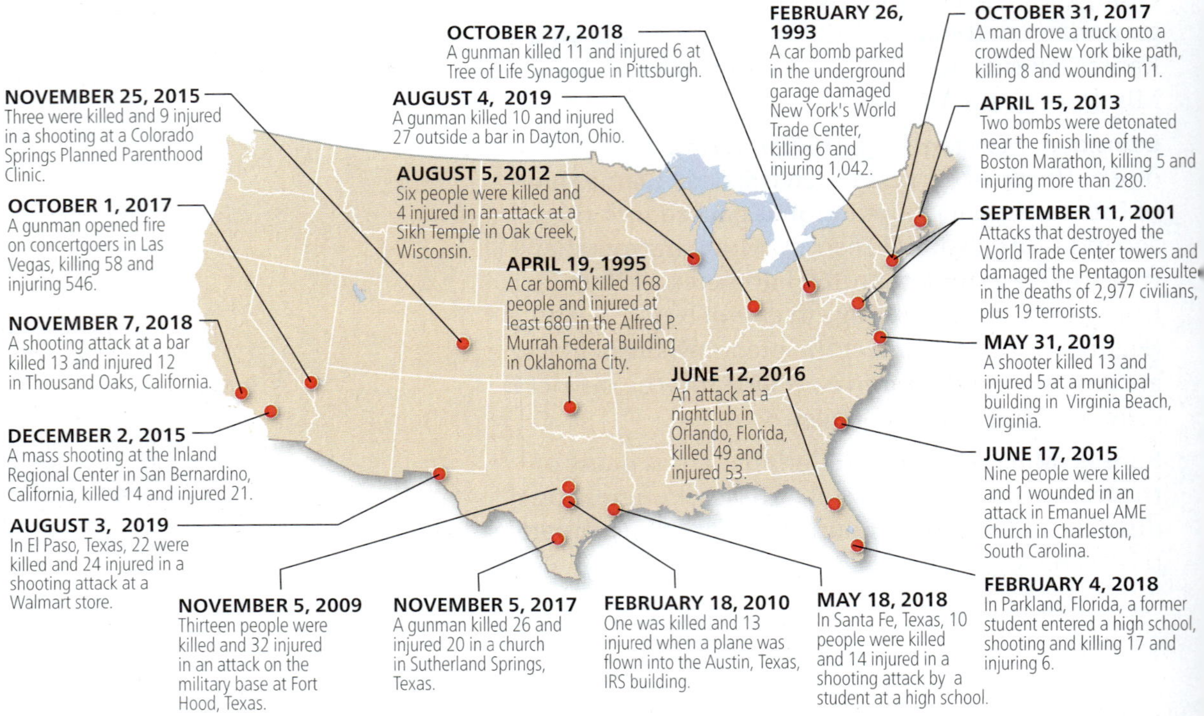

▶ 8.10.2 TERRORIST ATTACKS IN THE UNITED STATES
Terrorist attacks between 2009 and 2021 with at least 10 fatalities or injuries, plus three earlier attacks with especially high fatalities.
Source: Center for Systemic Peace

8.10.2 World Terrorism Trends

▲ 8.10.3 TERRORIST ATTACKS WITH AT LEAST 100 FATALITIES, 2000–2021

During the 1990s and the first years of the twenty-first century, the leading terrorist organization was al-Qaeda (Figure 8.10.3). Al-Qaeda was responsible for the September 11, 2001, attack on the World Trade Center and Pentagon.

Al-Qaeda's leader Osama bin Laden issued a declaration of war against the United States in 1996. Bin Laden argued that Muslims have a duty to wage a holy war against U.S. citizens because the United States was responsible for protecting the State of Israel and for maintaining the Saud royal family as rulers of Saudi Arabia. A U.S. military special operations unit killed bin Laden in 2011. Since then, al-Qaeda's influence has waned.

Leading Terrorist Organizations

The Institute for Economics & Peace has documented the number of deaths from terrorism in the world between 2000 and 2019. The number of deaths from terrorism increased from around 5,000 in 2000 to around 32,000 in 2014, before declining to around 14,000 in 2019 (Figure 8.10.4).

In recent years, the four most active terrorist organizations have been the Taliban, Boko Haram, the Islamic State, and Al-Shabaab. The four organizations operate in different regions of the world.

Taliban. The Taliban (which means "religious students") gained control of Afghanistan's government in 1996 and imposed very strict laws inspired by Islamic values as the Taliban interpreted them. For example, the Taliban banned playing music, watching TV, and surfing the internet. Men were beaten for shaving their beards, and women were stoned for committing adultery. Gay men were buried alive, and prostitutes were hanged in front of audiences. A NATO coalition overthrew the Taliban in 2001 and replaced it with a democratically elected government. However, the Taliban regained control of Afghanistan in 2021 and reimposed discriminatory practices against women and LGBTQ people.

Boko Haram. Boko Haram, which is Arabic for "Western education is forbidden," was founded in 2002 in Nigeria. It seeks to transform Nigeria into an Islamic state, and it opposes adoption of Western cultural practices. In recent years, it has expanded its attacks into Chad, Cameroon, and Niger.

Islamic State. The Islamic State originated in 1999 and became an affiliate of al-Qaeda in 2004. The two organizations split apart in 2014. Members of the Islamic State are Sunni Muslims who seek to impose strict religious laws throughout Southwest Asia. They have maintained control of territory through human rights violations, such as beheading, massacres, and torture. Over the years, the Islamic State has controlled considerable territory in northern Iraq and eastern Syria, but its activities have declined since 2019, when the United States killed its leader Abu Bakr al-Baghdad.

Al-Shabaab. Al-Shabaab originated in 2006 as an affiliate of al-Qaeda based in Somalia and Kenya. Its goal is to make Somalia an Islamic state.

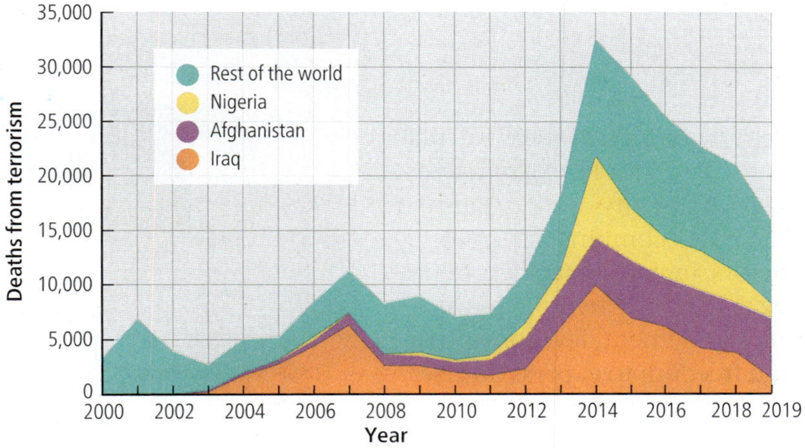

▲ 8.10.4 DEATHS FROM TERRORISM

States in War & Peace

- **Explain differences among regime types, and identify suspect states.**

The governments of some states are better able than those in other states to provide the leadership needed to promote peace and prosperity. Governments that are corrupt, repressive, and embroiled in wars may be less able to respond effectively to economic challenges and emergencies.

▼ 8.11.1 REGIME TYPES

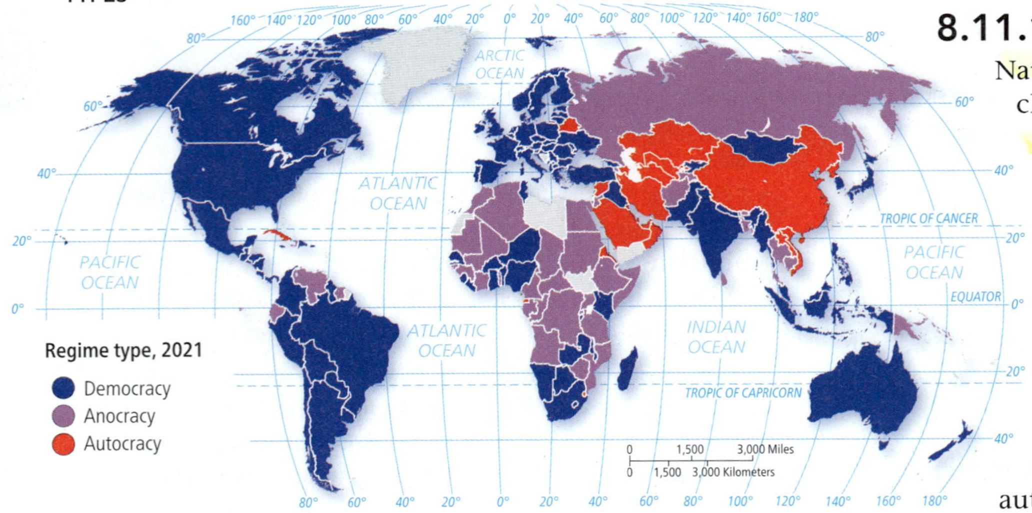

8.11.1 Regime Types

National governments can be classified as democratic, autocratic, or anocratic (Figure 8.11.1).

- A **democracy** is a country in which citizens elect leaders and can run for office.
- An **autocracy** is a country that is run according to the interests of the ruler rather than the people.
- An **anocracy** is a country that is not fully democratic or fully autocratic but rather displays a mix of the two types.

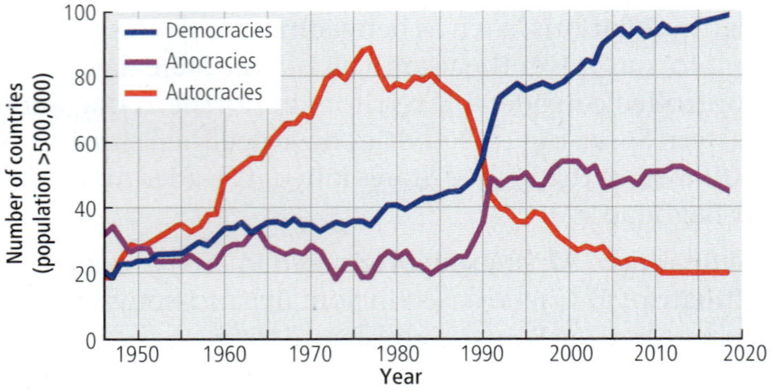

▲ 8.11.2 TREND TOWARD DEMOCRACY

The world has become more democratic (Figure 8.11.2). The Center for Systemic Peace cites these reasons: the replacement of increasingly irrelevant and out-of-touch monarchies with elected governments that broaden individual rights and liberties, the widening of participation in policymaking to all citizens through rights to vote and to serve in government, and the diffusion of democratic government structures created in Europe and North America to other regions.

State Fragility

The Fragile States Index, calculated by the Fund for Peace, measures the relative stability of every country. The index combines several factors, including fairness of the legal system, extent of youth unemployment, level of violence, and freedom to express diverse political views (Figure 8.11.3). The most fragile states are clustered in sub-Saharan Africa.

▼ 8.11.3 FRAGILE STATES INDEX

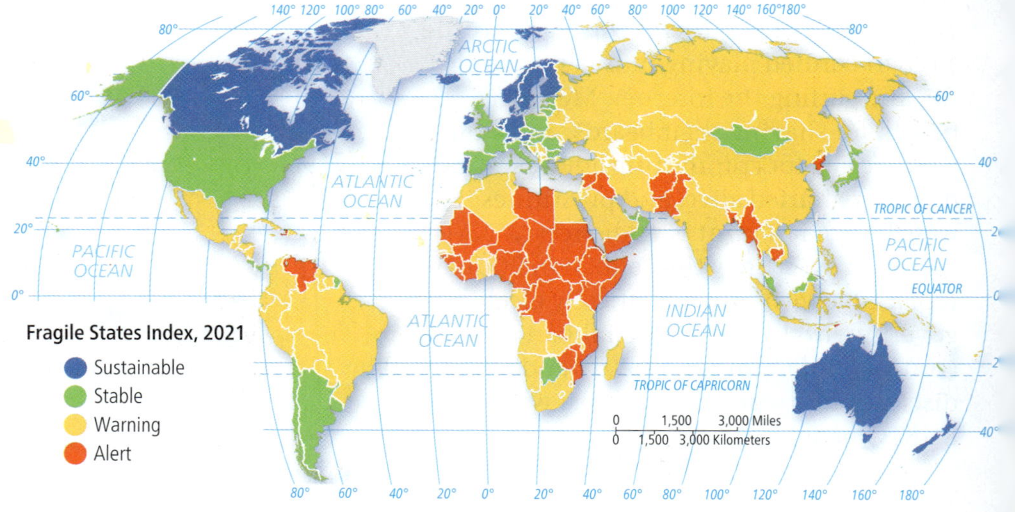

8.11.2 Nuclear States

A **weapon of mass destruction** is a nuclear, biological, chemical, or other weapon that can kill and bring significant harm to a large number of people or cause great damage to human-made structures, natural systems, or the biosphere. However, no treaty or international law contains an authoritative definition. The term was initially applied to nuclear weapons, but in recent years it has been expanded to encompass biological and chemical attacks as well.

The Treaty on the Non-Proliferation of Nuclear Weapons, which went into force in 1970, has been signed by 191 countries. The treaty was designed to prevent the spread of nuclear weapons and weapons technology and to promote the peaceful use of nuclear energy.

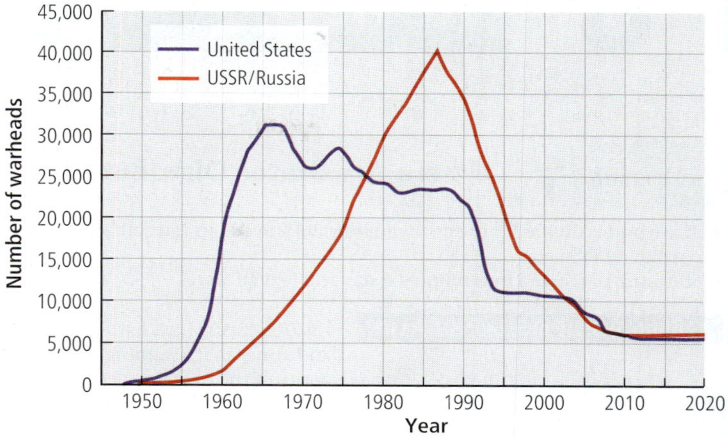

▲ **8.11.4** CHANGING NUMBER OF NUCLEAR WARHEADS

The United States and Russia together have more than 90 percent of the world's total of nuclear weapons. However, since the non-proliferation treaty went into effect, the two countries have destroyed a large percentage of their warheads (Figure 8.11.4). In addition to the United States and Russia, the other countries known to have nuclear warheads are China, France, the United Kingdom, Pakistan, India, and North Korea. Israel is suspected to have nuclear weapons but has not admitted to it, and it is not known if it has ever conducted a nuclear test.

Suspect States

A number of states that signed the Nuclear Non-Proliferation Treaty have been suspected of pursuing development of weapons of mass destruction, including nuclear and chemical weapons. The nuclear programs of Iran and North Korea have been of particular concern to the U.S and European governments.

Iran. Hostility between the United States and Iran dates from 1979, when a revolution forced abdication of Iran's pro-U.S. Shah Mohammad Reza Pahlavi. Iran's majority population, the Shiites, supported exiled fundamentalist Shiite Muslim leader Ayatollah Ruholiah Khomeini and proclaimed Iran an Islamic Republic. Militant supporters of the ayatollah seized the U.S. embassy in Iran on November 4, 1979, and held 52 Americans hostage until January 20, 1981. The United States and most other countries have opposed Iran's aggressive development of a nuclear program (Figure 8.11.5). Iran claimed that its nuclear program was for civilian purposes, but observers concluded that it was intended to develop weapons. Prolonged negotiations produced an agreement to degrade Iran's nuclear capabilities without resorting to yet another war in the region. President Trump pulled the United States out of the treaty in 2018, claiming that the treaty did not do enough to prevent Iran from continuing to pursue nuclear weapons.

▲ **8.11.5** IRAN'S NUCLEAR FACILITIES

North Korea. North Korea is one of the world's least economically developed and most isolated countries, and since 1948, it has been governed as a dictatorship in succession by Kim Il-sung, his son Kim Jong-il, and his grandson Kim Jong-un. North Korea builds and tests nuclear weapons and long-range missiles, even though the country lacks the ability to provide its citizens with adequate food, electricity, and other basic needs (Figure 8.11.6).

▶ **8.11.6** NORTH AND SOUTH KOREA
A nighttime satellite image recorded by the U.S. Air Force Defense Meteorological Satellite Program shows the illumination of electric lights in South Korea and the absence in North Korea.

CHAPTER

8 Review, Analyze, & Apply

KEY ISSUE 1 Where Are States Distributed?

- The world is divided into approximately 200 sovereign states that vary considerably in size.
- States are organized into unitary states and federal states.

THINKING GEOGRAPHICALLY In 2014, Russia invaded Crimea and annexed it, claiming that the majority of the Crimean people, who are ethnic Russians, supported the action (refer to Figure 8.4.5). Nearly every other country in the world continues to recognize Ukraine's sovereignty over Crimea. Should Crimea appear on maps (such as Figure 8.1.4) as part of Ukraine or as part of Russia? Why?

◀ **8.CR.1** BRIDGE ERECTED BY RUSSIA CONNECTING CRIMEA AND RUSSIA

KEY ISSUE 2 How Are States Created?

- Until relatively recently, the world was organized into city-states, empires, and kingdoms.
- Colonies once comprised much of the world but only a few colonies remain.
- During the last few centuries, nation-states have been created in many places, but none are perfect.

THINKING GEOGRAPHICALLY Puerto Rico is a commonwealth of the United States. Puerto Ricans are U.S. citizens but don't vote in U.S. elections or have a voting member of Congress. In a recent referendum, 52.5% of Puerto Ricans voted in favor of Puerto Rico becoming the fifty-first U.S. state, and 47.5% of Puerto Ricans voted for independence or a free association with the United States. What might be the advantages and challenges of Puerto Rico becoming a U.S. state or becoming an independent country?

▲ **8.CR.2** PUERTO RICO
Capitol Office Building, San Juan.

KEY ISSUE 3 Why Do States Have Distinctive Boundaries?

- Boundaries can be drawn using physical or cultural features.
- Five shapes of states are compact, perforated, elongated, prorupted, and fragmented.
- Boundaries dividing electoral districts within countries can be gerrymandered to favor one political party through cracking and packing.

THINKING GEOGRAPHICALLY The Arctic Ocean sea ice has receded in recent decades as a result of climate change. How might the reduction in sea ice affect claims by states to the Arctic region? How might states benefit or be harmed by the loss of the ice?

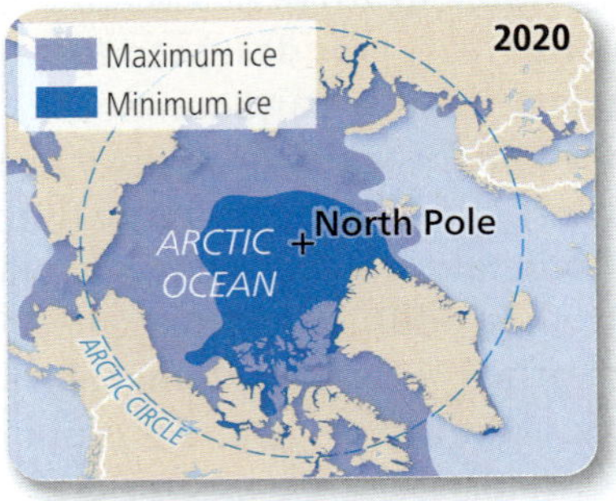

◀ **8.CR.3** RECEDING ARCTIC SEA ICE

KEY ISSUE 4 What Threats Do States Face?

- Terrorism by individuals and organizations has included the 9/11 attacks on the United States.
- The governance of states can be classified as democratic, anocratic, or autocratic.
- The number of democracies has been increasing.

THINKING GEOGRAPHICALLY For a few months in late 2020 and early 2021, the Center for Systemic Peace temporarily changed the classification of the United States from democracy to anocracy. What events in the United States at that time influenced the change?

◀ **8.CR.4** ATTACK ON U.S. CAPITOL, JANUARY 6, 2021

Explore

Europe's International Boundaries

Use Google Earth to view the international boundary formed by the Rhine River.

Search for **Le Jardin-des-Deux-Rives** (Two Rivers Garden).

1. In what country is the garden? What country is on the other side of the Rhine River?
2. Drop to Street View at the red pin marking the garden. Move the photo across the bridge. How do the border controls across the Rhine compare with those illustrated in Chapter 3 between the United States and Mexico?
3. Why do you think the border crossing at the Rhine differs from border crossings between the United States and Mexico?

▼ **8.CR.5** LE JARDIN-DES-DEUX-RIVES

Key Terms

Anocracy (8.11.1) A country that is not fully democratic or fully autocratic but rather displays a mix of the two types.

Antecedent boundary (8.6.1) A physical feature that serves as a boundary, so named because it existed before the present settlement.

Autocracy (8.11.1) A country that is run according to the interests of the ruler rather than the people.

Balance of power (8.2.2) A condition of roughly equal strength between opposing countries or alliances of countries.

Boundary (8.6.1) An invisible line that marks the extent of a state's territory.

City-state (8.3.1) A sovereign state composed of a city and its immediately surrounding countryside.

Colonialism (8.5.2) An attempt by one country to establish settlements and to impose its political, economic, and cultural principles in another territory.

Colony (8.5.1) A territory that is legally tied to a sovereign state rather than completely independent.

Compact state (8.8.1) A state in which the distance from the center to any boundary does not vary significantly.

Consequent boundary (8.6.2) A boundary drawn in order to separate groups based on ethnic, linguistic, religious, or economic differences.

Democracy (8.11.1) A country in which citizens elect leaders and can run for office.

Devolution (8.2.1) The transfer or delegation of power to a lower level, especially by a national government to local or regional administration.

Elongated state (8.8.2) A state with a long, narrow shape.

Federal state (8.2.1) An internal organization of a state that allocates most powers to units of local government.

Fragmented state (8.8.2) A state that includes several discontinuous pieces of territory.

Frontier (8.6.1) A zone separating two states in which neither state exercises political control.

Geometric boundary (8.6.2) A boundary based on a human construct such as a straight line or a parallel of latitude.

Gerrymandering (8.9.1) The process of redrawing legislative boundaries within a U.S. state for the purpose of benefiting the party in power.

Landlocked state (8.8.1) A state that does not have a direct outlet to the sea.

Microstate (8.1.2) A state that encompasses a very small land area.

Multinational state (8.4.2) A state that contains two or more cultural groups with traditions of self-determination that agree to coexist peacefully by recognizing each other as distinct nationalities.

Nation (8.3.2) A large group of people who are united by common cultural characteristics, such as language and ethnicity, or by shared history.

Nation-state (8.3.2) A state whose territory corresponds to that occupied by a particular nation.

Perforated state (8.8.1) A state that completely surrounds another one.

Prorupted state (8.8.2) An otherwise compact state with a large projecting extension.

Relic boundary (8.6.2) A boundary that no longer functions but can still be detected on the cultural landscape.

Self-determination (8.3.2) The concept that ethnicities have the right to govern themselves.

Sovereignty (8.2.1) Ability of a state to govern its territory free from control of its internal affairs by other states.

State (8.1.1) An area organized into a political unit and ruled by an established government that has control over its internal and foreign affairs.

Subsequent boundary (8.6.2) A boundary that is established after the settlement of an area and is changed to reflect developments such as war.

Superimposed boundary (8.6.2) A boundary drawn on an area by a conquering or colonizing power that ignores existing cultural patterns.

Terrorism (8.10.1) The threatened or actual use of illegal force and violence by a non-state actor to attain a political, economic, religious, or social goal through fear, coercion, or intimidation.

Unitary state (8.2.1) An internal organization of a state that places most power in the hands of central government officials.

Weapon of mass destruction (8.11.2) A nuclear, biological, chemical, or other weapon that can kill and bring significant harm to a large number of humans or cause great damage to human-made structures, natural structures, or the biosphere.

Geospatial Analysis

Log in to the Mastering Geography Study Area to access MapMaster 2.0.

Unity & Diversity in Europe

Most countries in Europe have joined economic and military alliances. But differences persist among the member countries. Add the ***NATO*** layer. Select the *Settings* icon from the *Legend*, and select *Show Political Labels*. Add the ***Gross National Income Per Capita*** layer and select *Split map window*. Zoom to Europe.

1. What differences in GNI per capita do you see between countries in the eastern countries of NATO and the western countries?
2. What historical factor accounts for the income difference between east and west? Refer to Figure 8.4.4 or view Former Warsaw Pact Members in MapMaster 2.0.

▶ **8.CR.6 NATO MEMBERS AND GROSS NATIONAL INCOME PER CAPITA IN EUROPE**

CHAPTER 9 Food & Agriculture

▲ Bamako Market, Mali.

All humans need food to survive. We have two choices in obtaining our food: buy it from others or produce it ourselves. For most of human history, people had to search for their food in the wild. Around 10,000 years ago, humans figured out how to deliberately grow their own food. In the modern world, most people buy their food.

With greater separation between the production of food and the consumption of it, humans are less aware of the connection between their food choices and the impacts of those choices on the environment. People may also be less aware of the connection between their food choices and the businesses that grow and distribute food.

KEY ISSUES

1 What Do People Eat?

People consume a wide variety of food, depending on their income, growing conditions in the physical environment, and cultural preferences. The U.N. estimates that approximately 780 million people are undernourished.

2 Where Does Food Come From?

Agriculture was invented around 10,000 years ago in multiple places. Today, the world can be divided into subsistence and commercial agricultural regions.

3 How Has Agriculture Become More Global?

Food production and trade have increased to meet the needs of a growing population. Some traditional agricultural practices have been lost in the transition.

4 How Can Agriculture Be More Earth-Friendly?

Agriculture faces issues of resource depletion. Organic farming has increased as a more Earth-friendly way to grow food.

LOCATIONS IN THIS CHAPTER

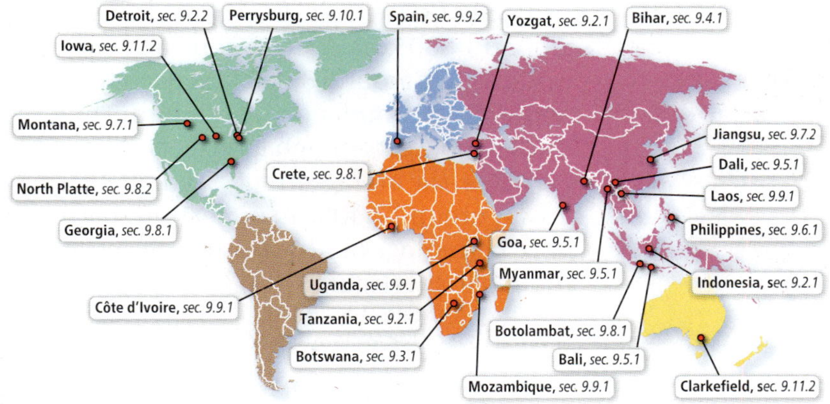

▲ **9.1.1 TYPICAL FOOD IN ZIMBABWE**
Maize (corn) is the most commonly grown crop and food source.

How Much We Eat

- **Explain variations in how much people eat.**

When you buy food in a supermarket, are you reminded of a farm? Not likely. The meat is carved into pieces that no longer resemble an animal and wrapped in paper or plastic film. Often the vegetables are canned or frozen. Food looks different in other places of the world (Figures 9.1.1 and 9.1.2).

9.1.1 Food Consumption

Consumption of food varies around the world, in both total amount and source of nutrients. The variation results from a combination of:

- **Level of development.** People in more developed countries tend to consume more food and from different sources than do people in developing countries.
- **Physical conditions.** Climate is important in influencing what can be most easily grown and therefore consumed in developing countries. In more developed countries, though, food is shipped long distances from locations with different climates.
- **Cultural preferences.** Some food preferences and avoidances are expressed without regard for physical and economic factors, as discussed in Chapter 4.

(a) (b)

▲ **9.1.2 TYPICAL MEAL IN ZIMBABWE**
(a) Maize is cooked into a thick porridge known as sadza in Zimbabwe. (b) Sadza is typically eaten in Zimbabwe mixed with tomatoes, onions, and chicken or local greens.

Food Security

The United Nations Food & Agricultural Organization (FAO) defines **food security** as physical, social, and economic access at all times to safe and nutritious food sufficient to meet dietary needs and food preferences for an active and healthy life. By this definition, around 9 percent of the world's inhabitants do not have food security.

Dietary energy consumption is the amount of food that an individual consumes. The unit of measurement of dietary energy is the kilocalorie (kcal), or Calorie in the United States. One gram (or ounce) of each food source delivers a kilocalorie level that nutritionists can measure.

According to the FAO, to maintain a moderate level of physical activity, an average individual needs to consume on a daily basis at least around 1,800 kcal. Average consumption worldwide is around 3,000 kcal per day, well above the recommended minimum. Thus, most people get enough food to survive. However, the average in sub-Saharan Africa is only around 2,400, an indication that a large percentage of Africans are not getting enough to eat (Figure 9.1.3).

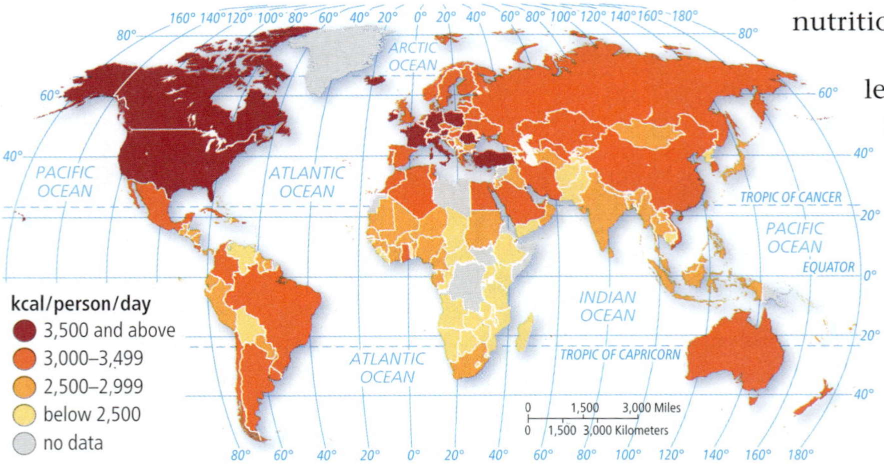

◀ **9.1.3 ADEQUACY OF AVERAGE DIET**

9.1.2 Undernourishment

The level of dietary energy consumption that is continuously below the minimum requirement for maintaining a healthy life and carrying out light physical activity is called **undernourishment**. The U.N. estimates that approximately 780 million people in the world are undernourished, one-third of them in India and China, the world's two most populous countries (Figure 9.1.4). Overall, 98 percent of the world's undernourished people are in developing countries.

As a percentage of a region's population, undernourishment is most prevalent in sub-Saharan Africa and South Asia. One-fifth of the population in sub-Saharan Africa and one-sixth in South Asia are undernourished.

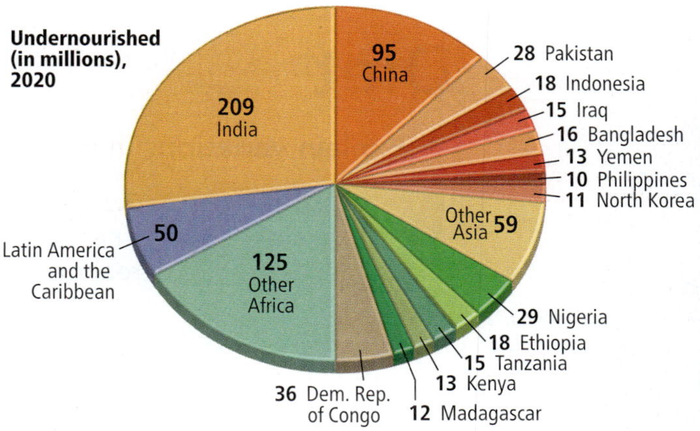

▲ **9.1.4 DISTRIBUTION OF UNDERNOURISHMENT**
The number for China is for 2005. The United Nations has been unable to calculate a more recent number.

The world as a whole has made progress in reducing hunger during the twenty-first century. Between 2000 and 2020, the number of undernourished people declined from 924 million to around 780 million, and the percentage of people who were undernourished declined from 25 percent in 2000 to 9 percent. East Asia, led by China, has had by far the largest decrease in the number and percentage of undernourished people.

Food Waste

Given the large number of undernourished people, especially challenging for world food security is the amount of food that is wasted. The U.N. reports that 17 percent of food produced globally is discarded. People in their homes account for 61 percent of food waste, food service 26 percent, and grocery retailers 13 percent, according to the U.N.

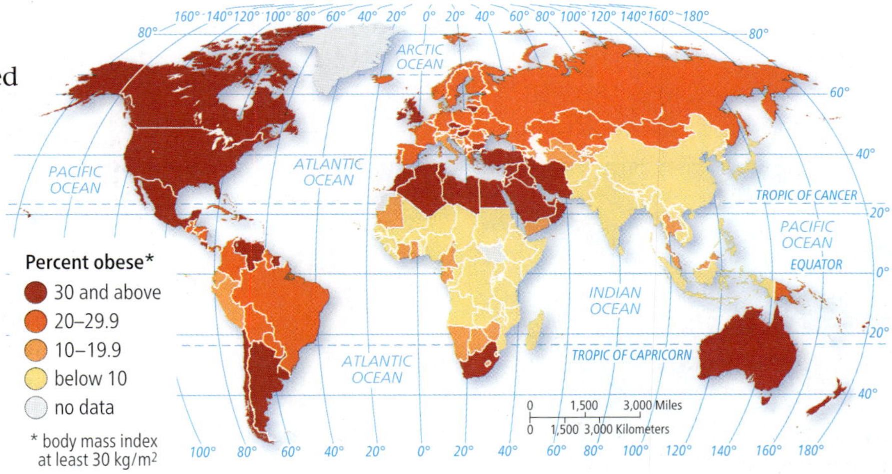

▲ **9.1.5 OBESITY**

Obesity

Undernourishment is a common cause of sickness in developing countries, whereas overconsumption of food is a major health concern in developed countries. The average American is consuming 3,800 kcal per day, the world's highest level and more than twice the recommended minimum. In fact, obesity is more prevalent than hunger in the United States as well as in other developed countries (Figure 9.1.5).

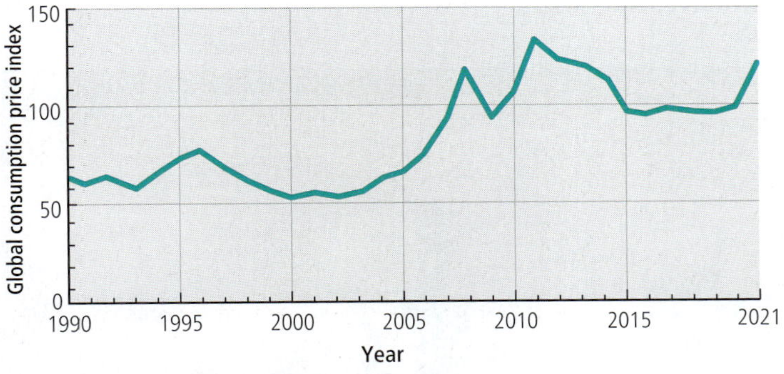

▲ **9.1.6 GLOBAL FOOD PRICE INDEX**

Food Prices

The greatest challenge to world food security in the twenty-first century is the cost of food rather than food supply. People on average spend less than 20 percent of their income on food in developed countries, compared with more than 40 percent in much of sub-Saharan Africa (Figure 9.1.6).

Diet

- **Explain variations in what people eat.**

Most humans derive most of their kilocalories through consumption of **cereal grain** (or simply cereal), which is a grass that yields grain for food. **Grain** is the seed from a cereal grass (Figure 9.2.1).

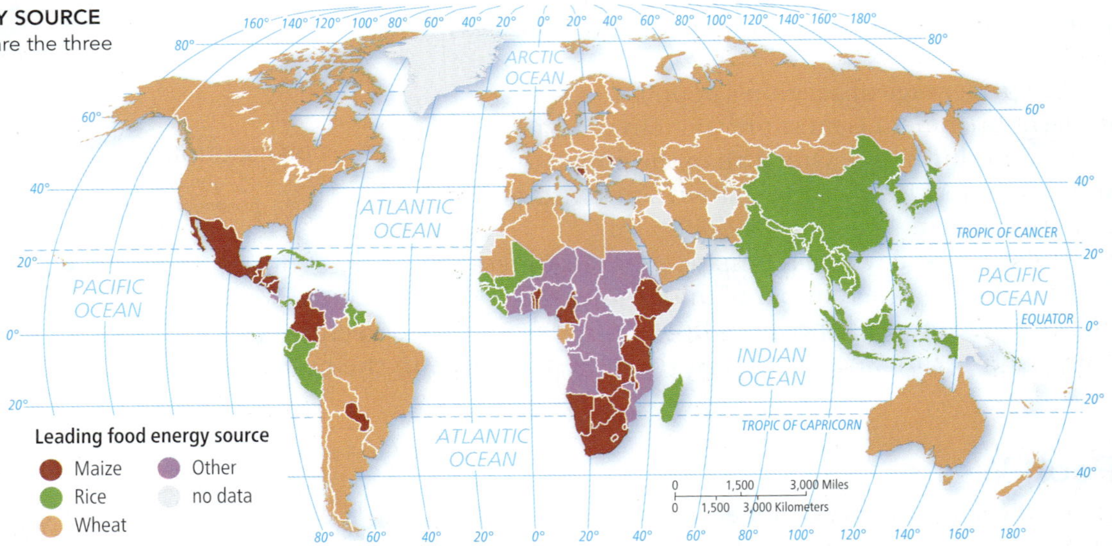

▶ **9.2.1 DIETARY ENERGY BY SOURCE**
Wheat, rice, and maize (corn) are the three main sources of kilocalories.

9.2.1 Source of Nutrients

The three leading cereal grains—wheat (Figure 9.2.2), maize (corn in North America, Figure 9.2.3), and rice (Figure 9.2.4)—together account for nearly 90 percent of all grain production and more than 40 percent of all dietary energy consumed worldwide. Other significant sources of nutrients are derived from yams, sorghum, cassava, millet, potatoes, and beans.

◀ **9.2.2 HARVESTING WHEAT**
Yozgat, Turkey.

▼ **9.2.3 HARVESTING MAIZE (CORN)**
Tanzania.

▼ **9.2.4 HARVESTING RICE**
Indonesia.

9.2.2 Meat Consumption

Protein is a nutrient needed for growth and maintenance of the human body. Many food sources provide protein of varying quantity and quality. One of the most fundamental differences between subsistence and commercial agriculture regions is the primary source of protein. In commercial agriculture regions, the leading source of protein is meat products, including beef, pork, and poultry.

Consumption of protein from meat exceeds 3 ounces per person per day in Europe, the United States, and Australia, compared with less than 1 ounce per day in many countries of South Asia and sub-Saharan Africa (Figure 9.2.5).

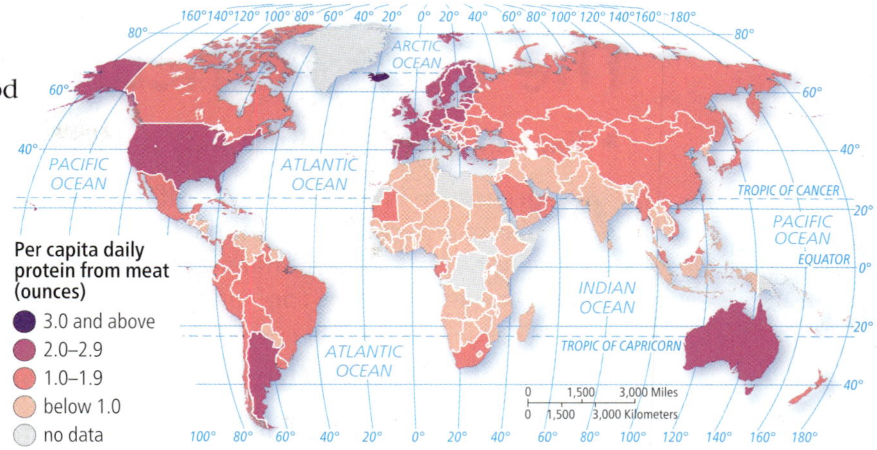

▲ **9.2.5 PROTEIN FROM MEAT**
Ounces per person per day.

▼ **9.2.6 DOUBLE CHEESEBURGER**
Around 430 calories.

Sustainable Food Choices

Choices to consume specific foods are made several times a day. These choices can be more or less sustainable.

Meat or vegetarian/vegan. Around 8 percent of people in the world are vegetarian. The percentage varies widely by country. For example, more than 30 percent of people in India identify as vegetarians, compared with only 5 percent in the United States, which has one of the highest rates of meat consumption per capita (Figure 9.2.6).

Meat is a great source of zinc, iron, and other omega-3s that are important for muscle development and health. On the other hand, meat consumption is associated with chronic illnesses, including hypertension, cancer, and cardiovascular disease.

▼ **9.2.7 DIRTY DOZEN AND CLEAN FIFTEEN**

Dirty Dozen & Clean Fifteen. The Environmental Working Group ranks several dozen types of produce according to level of pesticides, based on tests conducted by the U.S. Department of Agriculture. The 12 most pesticide-contaminated fruits and vegetables are called "the Dirty Dozen," and the 15 least contaminated are "the Clean Fifteen." Strawberries, spinach, and kale have been found to be the most contaminated, whereas avocados, corn on the cob, and pineapple are the least contaminated (Figure 9.2.7).

◀ **9.2.8 GLOBAL SOURCING**
Supermarket, Detroit.

Global sources or local sources. Most food in developed countries like the United States travels long distances to reach supermarkets and restaurants. At the same time, more people are patronizing farmers markets and "farm-to-table" restaurants. Consuming globally makes it easier to eat a balanced diet and obtain food that can't be grown locally (Figure 9.2.8). On the other hand, many foods transported long distance lose their nutritional value.

The First Agriculture Revolution

- **Explain how agriculture may have originated.**

Agriculture is deliberate modification of Earth's surface through cultivation of plants and raising of animals for sustenance or economic gain. Agriculture originated when humans domesticated plants and animals for their use. The word *cultivate* means "to care for," and a **crop** is any plant cultivated by people.

▼ 9.3.1 HUNTING & GATHERING
San people, Botswana.

9.3.1 Hunters & Gatherers

Before the invention of agriculture, all humans probably obtained the food they needed for survival through hunting for animals, fishing, or gathering plants (such as berries, nuts, fruits, and roots). Hunters and gatherers lived in small groups of usually fewer than 50 persons because a larger number would quickly exhaust the available resources within walking distance.

These groups traveled frequently, establishing new home bases or camps. The direction and frequency of migration depended on the movement of game and the seasonal growth of plants at various locations. We can assume that groups communicated with each other concerning matters such as hunting rights and intermarriage. For the most part, they kept the peace by steering clear of each other's territory.

It was once thought that the labor was strictly divided between men and women. The men hunted game or fished, while the women collected berries, nuts, and roots. Recent archaeological evidence, however, suggests that women participated in the hunting. The evidence also suggests that the hunters and gatherers got most of their calories from gathering food for a few hours each day, rather than from hunting.

Today, perhaps a quarter-million people, or less than 0.005 percent of the world's population, still survive by hunting and gathering rather than by agriculture. Contemporary hunting and gathering societies are isolated groups that live on the periphery of world settlement, but they provide insight into human customs that prevailed in prehistoric times before the invention of agriculture (Figure 9.3.1).

Agricultural Revolution

The time when human beings first domesticated plants and animals and no longer relied entirely on hunting and gathering is known as the **first agricultural revolution**. Geographers and other scientists believe that the first agricultural revolution occurred around the year 8000 B.C. because the world's population began to grow at a more rapid rate than it had in the past. By growing plants and raising animals, human beings created larger and more stable sources of food, so more people could survive.

Scientists do not agree on whether the first agricultural revolution originated primarily because of environmental factors or cultural factors. Probably a combination of both factors contributed:

- **Environmental factors.** The first domestication of crops and animals coincided with climate change that marked the end of the last ice age. At that time, permanent ice cover receded from Earth's mid-latitudes to the polar regions, resulting in a massive redistribution of humans, other animals, and plants.
- **Cultural factors.** A preference for living in a fixed place instead of as nomads may have led hunters and gatherers to build permanent settlements and to store surplus food there. In gathering wild vegetation, people inevitably cut plants and dropped berries, fruits, and seeds. These hunters probably observed that, over time, damaged or discarded food produced new plants. They may have deliberately cut plants or dropped berries on the ground to see if they would produce new plants. Subsequent generations learned to pour water over the site and to introduce manure and other soil improvements. Over thousands of years, plant cultivation apparently evolved from a combination of accident and deliberate experiment.

9.3.2 Agriculture Hearths

Scientists agree that agriculture originated in multiple hearths (Figure 9.3.2). Even within these hearths, scientists have determined that agriculture was invented independently by multiple groups.

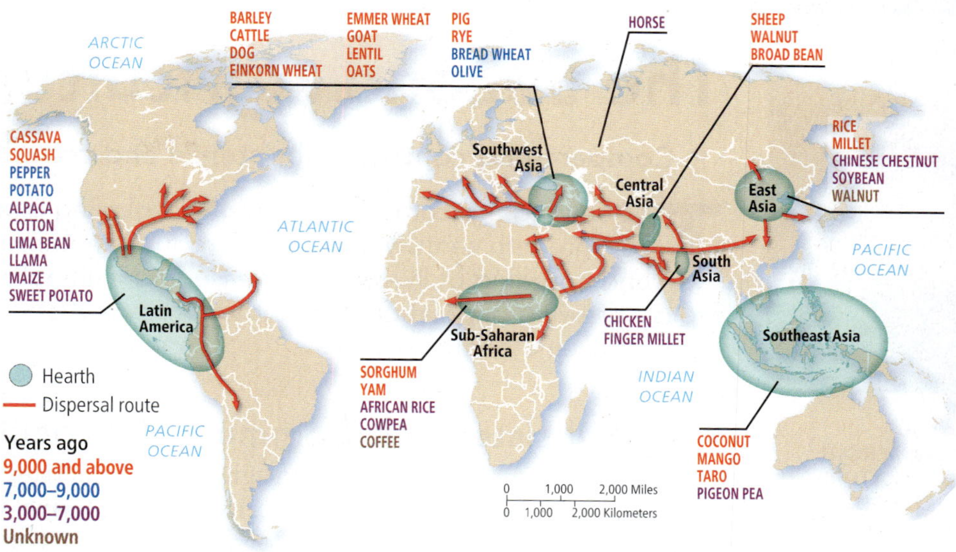

▲ 9.3.2 AGRICULTURAL HEARTHS

- **Southwest Asia.** The earliest domesticated crops trace back to around 10,000 years ago in Southwest Asia and are thought to have been barley, wheat, lentil, and olive crops. Indo-European languages may also have diffused from this hearth, as discussed in Chapter 5.
- **East Asia.** Rice crops are now thought to have been domesticated in East Asia more than 10,000 years ago.
- **Central & South Asia.** Chickens are thought to have been domesticated and diffused from South Asia around 4,000 years ago. Apples are considered to have been domesticated in Central Asia.
- **Sub-Saharan Africa.** Sorghum crops were domesticated in central Africa around 8,000 years ago. Yam crops may have been domesticated even earlier in that hearth.
- **Latin America.** Two important hearths of crop domestication are thought to have emerged in Mexico and Peru around 4,000 to 5,000 years ago. Mexico is considered a hearth for bean and maize crops, and Peru for potato crops.

Diffusion of Agriculture

The multiple origins of agriculture means that, from earliest times, people have produced food in distinctive ways in different regions. This diversity derives from a unique legacy of wild plants, climatic conditions, and cultural preferences in each place. Improved communications in recent centuries have encouraged the diffusion of some plants to varied locations around the world. Many plants and animals thrive across a wide portion of Earth's surface, not just in their place of original domestication. It wasn't until the sixteenth century, for example, that wheat, oats, and barley were introduced to the Western Hemisphere and maize to the Eastern Hemisphere. The transfer of plants and animals, as well as people, culture, and technology, between the Western Hemisphere and Europe occurred as a result of European colonization and trade and is known as the **Columbian Exchange** (Figure 9.3.3).

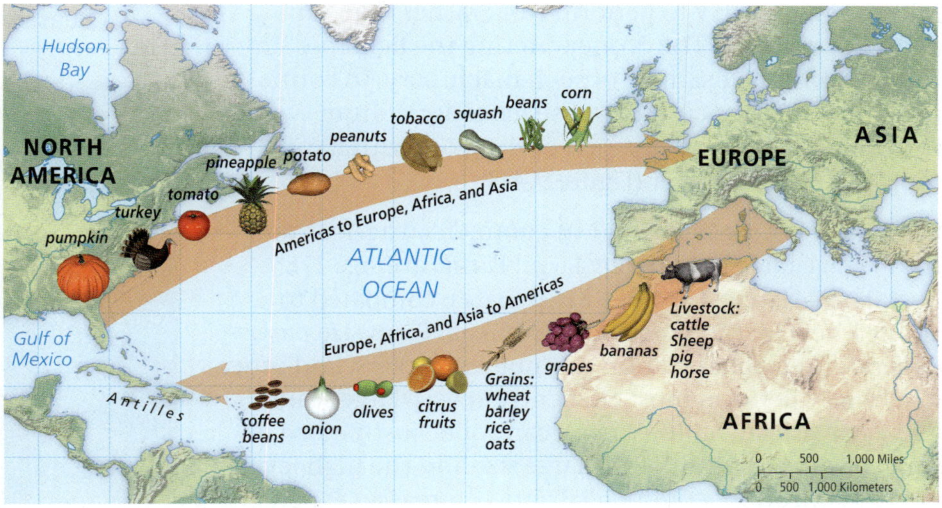

▲ 9.3.3 COLUMBIAN EXCHANGE

The Second Agricultural Revolution

- **Compare and contrast subsistence and commercial agriculture.**

The **second agricultural revolution**, beginning in the United Kingdom in the seventeenth century, increased productivity through improvement of crop rotation and breeding of livestock. Increased agricultural productivity helped feed the rapidly growing population in countries in stage 2 of the demographic transition during the nineteenth century.

▲ 9.4.1 COMMERCIAL AGRICULTURE TECHNOLOGY
A drone waters a field.

9.4.1 Subsistence & Commercial Agriculture

The most fundamental difference in agricultural practices is between subsistence agriculture and commercial agriculture. **Subsistence agriculture**, found in developing countries, is the production of food primarily for consumption by the farmer's family. **Commercial agriculture**, found primarily in developed countries, is the production of food primarily for sale off the farm. Commercial agriculture is distinct from subsistence agriculture in that commercial agriculture lowers the percentage of farmers and farm laborers needed to produce food, requires a larger farm, and uses more machinery and technology (Figure 9.4.1).

Farmers: Fewer grow more. All people must secure food in order to survive. Subsistence farmers produce the food that they and their families require. In commercial agriculture, relatively few people are engaged in farming. Most people buy food with money earned by working in factories or offices or by performing other services (Figure 9.4.2).

Technology: Agriculture's game changer. In commercial agriculture, a small number of farmers can feed many people because commercial agriculture relies on machinery to perform work rather than on people or animals. The drone watering the field in Figure 9.4.1 is an example of such machinery. In contrast to commercial agriculture, subsistence agriculture requires that farmers do much of the work with hand tools and animal power (Figure 9.4.3).

Farm size: Is bigger better? In commercial agriculture, the average farm is relatively large. Farms average 178 hectares (441 acres) in the United States, compared to about 1 hectare (2.5 acres) in South Asia. Farm size partly depends on mechanization. Combines, pickers, and other machinery perform most efficiently at very large scales, and their considerable expense cannot be justified on a small farm. As a result of the large size and the high level of mechanization, commercial agriculture is an expensive business.

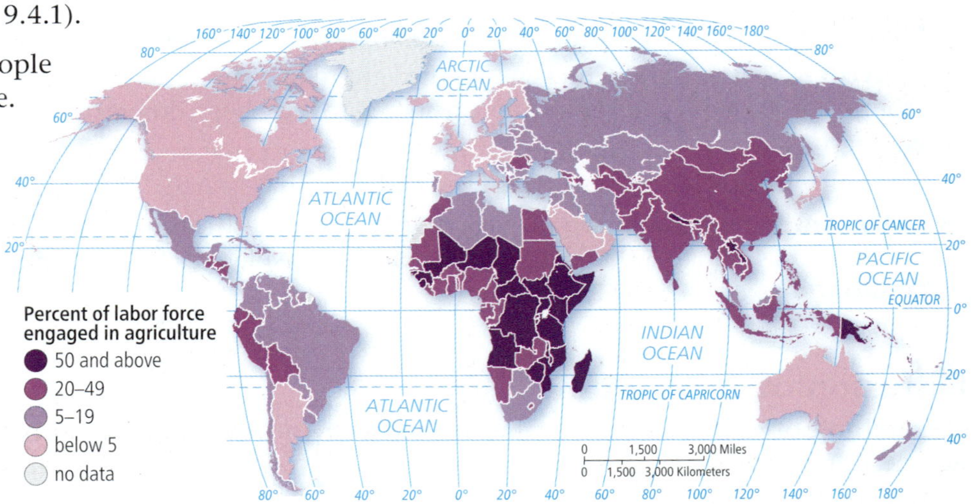

▲ 9.4.2 PERCENT OF LABOR FORCE IN AGRICULTURE

▼ 9.4.3 SUBSISTENCE AGRICULTURE TECHNOLOGY
Demonstrating the use of a row-scoring rake, Bihar, India.

9.4.2 Agriculture Regions

People have been able to practice agriculture in a wide variety of places. The most widely used map of world agricultural regions is based on work done by geographer Derwent Whittlesey in 1936. Whittlesey identified 11 main agricultural regions, plus an area where agriculture was nonexistent. Whittlesey's 11 regions are divided into five that are subsistence and six that are commercial. The subsistence agriculture regions included three that are intensive and two that are extensive. The commercial agriculture regions included four that are intensive and two that are extensive.

Subsistence Agriculture Regions

Intensive Subsistence

- **Wet rice dominant:** The large population concentrations of East Asia and South Asia.
- **Wet rice not dominant:** The large population concentrations of East Asia and South Asia, where growing rice is difficult.
- **Plantation:** The tropical and subtropical regions of Latin America, sub-Saharan Africa, South Asia, and Southeast Asia.

Extensive Subsistence

- **Pastoral nomadism:** The drylands of Southwest Asia & North Africa, Central Asia, and East Asia.
- **Shifting cultivation:** The tropical regions of Latin America, sub-Saharan Africa, and Southeast Asia.

Commercial Agriculture Regions

Intensive Commercial

- **Mixed crop and livestock:** The U.S. Midwest and central Europe.
- **Dairy:** Near population clusters in the northeastern United States, southeastern Canada, and northwestern Europe.
- **Mediterranean:** Lands surrounding the Mediterranean Sea, the western United States, the southern tip of Africa, and Chile.
- **Commercial gardening:** The southeastern United States and southeastern Australia.

Extensive Commercial

- **Grain:** The north-central United States, south-central Canada, and Eastern Europe.
- **Livestock ranching:** The dry lands of western North America, southeastern Latin America, Central Asia, sub-Saharan Africa, and the South Pacific.

Agriculture & Climate

Similarities between agriculture regions and climate regions are striking (Figure 9.4.4). For example, pastoral nomadism is the predominant type of agriculture in Southwest Asia & North Africa, which has a dry climate, whereas shifting cultivation is the predominant type of agriculture in sub-Saharan Africa, which has a tropical climate. Note the division between southeastern China (warm mid-latitude climate, intensive subsistence agriculture with wet-rice dominant) and northeastern China (cold mid-latitude climate, intensive subsistence agriculture with wet rice not dominant). In the United States, much of the West is distinguished from the rest of the country according to climate (dry) and agriculture (livestock ranching).

▼ 9.4.4 AGRICULTURE REGIONS AND CLIMATE REGIONS

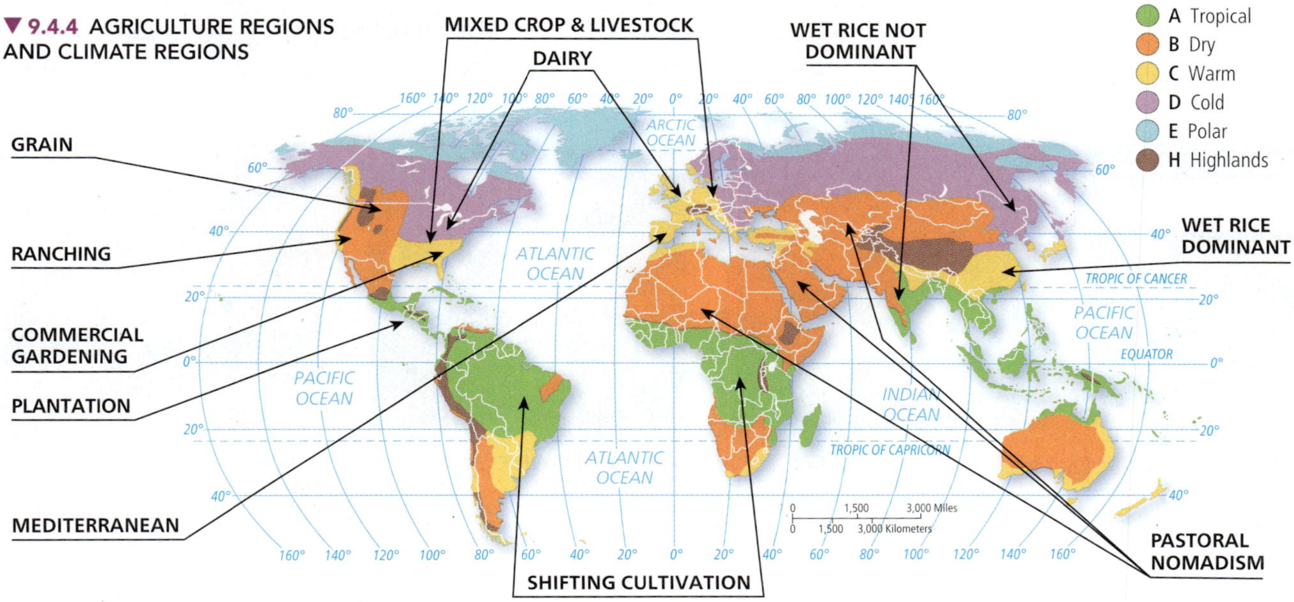

Comparing Subsistence & Commercial Agriculture

- **Compare the most common subsistence and commercial agriculture regions.**

The form of agriculture most common in the world is intensive subsistence agriculture in developing regions and mixed crop and livestock in developed regions.

9.5.1 Intensive Subsistence Agriculture

In developing countries, many people produce food for their own consumption. This is especially true in **intensive subsistence agriculture**, the most widely practiced form of agriculture in densely populated East Asia, South Asia, and Southeast Asia.

The intensive agriculture region of Asia can be divided between areas where wet rice dominates and areas where it does not. The term **wet rice** refers to rice planted on dry land in a nursery and then moved as seedlings to a flooded field to promote growth. Wet rice occupies a relatively small percentage of Asia's agricultural land but is the region's most important source of food (Figure 9.5.1).

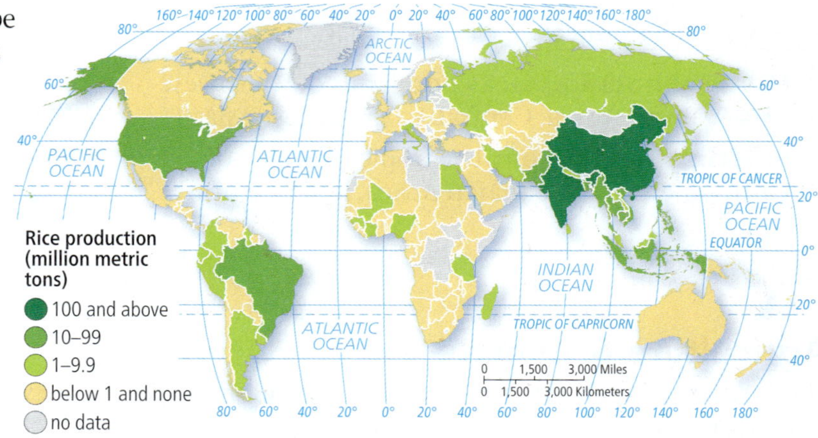

▲ 9.5.1 RICE PRODUCTION

Growing Rice

Successful production of large yields of rice is an elaborate process that is time-consuming and done mostly by hand. Four main steps are involved; as the name implies, all four steps are intensive. First, the field is prepared, typically with a plow drawn by water buffalo or oxen (Figure 9.5.2). Flat land is needed to grow rice, so hillsides are terraced. Second, the field is flooded with water (Figure 9.5.3). The flooded field is called a **sawah** in Indonesia and is sometimes incorrectly called a **paddy**, which is actually the Malay word for wet rice. Third, rice seedlings grown for the first month in a nursery are transplanted into the flooded field (Figure 9.5.4). Fourth, rice plants are harvested with knives. The chaff (husks) are separated from the seeds by threshing (beating) them on the ground (Figure 9.5.5). The threshed rice is placed in a tray, and the lighter chaff is winnowed (allowed to be blown away by the wind).

▲ 9.5.2 GROWING RICE: PREPARING THE FIELD
Water buffalo in the rice paddy, Goa, India.

▲ 9.5.3 GROWING RICE: FLOODING THE FIELD
Bali.

▲ 9.5.4 GROWING RICE: TRANSPLANTING
Rice plants are transplanted by hand, Dali, China.

▲ 9.5.5 GROWING RICE: HARVESTING
Rice is harvested and threshed by hand, Myanmar.

9.5.2 Intensive Commercial Agriculture

The two most widely practiced forms of intensive commercial agriculture are mixed crop and livestock and dairying.

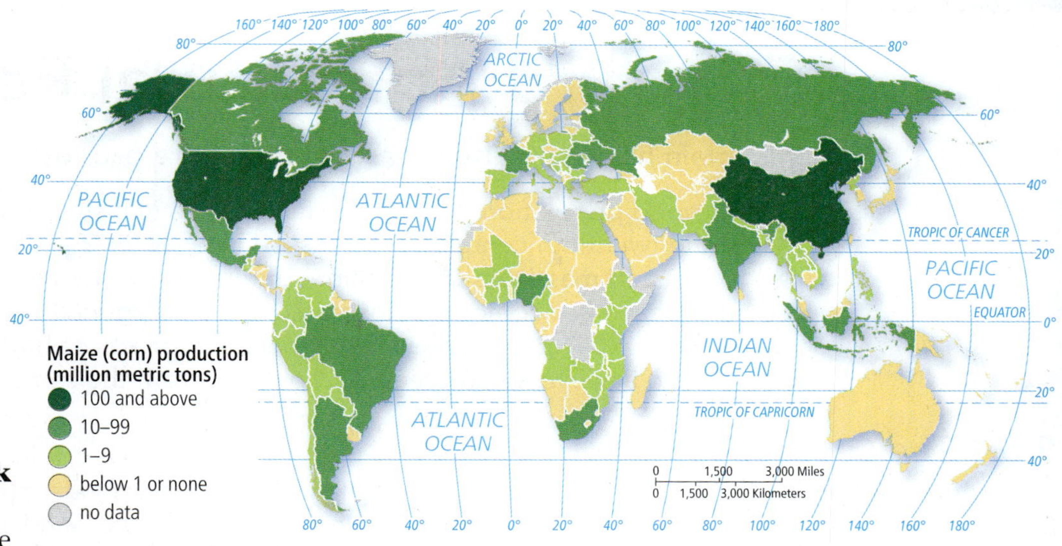

▲ 9.5.6 CORN PRODUCTION

Mixed Crop & Livestock

Mixed crop and livestock farming is the most common form of agriculture in much of Europe and in the United States between the Appalachians and 98° west longitude (a boundary between wetter and drier climates). Maize (corn) is the crop most frequently planted, followed by soybeans (Figure 9.5.6) . Crops are typically rotated or varied to help maintain fertility because various crops deplete the soil of some nutrients but restore others.

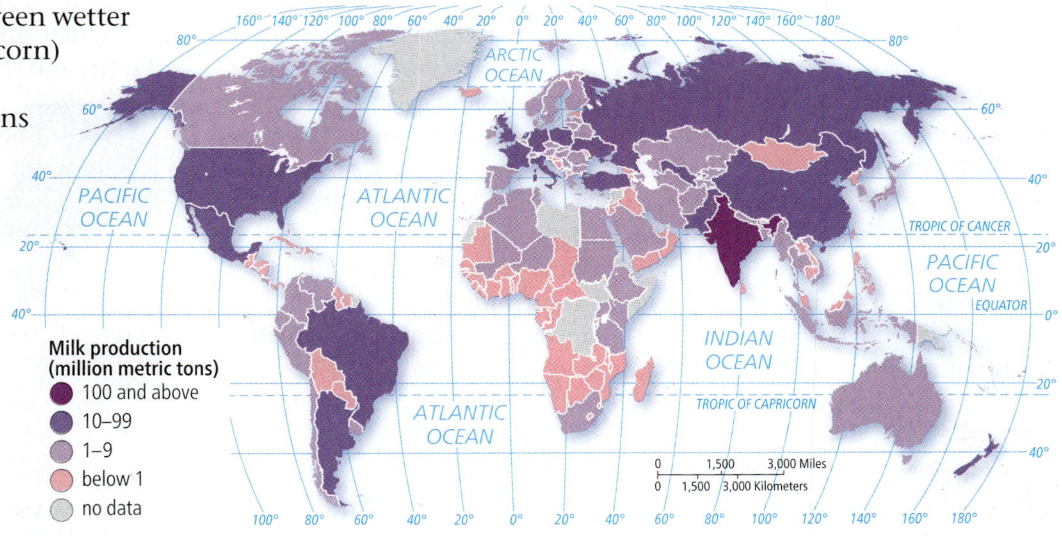

▲ 9.5.7 MILK PRODUCTION

Dairy Farming

The most important intensive commercial agriculture practiced near large urban areas in developed countries is **dairy farming**, a form of commercial agriculture that specializes in the production of milk and other dairy products (Figure 9.5.7). Dairy farms must be closer to their markets than other products because milk is highly perishable. The ring surrounding a city from which milk can be supplied without spoiling is known as the **milkshed**.

Traditionally, most milk was produced and consumed in developed countries (Figure 9.5.8). However, the share of the world's dairy farming conducted in developing countries has risen dramatically in recent years, surpassing the total in developed countries. Rising incomes permit urban residents in developing countries to buy more milk products.

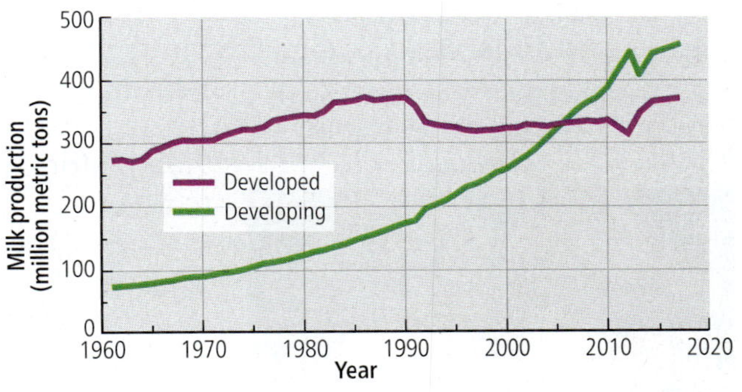

▲ 9.5.8 DAIRY PRODUCTION BY REGION

The Third Agricultural Revolution

- **Explain how agriculture has become more productive in recent years.**

The world population grew at the fastest rate in history during the second half of the twentieth century, as discussed in Chapter 2. At that time, many experts forecasted massive global famine, but in the twenty-first century, their dire predictions have not come true. Instead, increased agricultural productivity has resulted in an expansion of food supply. New agricultural practices have permitted farmers worldwide to achieve much greater crop yields from the same amount of land.

9.6.1 The Green Revolution

For hundreds if not thousands of years, subsistence farming yielded enough food for people living in rural villages to survive, assuming that no drought, flood, or other natural disaster occurred. Suddenly, in the late twentieth century, subsistence farming was faced with the dual challenge of providing enough food for the rapidly increasing population as well as for the growing number of urban residents who no longer grew their own food.

The **third agricultural revolution**, also called the **green revolution**, involved the invention and rapid diffusion of more productive agricultural techniques during the late twentieth century. Because of the green revolution, agricultural productivity at a global scale has increased faster than population growth. The green revolution involves two main practices: the introduction of new higher-yield seeds and the expanded use of fertilizers.

According to economist Ester Boserup, subsistence farmers have been able to increase the supply of food through intensification of production, achieved in two ways. First, land is left fallow for shorter periods. Second, new farming methods are adopted. Plows replace axes and sticks, and more weeding is done, more manure is applied, more terraces are carved out of hillsides, and more irrigation ditches are dug. Farmers adopt **crop rotation**, which varies the crop planted in a field each year to avoid exhausting the nutrients in the soil, and they practice **double cropping**, which allows for the same field to be harvested twice a year.

▲ 9.6.1 TESTING NEW VARIETIES OF RICE AT THE INTERNATIONAL RICE RESEARCH INSTITUTE (IRRI)

Miracle Seeds

Scientists began an intensive series of experiments during the 1950s to develop a higher-yield form of wheat. A decade later, the "miracle wheat seed" was ready. Shorter and stiffer than traditional breeds, the new wheat was less sensitive to variation in day length, responded better to fertilizers, and matured faster. The Rockefeller and Ford foundations sponsored many of the studies on the miracle wheat seed, and the program's director, Dr. Norman Borlaug, won the Nobel Peace Prize in 1970 for his contributions to its development.

The International Rice Research Institute (IRRI), established in the Philippines by the Rockefeller and Ford foundations, worked to create a miracle rice seed. During the 1960s, their scientists introduced a hybrid of Indonesian rice and Taiwan dwarf rice that was hardier than the traditional rice and that increased rice crop yields (Figures 9.6.1 and 9.6.2). More recently, scientists have developed new high-yield maize (corn).

▼ 9.6.2 MIRACLE RICE
Ordinary rice (left) and miracle rice (right) at the IRRI.

Both of the miracle seeds were diffused rapidly around the world. India's wheat production more than doubled in five years. After importing 10 million tons of wheat annually in the mid-1960s, India had a surplus of several million tons by 1971. Other Asian and Latin American countries recorded similar productivity increases. The green revolution was largely responsible for preventing a food crisis in these regions during the 1970s and 1980s. But will these scientific breakthroughs continue to keep up with human population growth in the years ahead?

9.6.2 Commercial Agriculture Productivity

Productivity has also increased among commercial farmers in recent years. New seeds, fertilizers, pesticides, mechanical equipment, and management practices have enabled commercial farmers to obtain greatly increased yields per area of land.

The experience of dairy farming in the United States demonstrates the growth in productivity. The number of dairy cows in the United States decreased from 10.8 million to 9.4 million between 1980 and 2020 (Figure 9.6.3). But milk production increased from 58 to 101 million metric tons. Thus, yield per cow increased 93 percent during this 40-year period, from 5.4 to 10.8 metric tons per cow.

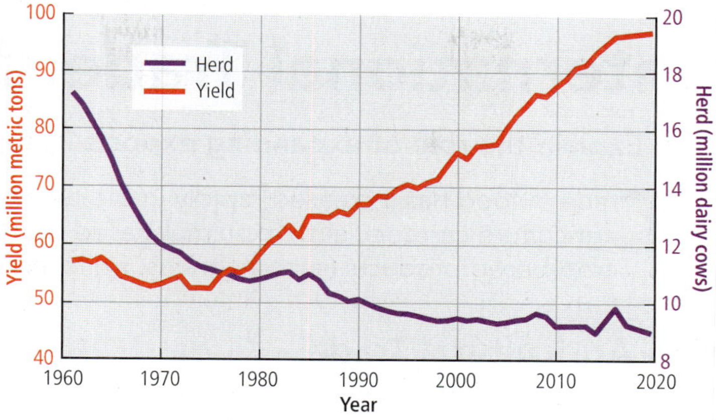

▲ 9.6.3 CHANGING U.S. DAIRY PRODUCTIVITY

The von Thünen Model

The use of land in developed countries is determined primarily by market forces of supply and demand. The value of the land affects the form of commercial agriculture practiced on it. Agricultural land is lost altogether to some other use if someone buys it for another purpose.

Because the purpose of commercial farming is to sell produce off the farm, the distance from the farm to the market influences the farmer's choice of crop to plant. Geographers use the von Thünen model to help explain the importance of proximity to market in the choice of crops on commercial farms. Johann Heinrich von Thünen, an estate owner in northern Germany, first proposed the model in 1826, in a book titled *The Isolated State*. He based his model on his experience as the owner of a large estate in northern Germany during the early nineteenth century and his observations on surrounding agriculture. He found that specific crops were grown in different rings around the cities in the area (Figure 9.6.4).

According to the von Thünen model (which geographers later modified), a commercial farmer initially considers which crops to cultivate and which animals to raise based on the location of markets where crops are sold. Then, the spatial arrangement of different crops is mapped out. In choosing an agricultural enterprise, the farmer compares two costs: the cost of the land and the cost of transporting products to market.

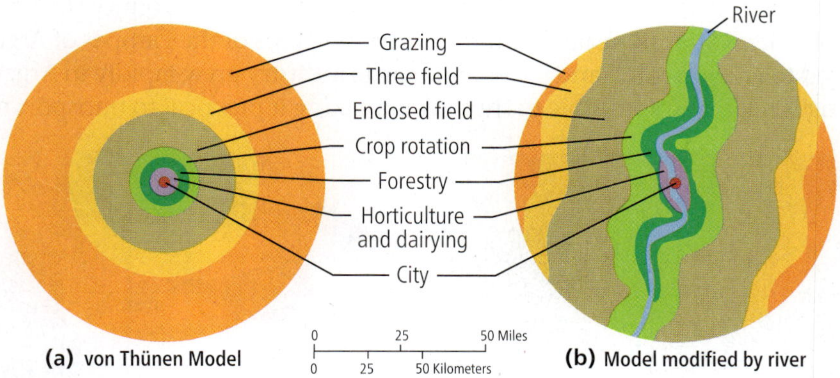

▲ 9.6.4 VON THÜNEN MODEL

(a) According to the von Thünen model, in the absence of topographic factors, different types of farming are conducted at different distances from a city, depending on the cost of transportation and the value of the product. (b) von Thünen recognized that his model would be modified by site factors, such as a river in this sketch. In the nineteenth century, when von Thünen created his model, many farmers depended on shipping to the market by boat. Consequently, agricultural uses that sought highly accessible locations, such as horticulture and dairying, were more likely to be located near the river than were fields used for grazing.

International Trade: Meat

- **Explain the role of expanding exports in supplying food.**

Trade in food has increased rapidly in the twenty-first century. International trade in agriculture can specialize in animal-based products (such as meat and dairy) or in crop-based products (such as grains, fruits, and vegetables).

9.7.1 Growth in Agricultural Trade

Trade in food has increased rapidly in the twenty-first century. Total food exports from all countries have increased from $331 billion in 2000 to $1.2 trillion in 2019 (Figure 9.7.1). In general, agricultural products are being imported into Africa and Asia and exported from Latin America and the South Pacific (Figure 9.7.2). Countries benefit from the revenues, and importing countries meet the food needs of their people. However, increased food trade is generating challenges for both exporting and importing countries.

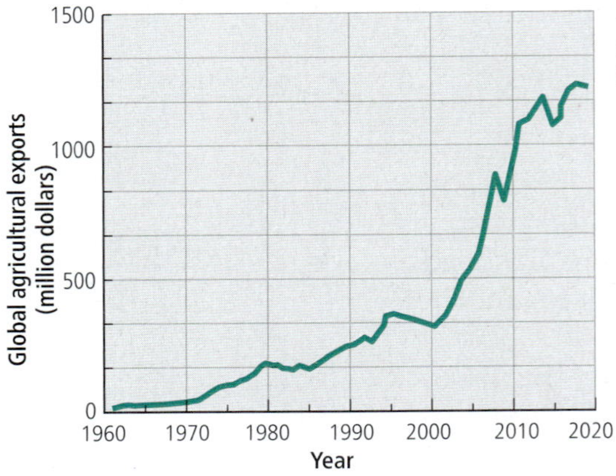

◀ 9.7.1 WORLDWIDE GROWTH IN FOOD EXPORTS

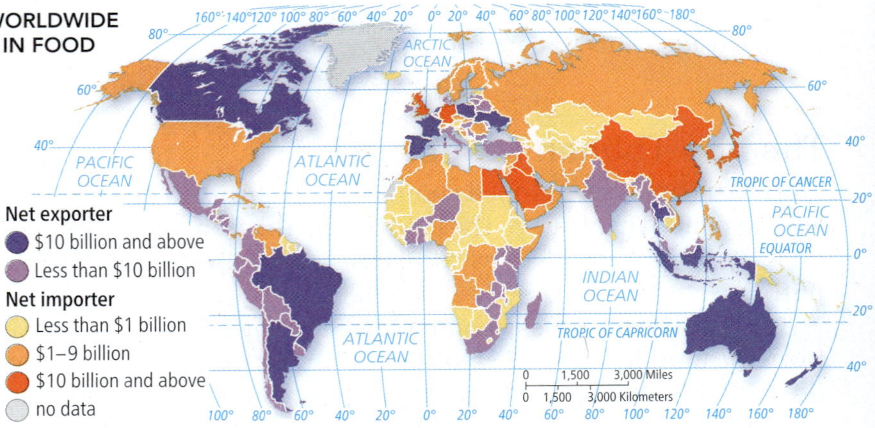

▲ 9.7.2 TRADE IN AGRICULTURE

Livestock Ranching

Ranching is the commercial grazing of livestock over an extensive area (Figure 9.7.3). Commercial ranching is a prominent type of agriculture in the western United States as well as in a handful of other countries.

China is the leading producer of meat, ahead of the United States, and Brazil is third in meat production (Figure 9.7.4). China passed the United States as the world's leading meat producer in 1990 and now produces twice as much as does the United States. In South America, large portions of the Pampas of Argentina, southern Brazil, and Uruguay are devoted to grazing cattle and sheep. The cattle industry grew rapidly in Argentina in part because the land devoted to ranching is relatively accessible to the ocean, making it practical to transport meat to overseas markets.

▲ 9.7.3 CATTLE RANCHING
Montana.

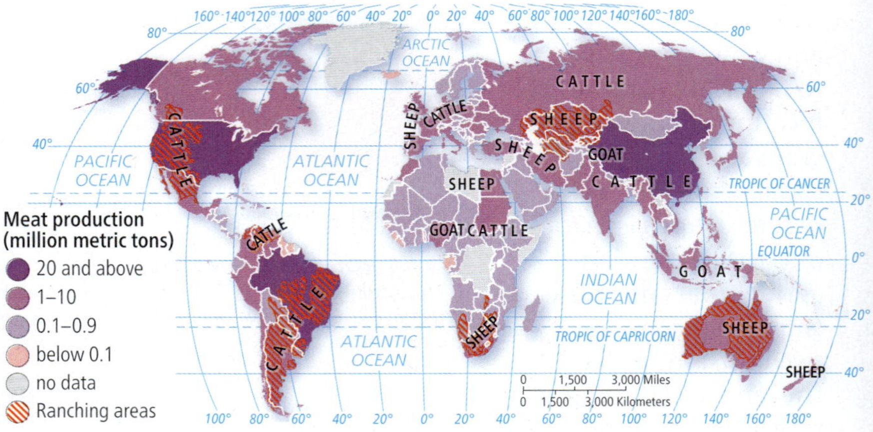

▲ 9.7.4 MEAT PRODUCTION

9.7.2 Fishing

Water-based food is acquired in two ways. **Fishing** is the capture of wild fish and other seafood living in a body of water. **Aquaculture** is the cultivation of fish and seafood under controlled aquatic conditions.

The world's oceans are divided into 16 major fishing regions, including six in the Atlantic Ocean, six in the Pacific Ocean, two in the Indian Ocean, one in the Mediterranean Sea, and one in the Antarctic Ocean (Figure 9.7.5). The fishing regions with the largest yields are Pacific Northwest and Pacific Western Central.

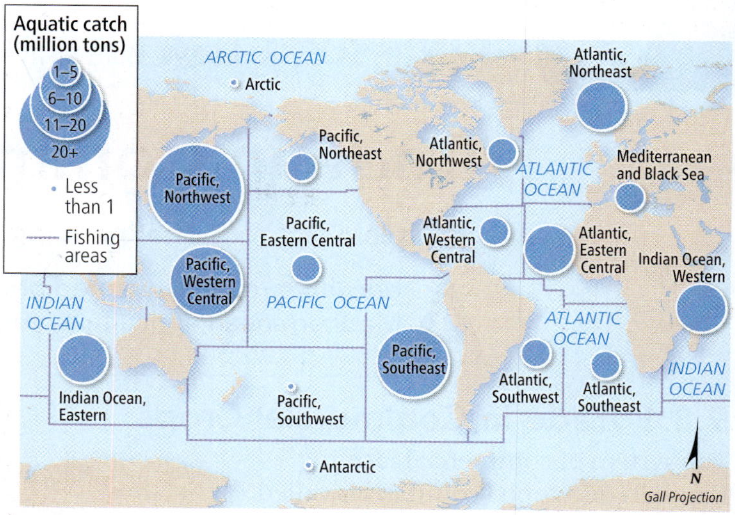

▲ 9.7.5 MAJOR FISHING REGIONS

Fish Production

China is responsible for one-third of the world's yield of fish (Figure 9.7.6). Worldwide fish consumption has increased more rapidly than human population growth. During the past half-century, per capita consumption of fish has nearly doubled in both developed and developing countries (Figure 9.7.7). Still, fish and seafood account for only 1 percent of all calories consumed by humans.

Fish production has increased even more rapidly than fish consumption (Figure 9.7.8). The growth results entirely from expansion of aquaculture (Figure 9.7.9). A comparison of Figures 9.7.7 and 9.7.8 shows that production of fish is around one-fourth higher than human consumption of it. Around three-fourths of the fish are consumed directly by humans. The remainder is converted to fish meal and fed to poultry and hogs.

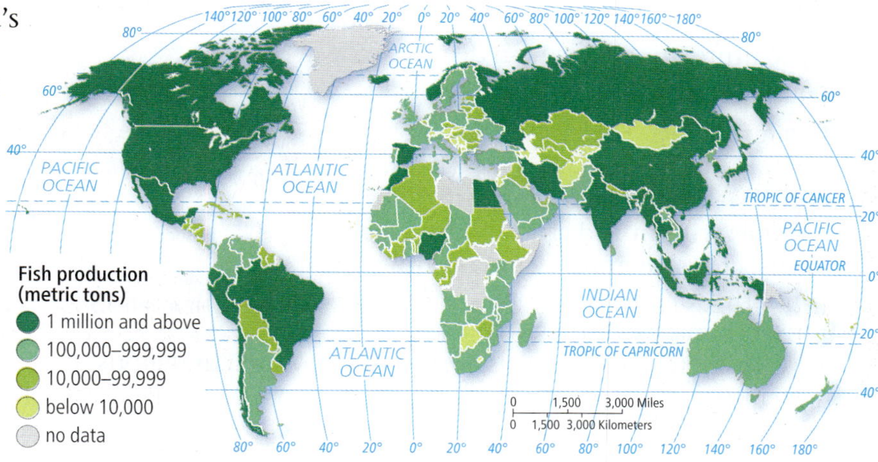

▲ 9.7.6 FISH PRODUCTION

Overfishing

The population of some fish species in the oceans and lakes has declined because of **overfishing**, which is capturing fish faster than they can reproduce. Overfishing has been particularly acute in the North Atlantic and Pacific oceans. Overfishing has reduced the population of tuna and swordfish by 90 percent in the past half-century, for example. The U.N. estimates that one-quarter of fish stocks have been overfished and one-half fully exploited, leaving only one-fourth underfished.

▼ 9.7.7 GROWTH IN HUMAN CONSUMPTION OF FISH

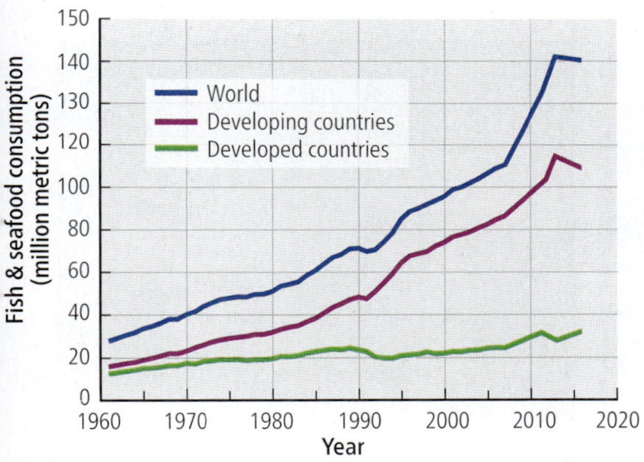

▼ 9.7.8 GROWTH IN FISH PRODUCTION

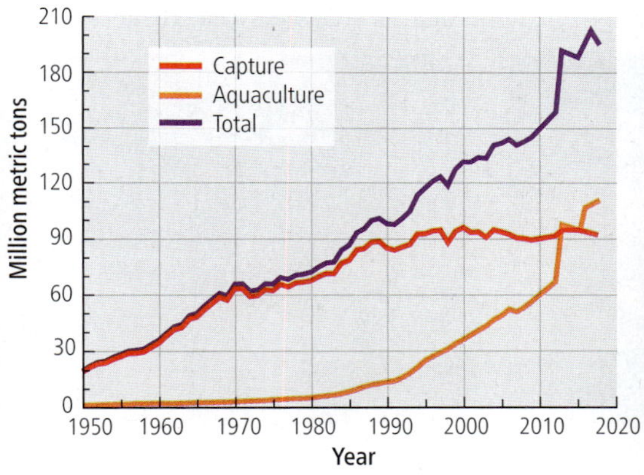

▼ 9.7.9 AQUACULTURE
Jiangsu, China.

Trade in Commercial Crops

- **Explain the role of expanding exports to the availability of food.**

Agriculture regions devoted primarily to trade in crops include grain farming, Mediterranean agriculture, commercial gardening, and plantation agriculture.

9.8.1 Trade in Commercial Crops

The system of commercial farming found in developed countries is called **agribusiness** because farming is integrated into a large food-production industry. Different types of commercial agriculture predominate in distinctive regions, depending on climate.

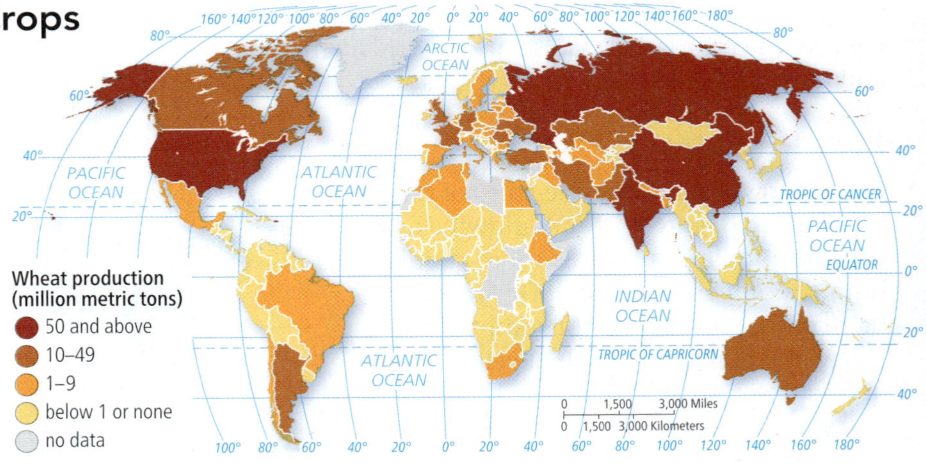

▲ 9.8.1 WHEAT PRODUCTION

Grain farming. The major crop in agribusiness is grain, such as wheat, corn, oats, barley, rice, and millet. In contrast with mixed crop and livestock farms, crops on grain farms are grown primarily for consumption by humans rather than by livestock.

Commercial grain farms are generally located in regions that are too dry for mixed crop and livestock agriculture. Wheat is the world's leading export crop (Figure 9.8.1). In agribusiness, wheat is a valuable commodity because it can generally be sold for a higher price than other grain and can be shipped profitably from remote farms to markets.

Mediterranean agriculture. Mediterranean agriculture is practiced primarily on the lands that border the Mediterranean Sea as well as along the Pacific coast of North and South America (Figure 9.8.2). Tree crops, as well as horticulture (the growing of fruits, vegetables, and flowers), form the commercial base of Mediterranean agriculture.

Commercial gardening. The predominant type of agribusiness in the southeastern United States is **commercial gardening**, especially fruits and vegetables (Figure 9.8.3). The region has a long growing season and humid climate, and it is accessible to the large number of consumers in the northeast United States. Commercial gardening is frequently called **truck farming**, from the Middle English word *truck*, meaning *barter* or "exchange of commodities."

Plantation farming. A **plantation** is a large commercial farm in a developing country that specializes in one or two crops. Most plantations are located in the tropics and subtropics of Latin America, Africa, and Asia. Although generally situated in developing countries, plantations are often owned or operated by European or North American agribusiness corporations, and they grow crops for sale primarily to developed countries (Figure 9.8.4).

▼ 9.8.2 MEDITERRANEAN AGRICULTURE
Olive farm, Crete, Greece.

▼ 9.8.3 COMMERCIAL GARDENING
Peach farm, Georgia, United States.

▼ 9.8.4 PLANTATION FARMING
Sugarcane plantation, Botolambat, Indonesia.

9.8.2 Government Policies

▲ **9.8.5 CORN DESTINED FOR EXPORT**
North Platte, Nebraska.

In many countries, government policies have aggravated food supply, cost, and distribution problems. To make food affordable for urban residents, some governments keep agricultural prices low. Constrained by price controls, farmers are unable to sell their commodities at a profit and therefore have little incentive to increase production.

The U.S. government has three agriculture policies designed to improve the financial position of farmers:

- Farmers are encouraged to avoid crops that are in excess supply. Because soil erosion is a constant threat, the government encourages planting fallow crops, such as clover, to restore nutrients to the soil and to help hold the soil in place. These crops can be used for hay or forage for pigs or to produce seeds for sale.
- When certain commodity prices are low, the government subsidizes farmers. The government sets a target price for a commodity and pays farmers the difference between the price they receive in the market and the target price. Target prices are calculated to give farmers the same price for the commodity today as they received in the past, adjusted for inflation.
- The government buys surplus production and sells or donates it to foreign governments (Figure 9.8.5). In addition, low-income Americans receive food stamps in part to stimulate their purchase of additional food.

The United States has averaged about $20 billion a year on farm subsidies in recent years. Subsidy payments are lower in years when market prices rise and production is down, typically as a result of poor weather conditions in the United States or political problems in other countries.

More farmers receive subsidies in Europe, and they receive more than American farmers. The high subsidies are a legacy of a long-standing commitment by the European Union to maintain agriculture in its member states, especially in France. Supporters point to the preservation of rural village life in parts of Europe, while critics charge that the subsidies create needlessly high food prices.

Government policies in developed countries point out a fundamental irony in worldwide agricultural patterns: In developed regions such as North America and Europe, farmers are encouraged to grow less food, whereas developing countries struggle to increase food production to match the rate of growth in the population.

Drug Trade

The export crops grown in some developing countries, especially in Latin America and Asia, are those that can be converted to drugs, especially cocaine, heroin, and marijuana. Cocaine is derived from the leaves of the coca plant, and most of the coca crops are grown in Colombia or the neighboring countries Peru and Bolivia (Figure 9.8.6). Heroin is derived from raw opium gum, which is produced by the opium poppy plant. Afghanistan is the source of more than 90 percent of the world's opium; most of the remainder is grown in Myanmar and Laos. Marijuana, produced from the Cannabis sativa plant, is cultivated widely around the world. The overwhelming majority of the marijuana that reaches the United States is grown in Mexico.

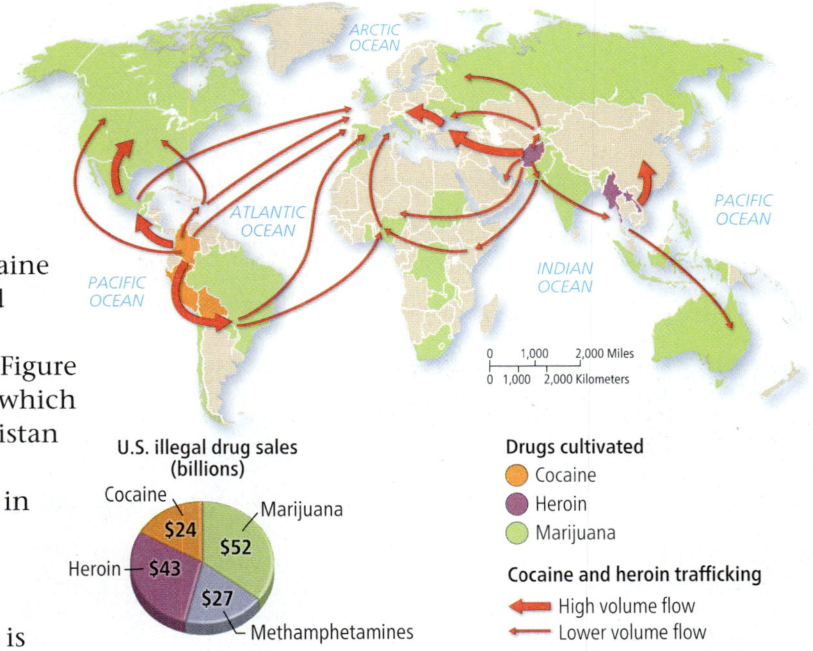

▲ **9.8.6 DRUG TRADE**

Threatened Subsistence Agriculture Practices

- **Understand reasons for the decline in some traditional subsistence agricultural practices.**

Extensive forms of subsistence agriculture that require access to a large amount of land are declining. These include shifting cultivation and pastoral nomadism.

9.9.1 Shifting Cultivation

A form of extensive subsistence agriculture in which farmers shift frequently from one field to another is known as **shifting cultivation**. It is practiced in much of the world's tropical (or A) climate regions, which have relatively high temperatures and abundant rainfall (refer to Figure 9.4.4). It is practiced by roughly 250 million people across 36 million square kilometers (14 million square miles), especially in the tropical rain forests of Latin America, sub-Saharan Africa, and Southeast Asia.

Two distinctive features of shifting cultivation are:

- **Slash & burn.** Farmers clear land for planting by slashing vegetation and burning the debris. Shifting cultivation is sometimes called *slash-and-burn agriculture* (Figure 9.9.1).
- **Frequent relocation.** Farmers grow crops on a cleared field for only a few years, until soil nutrients are depleted, and then leave it fallow (with nothing planted) for many years until the soil recovers.

Future of Shifting Cultivation

Land devoted to shifting cultivation is declining in the tropics at the rate of about 75,000 square kilometers (30,000 square miles), or 0.2 percent, per year according to the United Nations. The amount of Earth's surface covered by tropical rain forests today is less than half its original area, and until recent years, the World Bank supported deforestation with loans to finance development schemes that required clearing tropical rain forests.

Shifting cultivation is being replaced by logging, cattle ranching, and the cultivation of cash crops for agribusiness. Selling timber to builders and raising beef cattle for fast-food restaurants are more lucrative development strategies than maintaining shifting cultivation.

▼ **9.9.1 SHIFTING CULTIVATION**
(a) Slash and let dry (Laos); (b) burn during the dry season (Laos); (c) plant at the start of rainy season, tend, weed, water (Côte d'Ivoire); (d) harvest (rice, Mozambique); (e) consume (posho, made from cornmeal, Uganda).

(a)

(b)

(c)

(d)

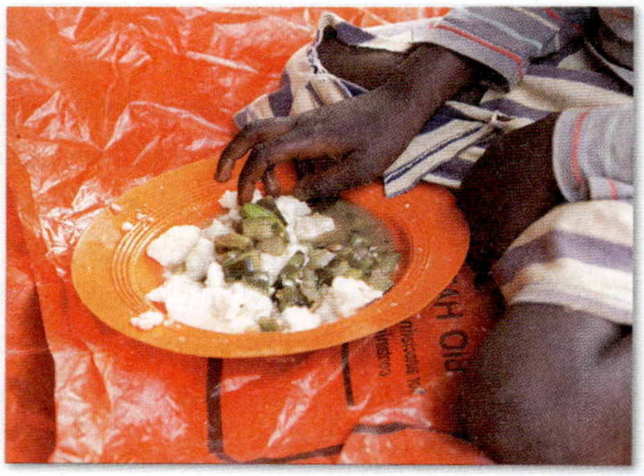

(e)

9.9.2 Pastoral Nomadism

A form of subsistence agriculture called **pastoral nomadism** is based on the herding of domesticated animals in dry climates, where growing crops for food is challenging (Figure 9.9.2). Pastoral nomads, unlike other subsistence farmers, depend primarily on animals for survival. The animals provide milk, and their skins and hair are used for clothing and tents. The animals of pastoral nomads are usually not slaughtered, although dead ones may be consumed.

▲ 9.9.2 PASTORAL NOMADISM
East Africa.

Nomads select the type and number of animals for the herd according to local cultural and physical characteristics. The choice depends on the relative prestige of animals and the ability of species to adapt to a particular climate and vegetation.

Agricultural experts once regarded pastoral nomadism as a stage in the evolution of agriculture—between the hunters and gatherers who migrated across Earth's surface in search of food and sedentary farmers who cultivated grain in one place. Because they had domesticated animals but not plants, pastoral nomads were considered more advanced than hunters and gatherers but less advanced than settled farmers.

Pastoral nomadism is now generally recognized as an offshoot of sedentary agriculture, not as a precursor of it. It is simply a practical way of surviving on land that receives too little rain for cultivation of crops.

Today, pastoral nomadism, a victim of modern technology, is a declining form of agriculture. Before recent transportation and communication inventions, pastoral nomads played an important role as carriers of goods and information across the sparsely inhabited drylands. They used to be the most powerful inhabitants of the drylands, but now, with modern weapons, national governments can control nomadic peoples more effectively and limit their movement within their traditional territories.

Governments force these communities to give up pastoral nomadism because officials believe the land can be used more efficiently for other uses. Land that can be irrigated is converted from nomadic to sedentary agriculture. In some instances, the mining and petroleum industries now operate in drylands formerly occupied by pastoral nomads. Some nomads are encouraged to try sedentary agriculture or to work for mining or petroleum companies. Others are still allowed to move about, but only within fixed boundaries. In the future, pastoral nomadism will be increasingly confined to areas that cannot be irrigated or that lack valuable raw materials.

Transhumance

Pastoral nomads do not wander randomly across the landscape but have a strong sense of territoriality. Every group controls a piece of territory and will invade another group's territory only in an emergency or if war is declared. The goal of each group is to control a territory large enough to contain the forage and water needed for survival. The actual amount of land a group controls depends on its wealth and power. The precise migration patterns evolve from intimate knowledge of the area's physical and cultural characteristics.

Some pastoral nomads, as well as other livestock growers, practice **transhumance**, which is seasonal migration of livestock between mountains and lowland pasture areas (Figure 9.9.3). Sheep or other animals may pasture in alpine meadows in the summer and be herded back down into valleys for winter pasture.

▲ 9.9.3 TRANSHUMANCE
Pyrenees Mountains, Spain.

Agriculture & Resource Depletion

- **Describe reasons for the loss of agricultural land.**

Both subsistence and commercial agriculture face challenges to produce more food for a growing world while protecting Earth's resources. The increasingly integrated global agricultural system has provided more food for the world's 8 billion humans but at a cost of negatively impacting natural resources, especially water and land.

9.10.1 Loss of Agricultural Land

Historically, world food production increased primarily by expanding the amount of land devoted to agriculture. In recent decades, population has increased much more rapidly than agricultural land (Figure 9.10.1). People believed that good agricultural land would always be available for willing pioneers. Today few scientists believe that further expansion of agricultural land can feed the growing world population. Around 38 percent of the world's land area is used for agriculture (Figure 9.10.2).

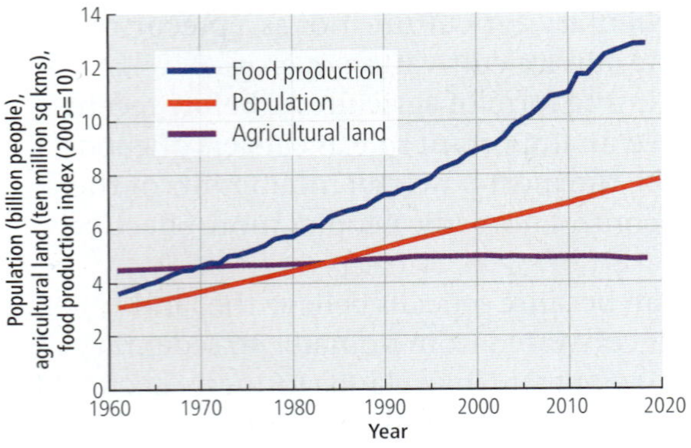

▲ 9.10.1 WORLD POPULATION GROWTH, AGRICULTURAL LAND, AND FOOD PRODUCTION
The food production index is set at 10 in the year 2005.

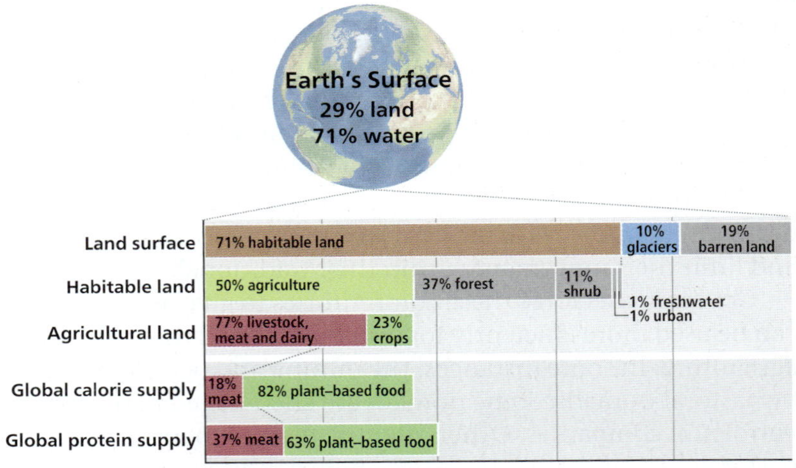

▲ 9.10.2 GLOBAL LAND USE FOR FOOD PRODUCTION

Urban Growth

Loss of farmland to urban growth is especially severe at the edge of the string of large metropolitan areas along the East Coast of the United States. As urban areas grow in population and expand in land area, farms on the periphery are forced out. A serious problem in the United States has been the loss of 200,000 hectares (500,000 acres) of the most productive farmland, known as **prime agricultural land**, as urban areas sprawl into the surrounding countryside. The von Thünen model shows the positive influence of proximity to urban markets for farmers, but the expansion of urban areas is reducing agricultural land (Figure 9.10.3).

◀ 9.10.3 URBAN DEVELOPMENT ENCROACHES ON FARM LAND
Perrysburg, Ohio.

9.10.2 Agriculture & Water Supply

Humans use around 9 billion cubic meters of water per year, or around 1,400 cubic meters per capita. The heaviest demand is for agriculture, followed by industry and municipal sewage systems (Figure 9.10.4). This is because the amount of land being used for agriculture has expanded into areas that do not have reliable rainfall and therefore rely on sophisticated irrigation methods.

Irrigation is the process of applying controlled amounts of water to plants at needed intervals using either channels or drips within field rows or high-tech sprinkler systems.

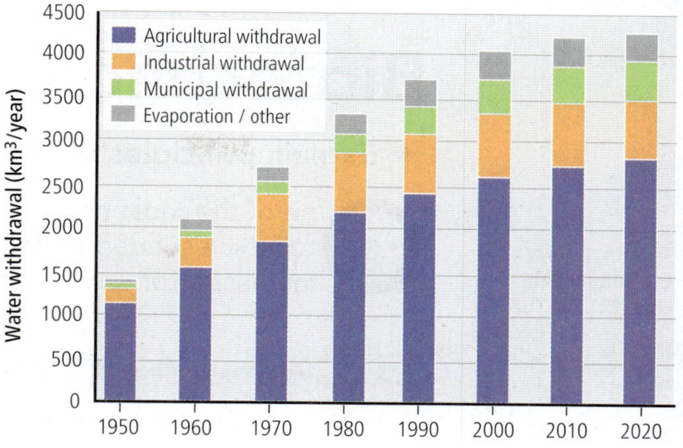

▲ **9.10.4** USES THAT DEMAND WATER

Desertification

Lack of water is causing stress on agriculture in many regions and is forcing people to abandon their land. Human actions that cause land to deteriorate to a desert-like condition is called **desertification**, or, more precisely, semiarid land degradation (Figure 9.10.5).

For example, hundreds of thousands have been forced to move from the Sahel region of northern Africa because of drought conditions. The people of the Sahel have traditionally been pastoral nomads, a form of agriculture adapted to dry lands but effective only at low population densities. The capacity of the Sahel to sustain human life—never very high—has declined recently because of population growth and several years of unusually low rainfall. Consequently, many of these nomads have been forced to move into cities and rural camps, where they survive on food donated by the government and international relief organizations.

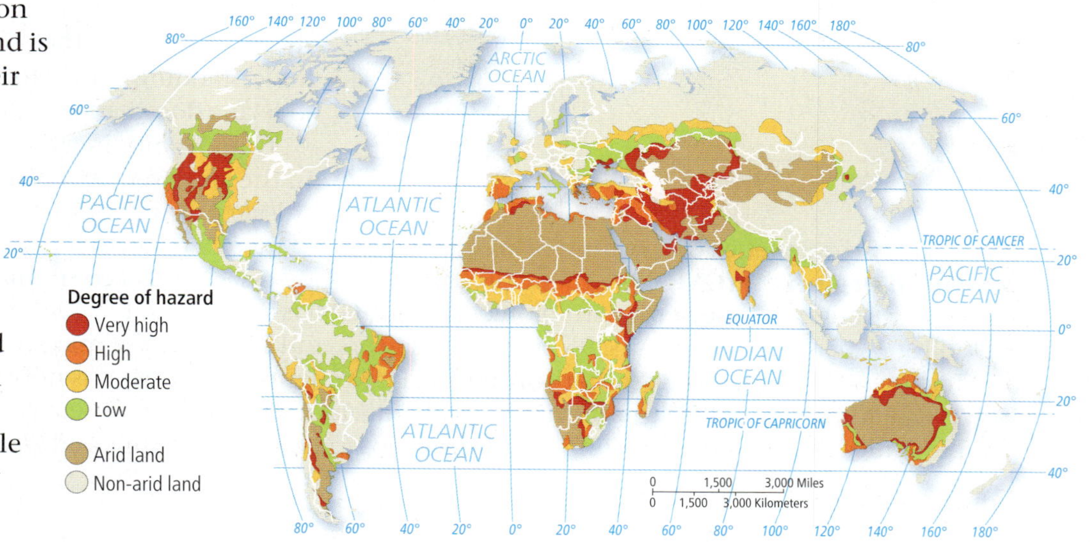

▲ **9.10.5** DESERTIFICATION
Also known as semiarid land degradation.

California Agriculture & Water

The state of California and its neighboring states in the U.S. Southwest have grown rapidly despite limited supplies of water. California farmers produce one-third of U.S. vegetables and two-thirds of the country's fruits and nuts. It takes a lot of water to grow these fruits and vegetables. Figure 9.10.6 has examples of the amounts of California water that is used to grow some of the state's fruits and vegetables. Through consumption of fruits and vegetables, the average American consumes around 40 gallons of California water per day.

Fruits & nuts	Gallons
1 apple, peach, pear, or plum or 1/4 melon	7.0
5 strawberries	3.0
1 almond	1.0
1 walnut	5.0
3 grapes	1.0
1 lemon, orange, grapefruit, or clementine	20
1 avocado	40
1 broccoli or cauliflower floret	0.5
Lettuce, cabbage, spinach (salad portion)	1.0
1 carrot or celery stalk	0.5
1 slice tomato, onion, or potato	0.5

▲ **9.10.6** AMOUNT OF WATER NEEDED TO GROW SELECTED FRUITS AND VEGETABLES IN CALIFORNIA

Biotechnology & Sustainability

- Explain principles of organic farming.

One of the most challenging issues in contemporary agriculture is the extent to which seeds should be genetically modified. Alternatively, demand has increased for organic crops that have not been genetically modified.

▼ 9.11.1 GMO CROPS IN THE UNITED STATES

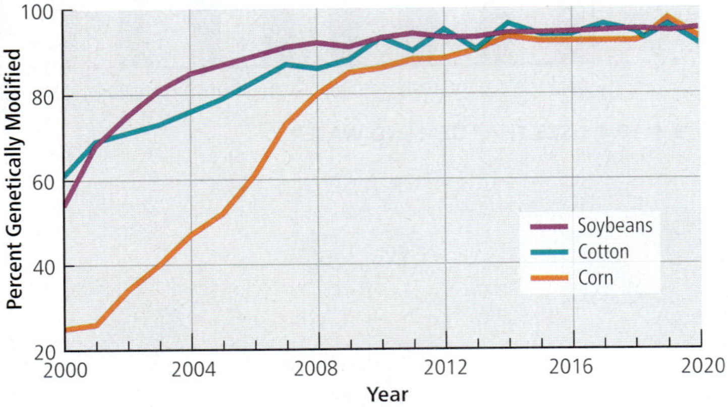

9.11.1 Genetically Modified Organisms

Farmers have been manipulating crops and livestock for thousands of years. The very essence of agriculture is to deliberately manipulate nature. Humans control selective reproduction of plants and animals in order to produce a larger number of stronger, hardier breeds. Beginning in the nineteenth century, the science of genetics expanded understanding of how to manipulate plants and animals to secure dominance of the most favorable traits.

Genetic modification, which became widespread in the late twentieth century, marks a sharp break with the agricultural practices of the past several thousand years. Today the genetic composition of an organism is not merely studied, it is actually altered. A **genetically modified organism (GMO)** is a living organism that possesses a novel combination of genetic material obtained through the use of modern biotechnology. GMOs are especially widespread in the United States, including 94 percent of soybeans, 96 percent of cotton, and 92 percent of maize (Figure 9.11.1).

Most medical studies suggest that consuming large quantities of GMOs is not injurious to human health. However, GMOs may reduce the effectiveness of antibiotics and growing them can require inputs of fertilizers and pesticides that may destroy long-standing ecological balances in soils and water.

GMO Labeling

Many countries, including most European countries, China, and India, require GMO labeling (Figure 9.11.2). The debate in the United States has been especially fierce concerning whether GMO food should be labeled. In the United States, the food industry opposes GMO labeling, but voters are divided on the issue.

Some consumers are questioning the health risks in consuming biotech food produced through heavy application of chemicals. The possible long-term adverse environmental impacts of agricultural biotechnology is also a concern.

Another controversial agricultural practice is injecting antibiotics in animals to increase their weight or make them grow faster. Public health officials fear that antibiotic use in animals promotes the development of hard-to-treat antibiotic-resistant superbugs that make people sick. As a result, the practice was banned in Europe in 2006 and in the United States in 2017.

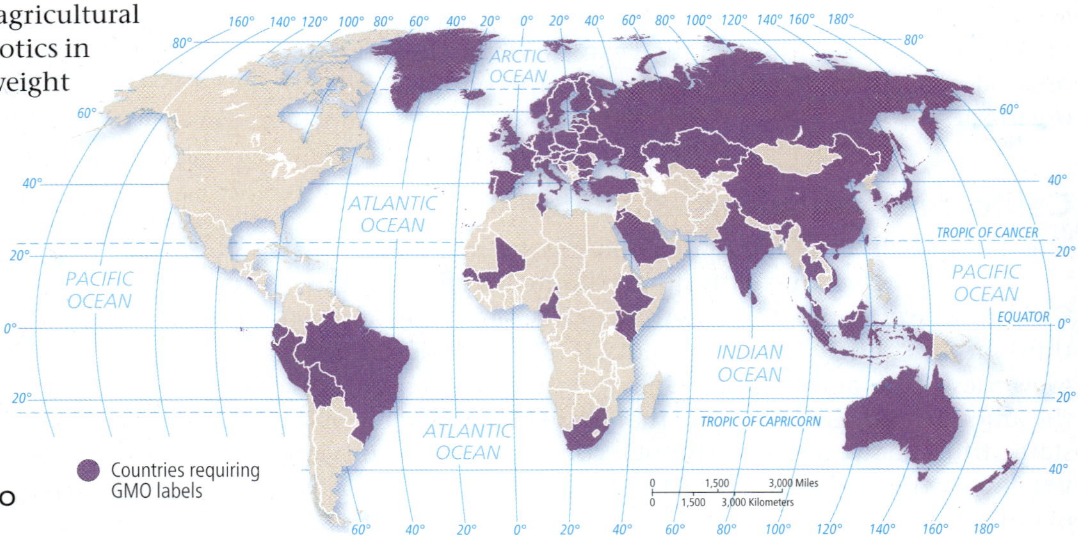

▶ 9.11.2 COUNTRIES WITH GMO LABELING LAWS

9.11.2 Organic Farming

Organic farming is agriculture that depends on the use of naturally occurring substances to grow crops and raise livestock while prohibiting or strictly limiting synthetic substances, such as herbicides, pesticides, and growth hormones. Worldwide, the Research Institute of Organic Agriculture classified 72 million hectares (178 million acres), or 1 percent of farmland, as organic in 2019 (Figure 9.11.3). Growth has been especially rapid in Australia.

Organic farming is sensitive to the complexities of biological and economic interdependencies between crops and livestock. Growing crops and raising livestock are integrated as much as possible at the level of the individual farm. This integration reflects a return to the historical practice of mixed crop and livestock farming, in which growing crops and raising animals were regarded as complementary activities on the farm. This was the common practice for centuries, until the mid-1900s, when technology, government policy, and economics encouraged farmers to become more specialized.

In organic farming, crops are grown without application of herbicides and pesticides to control weeds. GMO seeds are not used. In organic farming, animals consume crops grown on the farm and are not confined to small pens.

The moral and ethical debate over animal welfare is particularly intense regarding confined livestock production systems, where animals are reared in cages. Confining livestock leads to concentrated waste production that seeps into and pollutes surface waterways and groundwater particularly where the density of animals is high. Conversely, in organic farming, animals are not confined, so their manure is distributed over land and can contribute to soil fertility (Figure 9.11.4).

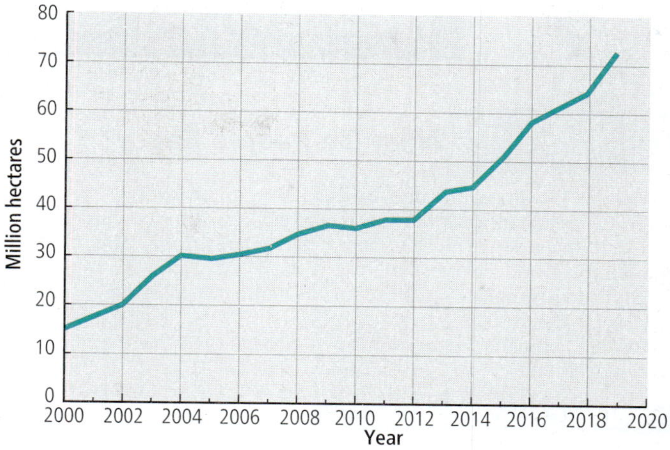

▲ **9.11.3 ORGANIC FARMLAND WORLDWIDE**

▲ **9.11.4 ORGANIC FREE RANGE CHICKENS**
Clarkefield, Australia.

Sustainable Land Management

Conservation tillage is a method of soil cultivation that reduces soil erosion and runoff. Under conservation tillage, some or all of the previous harvest is left on the fields through the winter. **No tillage**, as the name implies, leaves all of the soil undisturbed, and the entire residue of the previous year's harvest is left untouched on the fields (Figure 9.11.5). **Ridge tillage** is a system of planting crops on ridge tops. Crops are planted on 10- to 20-centimeter (4- to 8-inch) ridges that are formed during cultivation or after harvest. A crop is planted on the same ridges, in the same rows, year after year.

▶ **9.11.5 CONSERVATION TILLAGE**
Soybean field, Iowa.

CHAPTER

9 Review, Analyze, & Apply

KEY ISSUE 1 What Do People Eat?

- What people eat is influenced by a combination of level of development, cultural preferences, and environmental constraints.
- Most humans derive most of their dietary energy from grains, but the amount consumed varies widely.
- People in developed countries consume more protein through animal products like meat and dairy.
- One in 10 humans is undernourished, though the percentage has been declining.

THINKING GEOGRAPHICALLY

Pizza is the most popular food in the United States. Is it a favorite for you and your family? What are the principal ingredients in your favorite pizza? Are these ingredients grown locally, or are they imported?

◀ **9.CR.1** STUDENT HAVING LUNCH

KEY ISSUE 2 Where Does Food Come From?

- The first agricultural revolution was the time when human beings first domesticated plants and animals and no longer relied entirely on hunting and gathering.
- The second agricultural revolution increased productivity through improvement of crop rotation and breeding of livestock.
- Agriculture is now divided between subsistence agriculture in developing countries and commercial agriculture in developed countries.
- The third agricultural revolution involved the invention and rapid diffusion of more productive agricultural techniques.

THINKING GEOGRAPHICALLY

How does the diffusion of agriculture compare with the diffusion of the Indo-European language family (refer to Figures 5.4.4 and 5.4.5)? What might account for similarities and differences in the two distributions?

◀ **9.CR.2** MINARE, TURKEY

KEY ISSUE 3 How Has Agriculture Become More Global?

- International trade in food has increased rapidly.
- Crops, land animals, and fish are increasingly grown, raised, and caught in one country and sold in other countries.
- Converting land to produce food for trade has threatened the continuation of traditional subsistence agricultural practices like shifting cultivation and pastoral nomadism.

THINKING GEOGRAPHICALLY

▲ **9.CR.3** BANANAS FROM LATIN AMERICA IN DETROIT GROCERY STORE

Global trade has made possible shipping food long distances. The bananas in this photograph traveled 5,000 kilometers (3,000 miles) before reaching a supermarket shelf. What impact might the globalization of agriculture have on the nutritional value of food?

KEY ISSUE 4 How Can Agriculture Be More Earth-Friendly?

- Agriculture faces challenges of protecting land and water resources from environmental degradation.
- Organic farming depends on the use of naturally occurring substances while prohibiting or strictly limiting synthetic substances.

THINKING GEOGRAPHICALLY

What might be some of the obstacles to introducing or expanding organic food service at your university or school?

◀ **9.CR.4** SALAD BAR, UNIVERSITY OF CALIFORNIA, BERKELEY

Explore

Rice Fields in Asia

Use Google Earth to examine how rice is grown in Asia. ***Search*** for **Pura Desa Banjar Calo, Indonesia.** ***Zoom*** in to around 800 meters. Move the image so that ***Pura Desa Banjar Calo*** is on the far left of the screen.

1. Immediately to the east of Pura Desa Banjar Calo is patch of dense forest. Describe the landscape to the east of the forest patch. What do the many parallel features represent?
2. Search for Jl. Melati, Indonesia, which is the north-south road to the east of the parallel features in question 1. Drag Street View to Jl. Melati. Zoom in on the west side of the road. Why is the west side of the road wet?

▶ **9.CR.5** RICE FIELD, INDONESIA

Key Terms

Agribusiness (9.8.1) Commercial agriculture characterized by the integration of different steps in the food-processing industry, usually through ownership by large corporations.

Agriculture (9.3.1) The deliberate effort to modify a portion of Earth's surface through the cultivation of crops and the raising of livestock for sustenance or economic gain.

Aquaculture (9.7.2) The cultivation of seafood under controlled conditions.

Cereal grain (9.2.1) A grass that yields grain for food.

Columbian Exchange (9.3.2) The transfer of plants and animals, as well as people, culture, and technology, between the Western Hemisphere and Europe as a result of European colonization and trade.

Commercial agriculture (9.4.1) Agriculture undertaken primarily to generate products for sale off the farm.

Commercial gardening (9.8.1) See truck farming.

Conservation tillage (9.11.2) A method of soil cultivation that reduces soil erosion and runoff.

Crop (9.3.1) Any plant gathered from a field as a harvest during a particular season.

Crop rotation (9.6.1) The practice of rotating use of different fields from crop to crop each year to avoid exhausting the soil.

Dairy farming (9.5.2) A form of commercial agriculture that specializes in the production of milk and other dairy products.

Desertification (9.10.2) Degradation of land, especially in semiarid areas, primarily because of human actions such as excessive crop planting, animal grazing, and tree cutting. Also known as semiarid land degradation.

Dietary energy consumption (9.1.1) The amount of food that an individual consumes, measured in kilocalories (Calories in the United States).

Double cropping (9.6.1) Harvesting twice a year from the same field.

First agricultural revolution (9.3.1) The time when human beings first domesticated plants and animals and no longer relied entirely on hunting and gathering.

Fishing (9.7.2) The capture of wild fish and other seafood living in the waters.

Food security (9.1.1) Physical, social, and economic access at all times to safe and nutritious food sufficient to meet dietary needs and food preferences for an active and healthy life.

Genetically modified organism (GMO) (9.11.1) A living organism that possesses a novel combination of genetic material obtained through the use of modern biotechnology.

Grain (9.2.1) Seed of a cereal grass.

Green revolution (9.6.1) Rapid diffusion of new agricultural technology, especially new high-yield seeds and fertilizers.

Intensive subsistence agriculture (9.5.1) A form of subsistence agriculture characteristic of Asia's major population concentrations in which farmers must expend a relatively large amount of effort to produce the maximum feasible yield from a parcel of land.

Milkshed (9.5.2) The area surrounding a city from which milk is supplied.

Mixed crop and livestock farming (9.5.2) Commercial farming characterized by integration of crops and livestock; most of the crops are fed to animals rather than consumed directly by humans.

No tillage (9.11.2) A farming practice that leaves all of the soil undisturbed and the entire residue of the previous year's harvest left untouched on the fields.

Overfishing (9.7.2) Capturing fish faster than they can reproduce.

Paddy (9.5.1) The Malay word for wet rice, increasingly used to describe a flooded field.

Pastoral nomadism (9.9.2) A form of subsistence agriculture based on herding domesticated animals.

Plantation (9.8.1) A large farm in tropical and subtropical climates that specializes in the production of one or two crops for sale, usually to a more developed country.

Prime agricultural land (9.10.1) The most productive farmland.

Ranching (9.7.1) A form of commercial agriculture in which livestock graze over an extensive area.

Ridge tillage (9.11.2) A system of planting crops on ridge tops in order to reduce farm production costs and promote greater soil conservation.

Sawah (9.5.1) A flooded field for growing rice.

Second agricultural revolution (9.4.1) The increase in productivity through improvement of crop rotation, breeding of livestock, and other practices beginning in the United Kingdom in the seventeenth century.

Shifting cultivation (9.9.1) A form of subsistence agriculture in which people shift activity from one field to another; each field is used for crops for a few years and left fallow for a relatively long period.

Subsistence agriculture (9.4.1) Agriculture designed primarily to provide food for direct consumption by the farmer and the farmer's family.

Third agricultural revolution (9.6.1) See Green revolution.

Transhumance (9.9.2) Seasonal migration of livestock between mountains and lowland pasture areas.

Truck farming (9.8.1) Relatively small-scale production of fruits, vegetables, and other horticulture, so named for the Middle English word *truck*, meaning *barter* or "exchange of commodities."

Undernourishment (9.1.2) Dietary energy consumption that is continuously below the minimum requirement for maintaining a healthy life and carrying out light physical activity.

Wet rice (9.5.1) Rice planted on dry land in a nursery and then moved to a deliberately flooded field to promote growth.

Geospatial Analysis

Log in to the Mastering Geography Study Area to access MapMaster 2.0.

Agriculture & Climate in China

Eastern China is divided into three principal agricultural regions. How closely do they follow climate regions?

Add the ***Agricultural Regions*** layer. Select the *Settings* icon from the **Legend**, and select *Show Political Labels*. Add the ***Climate Zones*** data layer and select *Split map window*. Zoom to East Asia.

1. What is the principal form of agriculture in southeastern China? Hover over this area. What is the predominant climate?
2. What are the principal forms of agriculture in northeastern China? What is the most common climate?
3. What are the principal forms of agriculture in inland China? What is the predominant climate?
4. Why does the predominant agriculture vary among the climate regions?

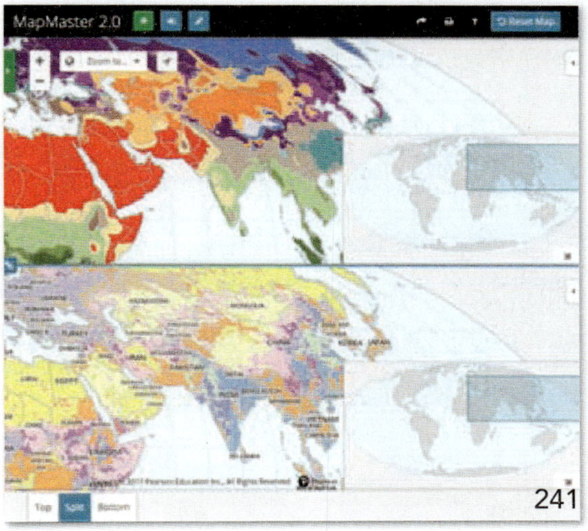

▶ **9.CR.6** AGRICULTURE REGIONS AND CLIMATE ZONES OF EAST ASIA

CHAPTER

10 Development

New housing under construction, Addis Ababa, Ethiopia.

The average person in the world is better off today than in the twentieth century. The average person possesses more wealth, has had more education, and will live longer than someone in the previous century. But not everyone is better off. Development is considered a process of improving people's prospects of leading long and healthy lives, acquiring knowledge, and obtaining adequate resources.

KEY ISSUES

1 How Does Development Vary Among Countries?

The United Nations divides countries according to their level of development. The U.N. Human Development Index is the basis for the allocation.

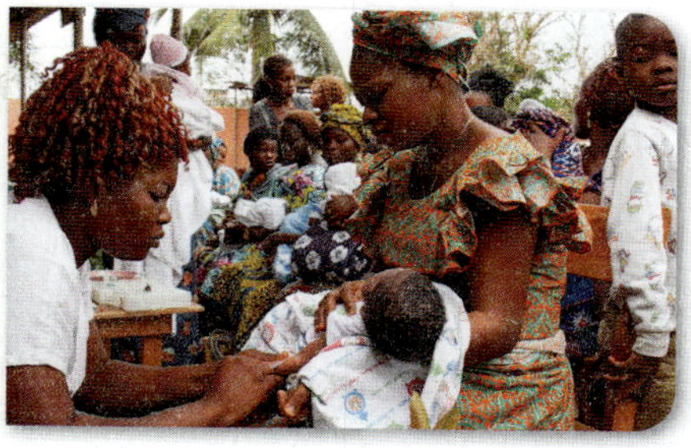

2 What Inequalities Are Found in Development?

According to U.N. development measures, the status of women is lower than that of men within every country. Economic inequalities also exist within every country.

3 How Do Countries Become More Developed?

Self-sufficiency and international trade are two paths to development. International trade has become more important, although countries have had increasing concerns with that development path.

4 How Can Development Be More Sustainable?

By most measures, the level of development worldwide has improved. The U.N. has adopted goals that emphasize sustainable development.

LOCATIONS IN THIS CHAPTER

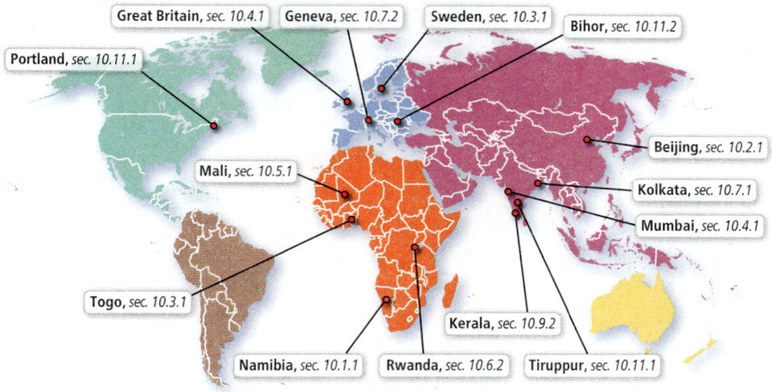

▲ **10.1.1 LOCAL CHILDREN MEET AN AMERICAN TOURIST**
Namibia.

Human Development

- **Measure a country's level of development.**

Earth's nearly 200 countries can be classified according to their level of development. The development process is continuous, involving never-ending actions to constantly improve the health and prosperity of people.

10.1.1 Development & Geography

Development is the process of improving the conditions of people through diffusion of knowledge and technology (Figure 10.1.1). The U.N. defines development as a process of enlarging people's ability to lead a long and healthy life, to acquire knowledge, and to have access to resources needed for a decent standard of living. With access to these three elements, people have greater opportunities to be creative and productive and to enjoy personal self-respect and guaranteed human rights.

Every place lies at some point along a continuum of development. Because many countries cluster at the high and low ends of the continuum of development, they can be divided into two groups:

- A **developed country**, also known as a more developed country (MDC) and referred to by the U.N. as a very high developed country, has progressed further along the development continuum.
- A **developing country**, also frequently called a less developed country (LDC), has made some progress toward development, although less than the developed countries. Recognizing that progress has varied widely among developing countries, the U.N. divides them into high, medium, and low developing.

Human Development Index

To determine the level of development of every country, the U.N. created the **Human Development Index (HDI)**. The HDI measures the level of development for a country through a combination of three factors: a decent standard of living, a long and healthy life, and access to knowledge. Each country gets an overall HDI score based on combining the three factors. The highest HDI possible is 1.0, or 100 percent (Figure 10.1.2). The U.N. has computed HDIs for countries every year since 1980, although it has occasionally modified the method of computation.

▶ **10.1.2 HUMAN DEVELOPMENT INDEX**
Developed countries are those with very high HDI scores. The other three classes are for developing countries.

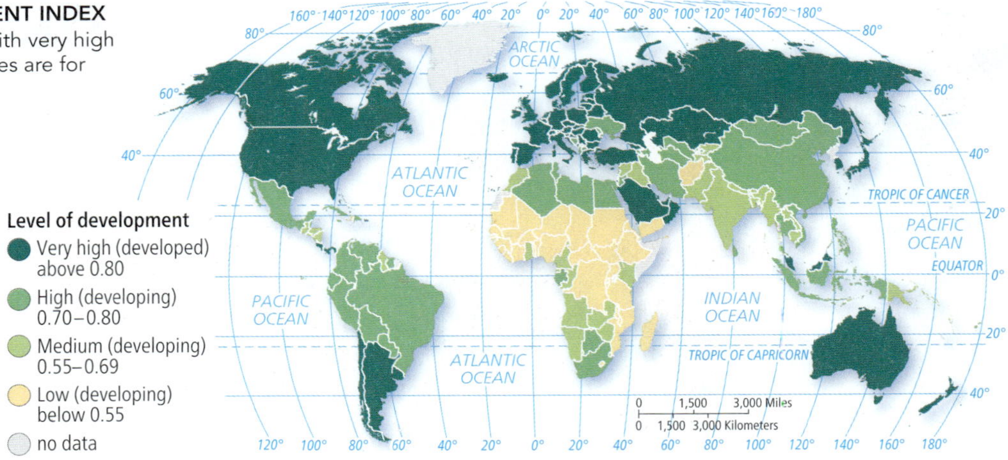

10.1.2 Development Regions

Geographers refer to the HDI scores to divide the world into two developed regions and seven developing regions (Figure 10.1.3). North America and Europe are the two developed regions. The seven developing regions are Latin America, East Asia, South Asia, Southeast Asia, Central Asia, Southwest Asia & North Africa, and sub-Saharan Africa. Each region has an overall HDI score, shown in Figure 10.1.3. Sub-Saharan Africa and South Asia are the regions with the two lowest HDI scores. As we will see in this chapter, regions sometimes have unexpectedly high or low scores in particular factors.

In addition to the nine regions, three other distinctive areas can be identified. Japan and South Korea are classified separately rather than included with the rest of East Asia because their level of development is much higher than that of their neighbors. The South Pacific is a much less populous area than the nine development regions; Australia (its most populous country) and New Zealand are developed, but the area's other countries are developing. Russia has an HDI score that places it on the boundary between very high developed and high developing. As a result, Russia is classified some years as developed and other years as developing.

▼ **10.1.3 DEVELOPMENT REGIONS**
The nine world regions are shown with their regional HDI score.

A Decent Standard of Living

- **Explain the HDI standard of living factor.**

Having enough wealth for a decent standard of living is key to development. The average individual in a developed country earns a much higher income than the average individual in a developing one. Geographers observe that people generate and spend their wealth in different ways in developed countries than in developing countries (Figure 10.2.1).

▲ 10.2.1 STARBUCKS
Beijing, China.

10.2.1 GNI PPP

The U.N. measures the standard of living in countries through a complex index called annual **gross national income per capita at purchasing power parity** (Figure 10.2.2). **Gross national income (GNI)** is the value of the output of goods and services produced in a country in a year, including money that leaves and enters the country. **Purchasing power parity (PPP)** is an adjustment made to the GNI to account for differences among countries in the cost of goods. For example, if a resident of country A has the same income as a resident of country B but must pay more for a Big Mac or a Starbucks latte, the resident of country B is better off.

Some studies refer to **gross domestic product (GDP)**, which is also the value of the output of goods and services produced in a country in a year. GDP does not account for money that leaves and enters the country.

By dividing GNI by total population, it is possible to measure the contribution made by the average individual toward generating a country's wealth in a year. For example, GNI in the United States was approximately $22 trillion in 2020, and the population was approximately 330 million, so GNI per capita was approximately $60,000. In 2020, per capita GNI was approximately $46,000 in developed countries compared to approximately $11,000 in developing countries.

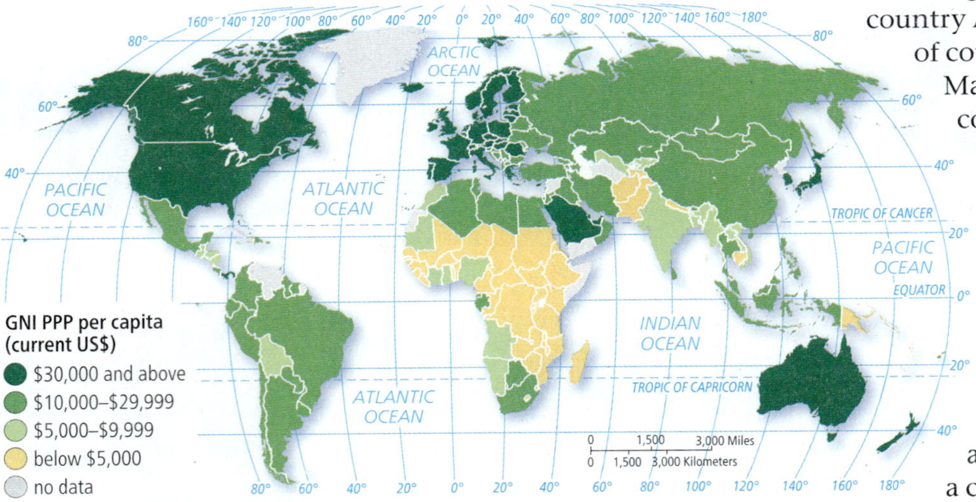

▲ 10.2.2 GNI PPP PER CAPITA

Per capita GNI—or, for that matter, any other single indicator—cannot measure perfectly the level of a country's development. The majority of people may be doing fairly well in a developing country, even with an average annual income of only a few thousand dollars. And not everyone is wealthy in a developed country with per capita GNI of $46,000. Per capita GNI measures average (mean) wealth, not the distribution of wealth. If only a few people receive much of the GNI, then the standard of living for the majority may be lower than the average figure implies. The higher the per capita GNI, the greater the potential for ensuring that all citizens can enjoy a comfortable life.

▼ 10.2.3 GNI BY REGION

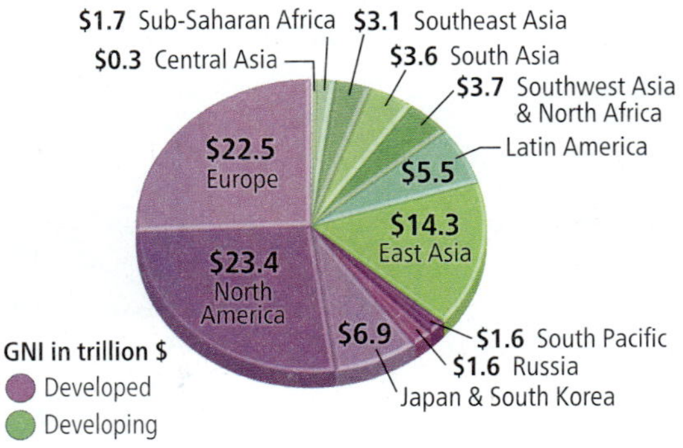

Share of GNI by Country & Region

Developed countries account for nearly two-thirds of the world's total GNI, compared with only one-sixth of the world's population (Figure 10.2.3). The developing countries of East Asia (primarily China) account for around one-sixth of the world's total GNI, nearly as much as the combined total of the other six developing regions.

10.2.2 Economic Structure

Average per capita income is higher in developed countries because people typically earn their living by different means than in developing countries. Jobs fall into three categories:

- The **primary sector** includes activities that directly extract materials from Earth through agriculture (as discussed in Chapter 9) and sometimes by mining, fishing, and forestry.
- The **secondary sector** includes manufacturers that process, transform, and assemble raw materials into useful products as well as industries that fabricate manufactured goods into finished consumer goods (discussed in Chapter 11).
- The **tertiary sector** involves the provision of goods and services to people in exchange for payment, such as retailing, banking, law, education, and government (discussed in Chapter 12).

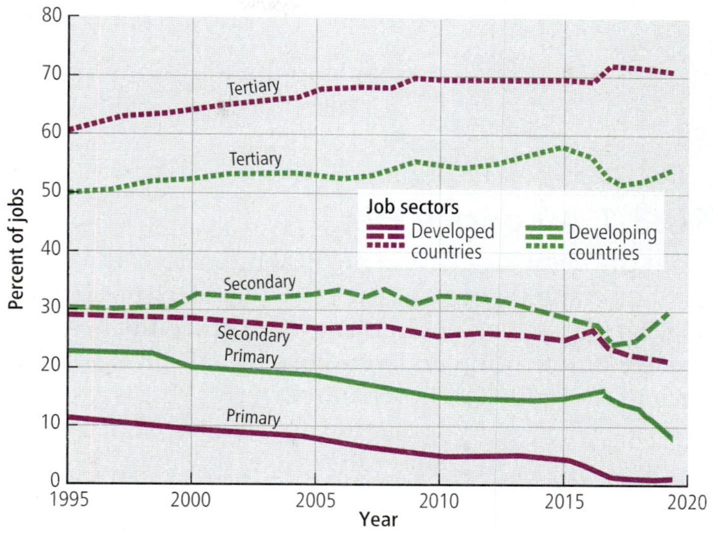

▲ 10.2.4 ECONOMIC STRUCTURE

The contribution to GNI among primary, secondary, and tertiary sectors varies between developed and developing countries (Figure 10.2.4). The share of GNI accounted for by the primary sector has decreased in developing countries, but it remains higher than in developed countries. The low share in developed countries indicates that a handful of farmers produce enough food for the rest of society. The share of GNI accounted for by the secondary sector decreased sharply in developed countries during the twentieth century and is now less than in developing countries. The share of GNI accounted for by the tertiary sector is relatively high in developed countries.

Productivity

Workers in developed countries are more productive than those in developing countries. **Productivity** is the value of a particular product compared to the amount of labor needed to make it (Figure 10.2.5). The World Bank measures a country's productivity by dividing the total GDP by the number of persons employed. GDP per employee exceeds $100,000 in North America and Europe, compared with around $10,000 in sub-Saharan Africa.

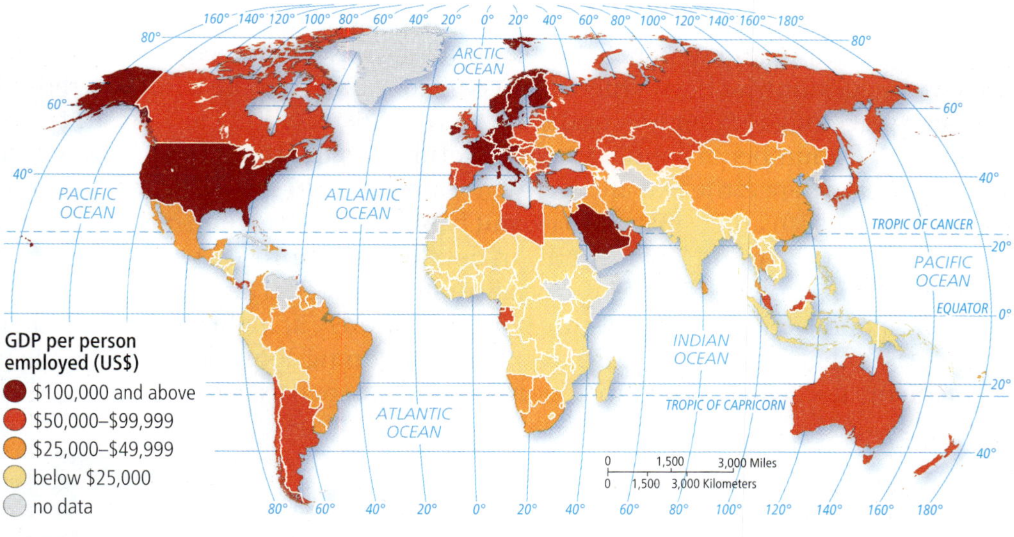

▲ 10.2.5 PRODUCTIVITY BY COUNTRY

Workers in developed countries produce more with less effort because they have access to more machines, tools, and equipment to perform much of the work. On the other hand, production in developing countries relies more on human and animal power (Figure 10.2.6). The larger per capita GNI in developed countries in part pays for the manufacture and purchase of machinery, which in turn makes workers more productive and generates more wealth.

▶ 10.2.6 PRODUCTIVITY IN A DEVELOPING COUNTRY
Plowing a field with animal power, India.

Access to Health & Knowledge

- **Identify HDI health and education factors.**

Development is about more than just wealth. The U.N. believes that development is also about people becoming healthier and wiser, not just wealthier.

10.3.1 A Long & Healthy Life

From the many health and medical indicators, the U.N. has selected life expectancy at birth as the contributor to the HDI (Figure 10.3.1). Life expectancy at birth was defined in Chapter 2 as the average number of years a newborn infant can expect to live at current mortality levels. A baby born this year is expected to live on average to age 73 worldwide, to 79 in developed countries, and to only 62 in sub-Saharan Africa (refer to Figure 2.5.4).

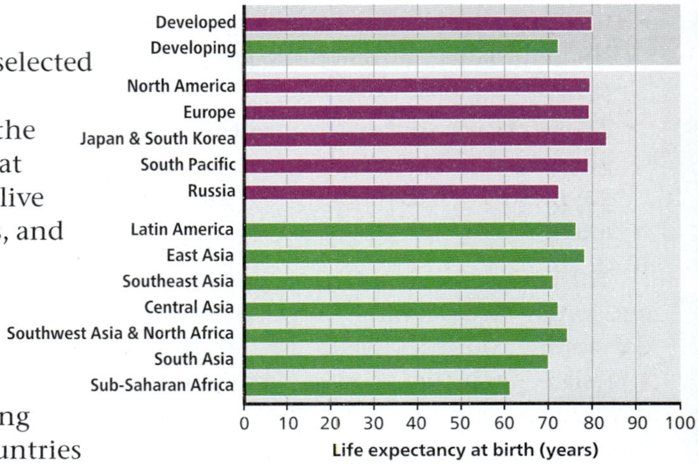

▲ 10.3.1 LIFE EXPECTANCY AT BIRTH BY REGION

Medical Services in Developed & Developing Countries

People are healthier in developed countries than in developing ones. When people in developed countries get sick, these countries possess the resources to care for them (Figure 10.3.2). Developed countries use part of their wealth to protect people who, for various reasons, are unable to work. In these countries, some public assistance is offered to those who are sick, elderly, of low income, disabled, orphaned, veterans of wars, widows, unemployed, or single parents.

Better health and welfare in developed countries permit people to live longer. With longer life expectancies, developed countries have a higher percentage of older people who have retired and receive public support and a lower percentage of children under age 15 who are too young to work and must also be supported by employed adults. The number of young people is six times higher than the number of older people in developing countries, whereas the two are nearly the same in developed countries (refer to Chapter 2).

Better health and welfare also permit babies to survive infancy in developed countries. About 97 percent of infants survive and 3 percent die in developing countries, whereas in developed countries more than 99.6 percent survive. The infant mortality rate is greater in developing countries for several reasons (Figure 10.3.3). In developing countries, babies are born prematurely at higher rates than in developed countries, which increases chances of early death. Babies in developing countries may also die from malnutrition or lack of medicine needed to survive infections or illnesses such as dehydration.

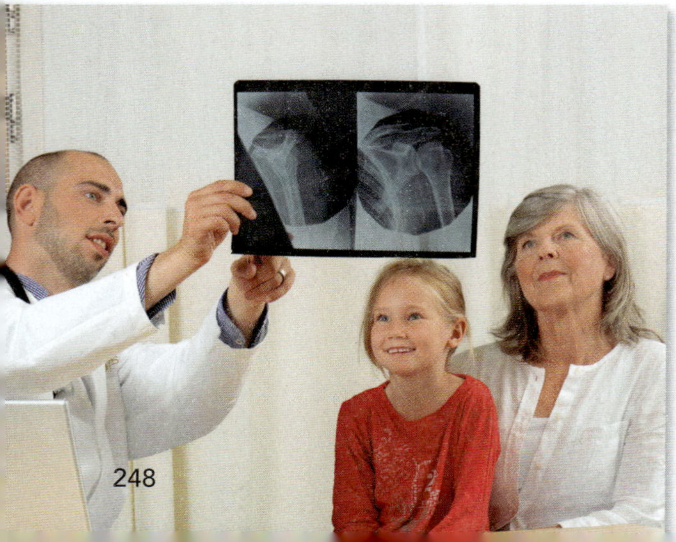

◀ 10.3.2 MEDICAL SERVICES: DEVELOPED COUNTRY
Sweden.

▶ 10.3.3 MEDICAL SERVICES: DEVELOPING COUNTRY
Togo.

10.3.2 Access to Knowledge

The U.N. considers years of schooling to be the most critical measure of the ability of an individual to gain access to knowledge needed for development. The assumption is that no matter how financially disadvantaged the school, the longer the pupils attend, the more likely they are to learn skills that will help them earn a living.

To form the access to knowledge component of HDI, the U.N. combines two measures of years of schooling:

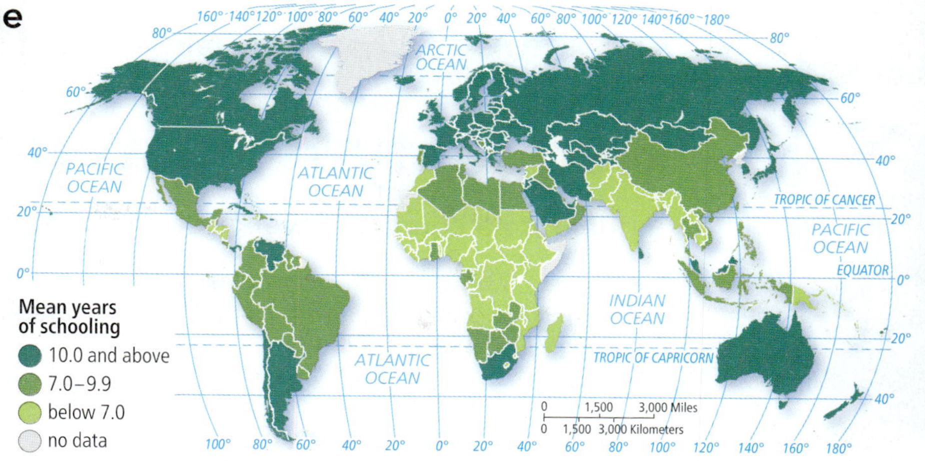

▲ **10.3.4 MEAN YEARS OF SCHOOLING**

- **Years of schooling for today's adults.** This measures the number of years that the average person age 25 or older in a country has spent in school (Figure 10.3.4). Adults have spent an average of 12.2 years in school in developed countries, compared to only 7.5 years in developing countries.
- **Expected years of schooling for today's youth.** This measures the number of years that the U.N. forecasts an average 5-year-old will spend in school. The U.N. expects that 5-year-olds in developed countries will spend an average of 16.3 years in school; in other words, roughly half of today's 5-year-olds will graduate from college in developed countries. On the other hand, the expected average is 10.1 years in sub-Saharan Africa.

Other Education Indicators

Other indicators can measure regional variations in access to knowledge:

- The **pupil/teacher ratio** is the number of enrolled students divided by the number of teachers. The fewer pupils a teacher has, the more likely that each student will receive effective instruction (Figure 10.3.5).
- The **literacy rate** is the percentage of a country's people who can read and write (Figure 10.3.6).

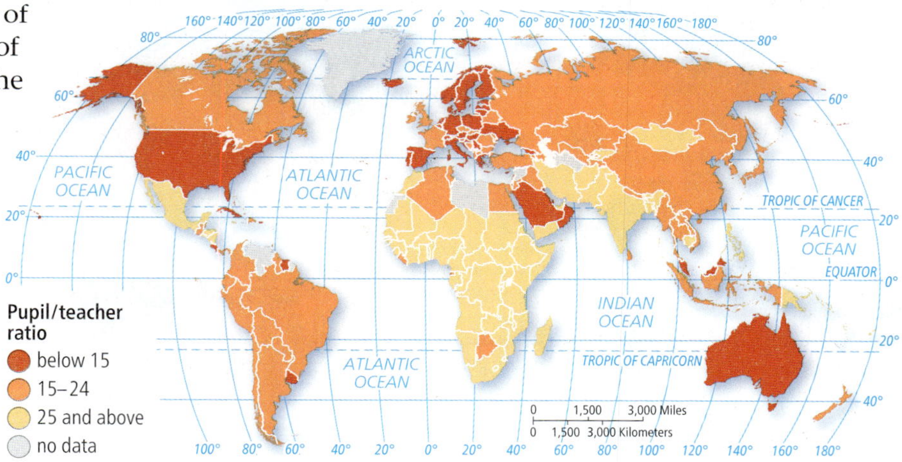

▲ **10.3.5 PUPIL/TEACHER RATIO**

Improved education is a major goal of many developing countries, but funds are scarce. Education may receive a higher percentage of GNI in developing countries, but those countries' GNI is far lower to begin with, so they spend far less per pupil than do developed countries. Many developing countries are unable to fund free public schools, forcing parents to pay high fees for their children's education.

Adding to the challenge of teaching and learning in developing countries, most books, newspapers, and magazines are published in developed countries, in part because more of their citizens can read and write and can afford to buy them. Developed countries dominate scientific and nonfiction publishing worldwide. (This text is an example.) Students in developing countries must learn technical information from books that usually are not in their native language.

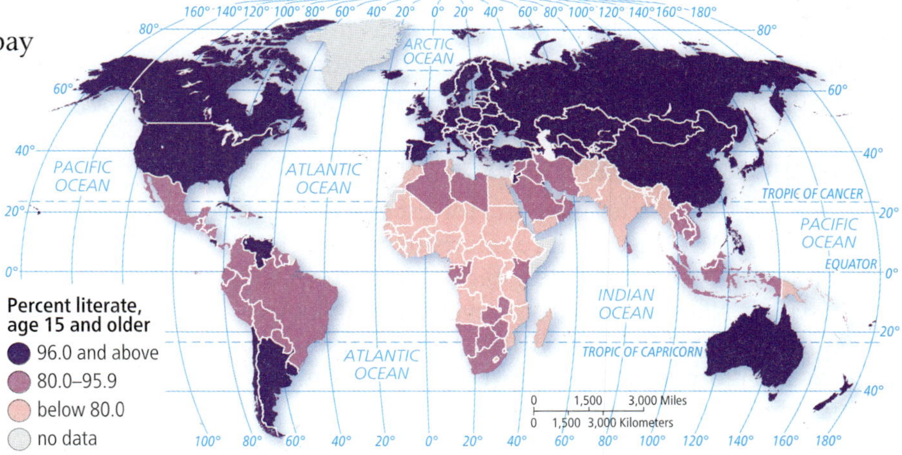

▲ **10.3.6 LITERACY RATE**

▲ **10.4.1 INEQUALITY IN INDIA**
Slums and modern apartment buildings, Mumbai.

Inequality-Adjusted HDI

- **Measure a country's level of development.**

The U.N. believes that every person should have access to decent standards of living, knowledge, and health. In reality, inequality occurs at the international, regional, and local levels.

10.4.1 Inequality Within a Country

To measure the extent of inequality, the U.N. has created the **Inequality-adjusted Human Development Index (IHDI)**. The IHDI modifies the HDI to account for inequality within a country. The highest levels of inequality are in sub-Saharan Africa and South Asia (Figure 10.4.1).

Under perfect equality, the HDI and the IHDI are the same. If the IHDI is lower than the HDI, the country has some inequality; the greater the difference in the two measures, the greater the inequality (Figure 10.4.2). For example, a country wherein only a few people have high incomes, college degrees, and good health care would have a lower IHDI than a country wherein differences in income, level of education, and access to health care are minimal.

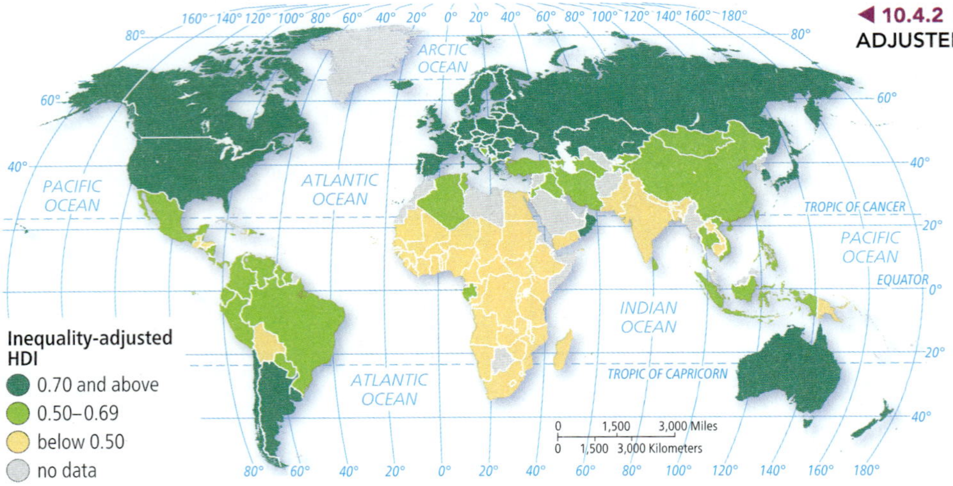

◀ **10.4.2 INEQUALITY-ADJUSTED HDI**

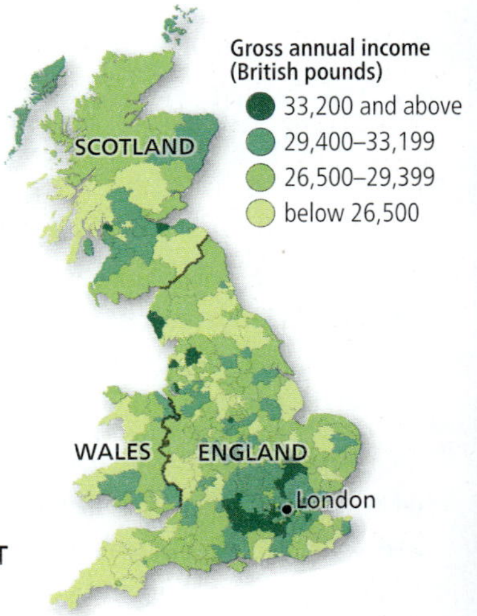

▶ **10.4.3 INEQUALITY WITHIN GREAT BRITAIN**

The inequality within a country may have regional dimensions. For example, in Great Britain incomes in Central London are twice as high as in some of the country's northern communities (Figure 10.4.3).

Regional-Scale Inequality

Developed countries have the lowest gap between HDI and IHDI, indicating a relatively modest level of inequality by worldwide standards (Figure 10.4.4).

The U.N. lacks data from several countries in Southwest Asia & North Africa, so they appear as gray in Figure 10.4.2, but the U.N. does estimate an overall IHDI score for the region, as shown in Figure 10.4.4.

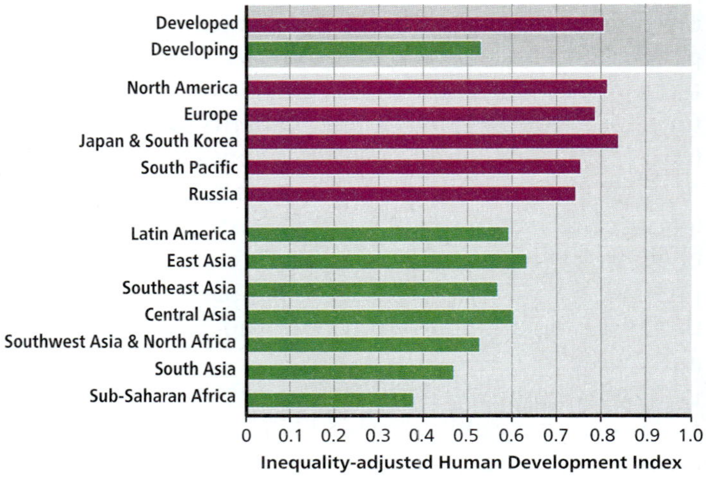

▲ **10.4.4 IHDI BY REGION**
South Asia and sub-Saharan Africa have the lowest scores and therefore the greatest degree of inequality.

10.4.2 Poverty

The U.N. conducts detailed studies to document the extent of poverty for the five-sixths of the world's people who live in developing countries. According to the U.N., 21 percent of people in developing countries live in poverty when it is defined as having an income of less than $1.90 per day. Poverty rates vary from 5 percent in East Asia to 43 percent in sub-Saharan Africa (Figure 10.4.5).

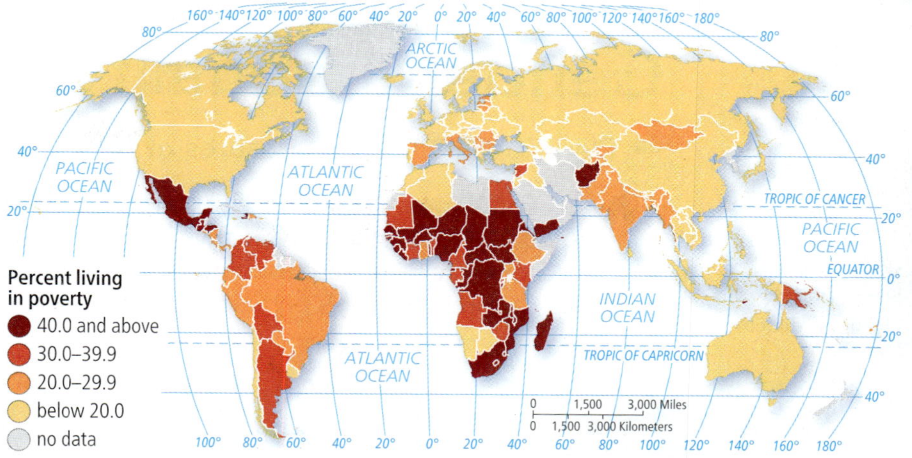

▲ **10.4.5 PERCENT LIVING IN POVERTY**

Multidimensional Poverty Index (MPI)

The U.N. considers living in poverty to have many dimensions in addition to limited income. The U.N.'s **Multidimensional Poverty Index (MPI)** measures the multiple deprivations in people's lives using the same three dimensions as the HDI (standard of living, health, and education). In addition to income, the MPI index for a developing country's standard of living combines the percentage of people with access to cooking fuel, adequate sanitation, safe drinking water, electricity, and decent housing. Education is a dimension of MPI calculated through a combination of average years of schooling and the percentage of children who attend school. The health dimension combines adequate nutrition and level of child mortality.

The overall MPI for developing countries is 22 percent, slightly higher than the monetary poverty rate. The percentage varies widely among developing regions—less than 10 percent in Central Asia, East Asia, and Latin America compared with 55 percent in sub-Saharan Africa and 29 percent in South Asia (Figure 10.4.6).

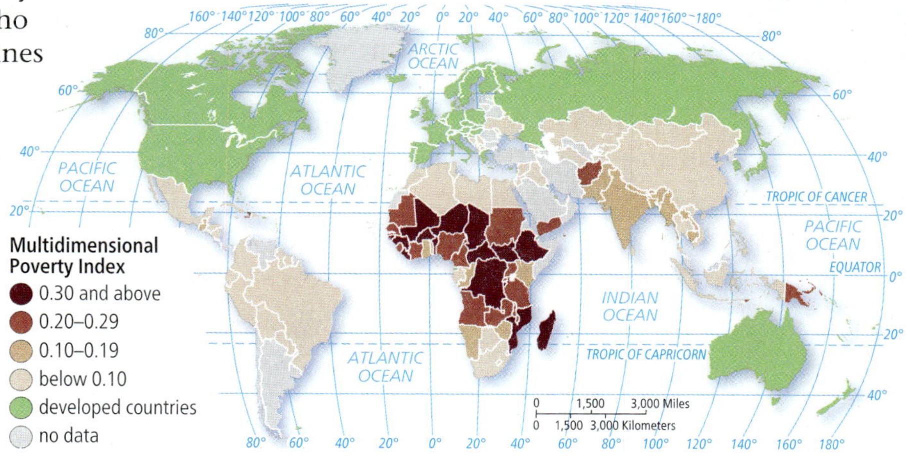

▲ **10.4.6 MULTIDIMENSIONAL POVERTY INDEX**

Regional Variations in MPI

The contribution of the three dimensions of standard of living, health, and education to calculating the MPI varies among countries and regions. A low standard of living accounts for around one-half of the index in sub-Saharan Africa, whereas poor health contributes around one-half in Central Asia. The three dimensions contribute around the same shares in Southwest Asia & North Africa, East Asia, and Latin America (Figure 10.4.7).

The U.N. concludes that most countries have made progress in reducing MPI. However, the COVID-19 pandemic is estimated to have halted progress in most countries by between 3 and 10 years.

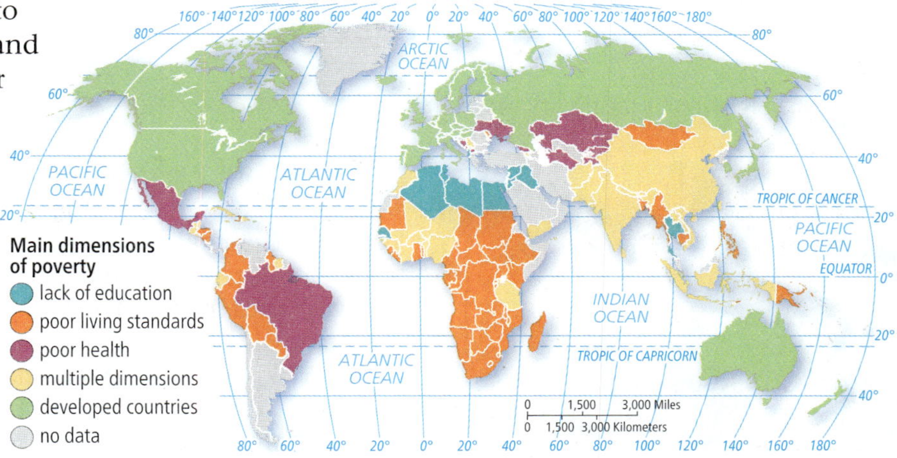

▲ **10.4.7 DIMENSIONS OF MPI**

Gender Inequality

- **Describe the U.N.'s measures of gender inequality.**

A country's overall level of development can mask inequalities in the status of men and women. An improved standard of living, access to knowledge, health, and a sustainable future are aspirations of all people. Yet long-standing cultural and legal obstacles can limit women's participation in development and access to its benefits.

▼ 10.5.1 GIRLS PUMP WATER FOR THEIR FAMILIES, MALI

10.5.1 Measuring Gender Inequality

The U.N., which uses several indexes to measure gender inequality, has not found a single country in the world where the women are treated as well as the men. At best, women have achieved near-equality with men in some countries, but in other countries, the level of development for women lags far behind the level for men. The U.N. argues that inequality between men and women is a major factor that keeps a country from achieving a higher level of development.

Gender Inequality Index (GII)

The **Gender Inequality Index (GII)** measures the gender gap in the level of achievement in three dimensions: reproductive health, empowerment, and the labor market. The GII uses similar methodology to the IHDI, as discussed in Section 10.4.1. The higher the GII, the greater the inequality between men and women. A score of 0 would mean that men and women fare equally, and a score of 1.0 would mean that women fare as poorly as possible in all measures.

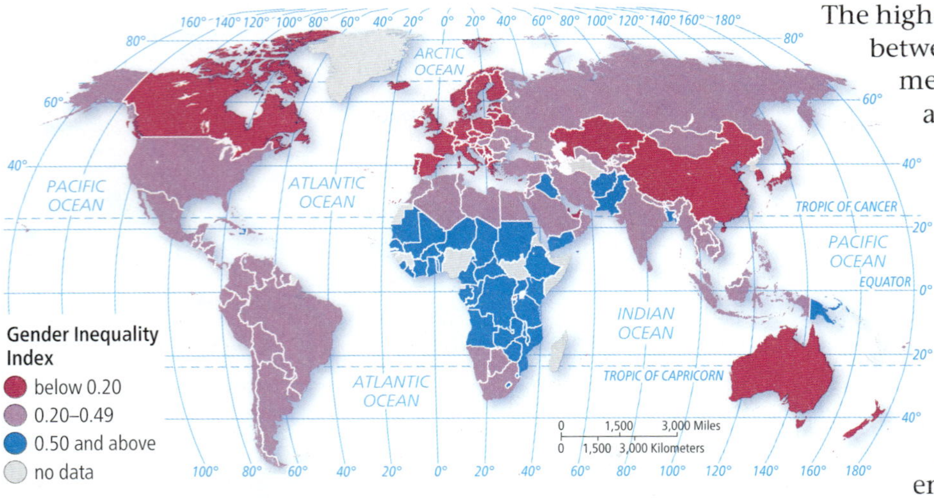

▲ 10.5.2 GENDER INEQUALITY INDEX

Sub-Saharan Africa, South Asia, and Southwest Asia & North Africa are the developing regions with the highest levels of gender inequality (Figure 10.5.1). Reproductive health is the largest contributor to gender inequality in these regions. South Asia and Southwest Asia & North Africa also have relatively poor female empowerment scores. The GII is .463 in developing countries compared with .173 in developed ones (Figure 10.5.2).

At the other extreme, 25 countries, including 19 in Europe as well as Canada, have GIIs less than 0.1, meaning that men and women are nearly equal. In general, countries with high HDIs have low GIIs and vice versa.

The United States has a GII rank of only 46, although it ranks sixteenth in the world in HDI. The U.N. points to two factors accounting for the relatively low U.S. GII ranking. Compared with other very high HDI countries, the United States has a much higher birth rate among teenage women (Figure 10.5.3) and a higher mortality rate among women during childbirth. The percentage of women in the national legislature is lower in the United States than in other high HDI countries.

◀ 10.5.3 TEENAGE MOTHER AND BABY, UNITED STATES

10.5.2 Components of the GII

The GII combines three sets of measures to come up with a composite score for gender inequality: empowerment, employment, and reproductive health.

Empowerment

In the context of gender inequality, empowerment refers to the ability of women to achieve improvements in their own status—that is, to achieve economic and political power. The empowerment dimension of GII is measured by two indicators: the percentage of seats held by women in the national legislature and the percentage of women who have completed some secondary school (high school).

In every country of the world, both developed and developing, fewer women than men hold positions of political power (Figure 10.5.4). Worldwide, 61 percent of women have completed some secondary (high) school, compared to 68 percent of men. The gap in education between girls and boys is especially high in South Asia.

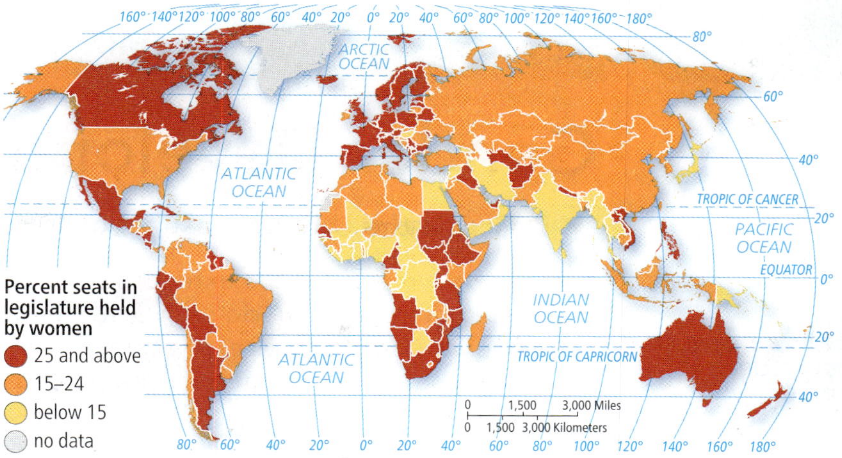

▲ 10.5.4 WOMEN IN THE NATIONAL LEGISLATURE

Employment

The **female labor force participation rate** is the percentage of women holding full-time jobs outside the home. Worldwide, 46 percent of women work outside the home, compared to 76 percent of men.

In general, women in developed countries are more likely than women in developing countries to hold full-time jobs outside the home. Figures vary widely among developing regions (Figure 10.5.5).

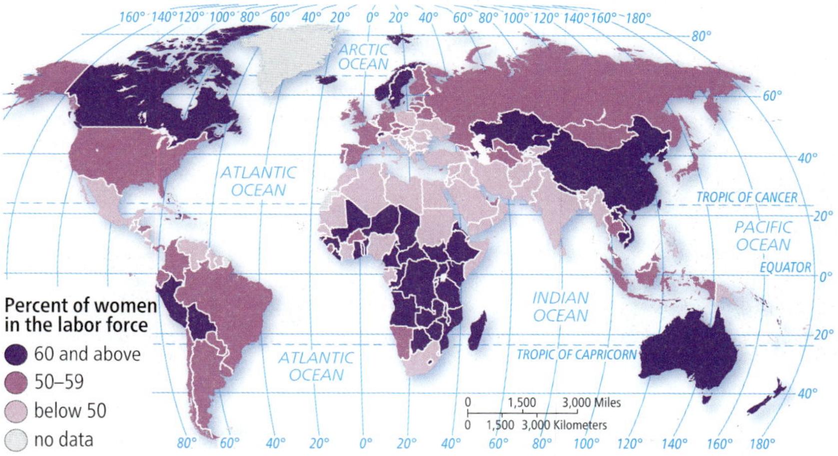

▲ 10.5.5 PERCENT OF WOMEN IN THE LABOR FORCE

Reproductive Health

The third component of the GII is reproductive health. The effects of reproduction are significantly greater for females by nature. Poor reproductive health is a major contributor to gender inequality in many countries. The reproductive health dimension of the GII is based on two indicators: adolescent fertility rate and maternal mortality rate.

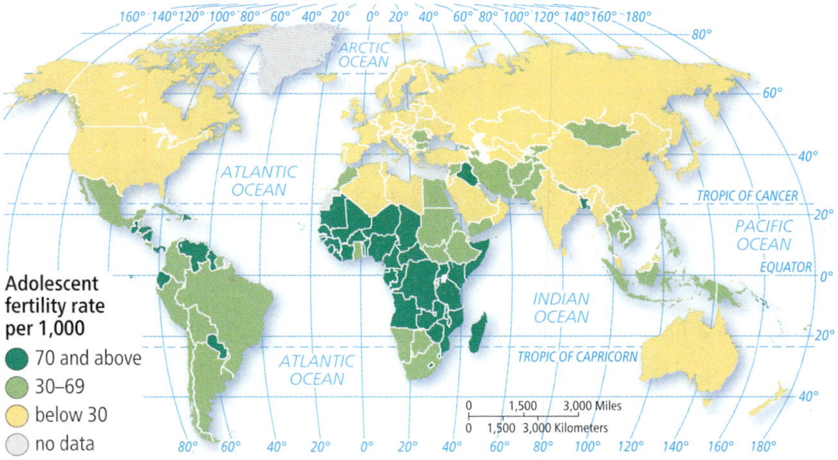

▲ 10.5.6 ADOLESCENT FERTILITY RATE

The **adolescent fertility rate** is the number of births per 1,000 women ages 15 to 19. The rate is 17 births per 1,000 women ages 15 to 19 in developed countries and 47 in developing countries. The teenage pregnancy rate is below 10 per 1,000 in most European countries, where most couples use some form of contraception. In sub-Saharan Africa, where gender inequality is high, contraceptive use is below 10 percent, and the teenage pregnancy rate is 105 per 1,000 (Figure 10.5.6).

The maternal mortality rate was defined in Chapter 2 as the number of women who die giving birth per 100,000 births. The ratio is 14 deaths of mothers per 100,000 live births in developed countries and 224 in developing countries. The highest rates (most deaths per births) are in sub-Saharan Africa (refer to Figure 2.6.1).

Gender Development

- **Describe the U.N.'s measures of gender development.**

The U.N. measures the extent to which women experience the three measures of HDI differently than do men. The Gender Development Index (GDI) measures the gender gap in the level of achievement for the three dimensions of the Human Development Index: income, education, and life expectancy.

▼ 10.6.1 GENDER DEVELOPMENT INDEX

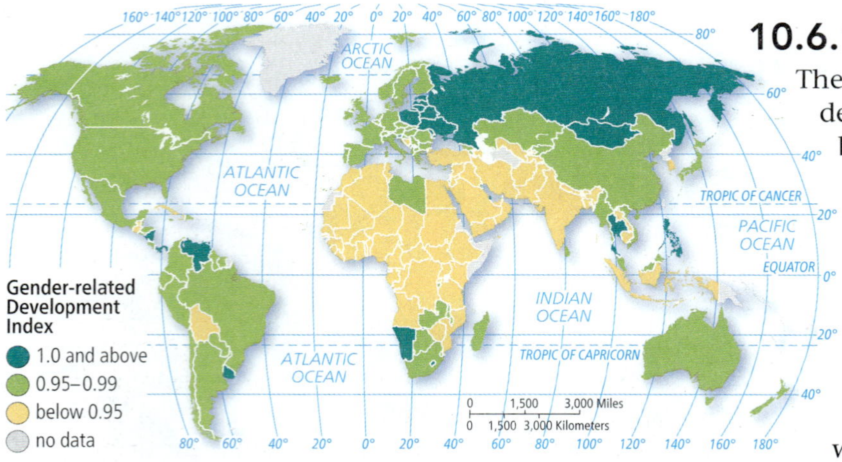

10.6.1 Gender Development Index

The GDI uses the same methodology as the HDI, described in Section 10.1.1. Countries are ranked based on their deviation from gender parity in the three dimensions of the HDI.

If females and males had precisely the same HDI scores, the GDI would be 1.000. In fact, the overall GDI in the world is .943, which means that the average HDI for all females in the world (.714) is 94.3 percent of the average HDI for all males (.757). The average GDI in developed regions is .981, indicating close to parity, whereas the average GDI in developing countries is .919 (Figure 10.6.1).

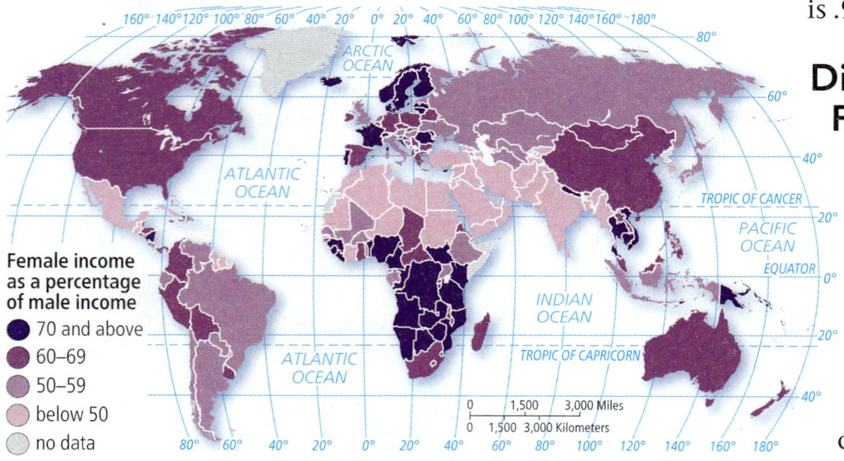

▲ 10.6.2 FEMALE INCOME AS A PERCENTAGE OF MALE INCOME

Disparities Between Females & Males

The three principal components of the GDI score can be examined in more detail.

Income. The U.N. has not found a single country in the world where the average income earned by women exceeds that earned by men (Figure 10.6.2). Worldwide, the average income of women is 60 percent that of men in developed countries and 49 percent in developing countries.

Education. Worldwide, women have completed an average of 8.1 years of schooling, compared with 9.2 years for men. In developed countries, the average education of women and men is virtually the same at around 12 years of schooling, whereas in developing countries women have completed an average of 6.9 years of schooling, compared with 8.3 years for men (Figure 10.6.3).

Life expectancy. The life expectancy at birth worldwide is 75 for females and 71 for males. The gap is 6 years in developed countries and 4 years in developing countries.

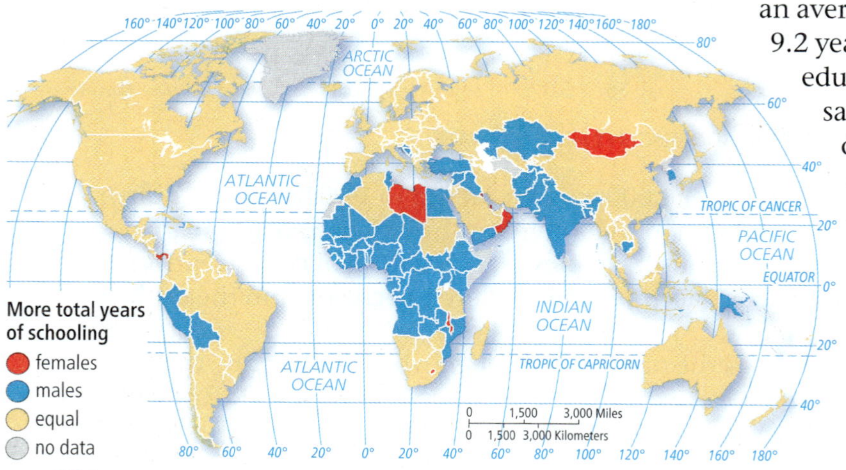

◀ 10.6.3 GENDER DIFFERENCES IN MEAN YEARS OF SCHOOLING

10.6.2 Gender Attitudes

Underlying the challenges faced by women, as demonstrated by the GII and GDI, is the prevalence of attitudes favoring men in many parts of the world. In recent years, the #MeToo movement has been a social movement against sexual abuse and discrimination. In India, the #IWillGoOut movement has demanded equal rights for women, and in Latin America, the #NiUnaMenos movement has highlighted violence against women.

To measure the strength of attitudes biased against women, the U.N. has constructed the **Gender Social Norms Index (GSNI)**. According to the GSNI, around one-half of the people in developed countries have biased attitudes against the aspirations of women, whereas in developing countries at least some bias against women is nearly universal (Figure 10.6.4).

Unlike other U.N. development indexes, the GSNI is based on surveys of people's attitudes rather than based on demographic or economic data. The GSNI is calculated for each country by obtaining a random sample of responses to seven statements.

GSNI Dimensions

The GSNI measures the extent to which people in different countries are biased against women in four dimensions: political power, education, the economy, and physical integrity. In addition to an overall GSNI, separate indexes are constructed for each of the four dimensions, and separate indexes are constructed for male and female respondents.

Political empowerment index. Attitudes toward political empowerment are based on responses to two statements: "Men make better political leaders than women do" and "Women have the same rights as men." Respondents are classified by the U.N. as "biased" if they strongly agree or agree with the first statement, and they are considered "biased" if they respond "not essential" to the second statement. Based on responses to these two questions, the GSNI classifies around three-fourths of people in developing countries and one-fourth of people in developed countries (including 40 percent in the United States) as biased against women's political aspirations (Figure 10.6.5).

▲ 10.6.5 WOMEN IN GOVERNMENT, RWANDA

Educational empowerment index. The GSNI bases the educational empowerment index on responses to this statement: "University is more important for a man than for a woman." Respondents are classified by the U.N. as "biased" if they strongly agree or agree with the statement. This statement had the lowest level of bias in both developing countries (around one-third) and in developed countries (around 5 percent).

Economic empowerment index. Attitudes toward economic empowerment are based on responses to two statements: "Men should have more right to a job than women" and "Men make better business executives than women do." Respondents are classified by the U.N. as "biased" if they strongly agree or agree with the statements. Based on responses to these two questions, the GSNI classifies around two-thirds of people in developing countries and one-fourth of people in developed countries (including one-sixth in the United States) as biased against women's economic aspirations.

Physical integrity index. Attitudes toward physical integrity are based on responses to these two statements: "My intimate partner is never violent with me" or "I am never violent with my intimate partner" and "I have reproductive rights" or "My intimate partner has reproductive rights." Respondents are classified by the U.N. as "biased" if they disagree with the statements. Based on responses to these two statements, the GSNI classifies around two-thirds of people in developing countries and one-third of people in developed countries as biased against women's physical integrity rights.

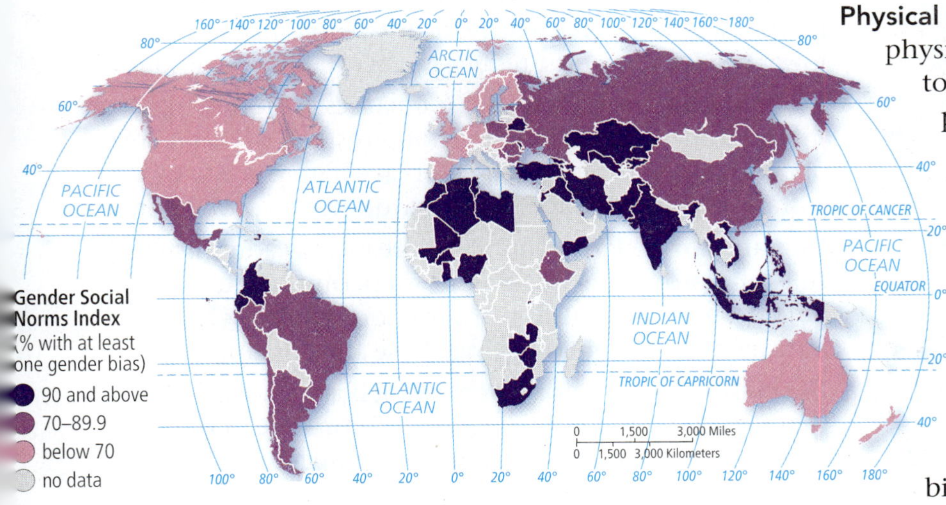

▲ 10.6.4 GENDER SOCIAL NORMS INDEX

▲ **10.7.1 SELF-SUFFICIENCY** India manufactured its own cars, called Maruti.

Two Paths to Development

- **Summarize the two paths to development.**

To promote development, developing countries choose one of two paths: self-sufficiency or international trade. Each path has important advantages and challenges.

10.7.1 Self-Sufficiency Path

In the self-sufficiency path, countries encourage domestic production of goods, discourage foreign ownership of businesses and resources, and protect their businesses from international competition (Figure 10.7.1). Key elements of the self-sufficiency path to development include the following:

- **Limits to imports.** A country may enact barriers that limit the import of goods from other places. Three widely used barriers include setting high taxes (tariffs) on imported goods to make them more expensive than domestic goods, fixing quotas to limit the quantity of imported goods, and requiring licenses in order to restrict the number of legal importers.
- **Insulation.** A country may nurse fledgling businesses to success by restricting or prohibiting large international corporations from doing business in the country. Such insulation from the potentially adverse impacts of decisions made by businesses and governments in developed countries encourages a country's fragile businesses to achieve independence.
- **Equal investment.** Investment is spread as equally as possible across all sectors of a country's economy and in all of its regions.
- **Equal income.** Incomes in the countryside keep pace with those in the city, and reducing poverty takes precedence over encouraging a few people to become wealthy consumers.

Shortcomings of Self-Sufficiency Path

The self-sufficiency path has a number of shortcomings:

▼ **10.7.2 BUREAUCRACY** Government official works at an outside desk, Kolkata, India.

- **Inefficient businesses.** Businesses could sell all they make at high government-controlled prices to customers culled from long waiting lists. With such assured control of the market, businesses have little incentive to improve quality, lower production costs, reduce prices, or increase production.
- **Lack of competitiveness.** Companies protected from international competition are not pressured to keep abreast of rapid technological changes or give high priority to sustainable development and environmental protection.
- **Corruption.** A large, complex bureaucracy administering rules and processing documents for permits (Figure 10.7.2) burdens businesses and grants unmonitored power to bureaucrats, thus encouraging abuse and corruption.
- **Black market.** Aspiring or existing businesses may find that struggling to produce goods and compete in the open market is less profitable than illegally importing goods and selling them at inflated prices on the black market.

10.7.2 International Trade Path

The international trade path of development calls for a country to identify its distinctive or unique economic assets. What animal, vegetable, or mineral resources does the country have in abundance that other countries are willing to buy? What product can the country manufacture and distribute at a higher quality and a lower cost than other countries? According to the international trade path, a country can develop economically by concentrating scarce resources on expansion of its distinctive local industries. The sale of these products in the world market brings into the country funds that can be used to finance other development.

Rostow's International Trade Model

A pioneering advocate of the international trade path was W. W. Rostow, who in the 1950s proposed a five-stage model of development. According to Rostow's international trade model, each country is in one of these five stages of development:

1. **Traditional society.** A traditional society has not yet started a process of development. It contains a very high percentage of people engaged in agriculture and a high percentage of national wealth allocated to what Rostow called "nonproductive" activities, such as the military and religion.
2. **Preconditions for takeoff.** An elite group initiates innovative economic activities. Under the influence of these well-educated leaders, the country starts to invest in new technology and infrastructure, such as water supplies and transportation systems. Support from international funding sources often emphasizes the importance of constructing new infrastructure. These projects will ultimately stimulate an increase in productivity.
3. **Takeoff.** Rapid growth is generated in a limited number of economic activities, such as textiles or food products. These few takeoff industries achieve technical advances and become productive, whereas other sectors of the economy remain dominated by traditional practices.
4. **Drive to maturity.** Modern technology, previously confined to a few takeoff industries, diffuses to a wide variety of industries, which then experience rapid growth comparable to the growth of the takeoff industries. Workers become more skilled and specialized.
5. **Age of mass consumption.** The economy shifts from production of heavy industry, such as steel and energy, to consumer goods, such as motor vehicles and refrigerators.

The World Trade Organization

To promote the international trade development path, countries representing 97 percent of world trade established the World Trade Organization (WTO) in 1995 (Figure 10.7.3). The WTO works to reduce barriers to international trade in two principal ways.

Reduce restrictions. Through the WTO, countries negotiate reduction or elimination of international trade restrictions on manufactured goods, such as government subsidies for exports, quotas for imports, and tariffs on both imports and exports. Also reduced or eliminated are restrictions on the international movement of money by banks, corporations, and wealthy individuals.

Enforcement. The WTO also promotes international trade by enforcing agreements. One country can bring to the WTO an accusation that another country has violated a WTO agreement. The WTO is authorized to rule on the validity of the charge and order remedies. The WTO also tries to protect intellectual property in the age of the internet. An individual or a corporation can also bring charges to the WTO that someone in another country has violated a copyright or patent, and the WTO can order illegal actions to stop.

▶ **10.7.3 WORLD TRADE ORGANIZATION HEADQUARTERS, GENEVA, SWITZERLAND**
Sculpture "Human Effort" by James Vibert.

World Trade

- **Analyze reasons for the triumph of the international trade development path.**

During the late twentieth century and into the twenty-first century, most countries embraced the international trade approach as the preferred alternative for stimulating development.

▼ 10.8.1 INTERNATIONAL TRADE AS A PERCENTAGE OF GDP

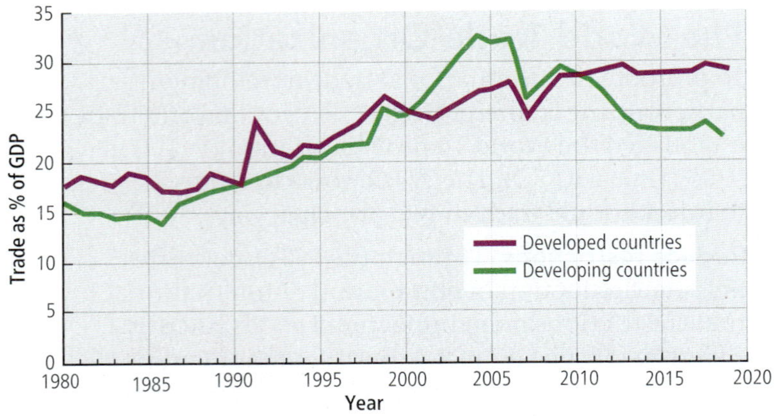

10.8.1 International Trade Triumphs

Trade increased more rapidly than wealth (as measured by GDP) during the late twentieth and early twenty-first centuries (Figure 10.8.1). The increase is a reflection of widespread acceptance of international trade as the best way to promote development, based on three observations:

- **Other countries' successes.** If Japan and South Korea could become more developed through international trade, why couldn't other countries?
- **Resource riches.** Developing countries contain many of the raw materials sought by manufacturers and producers in developed countries. In the past, European colonial powers extracted resources without paying for them. In a global economy, the sale of raw materials can help developing countries pay for development.
- **International competitiveness.** A country that concentrates on international trade benefits from exposure to the demands, needs, and preferences of consumers in other countries. To remain competitive, the takeoff industries constantly evaluate changes in international consumer preferences, marketing strategies, production engineering, and design technologies.

India Moves Away from Self-Sufficiency

For several decades after it gained independence from the United Kingdom in 1947, India was a leading example of the self-sufficiency strategy. India limited foreign companies from importing into the country through such measures as requiring an import license that was difficult to get, levying heavy taxes on imported goods, and limiting the ability to move currency in and out of the country.

India controlled its own companies by requiring government permission for many actions, such as introducing a new product, modernizing a factory, setting prices, and hiring and firing workers. The government itself owned many of the companies, including automakers and insurance firms.

India adopted many international trade policies during the 1990s. Foreign companies were allowed to set up factories and sell in India, taxes were reduced, and monopolies were eliminated. The result was a more substantial increase in GDP per capita than had been experienced under self-sufficiency (Figure 10.8.2).

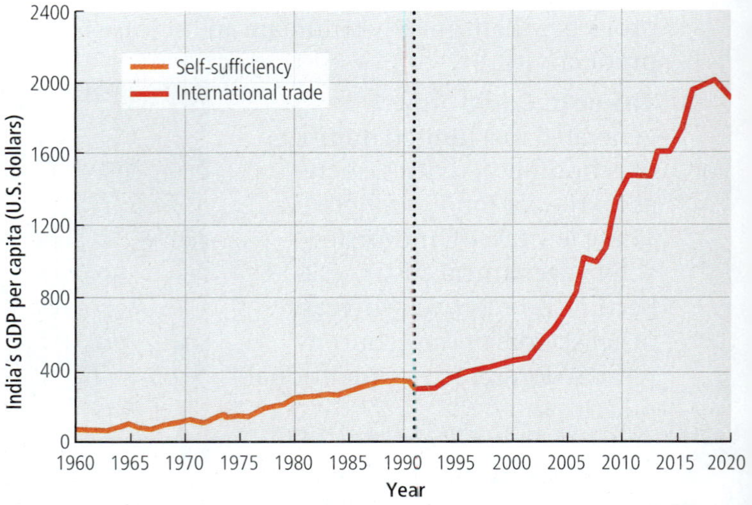

▲ 10.8.2 GDP PER CAPITA: INDIA

10.8.2 Core & Periphery

Immanuel Wallerstein, a U.S. social scientist, developed the **world-systems theory** to describe relationships between developed and developing countries. According to Wallerstein's world-systems theory, in an increasingly unified world economy, developed countries form an inner core area, whereas developing countries occupy peripheral locations (Figure 10.8.3).

Countries in Africa, Asia, and Latin America contain three-fourths of the world's population and nearly all of its population growth. However, these countries find themselves on a periphery, or outer edge, with respect to the wealthier core regions of North America, Europe, and Japan.

Since the original formulation of world systems theory, an increasingly important element has been the emergence of countries classified as semi-periphery. These are countries that are either intermediate in level of economic development or situated close to both core and periphery regions.

▲ 10.8.3 CORE & PERIPHERY: NORTH POLAR PROJECTION
When seen from the North Pole, developed regions form an inner core and developing regions an outer periphery.

North–South Split

The relationship between developed countries and developing countries appears on a traditional projection as a north-south split because most of the developed countries of the core (except for Australia and New Zealand) are in the Northern Hemisphere's temperate zone north of the Tropic of Cancer, whereas many developing countries are to the south (Figure 10.8.4). However, the north polar projection used in Figure 10.8.3 portrays more graphically the core–periphery relationship between developed and developing regions.

Global investment arrives from the core through hierarchical diffusion of decisions made by transnational corporations. Particular core and periphery regions have special connections. The development prospects of Latin America are tied to governments and businesses primarily in North America, those of Africa and Eastern Europe to Western Europe (Figure 10.8.5), and those of Asia to Japan and to a lesser extent Europe and North America. As countries such as China, India, and Brazil develop, relationships between core and periphery countries are changing, and the line between core and periphery countries may need to be redrawn.

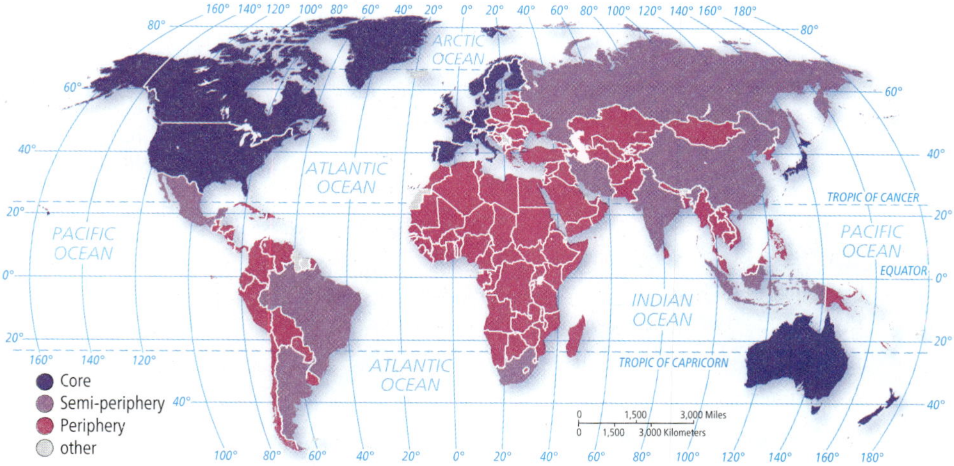

▲ 10.8.4 CORE & PERIPHERY: NORTH–SOUTH DIVIDE

▼ 10.8.5 CORE & PERIPHERY INTERACT
Used cars are being exported from a core country (Germany) to a periphery country (Benin).

Financing Development

- **Identify the main source for financing development.**

Developing countries lack money to fund development, so they obtain financial support from two primary sources: direct investment by transnational corporations and loans from banks and international organizations.

▼ 10.9.1 GROWTH IN FDI

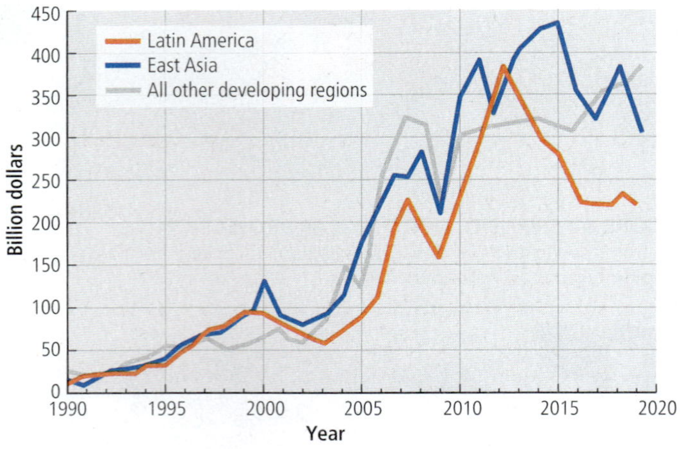

10.9.1 Foreign Direct Investment

Investment made by a foreign company in the economy of another country is known as **foreign direct investment (FDI)**. FDI has grown rapidly from $172 billion in 2002 to $1.7 trillion in 2019 (Figure 10.9.1).

Investment does not flow equally around the world (Figure 10.9.2). Only half of FDI went to developing countries while the other half went to developed countries. Half of all FDI destined for developing countries went to China. The major sources of FDI are transnational corporations that invest and operate in countries other than the one in which the company headquarters are located. Of the 500 largest transnational corporations in 2020, 343 had headquarters in developed countries, including 121 in the United States. China (including Hong Kong) was the location of 124 of the 157 with headquarters in developing countries.

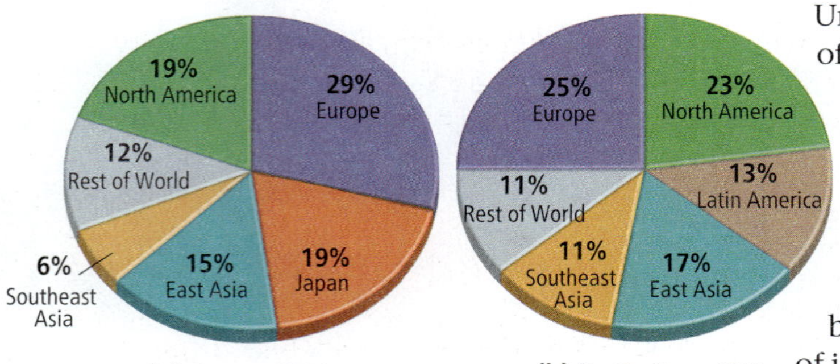

(a) Sources of FDI **(b)** Destinations of FDI

▲ 10.9.2 SOURCES AND DESTINATIONS OF FDI, 2019

Principal Lenders

The two major lenders to developing countries are the International Monetary Fund (IMF) and the World Bank.

IMF. The IMF provides loans to countries experiencing balance-of-payments problems that threaten expansion of international trade. IMF assistance is designed to help a country rebuild international reserves, stabilize currency exchange rates, and pay for imports without the imposition of harsh trade restrictions or capital controls that could hamper the growth of world trade.

World Bank. The World Bank includes the International Bank for Reconstruction and Development (IBRD) and the International Development Association (IDA). The IBRD provides loans to countries to reform public administration and legal institutions, develop and strengthen financial institutions, and implement transportation and social service projects. The IDA provides support to developing countries considered too risky to qualify for IBRD loans. The IBRD lends money raised from sales of bonds to private investors, whereas the IDA lends money from government contributions (Figure 10.9.3).

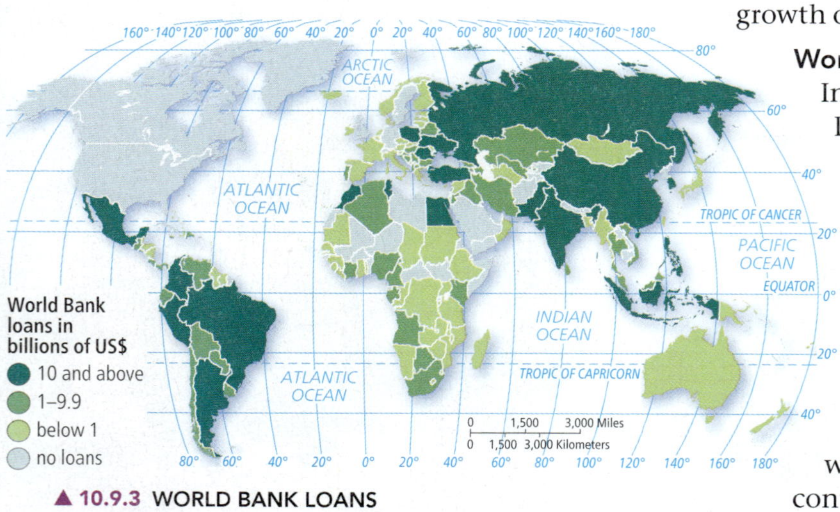

▲ 10.9.3 WORLD BANK LOANS

10.9.2 Loan Debt

Developing countries borrow money to build new infrastructure, such as hydroelectric dams, electric transmission lines, flood-protection systems, water supplies, roads, and hotels. The theory is that new infrastructure will make conditions more favorable for domestic and foreign businesses to open or expand. After all, no business wants to be located in a place that lacks paved roads, running water, and electricity.

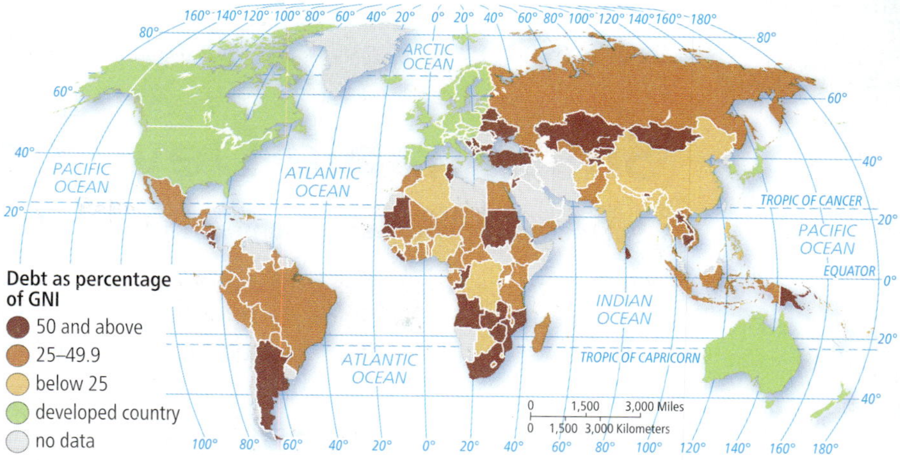

▲ 10.9.4 DEBT AS PERCENTAGE OF GNI

In principle, new or expanded businesses are attracted to an area because improved infrastructure will contribute additional taxes that the developing country will use in part to repay loans and in part to improve its citizens' living conditions. In reality, the World Bank itself has judged half the projects it has funded in Africa to be failures. Projects may not function as intended because of faulty engineering, or the money may be stolen through corruption.

Some countries have been unable to repay the interest on their loans, let alone the principal. Debt actually exceeds annual income in a number of countries (Figure 10.9.4). When countries cannot repay their debts, financial institutions in developed countries refuse to make further loans, so construction of needed infrastructure stops. The inability of many countries to repay loans also damages the financial stability of banks in developed countries.

Austerity Programs

To apply for debt relief, a country may be required to adopt an austerity program. Key elements of an austerity program include sharp cuts in government spending and taxes. The expectation of an austerity program is that lower taxes will encourage people and businesses to use their tax savings to buy more goods and services and thereby stimulate economic growth.

Austerity is imposed through a policy framework paper (PFP) that outlines a structural adjustment program. A **structural adjustment program** contains economic "reforms" or "adjustments," such as economic goals, strategies for achieving the objectives, and external financing requirements.

Reforms required of a developing country typically include spending only what it can afford, investing scarce resources where they will have the most impact, and encouraging a more productive private sector. Critics charge that poverty worsens under structural adjustment programs. By placing priority on reducing government spending and inflation, structural adjustment programs may result in cuts in health, education, and social services that benefit those living in poverty.

Microfinance for Development

Many would-be business owners in developing countries do not have the collateral to qualify for regular bank loans. An alternative source of loans is **microfinance**, which is the provision of small loans and other financial services to individuals and small businesses in developing countries that are unable to obtain loans from commercial banks (Figure 10.9.5). Microfinance companies provide small loans, typically in the range of just a few hundred dollars, to aspiring entrepreneurs whose income falls below the poverty line in developing countries. The five leading microfinance companies are 51Give (based in China), Bank Raykat Indonesia, BRAC (based in Bangladesh), Grameen Bank (based in Bangladesh), and Kiva (based in the United States).

▼ 10.9.5 MICROFINANCE
The Peermade Development Society, Kerala, India, provides microloans to women trying to start small businesses.

Sustainable Development

- **Describe progress in achieving the U.N.'s Sustainable Development Goals.**

To reduce disparities between developed and developing countries, all 193 U.N. members in 2015 adopted 17 **Sustainable Development Goals**. The Sustainable Development Goals replaced eight **Millennium Development Goals** adopted in 2002.

10.10.1 Sustainable Development Goals

Figure 10.10.1 displays the 17 U.N. Sustainable Development Goals. All U.N. members agreed to achieve these goals by 2030.

▼ 10.10.1 SUSTAINABLE DEVELOPMENT GOALS

Goal	Relevant Information
1. End poverty in all its forms everywhere.	The percentage of people living in poverty in the world has been cut in half since 1990, but 700 million people still live in extreme poverty, which the U.N. defines as earning less than $1.90 per day.
2. End hunger, achieve food security and improved nutrition, and promote sustainable agriculture.	The U.N. estimates 780 million people are undernourished. It wants to double agricultural productivity by 2030.
3. Ensure healthy lives, and promote well-being for all at all ages.	Progress has been made in reducing infant mortality and maternal mortality, but the U.N. hopes to reduce these rates further by 2030.
4. Ensure inclusive and quality education for all, and promote lifelong learning.	The U.N. wants to ensure that all children attend school, and it estimates that 57 million children currently do not attend school, more than half of whom are in sub-Saharan Africa.
5. Achieve gender equality, and empower all women and girls.	Gender inequalities have been reduced but persist in all regions, as discussed in Sections 10.5 and 10.6.
6. Ensure access to water and sanitation for all.	According to the U.N., 884 million people lack access to safe drinking water, and 4.2 billion lack access to safely managed sanitation systems.
7. Ensure access to affordable, reliable, sustainable, and modern energy for all.	According to the International Energy Agency, 940 million people lack access to electricity, and fossil fuels are the principal contributor to climate change.
8. Promote inclusive and sustainable economic growth, employment, and decent work for all.	The U.N. wants an increase in GDP per capita in developing countries of at least 7 percent per year.
9. Build resilient infrastructure, promote sustainable industrialization, and foster innovation.	In addition to electricity and sanitation issues cited in earlier goals, the U.N. also estimates that 1.5 billion people lack access to reliable phones.
10. Reduce inequality within and among countries.	Income inequality has increased since 1990 within and between countries. Refer to figures 1.10.4 and 10.4.2.
11. Make cities inclusive, safe, resilient, and sustainable.	Rapid urbanization has put pressure on water, sewerage, and health; The U.N. estimates that around 1 billion people live in what the U.N. defines as informal settlements. Refer to Section 13.6.1.
12. Ensure sustainable consumption and production patterns.	The U.N. estimates that 30 percent of food rots or spoils because of poor transportation or harvesting methods.
13. Take urgent action to combat climate change and its impacts.	It is not too late to reverse the recent sharp increase in carbon dioxide emissions that is producing climate change. Refer to Section 14.5.
14. Conserve and sustainably use the oceans, seas, and marine resources.	Progress has been made in water quality, but overfishing is a problem. Refer to Section 9.7.2.
15. Sustainably manage forests, combat desertification, halt and reverse land degradation, and halt biodiversity loss.	Drought, desertification, and deforestation are negatively impacting agriculture and biodiversity. Refer to Section 9.10.
16. Promote just, peaceful, and inclusive societies.	The World Economic Forum estimates that the global cost of corruption, bribery, theft, and tax evasion is $2.6 trillion annually.
17. Revitalize the global partnership for sustainable development.	Overall, financial aid has increased from developed countries to developing countries.

10.10.2 Measuring Progress in Development

Since the U.N. began measuring HDI in 1980, both developed and developing regions have made progress in the HDI. The gap in HDI between developed countries and developing countries has narrowed since 1980 (Figure 10.10.2). However, progress varies among the three factors contributing to the HDI.

- **GNI per capita.** Since 1980, GNI per capita has increased much more rapidly in developed countries than in developing countries (refer to Figure 1.10.4).
- **Life expectancy.** Since 1980, life expectancy has increased by around the same number of years in developed and developing countries (Figure 10.10.3).
- **Education.** Since 1980, mean years of education has increased by around the same number in developed and developing countries (Figure 10.10.4).

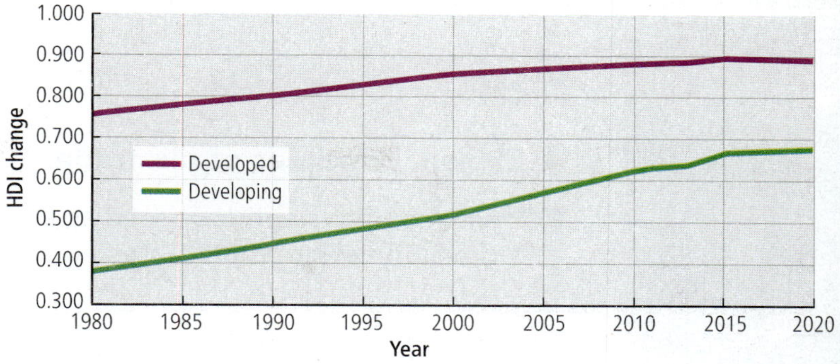

▲ **10.10.2** CHANGE IN HDI

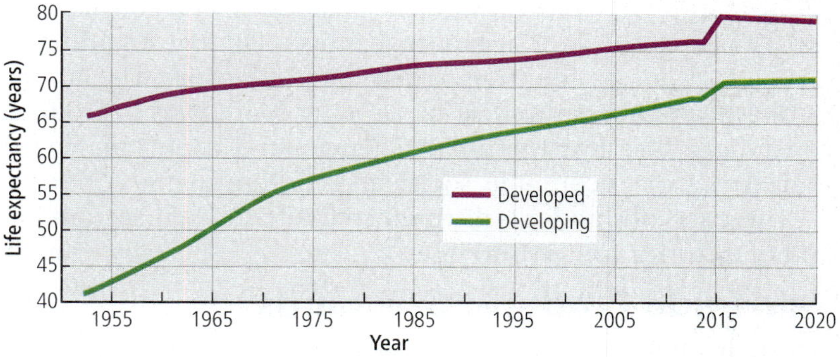

▲ **10.10.3** CHANGE IN LIFE EXPECTANCY

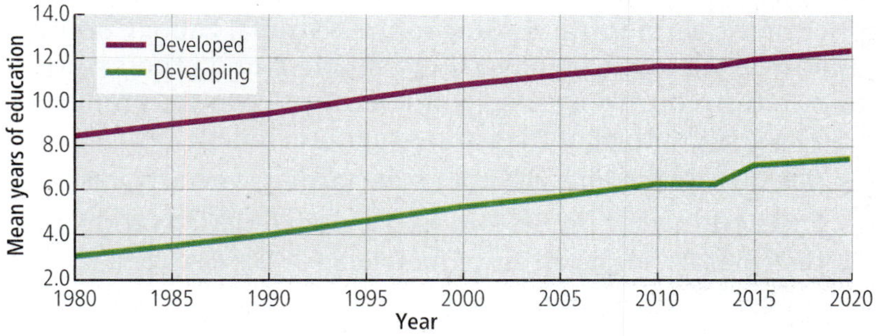

▲ **10.10.4** CHANGE IN EDUCATION

Planetary Pressures-Adjusted Human Development Index

The annual U.N. Human Development Report added a new index in 2020 called the **Planetary pressures-adjusted Human Development Index (PHDI)**. The PHDI modifies the HDI by accounting for two resource factors: the carbon dioxide emissions per person and the amount of raw materials extracted per person to meet demand for goods and services.

When the two resource factors are factored into the HDI score, most of the highly developed countries fall to lower ranks (Figure 10.10.5). For example, the United States ranks sixteenth in its level of development among the world's countries, according to its HDI, but falls to 62nd when factoring in the country's relatively high carbon dioxide emissions and consumption of raw materials. Canada's ranking drops by a similar level. The U.N. argues that further progress in human development is impossible without addressing environmental concerns.

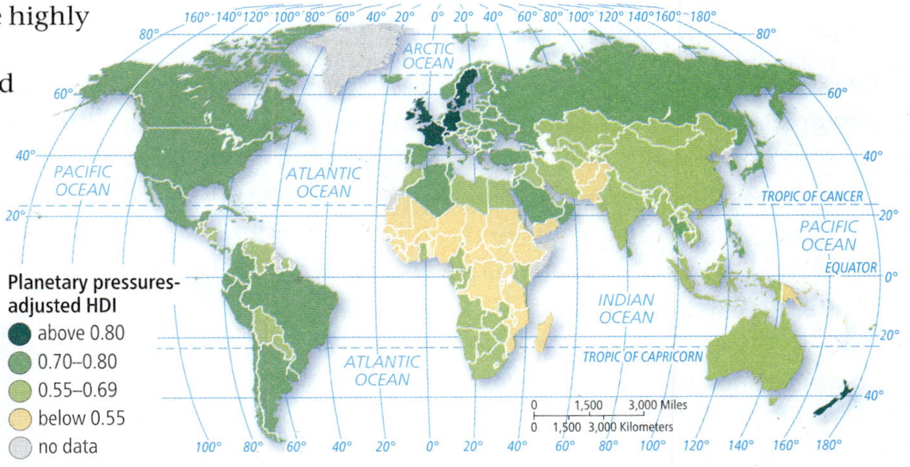

▲ **10.10.5** PLANETARY PRESSURES-ADJUSTED HUMAN DEVELOPMENT INDEX

▲ **10.11.1 FAIR TRADE TEXTILE FACTORY**
Tiruppur, India.

Fair Trade

- **Explain principles of fair trade.**

Fair trade has been proposed as a variation of the international trade model of development that promotes equity and sustainability.

10.11.1 Fair Trade

Fair trade is international trade that provides greater equity to workers and small businesses in developing countries. Fair trade products are made and traded according to standards that protect workers, businesses, and consumers.

The fair trade movement focuses primarily on products exported from developing countries to developed countries. The fair trade movement promotes sustainable development by offering higher pay and better working conditions in producing countries. Fair trade organizations, backed by consumers, raise awareness of deficiencies in conventional international production and trade, and they use fair trade to improve economic, social, and environmental conditions for producers and workers.

Three sets of standards distinguish fair trade: One set applies to producers, a second set to workers on farms and in factories, and a third set to consumers, especially those living in developed countries.

Fair Trade Producer Standards

Fair trade includes a set of producer standards designed to advance a number of economic, social, and environmental goals. Critics of international trade charge that only a tiny percentage of the price a consumer pays for a good reaches the individual in the developing country who is responsible for making or growing it. Someone in India sewing clothing for the U.S. market, for example, earns less than 1 percent of the retail price of the garment, according to the National Labor Committee (Figure 10.11.1). The rest goes to wholesalers, importers, distributors, advertisers, retailers, and others who did not actually make the item. In contrast, fair trade returns on average one-third of the price to the producer in the developing country.

The goals of the fair trade movement include:

- Raising the incomes of small-scale farmers and artisans by eliminating some of the intermediate steps.
- Distributing the profits and risks associated with production and sale of goods more fairly among producers, distributors, retailers, and financiers.
- Increasing the entrepreneurial and management skills of the producers.
- Promoting safe and sustainable farming methods and working conditions, such as by prohibiting the use of dangerous pesticides and herbicides and promoting the production of certified crops.

International fair trade organizations set standards and earn certification for implementing these principles, and they monitor, audit, and certify that practices comply with the standards.

Cooperatives

Many small-scale farmers and artisans join democratically managed cooperatives. A **cooperative** is an association of persons united voluntarily to meet their common economic, social, and cultural needs and aspirations through a jointly owned enterprise (Figure 10.11.2). Cooperatives offer several advantages:

- The cooperative can qualify for credit so that funds can be borrowed to buy equipment and invest in improving farms.
- Materials can be purchased at a lower cost.
- The people who grow or make the products democratically manage allocation of resources and assure safe and healthy working conditions.
- Profits are reinvested in the community instead of going to absentee corporate owners.

▶ **10.11.2 COOPERATIVELY OWNED GROCERY STORE**
Portland, Maine.

10.11.2 Fair Trade Worker Standards

Protection of workers' rights is not a high priority in the international trade development approach, according to its critics. Critics charge that:

- Oversight of workers' conditions by governments and international lending agencies is minimal.
- Workers allegedly work long hours in poor conditions for low pay.
- Children or forced labor may be in the workforce.
- Health problems may result from poor sanitation and injuries from inadequate safety precautions.
- Injured, ill, or laid-off workers are not compensated.

▲ 10.11.3 FAIR TRADE FOR WORKERS: DISTINCTIVE CLOTHING IN BIHOR, ROMANIA

In contrast fair trade requires that workers be:

- Paid fair wages, at least enough to cover food, shelter, education, health care, and other basic needs.
- Permitted to organize a union and to have the right to collective bargaining.
- Protected by high environmental and safety standards.

Fair Trade for Bihor

For many generations, women in Bihor, Romania, have practiced economic self-sufficiency in part by making distinctive clothing for their community's local consumption. To design high-priced clothing for international consumption, big fashion houses get much of their inspiration from local traditions. Dior, one of the world's major fashion houses, recently created an embroidered coat that copied one made in Bihor for more than a century. Dior priced its copy of a Bihor coat at $40,000.

The Bihor women fought back. With the help of a Romanian magazine, they made a video documenting Dior's theft and established a local company to create and sell their locally made clothes (Figure 10.11.3).

Fair Trade & Consumers

Most fair trade sales are in food, including coffee, tea, banana, chocolate, cocoa, juice, sugar, and honey products (Figure 10.11.4). In North America, fair trade products also include craft products such as decorative home accessories, jewelry, textiles, and ceramics. Ten Thousand Villages is the largest fair trade organization in North America, specializing in handicrafts.

Buying fair trade products helps consumers connect more directly with the producers of the food, clothing, and household items that they buy. Fair trade products do not necessarily cost the consumer more than conventionally grown or produced alternatives. Because fair trade organizations bypass exploitative intermediaries and work directly with producers, they are able to cut costs and return a greater percentage of the retail price to the producers.

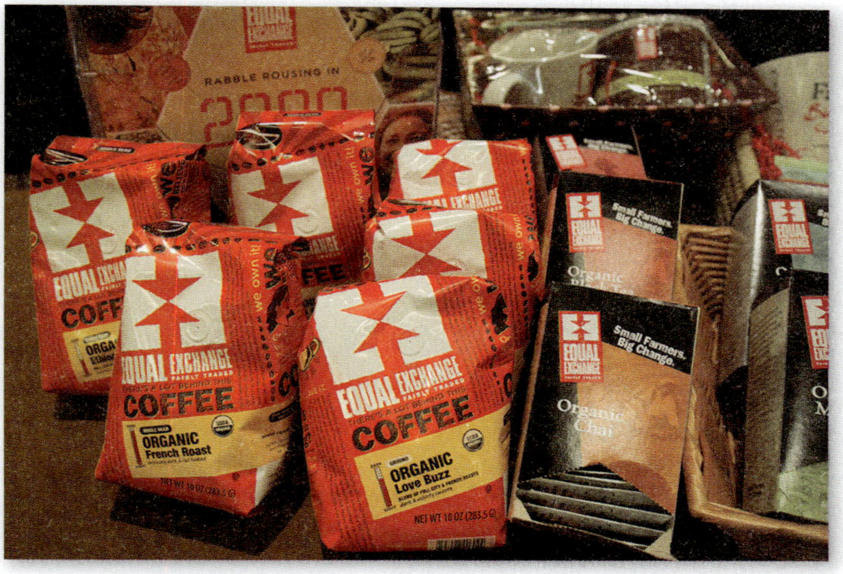

▲ 10.11.4 FAIR TRADE COFFEE

CHAPTER 10

Review, Analyze, & Apply

KEY ISSUE 1 How Does Development Vary Among Countries?

- The Human Development Index (HDI) measures the level of development of each country.
- HDI is based on three factors: a decent standard of living, a long and healthy life, and access to knowledge.

THINKING GEOGRAPHICALLY

Which indexes of development presented in this chapter (HDI, IHDI, MPI, GDI, GII, PHDI) do you think best measures a country's level of development? Which do you think is least successful at measuring a country's level of development? Why?

◀ **10.CR.1** SCHOOL, VRINDĀVAN, INDIA

KEY ISSUE 2 What Inequalities Are Found in Development?

- The Inequality-adjusted Human Development Index (IHDI) measures the extent of economic inequality in a country.
- The Multidimensional Poverty Index (MPI) measures the variables that contribute to poverty in a country.
- The Gender Inequality Index (GII) and Gender Development Index (GDI) measure the extent of inequality between men and women.

THINKING GEOGRAPHICALLY

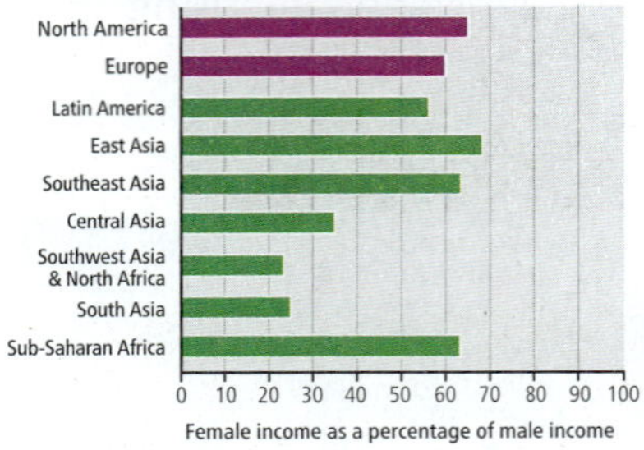

Sub-Saharan Africa and South Asia appear to have the highest levels of gender inequality. What factors discussed in earlier chapters might be contributing to the high level of gender inequality in these two regions? Why?

◀ **10.CR.2** FEMALE INCOME AS A PERCENTAGE OF MALE INCOME BY REGION

KEY ISSUE 3 How Do Countries Become More Developed?

- Two paths to development are self-sufficiency and international trade.
- International trade has been embraced by most countries in recent decades.
- The world-systems theory suggests that developed countries are located at the core of international trade and developing countries at the periphery.
- Development is financed through foreign direct investment as well as loans from international organizations.

THINKING GEOGRAPHICALLY

Although the international trade path has proved more successful at promoting development, developed countries like the United States and the United Kingdom have taken steps to reduce their commitment to international trade. What reason might be behind this movement?

◀ **10.CR.3** BREXIT SUPPORTER

KEY ISSUE 4 How Can Development Be More Sustainable?

- Progress has been made in achieving development in most regions, but substantial gaps in performance between developed and developing countries persist.
- The Planterary pressures-adjusted HDI (PHDI) modifies the HDI for a country's use of resources.
- Fair trade is an alternative to the international trade model that provides greater equity to workers, small businesses, and consumers.

THINKING GEOGRAPHICALLY

Do you make purchase decisions on whether the producer or the shop promotes fair trade? Why or why not?

◀ **10.CR.4** FAIR TRADE SHOP, CANTERBURY, UNITED KINGDOM

Explore

The World's Lowest HDI

Central African Republic has the world's lowest HDI score. Let's examine its development possibilities. In Google Earth, ***search*** for *Bayanga, Central African Republic*

1. Describe the image that Google Earth supplies for Bayanga.
2. Does Google Earth offer Street View for Bayanga? Why do you think this is the situation?
3. ***Zoom*** in to around 5 km. What transportation features are visible? How might these features help Bayanga become more economically developed?

▼ **10.CR.5** CENTRAL AFRICAN REPUBLIC

Key Terms

Adolescent fertility rate (10.5.2) The number of births per 1,000 women ages 15 to 19.

Cooperative (10.11.1) An association of persons united voluntarily to meet their common economic, social, and cultural needs and aspirations through a jointly owned enterprise.

Developed country (10.1.1) A country that has progressed relatively far along a continuum of development.

Developing country (10.1.1) A country that is at a relatively early stage in the process of development.

Development (10.1.1) A process of improvement in the conditions of people through diffusion of knowledge and technology.

Fair trade (10.11.1) A variation of international trade that provides greater equity to workers, small businesses, and consumers, focusing primarily on products exported from developing countries to developed countries.

Female labor force participation rate (10.5.2) The percentage of women holding full-time jobs outside the home.

Foreign direct investment (FDI) (10.9.1) Investment made by a foreign company in the economy of another country.

Gender Development Index (GDI) (10.6.1) An index constructed by the United Nations that modifies the Human Development Index to measure the gender gap in level of achievement in terms of income, education, and life expectancy.

Gender Inequality Index (GII) (10.5.1) An index constructed by the United Nations to measure the extent of each country's gender inequality in terms of reproductive health, empowerment, and the labor market.

Gender Social Norms Index (GSNI) (10.6.2) An index constructed by the United Nations to measure the strength of attitudes biased against women.

Gross domestic product (GDP) (10.2.1) The value of the total output of goods and services produced in a country in a year, not accounting for money that leaves and enters the country.

Gross national income (GNI) (10.2.1) The value of the output of goods and services produced in a country in a year, including money that leaves and enters the country.

Gross national income per capita at purchasing power parity (10.2.1) The value of the output of goods and services produced in a country in a year, including money that leaves and enters the country, adjusted to account for differences among countries in the cost of goods.

Human Development Index (HDI) (10.1.1) An index constructed by the United Nations to measure the level of development for a country through a combination of income, education, and life expectancy.

Inequality-adjusted Human Development Index (IHDI) (10.4.1) An index constructed by the United Nations that modifies the Human Development Index to account for inequality.

Literacy rate (10.3.2) The percentage of a country's people who can read and write.

Microfinance (10.9.2) Provision of small loans and financial services to individuals and small businesses in developing countries.

Millennium Development Goals (10.10.1) Eight goals adopted by the United Nations in 2002 to reduce disparities between developed and developing countries by 2015.

Multidimensional Poverty Index (MPI) (10.4.2) An index constructed by the United Nations that modifies the Human Development Index to measure the complexities of the lives of people living in poverty.

Planetary pressures-adjusted Human Development Index (PHDI) (10.10.2) An index constructed by the United Nations that modifies the Human Development Index to account for a country's carbon dioxide emissions per person and the amount of raw materials extracted per person to meet the country's demand for goods and services.

Primary sector (10.2.2) The portion of the economy concerned with the direct extraction of materials from Earth, generally through agriculture.

Productivity (10.2.2) The value of a particular product compared to the amount of labor needed to make it.

Pupil/teacher ratio (10.3.2) The number of enrolled students divided by the number of teachers.

Purchasing power parity (PPP) (10.2.1) The amount of money needed in one country to purchase the same goods and services in another country.

Secondary sector (10.2.2) The portion of the economy concerned with manufacturing useful products through processing, transforming, and assembling raw materials.

Structural adjustment program (10.9.2) Economic policies imposed on less developed countries by international agencies to create conditions that encourage international trade.

Sustainable Development Goals (10.10.1) Seventeen goals adopted by the United Nations in 2015 to reduce disparities between developed and developing countries by 2030.

Tertiary sector (10.2.2) The portion of the economy concerned with transportation, communications, and utilities, sometimes extended to the provision of all goods and services to people in exchange for payment.

World-systems theory (10.8.2) Theory developed by Immanuel Wallerstein that in an increasingly unified world economy developed countries form an inner core area whereas developing countries are found on the periphery and semi-periphery areas.

Geospatial Analysis

Log in to the Mastering Geography Study Area to access MapMaster 2.0.

HDI in Southwest Asia

The HDI for Southwest Asia & North Africa overall ranks in the middle among developing regions. The region's mid-range HDI masks wide variations among the region's individual countries and factors contributing to the HDI.

Add the ***Human Development Index (HDI)*** layer. Select the *Settings* icon from the *Legend*, and select *Show Political Labels.*

Add the ***Per Capita Gross Domestic Product*** layer, and select *Split map window*. Zoom to Southwest Asia and North Africa.

1. Do most of the countries with relatively high HDI have GDP that is relatively low or average?
2. Remove ***Per Capita Gross Domestic Product*** layer. Add ***Life Expectancy,*** and select *Split map window*. Do the countries with the highest HDI have the highest life expectancy? What might account for this situation?
3. Remove the ***Life Expectancy*** layer. Add the ***Literacy Rate*** layer, and select *Split map window*. Do the countries with the highest HDI have the highest literacy rate? What might account for this situation?

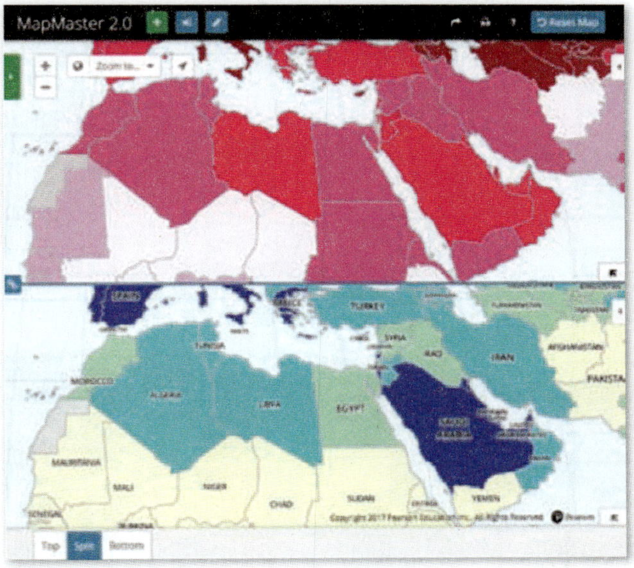

▶ **10.CR.6** HDI AND LITERACY RATES, SOUTHWEST ASIA & NORTH AFRICA

CHAPTER

11 Industry

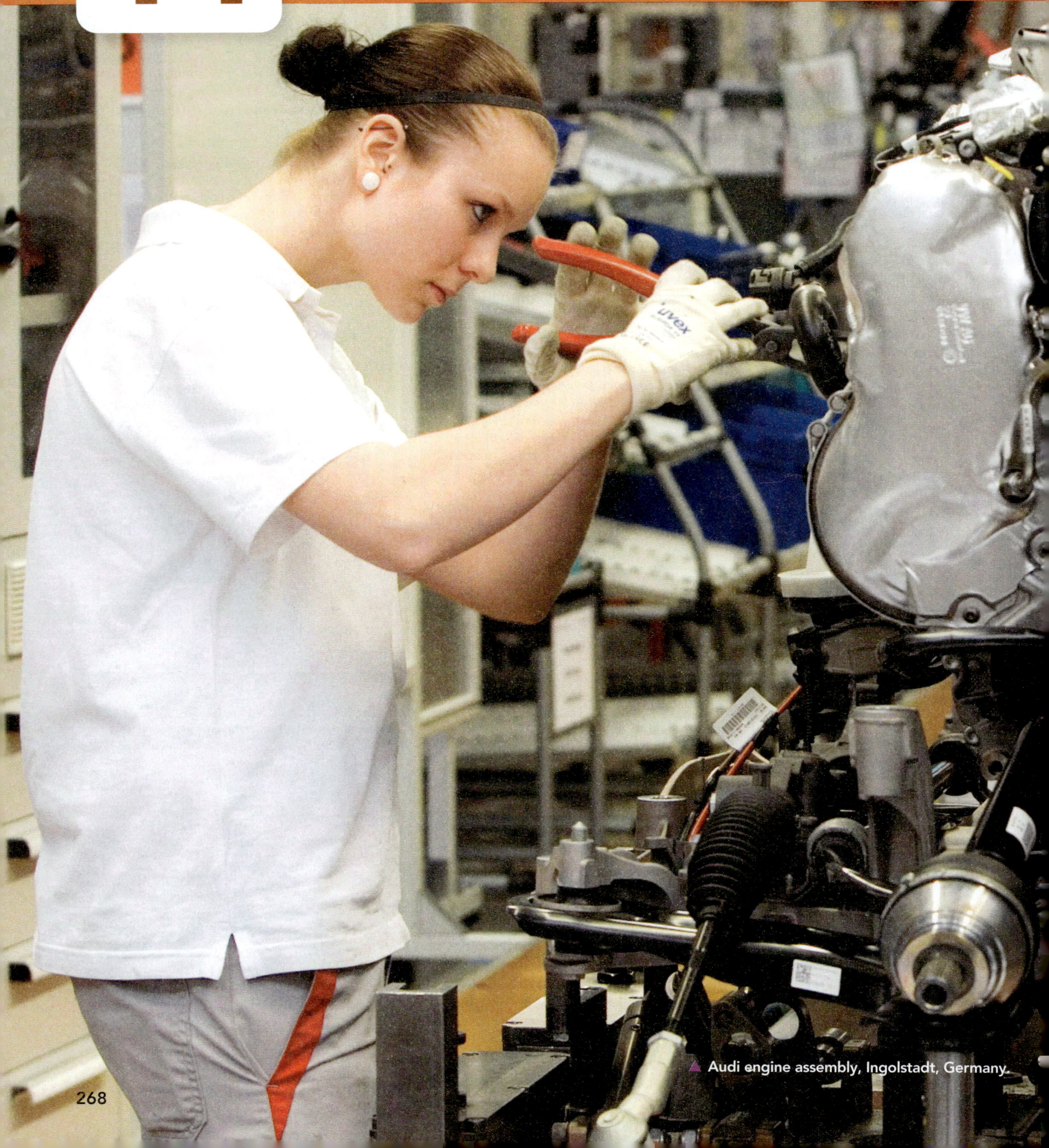

Audi engine assembly, Ingolstadt, Germany.

Industry was once highly clustered in a handful of communities within a handful of developed countries. But industry has diffused to many communities in an increasing number of developing countries. Manufacturing jobs are viewed as a special asset to communities and entire countries around the world. Local residents mourn when factories close and rejoice when new ones open.

KEY ISSUES

1 Where Is Industry Distributed?

Industry is clustered in three regions: Europe, East Asia, and North America. Europe was the first region to industrialize, and East Asia is the fastest growing.

2 Why Are Site Factors Important for Industry?

Site factors are related to the costs of production inside a factory. The three site factors that may vary among locations are labor, land, and capital.

3 Why Are Situation Factors Important for Industry?

Situation factors involve transporting materials to and from a factory. Some industries locate factories near inputs, whereas others choose locations near markets.

4 How Has the Distribution of Industry Changed?

Manufacturing is increasing in some developing countries, attracted by low-wage labor. China has become the world's leading industrial country.

LOCATIONS IN THIS CHAPTER

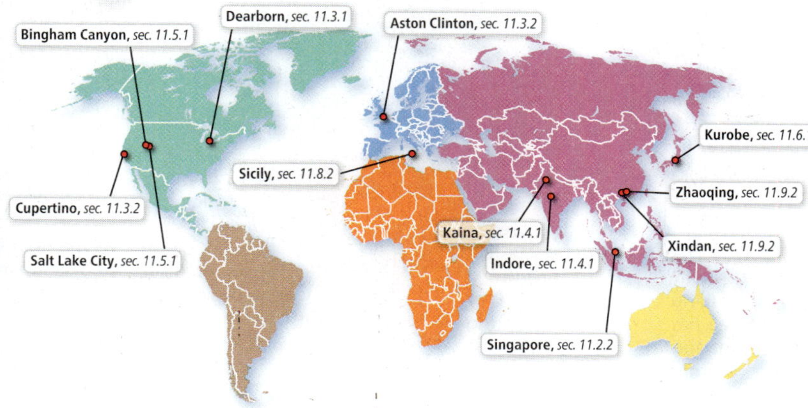

Industrial Regions

- **Describe the locations of the three principal industrial regions.**

The hearth of modern industry—meaning the manufacturing of goods in a factory—was in northern England and southern Scotland during the second half of the eighteenth century. From that hearth, industry diffused to Europe and to North America in the nineteenth century and to other regions in the twentieth century.

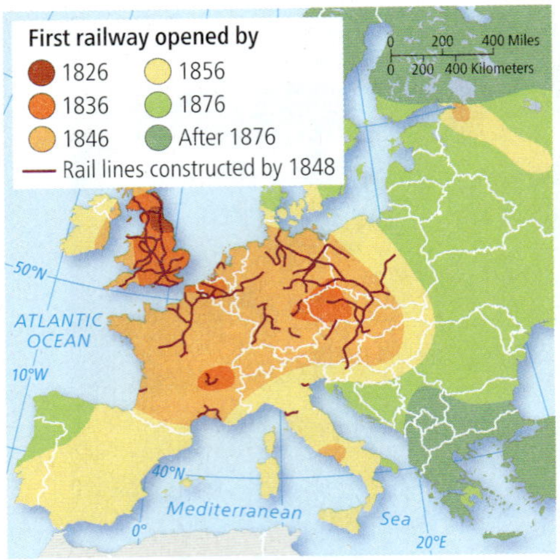

▲ 11.1.1 DIFFUSION OF THE INDUSTRIAL REVOLUTION ACROSS EUROPE

11.1.1 The Industrial Revolution

The Industrial Revolution was defined in Chapter 2 as a series of improvements in industrial technology that transformed the process of manufacturing goods. Prior to the Industrial Revolution, industry was geographically dispersed across the landscape. People made household tools and agricultural equipment in their own homes or obtained them in the local village. Home-based manufacturing was known as the **cottage industry** system.

The catalyst of the Industrial Revolution was technology, and several inventions transformed the way in which goods were manufactured. The invention most important to the development of factories was the steam engine, patented in 1769 by James Watt, a maker of mathematical instruments in Glasgow, Scotland. The large supply of steam power available from Watt's steam engines induced manufacturers to concentrate all their process steps in one building attached to a single power source. Transportation improvements played a critical role in diffusing the Industrial Revolution from its hearth in the United Kingdom to the European continent (Figure 11.1.1).

Europe: Nineteenth Century Industry Leader

Europe was the first region to industrialize, during the nineteenth century. Numerous industrial centers emerged in Europe as countries competed with each other for leadership (Figure 11.1.2).

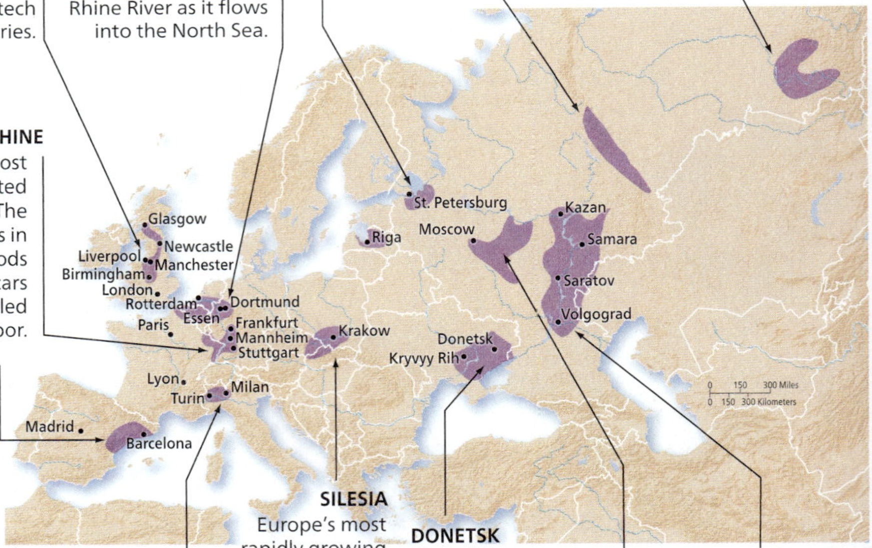

▶ 11.1.2 EUROPE'S INDUSTRIAL REGIONS

11.1.2 North America: Twentieth-Century Industry Leader

The United States became the world's leading industrial power in the late nineteenth century and retained that position through the twentieth century. North America's manufacturing was traditionally highly concentrated in northeastern United States and southeastern Canada (Figure 11.1.3). In recent years, manufacturing has relocated to southern U.S. states and to Mexico, lured by lower wages and legislation that has made it difficult for independent unions to organize factory workers (refer to Section 11.8).

▶ **11.1.3** NORTH AMERICA'S INDUSTRIAL AREAS

SOUTHEASTERN ONTARIO Canada's most important industrial area, central to the Canadian and U.S. markets and near the Great Lakes and Niagara Falls.

MOHAWK VALLEY Takes advantage of inexpensive electricity generated at nearby Niagara Falls.

NORTHERN CALIFORNIA A leading center for social media, biotech, software, and electronic devices.

NEW ENGLAND A cotton textile center in the early nineteenth century. Cotton was imported from southern states, and finished cotton products were shipped to Europe.

MIDDLE ATLANTIC The largest U.S. market, so attracts industries that need proximity to a large number of consumers and depend on foreign trade through one of this region's large ports.

SOUTHERN CALIFORNIA Now the country's largest area of clothing and textile production, the second-largest furniture producer, and a major food-processing center.

WESTERN GREAT LAKES Centered on Chicago, the hub of the nation's transportation network and now the center of steel production.

PITTSBURGH-LAKE ERIE The leading steel-producing area in the nineteenth century because of its proximity to Appalachian coal and iron ore.

East Asia: Twenty-First-Century Industry Leader

East Asia has become the most important industrial region in the world, beginning with Japan in the late twentieth century. In the twenty-first century, China is the world's leading manufacturing country by most measures (Figure 11.1.4).

▶ **11.1.4** EAST ASIA'S INDUSTRIAL AREAS

CHINA The world's leading manufacturer of many products, thanks to having the world's largest supply of low-cost labor and the world's largest market for consumer products. Manufacturers cluster in three areas along the east coast.

JAPAN An industrial power since the 1950s and 1960s, initially by producing goods that could be sold in large quantities at cut-rate prices to consumers in other countries, now by manufacturing high-quality electronic products.

SOUTH KOREA Has followed Japan's lead in focusing on export-oriented manufacturers, such as cars, electronics, and steel. The country is a leading producer of oceangoing ships.

Shipping Raw Materials & Products

▲ **11.2.1 SHIP BY AIR**
FedEx Express.

- **Explain why industries use different modes of transportation.**

Industries ship their raw materials and products in one of four ways: truck, train, boat, or airplane. Manufacturers seek the lowest-cost mode of transportation, but the most appropriate alternative depends on what is being shipped and how far it is being shipped.

11.2.1 Ship by Boat, Rail, Truck, or Air?

In general, the farther something is transported, the lower the cost per kilometer (or mile). Longer-distance transportation is cheaper per kilometer in part because manufacturers must pay workers to load goods on and off vehicles, whether the material travels 10 kilometers or 10,000 kilometers. The cost per kilometer decreases at different rates for each of the four modes because the loading and unloading expenses differ for each mode.

Airplanes for small & valuable packages. Airplanes are the most expensive transport option for all distances, so airplane shipment is usually reserved for speedy delivery of small-bulk, high-value packages (Figure 11.2.1).

Trucks for short-distance delivery. Trucks are most often used for short-distance delivery because they can be loaded and unloaded quickly and cheaply. Truck delivery is especially advantageous if the driver can reach the destination within one day before having to stop for an extended rest (Figure 11.2.2).

Trains for cross-country shipment. Trains are most often used to ship to destinations that take longer than one day to reach, such as between the East and West coasts of the United States. Loading trains takes longer than loading trucks, but once underway, trains aren't required to make daily rest stops like trucks (Figure 11.2.3).

Boats for shipping across oceans. Boats are attractive for transport over very long distances because the cost per kilometer is very low. Boats are slower than land-based transportation, but unlike trains or trucks, boats can cross oceans, such as to North America from Europe or Asia (Figure 11.2.4).

▲ **11.2.2 SHIP BY TRUCK**
The highest volumes of freight truck traffic in the United States are in the East, the country's most densely populated area.

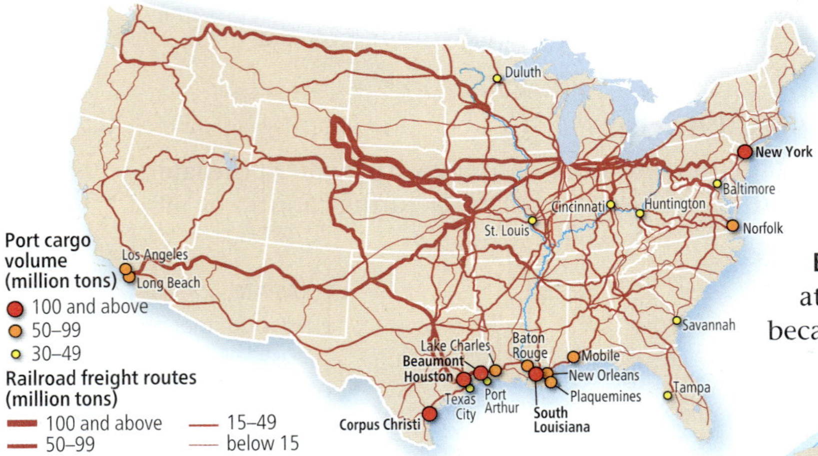

▲ **11.2.3 SHIP BY TRAIN**
Most long-distance rail traffic in the United States moves east-west.

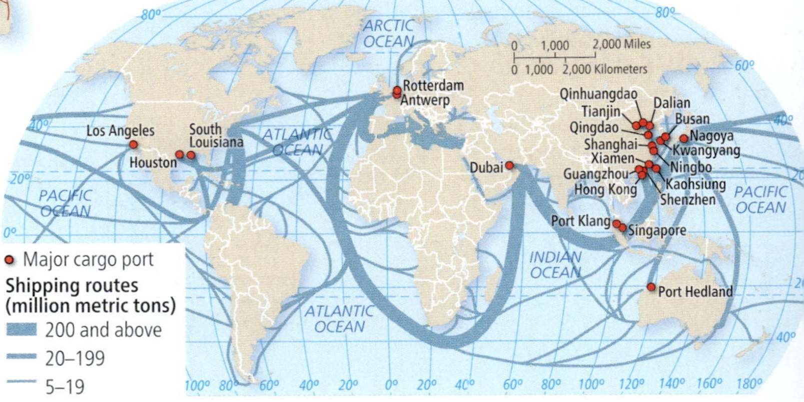

▶ **11.2.4 SHIP BY BOAT**
The world's largest cargo ports are in Asia, and the leading routes are from Asia to Europe and North America.

11.2.2 Break-of-Bulk Points

Mixed modes of delivery are often used. For example, air freight companies pick up packages in the afternoon and transport them by truck to the nearest airport. Late at night, planes filled with packages are flown to a central hub airport in the interior of the country, such as Memphis, Tennessee, or Louisville, Kentucky. The packages are transferred to other planes, flown to airports nearest their destination, transferred to trucks, and delivered the next morning.

Many companies that use multiple transport modes locate at a **break-of-bulk point**, which is a location where transfer among transportation modes is possible. Important break-of-bulk points include seaports and airports. For example, a steel mill in Gary, Indiana, may receive iron ore by ship via Lake Michigan and coal by train from Appalachia.

Containerization has facilitated transfer of packages between modes. Containers may be packed into a rail car, transferred quickly to a container ship to cross the ocean, and unloaded onto trucks at the other end (Figure 11.2.5). Large ships have been specially built to accommodate large numbers of rectangular box-like containers.

▲ 11.2.5 CONTAINERS LOADED ONTO A SHIP, SINGAPORE

Regardless of transportation mode, cost rises each time inputs or products are transferred from one mode to another. For example, workers must unload goods from a truck and then reload them onto a plane. The company may need to build or rent a warehouse to store goods temporarily after unloading from one mode and before loading to another mode. Some companies may calculate that the cost of one mode is lower for some inputs and products, whereas another mode may be cheaper for other goods.

Just-in-Time Delivery

Just-in-time delivery is the shipment of parts and materials to arrive at a factory shortly before they are needed. Just-in-time delivery is especially important for delivery of inputs, such as parts and raw materials, to manufacturers of fabricated products, such as cars and computers.

Under a just-in-time system, parts and materials arrive at a factory frequently, in many cases daily or even hourly (Figure 11.2.6). Suppliers of the parts and materials are told a few days in advance how much will be needed over the next week or two, and first thing each morning, they are told exactly what will be needed at precisely what time that day. To meet a tight timetable, a supplier of parts and materials must locate factories near its customers. If given only an hour or two of notice, a supplier has no choice but to locate a factory within around one hour of the customer.

Just-in-time delivery reduces the money that a manufacturer must dedicate to the storage of inventory. As a result, manufacturers also reduce the size of the factory because space does not have to be wasted on piling up a mountain of inventory. Leading computer manufacturers have eliminated inventory altogether. They build computers only in response to customer orders placed primarily over the internet or by telephone. Just-in-time delivery, however, means that producers have lower inventory to cushion against disruptions in the arrival of needed parts.

Four kinds of disruptions can result from reliance on just-in-time delivery:

- **Natural hazards.** Poor weather conditions can affect deliveries anywhere in the world. For example, blizzards can close highways, rail lines, and airports.
- **Traffic.** Deliveries may be delayed when traffic is slowed by accident, construction, or unusually heavy volume.
- **Labor unrest.** A strike at one supplier plant can shut down the entire production within a couple of days.
- **Pandemic emergency.** The COVID-19 pandemic disrupted production and delivery of many items, such as computer chips.

▼ 11.2.6 JUST-IN-TIME DELIVERY

Site Factors in Industry

- **Analyze the three types of site factors in industrial location.**

Geographers try to explain why one location may be more profitable for a factory than others. A company ordinarily faces two geographic costs: site factors and situation factors.

▼ 11.3.1 HOURLY WAGES

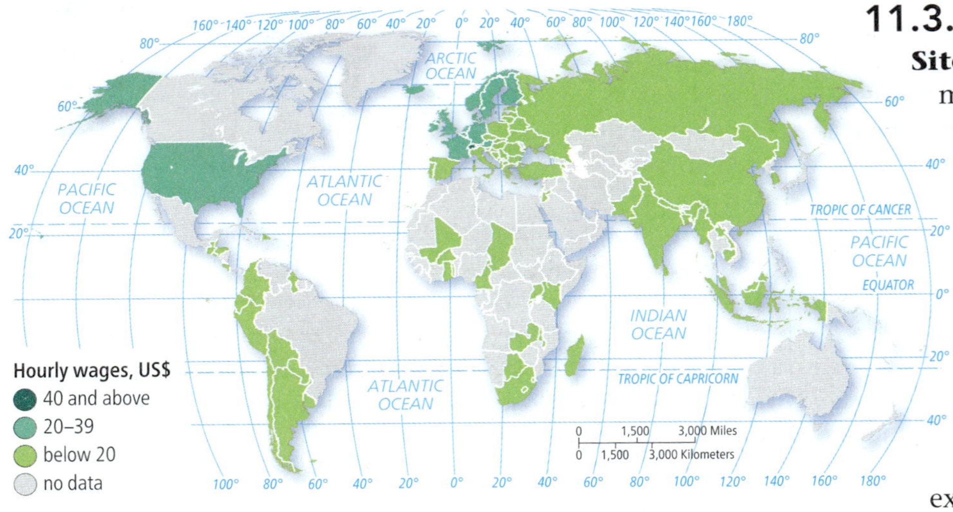

11.3.1 Site Factors: Labor

Site factors are related to the costs of manufacturing inside a plant. The three site factors that may vary among locations are labor, capital, and land.

The most important site factor on a global scale is labor. Minimizing labor costs is important for some industries, and the variation of labor costs around the world is large.

A **labor-intensive industry** is an industry in which wages and other compensation paid to employees constitute a high percentage of expenses. Labor constitutes an average of 11 percent of overall manufacturing costs in the United States, so a labor-intensive industry in the United States would have a much higher percentage than that. The reverse case, an industry with a much lower-than-average percentage of expenditures on labor, is considered capital intensive.

The average wage paid to manufacturing workers exceeds $20 per hour in many developed countries, compared with less than $1 per hour in many developing countries (Figure 11.3.1). For some manufacturers—but not all—the difference between paying workers $1 and $20 per hour is critical.

A labor-intensive industry is not the same as a high-wage industry. "Labor-intensive" is measured as a percentage, whereas "high-wage" is measured in dollars or other currencies. For example, motor vehicle workers are paid much higher hourly wages than textile workers, yet the textile industry is labor intensive, and the auto industry is not. Although auto workers earn relatively high wages, most of the value of a car is accounted for by the parts and the machinery needed to put together the parts. On the other hand, labor accounts for a large percentage of the cost of producing a towel or shirt compared with materials and machinery.

▼ 11.3.2 FORDIST PRODUCTION
Ford Motor Company, Dearborn, Michigan, 1913.

Fordist Production & Post-Fordist Production

Manufacturing centers in North America and Europe still attract industries that depend on skilled labor. Traditionally, factories assigned each worker one task to perform repeatedly. Some geographers call this approach **Fordist production**, or mass production, because Ford Motor Co. was one of the first companies to organize its production this way early in the twentieth century (Figure 11.3.2).

Many industries now follow a lean, or flexible, production approach. The term **post-Fordist production** is sometimes used to describe flexible production, in contrast with Fordist production. Post-Fordist flexible production organizes workers into teams that perform a variety of tasks and solve problems through consensus. Increasingly, factory workers need skills such as computer literacy, and many now have college degrees.

11.3.2 Site Factors: Capital

Manufacturers typically borrow capital—the funds to establish new factories or modernize existing ones. The U.S. motor vehicle industry concentrated in Michigan early in the twentieth century largely because that region's financial institutions were more willing than eastern banks to lend money to the industry's pioneers.

The most important factor in the clustering of high-tech industries in California's Silicon Valley—even more important than proximity to skilled labor—has been the availability of capital (Figure 11.3.3). Financial institutions in Silicon Valley have long been willing to provide money for new software and communications firms, even though lenders elsewhere have hesitated. High-tech industries have been risky propositions—roughly two-thirds of them fail—but Silicon Valley financial institutions have continued to lend money to engineers who have good ideas so that they can buy the software, communications, and networks they need to get started. One-fourth of all capital in the United States is spent on new industries in Silicon Valley.

▲ **11.3.3 APPLE HEADQUARTERS**
Cupertino, California.

The ability to borrow money has become a critical factor in the distribution of industry in developing countries. Financial institutions in many developing countries are short of funds, so new industries must seek loans from banks in developed countries. But enterprises may not get loans if they are located in a country that is perceived to have an unstable political system, a high debt level, or ill-advised economic policies.

Site Factors: Land

Land suitable for constructing a factory can be found in many places. If considered to encompass energy and other natural resources in addition to terra firma, "land" is an especially critical site factor.

In the past, most factories located inside cities. To get the necessary space in cities, early factories were typically multistory. Raw materials were hoisted to the upper floors to make smaller parts, which were then sent downstairs on chutes and pulleys for final assembly and shipment.

Contemporary factories operate most efficiently when laid out in one-story buildings. Raw materials are typically delivered at one end and moved through the factory on conveyors or forklift trucks. Products are assembled in logical order and shipped out at the other end. The land needed to build one-story factories is now more likely to be available in suburban and rural locations. Also, land is much cheaper in suburban and rural locations than near the center of a city.

In addition to providing enough space for one-story buildings, locations outside cities are also attractive because they facilitate delivery of inputs and shipment of products (Figure 11.3.4). In the past, when most material moved in and out of a factory by rail, a central location was attractive because rail lines converged there. With trucks now responsible for transporting most inputs and products, proximity to major highways is more important for a factory. Especially attractive is the proximity to the junction of a long-distance route and the beltway, or ring road, that encircles most cities. Thus, factories cluster in industrial parks located near suburban highway junctions.

▶ **11.3.4 ONE-STORY FACTORY IN A RURAL AREA**
Aston Clinton, United Kingdom.

11.4 **KEY ISSUE 2** Why are site factors important for industry?

▲ **11.4.1** COTTON SPINNING
Indore, India.

Changing Site Factors: Clothing

- **Explain the distribution of clothing production.**

Production of **textiles** (woven fabrics) and **apparel** (clothing) is a prominent example of a labor-intensive industry that generally requires less-skilled, low-cost workers. The textile and apparel industry accounts for 6 percent of the dollar value of world manufacturing but a much higher 14 percent of world manufacturing employment, an indicator that it is a labor-intensive industry. More than three-fourths of the employees in the world's textile and apparel industries are women.

▲ **11.4.2** COTTON WEAVING
Kalna, India.

11.4.1 Textile & Apparel Production

Textile and apparel production involves three principal steps: spinning of fibers to make yarn (Figure 11.4.1), weaving or knitting of yarn into fabric (Figure 11.4.2), and assembly of fabric into products. Spinning, weaving, and sewing are all labor intensive compared to other industries, but the importance of labor varies somewhat among them. As a result, their global distributions are not identical because the three steps are not equally labor intensive.

Spinning

Fibers can be spun from natural or synthetic elements. The principal natural fiber is cotton, and various synthetics now account for three-fourths of world thread production. Because it is a labor-intensive industry, spinning is done primarily in low-wage countries. China produces 41 percent and India 23 percent of the world's cotton thread (Figure 11.4.3).

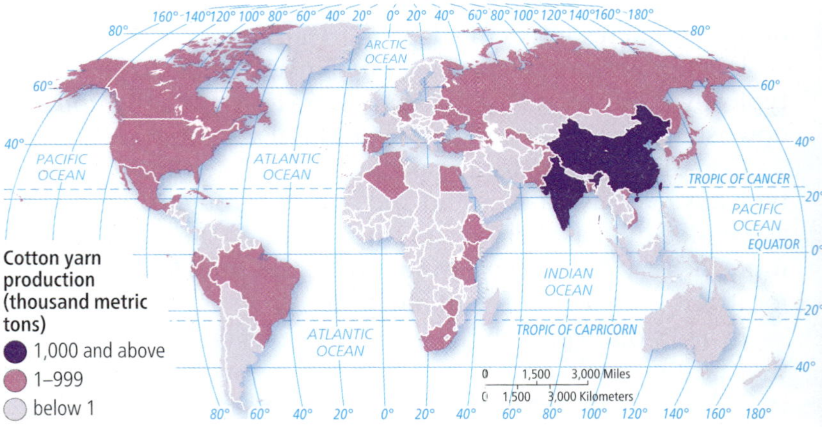

▲ **11.4.3** DISTRIBUTION OF COTTON YARN PRODUCTION

Weaving

For thousands of years, fabric has been woven or laced together by hand on a loom, which is a frame on which two sets of threads are placed at right angles to each other. One set of threads, called the warp, is strung lengthwise. A second set of threads, called the weft, is carried widthwise in a shuttle that is inserted over and under the warp. For mechanized weaving, labor constitutes a high percentage of the total production cost.

Consequently, weaving is highly clustered in low-wage countries (Figure 11.4.4). Despite their remoteness from European and North American markets, China and India have become the dominant fabric producers because their lower labor costs offset the expense of shipping inputs and products long distances. China accounts for nearly 60 percent of the world's woven cotton fabric production, and India accounts for 30 percent.

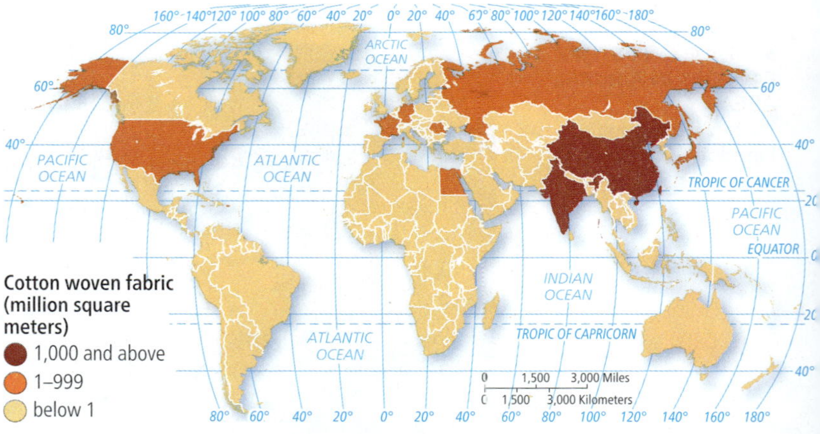

▶ **11.4.4** DISTRIBUTION OF COTTON WEAVING PRODUCTION

11.4.2 Assembly of Clothing

Sewing by hand is a very old human activity. Needles made from animal horns or bones date back tens of thousands of years, and iron needles date from the fourteenth century. The first functional sewing machine was invented by French tailor Barthelemy Thimonnier in 1830. Isaac Singer manufactured the first commercially successful sewing machine in the United States during the 1850s.

▲ 11.4.5 MADE IN CHINA

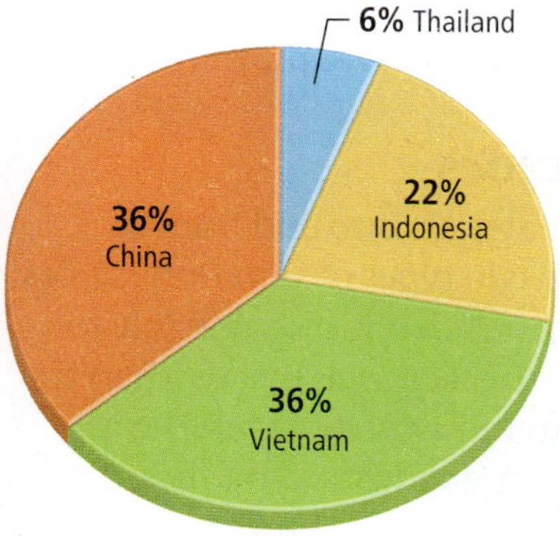

▲ 11.4.6 DISTRIBUTION OF NIKE SHOE FACTORIES

Textiles are cut and sewn to be assembled into four main types of products: garments, carpets, home products such as bed linens and curtains, and industrial items such as headliners for motor vehicles. Developed countries play a larger role in assembly than they do in spinning and weaving because most of the consumers of assembled products are located in developed countries. Nonetheless, most clothing is assembled in developing countries (Figure 11.4.5).

Overall production costs are generally lower in developing countries because substantially lower labor costs than in developed countries offset higher shipping and taxation costs (Figure 11.4.6). Most of the cost of apparel sold in developed countries is markup by the retailer (Figure 11.4.7). Workers in developing countries earn only around 1 percent of the final cost to the consumer.

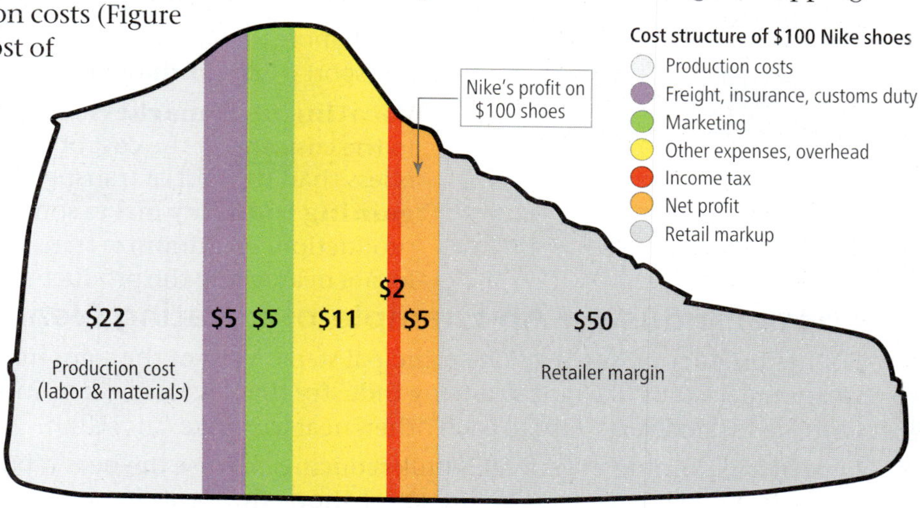

▶ 11.4.7 COST STRUCTURE OF $100 NIKE SHOES

Trade in Clothing

While most textiles and apparel are manufactured in developing countries, a large percentage of the customers are located in developed countries, especially in Europe and North America. Consequently, there are substantial flows of clothing shipped from East Asia and Southeast Asia to Europe and North America (Figure 11.4.8).

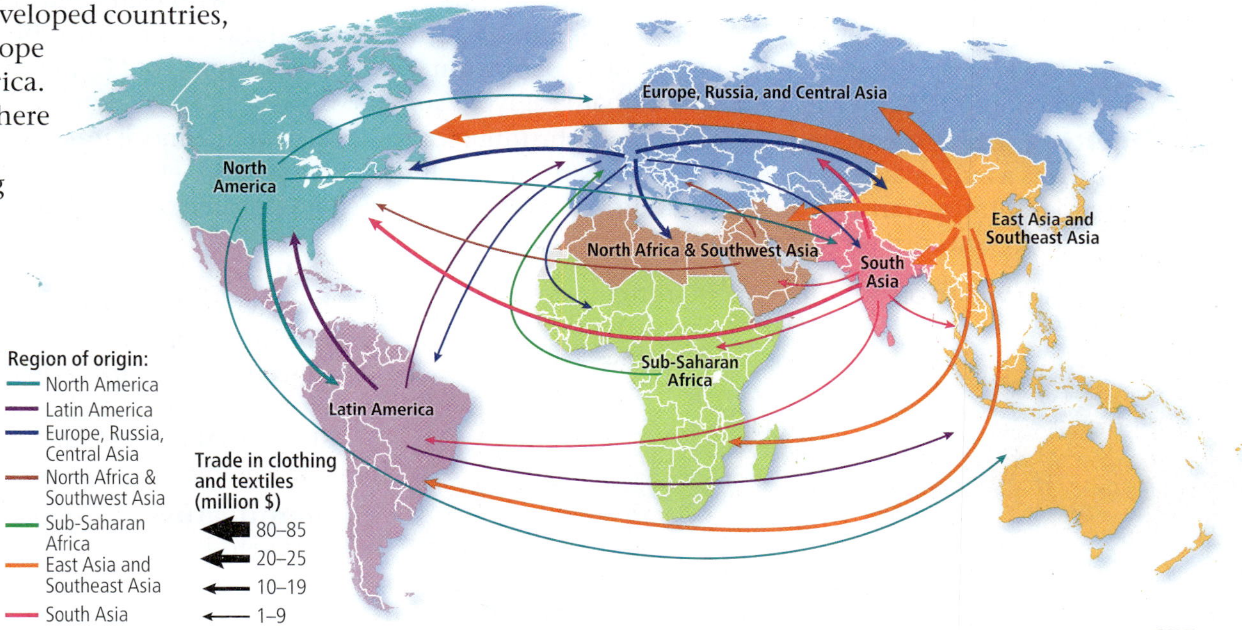

▶ 11.4.8 INTERNATIONAL TRADE IN CLOTHING & TEXTILES

Situation Factors: Locating Near Inputs

- **Explain why some industries locate near their inputs.**

Manufacturers buy from suppliers of inputs, such as minerals, materials, energy, machinery, and supporting services. They sell to companies and individuals who purchase the products. The farther something is transported, the higher the cost. So to minimize transportation costs, some manufacturers locate factories close to their sources of inputs, and some manufacturers locate factories close to their markets.

▼ 11.5.1 COPPER PRODUCTION NEAR SALT LAKE CITY, UTAH

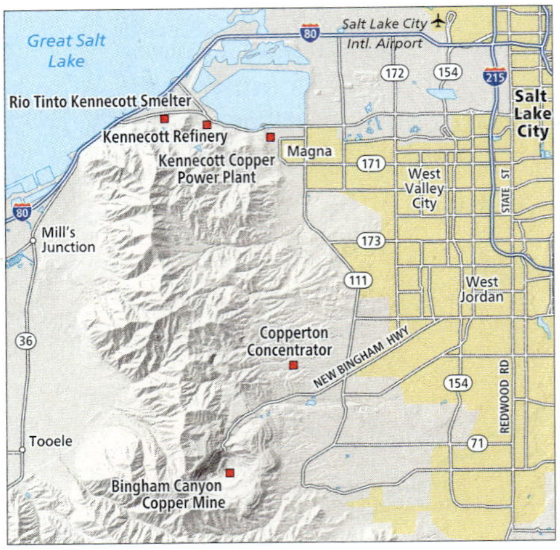

11.5.1 Two Types of Situation Factors

Location factors relating to the transportation of materials into and from a factory are known as **situation factors**. There are two types of situation factors:

- **Locating near inputs.** The optimal plant location is as close as possible to inputs if the cost of transporting raw materials to the factory is greater than the cost of transporting the product to consumers. An industry in which the inputs weigh more than the final products is a bulk-reducing industry. To minimize transport costs, a **bulk-reducing industry** locates near its sources of inputs.
- **Locating near markets.** The optimal plant location is as close as possible to the customer if the cost of transporting raw materials to the factory is less than the cost of transporting the product to consumers. A **bulk-gaining industry** makes something that gains volume or weight during production. To minimize transport costs, a bulk-gaining industry needs to locate near where the product is sold.

Copper Industry: An Example of Locating Near Inputs

Copper production involves five principal steps. Most of the steps in the production of copper illustrate the location decisions of a bulk-reducing industry that locates near its inputs. Figure 11.5.1 shows that the first four steps are located very close to each other, near Salt Lake City, Utah.

1. **Mining.** Mining in general is bulk reducing because the heavy, bulky ore extracted from mines is mostly waste, known as gangue. Figure 11.5.1 shows the location of Bingham Canyon Copper Mine (Figure 11.5.2).
2. **Concentration.** The ore is crushed and ground into fine particles, mixed with water and chemicals, filtered, and dried. Concentration mills are near copper mines because concentration transforms the heavy, bulky copper ore into a product of much higher value per weight. Figure 11.5.1 shows that Copperton Concentrator is located only 10 kilometers from the Bingham Canyon Copper Mine.
3. **Smelting.** The concentrated copper becomes the input for smelters, which remove more impurities. Because smelting is a bulk-reducing industry, smelters are built near their main inputs—the concentration mills—again to minimize transportation costs. Rio Tinto Kennecott Smelter is located only 20 kilometers from Copperton Concentrator (Figure 11.5.3).
4. **Refining.** The purified copper produced by smelters is treated at refineries to produce copper cathodes, about 99.99 percent pure copper. Most refineries are located near smelters. Figure 11.5.1 shows that Kennecott Refinery is only 2 kilometers from the Rio Tinto smelter.
5. **Manufacturing.** Copper that is ready for use in other products is produced in foundries. This step is not bulk reducing, so foundries are located near markets on the East and West coasts rather than near inputs.

▼ 11.5.2 BINGHAM CANYON COPPER MINE

▼ 11.5.3 RIO TINTO KENNECOTT SMELTER

11.5.2 Mineral Resources

Minerals are especially important inputs for many industries. Earth has 92 natural elements, but about 99 percent of the crust is composed of eight of them (Figure 11.5.4). The eight most common elements combine with less abundant elements to form approximately 3,000 different minerals, all with their own properties of hardness, color, and density as well as spatial distribution. Many of these minerals have important industrial uses.

Minerals are either nonmetallic or metallic:

- **Nonmetallic minerals.** By weight, more than 90 percent of the minerals that humans use are nonmetallic. Important nonmetallic minerals include building stones, gemstones such as diamonds, and minerals used in the manufacture of fertilizers such as nitrogen, phosphorus, potassium, calcium, and sulfur.
- **Metallic minerals.** Metallic minerals have properties that are especially valuable for fashioning machinery, vehicles, and other essential elements of contemporary society. They are to varying degrees malleable (able to be hammered into thin plates) and ductile (able to be drawn into fine wire) and are good conductors of heat and electricity.

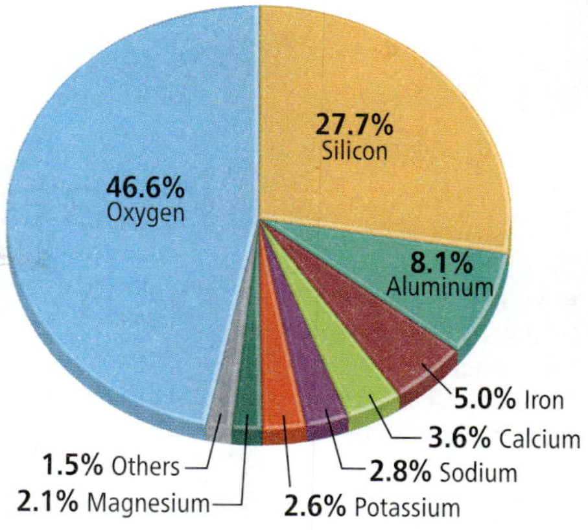

▲ 11.5.4 ELEMENTS IN EARTH'S CRUST

Mineral resources are not distributed uniformly across Earth. China and North America have the most abundant reserves of minerals critical for industry that are expected to be in short supply in the years ahead (Figure 11.5.5). Few important minerals are found in Europe, Central Asia, and Southwest Asia & North Africa.

▼ 11.5.5 DISTRIBUTION OF MINERALS

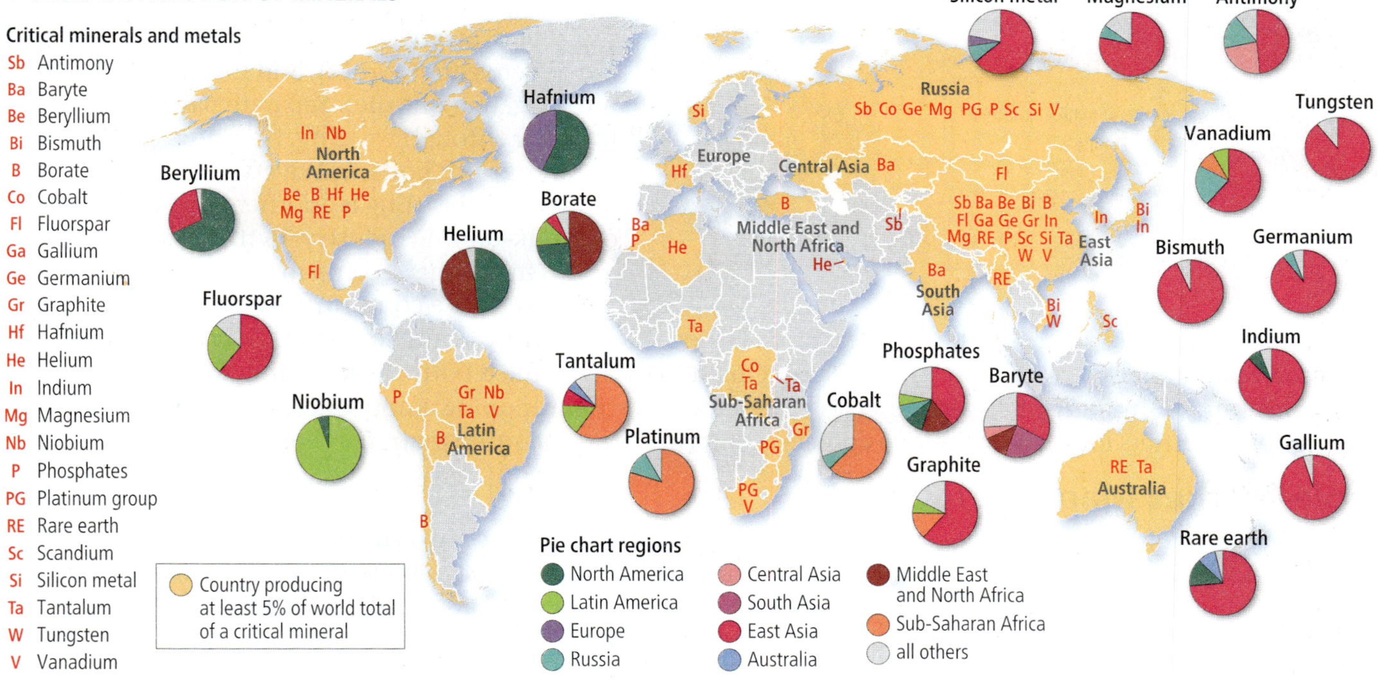

Alloys

Many metals are capable of combining with other metals to form alloys with distinctive properties that are important for industry. Alloys are known as ferrous or nonferrous. A ferrous alloy contains iron, and a nonferrous one does not:

- **Ferrous alloys.** The word *ferrous* comes from the Latin for "iron." Iron is extracted from iron ore, by far the world's most widely used ore. Humans began fashioning tools and weapons from iron 4,000 years ago. Important metals used to make ferrous alloys include chromium, manganese, molybdenum, nickel, tin, titanium, and tungsten.
- **Nonferrous alloys.** Important metals utilized to manufacture products that don't contain iron and steel include aluminum, copper, lead, lithium, magnesium, zinc, precious metals (silver, gold, and the platinum group), and rare earth metals.

Situation Factors: Locating Near Customers

- **Explain why some industries locate near their customers.**

For many industries, the optimal location is close to customers. Proximity to markets is a critical locational factor for three types of industries: bulk-gaining industries, single-market manufacturers, and manufacturers of perishable products.

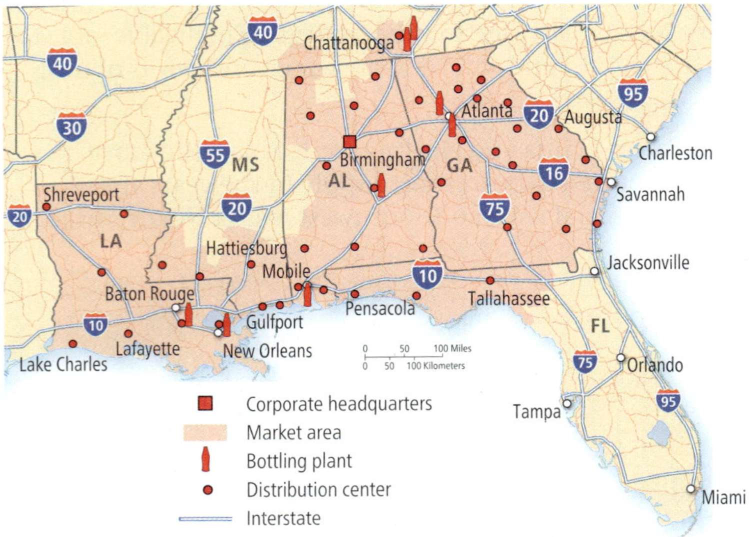

▼ **11.6.1 BULK-GAINING INDUSTRY** Coca-Cola Bottling Company UNITED operates numerous bottling plants near its customers, who are all located in the southeastern United States.

11.6.1 Bulk-Gaining Industries

A prominent example of a bulk-gaining industry is the fabrication of parts and machinery from steel and other metals. A fabricated-metal factory combines previously manufactured parts with metals such as steel and transforms them into a more complex product. Fabricators shape individual pieces of metal using such processes as bending, forging (hammering or rolling metal between two dies), stamping (pressing metal between two dies), and forming (pressing metal against one die). Separate parts are joined together through welding, bonding, and fastening with bolts and rivets.

Beverage bottling is another good example of a bulk-gaining industry. In this case, the bulk gain is in weight. A company like Coca-Cola manufactures a concentrated syrup (which is a bulk-reducing industry) and sells it to bottlers, such as Coca-Cola Bottling Company UNITED, which mix the syrup with water. Because water is the principal ingredient in cola, the filled bottle or can is much heavier than when it was empty. Shipping filled containers is more expensive than shipping empty ones, so to minimize shipping costs, bottlers locate near their customers rather than near the syrup manufacturer's plant (Figure 11.6.1).

Single-Market Manufacturers

A single-market manufacturer is a specialized manufacturer with only one or two customers. The optimal location for these factories is often in close proximity to the customers. An example of a single-market manufacturer is a producer of buttons, zippers, clips, pins, or other specialized components attached to clothing. The clothing manufacturer may need additional supplies of these pieces on very short notice. The world's largest manufacturer of zippers, YKK, for example, has factories in 71 countries in order to be near its customers, the manufacturers of clothing (Figure 11.6.2).

▼ **11.6.2 SINGLE-MARKET MANUFACTURER** YKK Zipper factory, Kurobe, Japan.

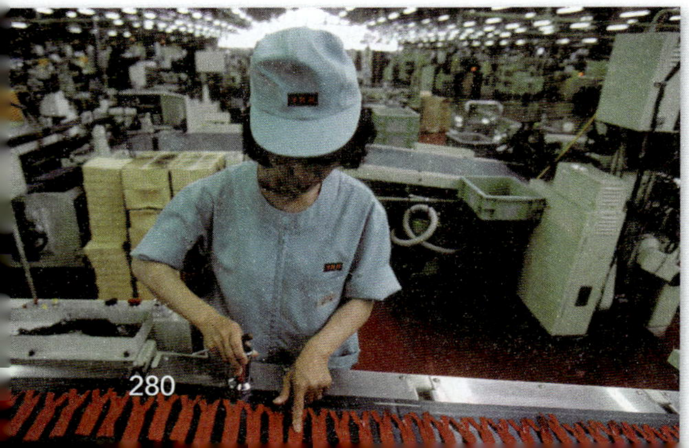

Manufacturers of Perishable Products

To deliver their products to consumers as rapidly as possible, perishable-products industries must be located near their markets. Because few people want stale bread or sour milk, food producers such as bakers and milk bottlers must locate near their customers to ensure rapid delivery. Processors of fresh food into frozen, canned, and preserved products can, however, locate far from their customers. Cheese and butter, for example, are manufactured in Wisconsin because rapid delivery to the urban markets is not critical for products with a long shelf life and the area is well suited agriculturally for raising dairy cows.

11.6.2 Motor Vehicle Production & Sales

The motor vehicle is a prominent example of a fabricated metal product that is likely to be built near its market. Around 90 million new vehicles are sold annually worldwide. China accounts for 29 percent of those sales, other Asian countries 20 percent, Europe 23 percent, North America 23 percent, and the rest of the world 6 percent (Figure 11.6.3).

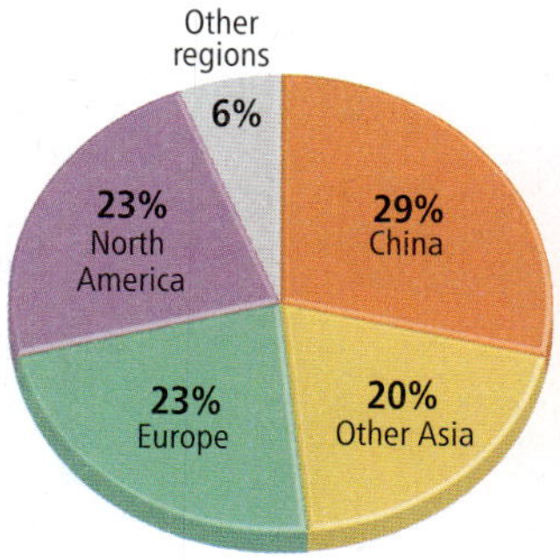

▲ **11.6.3 MOTOR VEHICLE SALES**

Not surprisingly, in view of the importance of producing vehicles near their markets, the regional distribution of vehicle production is extremely close to that for sales (Figure 11.6.4). China has 28 percent of world production, the rest of Asia 28 percent, Europe 20 percent, North America (including Mexico) 18 percent, and the rest of the world 5 percent. For example, around 76 percent of vehicles sold in North America are produced in North America. Around 11 percent come from Japan, 7 percent from Europe, and 5 percent from South Korea. Similarly, most vehicles sold in Europe are assembled in Europe, most vehicles sold in Japan are assembled in Japan, and most vehicles sold in China are assembled in China.

While most vehicles are produced near where they are sold, the nationality of the vehicle producers is less likely to be local. Eight companies sold at least 5 million vehicles in 2020 and together accounted for 70 percent of the world's sales. These included two based in North America (Ford and GM); three based in Europe (Germany's Volkswagen; France's Renault, which controls Nissan and Mitsubishi; and Stellantis); and three based in East Asia (Japan's Toyota and Honda and South Korea's Hyundai). These carmakers operate assembly plants in all three major industrial regions.

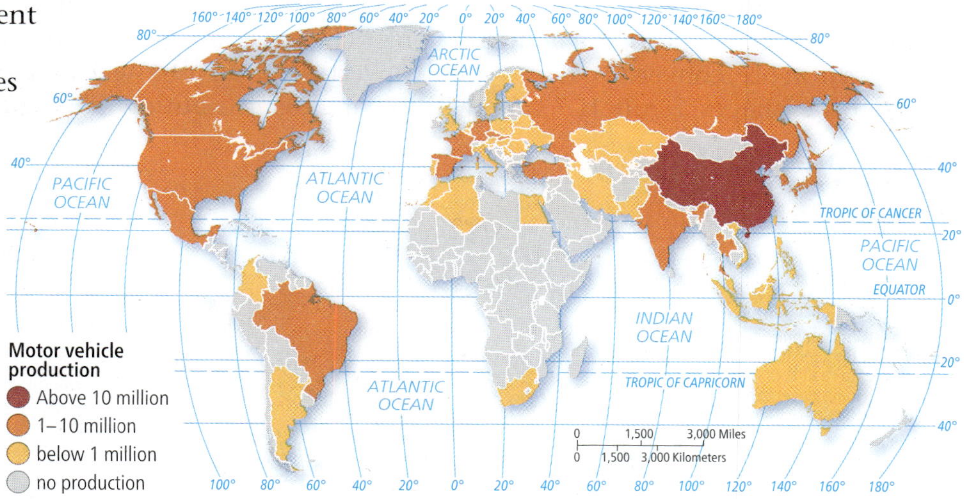

▲ **11.6.4 MOTOR VEHICLE PRODUCTION**

Distribution of Motor Vehicle Production in North America

Within North America, most motor vehicle production is clustered in the corridor known as Auto Alley, formed by north-south interstate highways 65 and 75 between Michigan and Alabama, with an extension into southwestern Ontario (Figure 11.6.5). A second cluster has formed in central Mexico. Auto Alley is a convenient location for factories to ship vehicles to most U.S. customers in just one day.

Also note in Figure 11.6.5 the location of engine plants in close proximity to their customers, the final assembly plants. An engine plant is a good example of a single-market manufacturer.

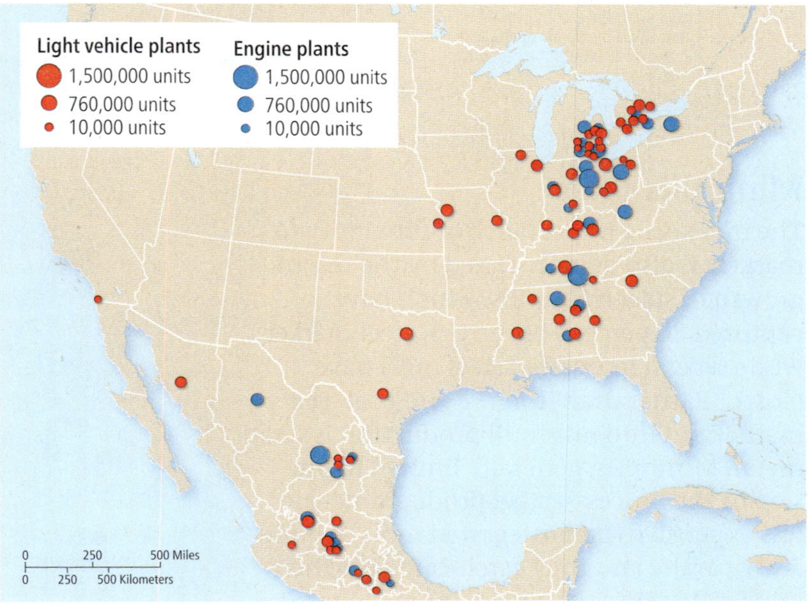

▲ **11.6.5 MOTOR VEHICLE ASSEMBLY PLANTS AND ENGINE PLANTS IN NORTH AMERICA**

Changing Situation Factors: Steel

- **Summarize changes in the distribution of the steel industry.**

The two principal inputs in steel production are iron ore and coal. Because of the need for large quantities of these two bulky inputs, steelmaking traditionally clustered near the sources of these two key inputs. More recently, steel production has relocated to be closer to markets.

11.7.1 U.S. Steel Plants

Steel is an alloy of iron that is manufactured by removing impurities in iron, such as silicon, phosphorus, sulfur, and oxygen, and adding desirable elements, such as manganese and chromium. An integrated steel mill processes iron ore, converts coal into coke, converts the iron into steel, and forms the steel into sheets, beams, rods, or other shapes. Figure 11.7.1 shows the changing distribution of integrated steel mills in the United States.

▶ 11.7.1 INTEGRATED STEEL MILLS IN THE UNITED STATES

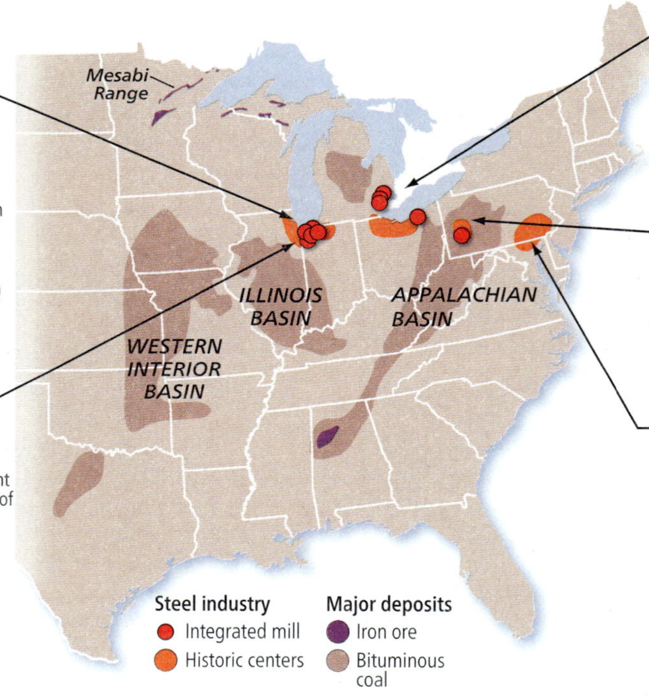

EARLY 1900s
Most new steel mills were located near the southern end of Lake Michigan. The main raw materials continued to be iron ore and coal, but changes in steelmaking required more iron ore in proportion to coal. Thus, new steel mills were built closer to the Mesabi Range to minimize transportation cost. Coal was available from nearby southern Illinois, as well as from Appalachia.

LATE 1900s
Most U.S. steel mills closed. The survivors are mostly around the southern Great Lakes. Proximity to markets has become more important than the traditional situation factor of proximity to inputs. Coastal plants provide steel to large East Coast population centers, and southern Great Lakes plants are centrally located to distribute their products countrywide.

LATE 1800s
Steel mills were built around Lake Erie. The shift westward from Pennsylvania was influenced by the discovery of rich iron ore in the Mesabi Range in northern Minnesota. The ore was transported through the Great Lakes. Coal was shipped from Appalachia by train.

MID–1800s
Steel concentrated around Pittsburgh because iron ore and coal were both mined there. The area no longer has steel mills, but it remains the center for steel industry research and administration.

MID–1900s
Most new U.S. steel mills were located near the East and West coasts, including Baltimore, Los Angeles, and Trenton, New Jersey. Iron ore increasingly came from other countries, especially Canada and Venezuela, and locations near the Atlantic and Pacific oceans were more accessible to those foreign sources. Scrap iron and steel—widely available in the large metropolitan areas of the East and West coasts—became an important input in the steel-production process.

Minimills

The increasing importance of proximity to markets is also demonstrated by the recent growth of steel minimills, which have captured 80 percent of the U.S. steel market, while several integrated steel mills have closed. Rather than iron ore and coal, the main input into minimill production is scrap metal. Minimills, generally limited to one step in the process—steel production—are less expensive than integrated mills to build and operate, and they can locate near their markets because their main input—scrap metal—is widely available (Figure 11.7.2).

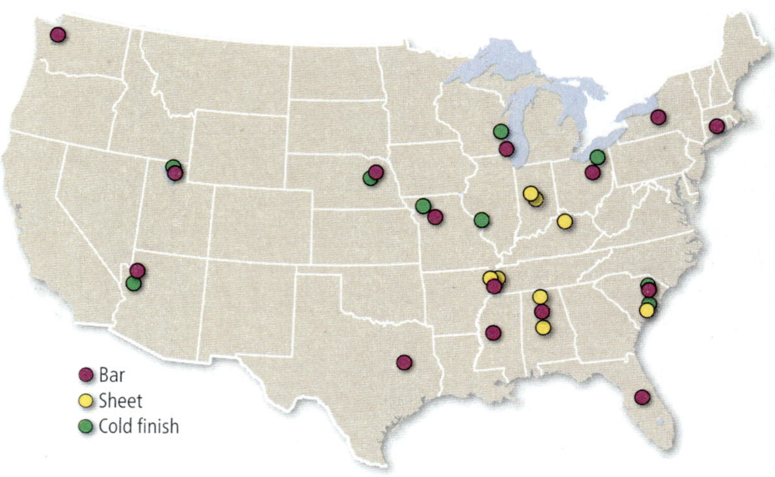

▲ 11.7.2 NUCOR CORPORATION MINIMILL PLANTS

11.7.2 World Steel Production

The changing distribution of world manufacturing can be seen clearly in steel production. Situation factors are still important, but two changes in these factors have altered the distribution of steel producers:

- Changes in the relative importance of the main inputs.
- Increasing importance of proximity to markets.

In 1980, 81 percent of world steel was produced in developed countries and 19 percent in developing countries (Figure 11.7.3). Three countries—the Soviet Union, Japan, and the United States—produced 50 percent of the world's steel in 1980 (Figure 11.7.4).

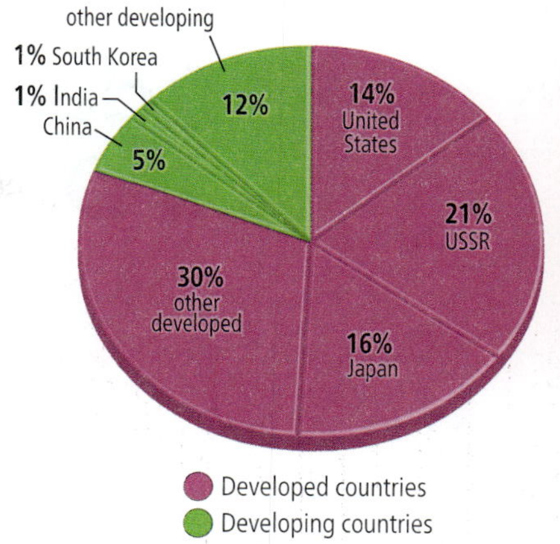

▲ **11.7.3 DISTRIBUTION OF STEEL PRODUCTION, 1980**

Changing Distribution of Steel Production

Between 1980 and 2020, the share of world steel production declined to 26 percent in developed countries and increased to 74 percent in developing countries (Figure 11.7.5). China alone produced more than 50 percent of the world's steel in 2020 (Figure 11.7.6). China's steel industry has grown in part because of access to the primary inputs iron ore and coal. However, the principal factor in recent years has been increased demand by growing industries in China that use a lot of steel, such as for production of motor vehicles.

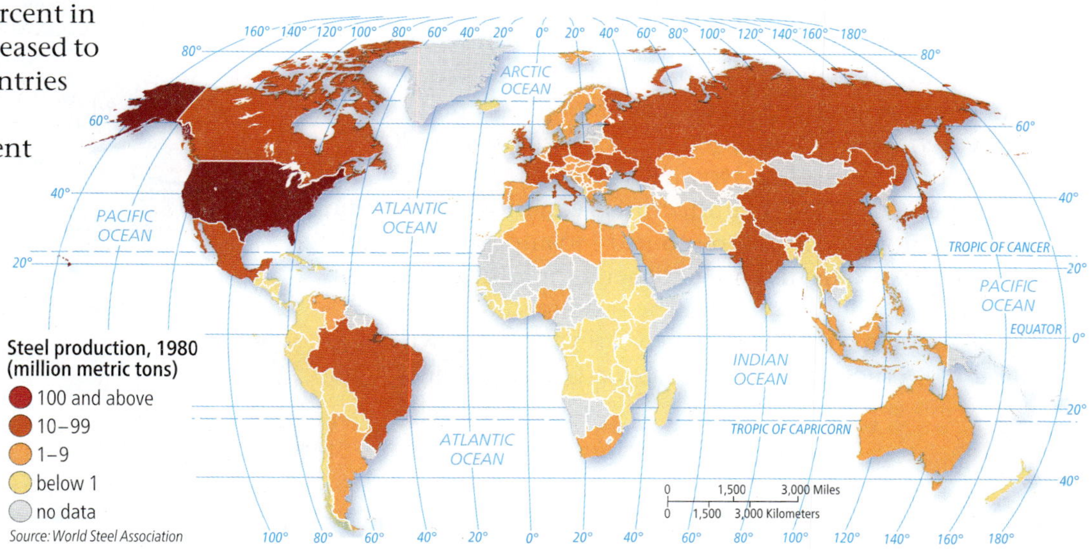

▲ **11.7.4 STEEL PRODUCTION, 1980**

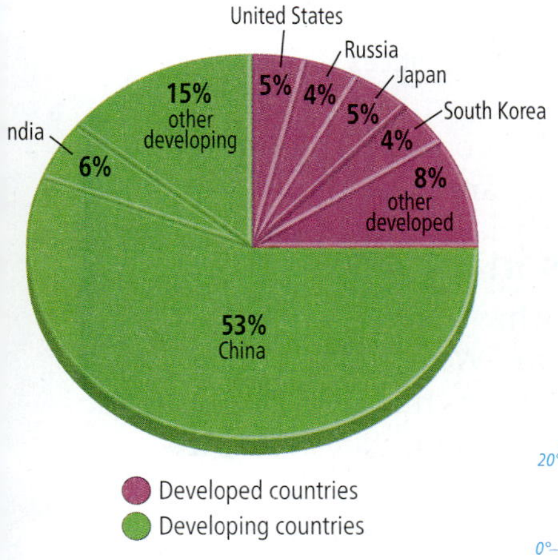

▲ **11.7.5 DISTRIBUTION OF STEEL PRODUCTION, 2020**

▼ **11.7.6 STEEL PRODUCTION BY COUNTRY, 2020**

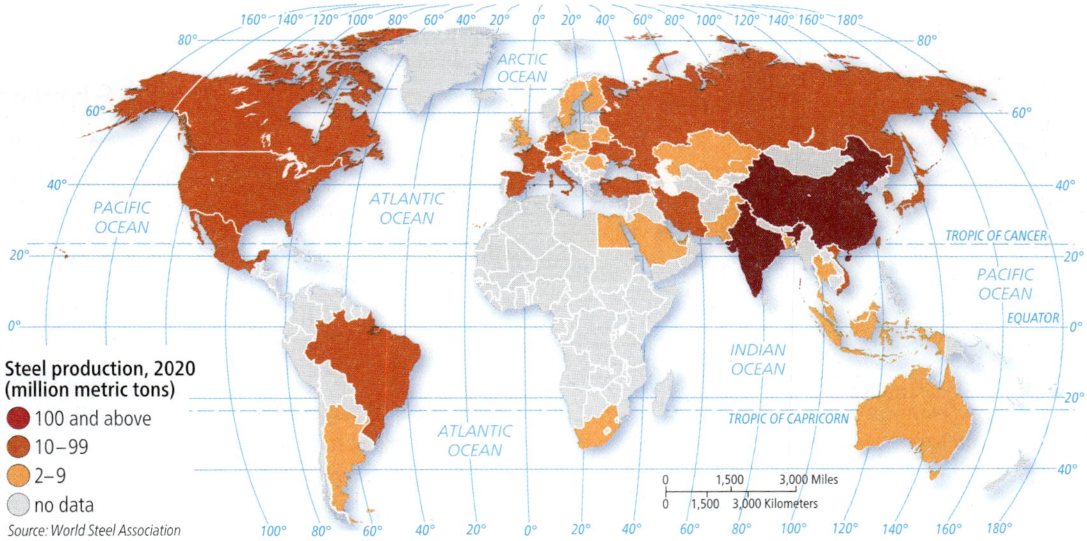

Industrial Change in Developed Regions

- **Explain industrial location changes within developed regions.**

In developed regions, industry is shifting away from the traditional industrial areas of the northeastern United States and northwestern Europe. Mexico and Central Europe attract industries concerned with both a site factor (low-cost labor) and a situation factor (proximity to consumers with high incomes).

11.8.1 Intraregional Shifts in North America

Trade agreements have eliminated most barriers to moving finished products and inputs within North America. Under the United States–Mexico–Canada Agreement (USMCA), which was implemented in 2020, and its predecessor, the North American Free Trade Agreement (NAFTA), manufacturers can move inputs and finished products from one country to another without having to pay a tariff or meet other requirements, as long as most of the inputs in the products are made somewhere in North America. Nearly all the growth of motor vehicle production in North America, for example, has been located in Mexico rather than the United States or Canada (Figures 11.8.1 and 11.8.2).

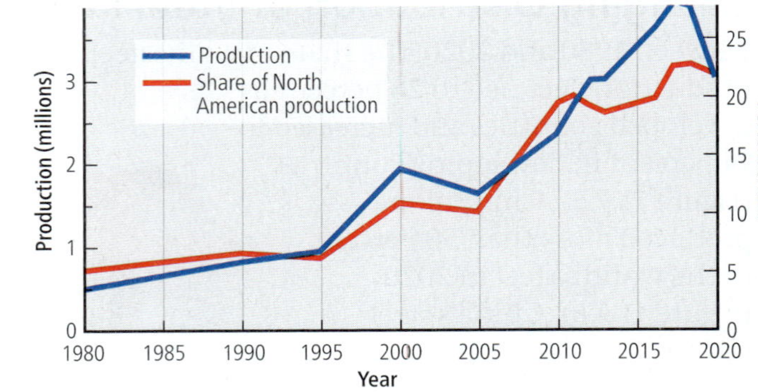

▲ 11.8.1 VEHICLE PRODUCTION IN MEXICO

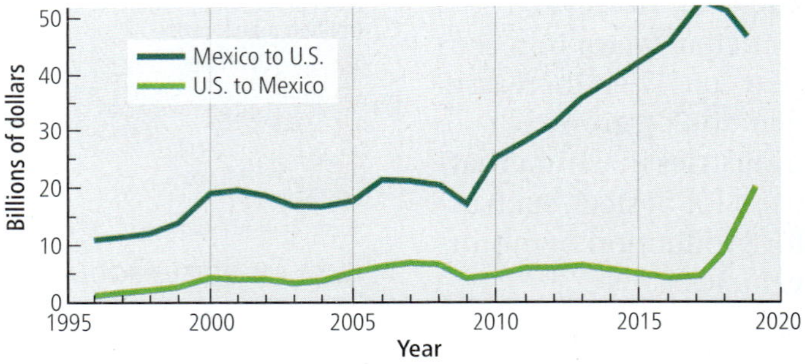

▲ 11.8.2 U.S.-MEXICO TRADE IN VEHICLES

Intraregional Shifts in the United States

In the United States, industry has shifted from the Northeast toward the South and West (Figure 11.8.3). The principal lure for many manufacturers has been right-to-work laws in many southern and some western states. A **right-to-work law** requires a factory to maintain an "open shop" and prohibits a "closed shop." In a closed shop, a company and a union agree that everyone must join the union to work in the factory. In an open shop, a union and a company may not negotiate a contract that requires workers to join a union as a condition of employment. Right-to-work laws make it much more difficult for unions to organize factory workers, collect dues, and bargain with employers from a position of strength.

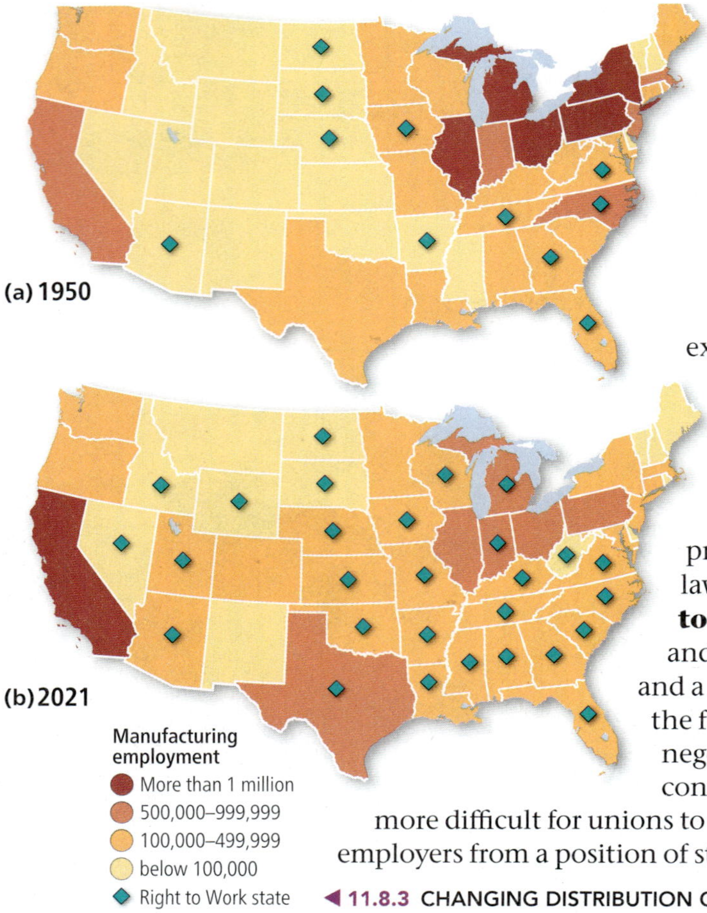

◀ 11.8.3 CHANGING DISTRIBUTION OF U.S. INDUSTRY

11.8.2 Intraregional Shifts in Europe

Manufacturing has diffused from traditional industrial centers in northwestern Europe toward Southern and Eastern Europe. Government policies have encouraged relocation toward economically distressed peripheral areas.

During the late twentieth century, the Western European country with the most rapid manufacturing growth was Spain, especially after its admission to the European Union in 1986. However, investment in Spain has slowed in recent years.

Since the fall of communism in the early 1990s, industry has increased relatively rapidly in Central Europe, especially Czechia and Poland (Figure 11.8.4). Central Europe as a whole increased its share of Europe's industry from 4 percent in 2000 to 10 percent in 2020. Central Europe offers manufacturers two important site and situation factors: low-wage but relatively skilled labor and proximity to Western Europe markets. Ireland has also seen a relatively strong increase in manufacturing.

Conversely, the share of Europe's manufacturing located in Germany and the United Kingdom declined relatively rapidly. The two countries accounted for 36 percent of Europe's industry in 2020, compared with 46 percent in 2000. Manufacturers are able to locate in Czechia and Poland and easily ship products to consumers in Germany. The decline in the share of Europe's manufacturing in the United Kingdom has been accelerated by that country's departure from the EU.

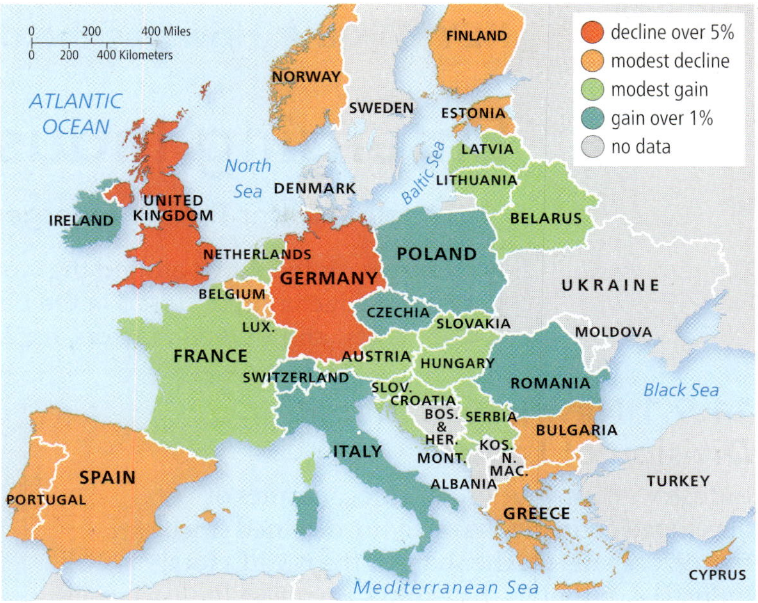

▲ **11.8.4** CHANGE IN SHARE OF EUROPE'S MANUFACTURING, 2000–2020

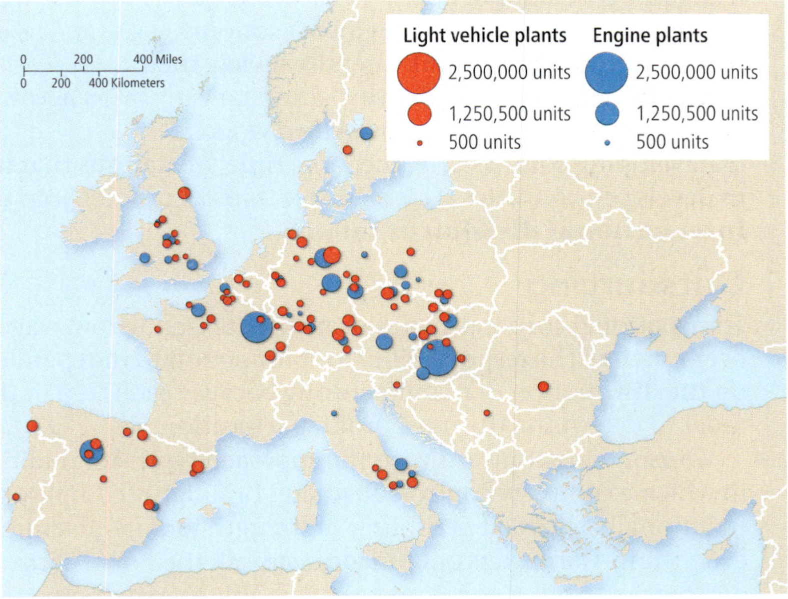

▲ **11.8.5** MOTOR VEHICLE ASSEMBLY & ENGINE PLANTS

Changing Distribution of Europe's Motor Vehicle Industry

Motor vehicle production is an example of an industrial sector that has grown in different areas of Europe at different times (Figure 11.8.5). The continent has 56 assembly plants that date from 1970 or earlier. These include 30 in West Germany (before its reunification with East Germany in 1990), 9 in the United Kingdom, and 8 in France. Only 7 of the 45 assembly plants dating from 1970 or earlier are in Central Europe. In contrast, of the 39 assembly plants opened since 1970, only 12 are in France, Germany, and the United Kingdom, compared with 17 in Central Europe. A number of plants in Italy and the United Kingdom have been closed (Figure 11.8.6).

▶ **11.8.6** WORKERS OUTSIDE ASSEMBLY PLANT IN SICILY, ITALY, AFTER ANNOUNCEMENT OF ITS CLOSURE IN 2011

Emerging Industrial Regions

- **Explain reasons for the emergence of new clusters of industries.**

Industry is on the move around the world. Industry has grown in locations where relatively little manufacturing occurred in the past. Reasons include site factors, especially lower labor costs, and situation factors, especially proximity to growing markets.

11.9.1 New International Division of Labor

Industry's contribution to the economies of developed countries has steadily declined since the 1970s, whereas the share of the world's total industrial output in the world's emerging industrial regions has increased (Figure 11.9.1). Transnational corporations have been especially aggressive in using low-cost labor in developing countries. To remain competitive in the global economy, transnational corporations carefully review their production processes to identify steps that can be performed by low-paid, low-skilled workers in developing countries. At the same time, operations that require highly skilled workers remain in factories in developed countries. This selective transfer of some jobs to developing countries is known as the **new international division of labor**.

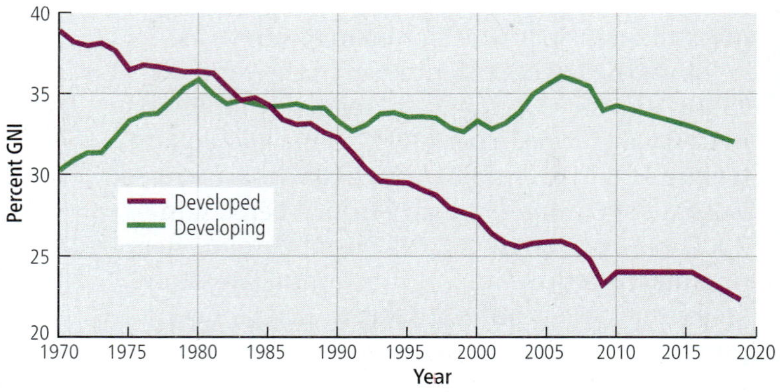

▲ 11.9.1 MANUFACTURING VALUE ADDED AS PERCENTAGE OF GNI

Outsourcing

Transnational corporations allocate production to low-wage countries through **outsourcing**, which is turning over much of the responsibility for production to independent suppliers. Outsourcing has had a major impact on the distribution of manufacturing because each step in the production process is now scrutinized closely in order to determine the most profitable location.

Outsourcing contrasts with the approach typical of traditional mass production, called **vertical integration**, in which a company controls all phases of a highly complex production process. Vertical integration was traditionally regarded as a source of strength for manufacturers because it gave them the ability to do and control everything. Carmakers once made nearly all their own parts, for example, but now most of this operation is outsourced to other companies that are able to make better-quality parts at a lower cost.

BRIC Countries

Much of the world's future growth in manufacturing is expected to cluster in a handful of countries known as BRIC, which is an acronym coined by the investment banking firm Goldman Sachs for Brazil, Russia, India, and China. South Africa is sometimes included, and the group is then known as BRICS.

The four BRIC countries together encompass 25 percent of the world's land area and 41 percent of the world's inhabitants, but they account for only 21 percent of world GDP (Figure 11.9.2). The International Monetary Fund ranks their economies second (China), fifth (India), ninth (Brazil), and twelfth (Russia). China is expected to pass the United States as the world's largest economy in the 2020s, and India is expected to become second around 2050.

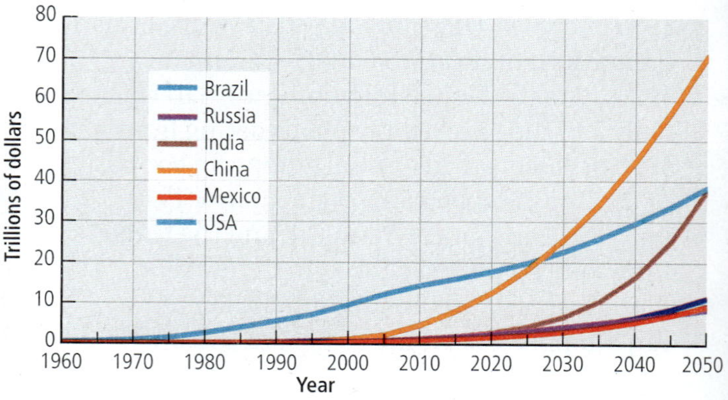

▲ 11.9.2 GDP FOR BRIC COUNTRIES, MEXICO, AND THE UNITED STATES

11.9.2 World Industrial Distribution

Three countries account for around one-half of the world's total output of industrial goods: 20 percent of the world's industrial output is in China, 18 percent is in the United States, and 10 percent is in Japan (Figure 11.9.3). China is the first country to achieve an annual output of more than $2 trillion.

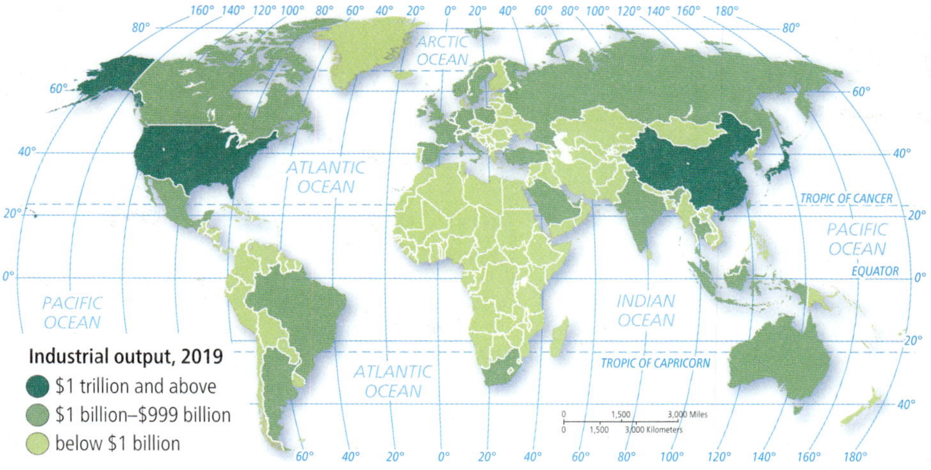

▲ 11.9.3 INDUSTRIAL OUTPUT

China's Situation Factors

China's industrial leadership rests on a combination of situation and site factors. China has important assets in both types of situation factors: proximity to inputs and proximity to markets.

Locating near customers. Because it is the world's most populous country, China is attractive to industries that wish to locate near customers. For example, nearly one-third of the world's motor vehicles are produced and sold in China (refer to Figures 11.6.3 and 11.6.4). The percentage of the 1.4 billion Chinese people who can afford products such as cars has increased rapidly in the twenty-first century. The income per capita in China has risen from around $3,000 in 2000 to $15,000 in 2020. In comparison, developing countries other than China have had more modest increases in income per capita, from $1,500 in 2000 to $9,000 in 2020 (Figure 11.9.4).

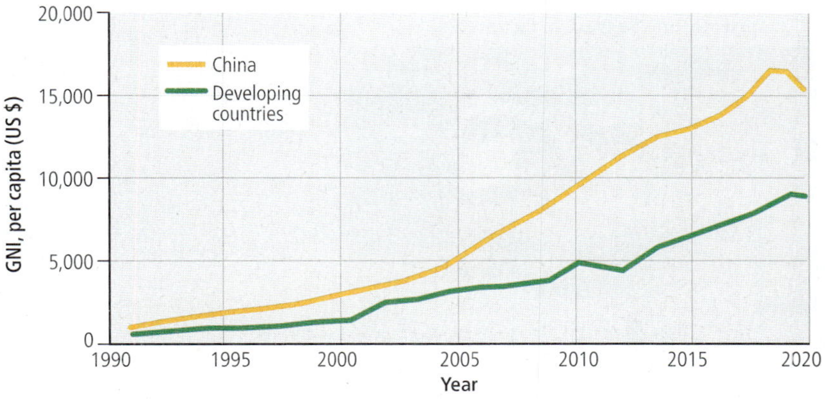

▲ 11.9.4 INCOME PER CAPITA, CHINA AND OTHER DEVELOPING COUNTRIES

Locating near inputs. China is an attractive location for industries that need to be near raw materials. China is the leading producer of nearly every critical mineral and metal needed for manufacturing, especially those needed for production of electronics.

▼ 11.9.5 FITNESS EQUIPMENT FACTORY
Xindan, China.

China's Site Factors

China's industrial leadership derives not only from the situation factor of mineral and metal abundance but also from all three site factors: land, labor, and capital.

Land. At a national scale, China's industries are clustered in the eastern half of the country (refer to Figure 11.1.4). Chinese companies that export most of their products are more likely to select locations near a major port on the east coast of their country (Figure 11.9.5).

Labor. In China, labor is abundant, given the country's large population. Wages average around $7 per hour in China, compared with $30 to $40 in developed countries (refer to Figure 11.3.1).

Capital. In China, much of the capital invested to develop new and expanding industrial sectors such as electronics comes from the government. Capital also comes from foreign companies that are required to invest in Chinese companies in order to produce and sell in China. For example, the leading motor vehicle producers in China are mostly joint ventures between Chinese companies and international companies like GM and Volkswagen. Availability of capital has been a key factor in China becoming the dominant producer of electric vehicles (Figure 11.9.6).

▼ 11.9.6 ELECTRIC VEHICLE FACTORY
Zhaoqing, China.

CHAPTER 11 Review, Analyze, & Apply

KEY ISSUE 1 Where Is Industry Distributed?

- Most of the world's industry is clustered in three regions: Europe, North America, and East Asia.
- The concept of manufacturing goods in a factory originated with the Industrial Revolution in the United Kingdom.
- Products are shipped by boat, rail, truck, or air. The optimal shipping choice depends on the distance something is being transported.

THINKING GEOGRAPHICALLY

What might be the advantages and challenges of utilizing drones for just-in-time delivery to factories?

▲ 11.CR.1 DRONE DELIVERING A PACKAGE INSIDE A FACTORY

KEY ISSUE 2 Why Are Site Factors Important for Industry?

- A company ordinarily considers a combination of site and situation factors in deciding where to locate a factory or to find suppliers.
- Site factors are related to the costs of production inside a factory.
- The three site factors are labor, capital, and land.

THINKING GEOGRAPHICALLY

Check your clothing labels. Do you have any clothing made in a developed country of Europe or North America? What might account for that clothing being made in a developed country rather than in a developing country?

▲ 11.CR.2 HARRIS TWEED, MADE IN THE ISLE OF HARRIS, SCOTLAND

KEY ISSUE 3 Why Are Situation Factors Important for Industry?

- Situation factors involve transporting inputs into a factory and transporting manufactured goods to the markets.
- Manufacturers tend to locate near their source of inputs if their product is bulk-reducing.
- Manufacturers tend to locate near their markets if their product is bulk-gaining, shipped to only one customer, or highly perishable.

THINKING GEOGRAPHICALLY

▲ 11.CR.3 VEHICLE IDENTIFICATION NUMBER

Every car has a 17-digit Vehicle Identification Number (VIN) that tells a lot about the vehicle, such as where the vehicle was assembled. Enter the VIN in a search engine for a vehicle owned by you, a family member, or a friend to find out where it was made. Locate the region of assembly for the vehicle. Was the location where the vehicle was assembled important to the purchaser of the vehicle? Why or why not?

KEY ISSUE 4 How Has the Distribution of Industry Changed?

- The location of manufacturing has changed. Industry is increasing in some developing countries and in some areas within developed regions.
- Site factors, especially labor costs, have stimulated industrial growth in new regions, especially the BRICS countries.
- China has become the world's leading industrial country.

THINKING GEOGRAPHICALLY

▲ 11.CR.4 BICYCLE FACTORY, BURKINA FASO, A LANDLOCKED STATE

Refer to Section 8.8.1 about landlocked states and Section 8.11.1 about regime types. How might the distribution of landlocked states and regime types impact the ability of African countries to emerge as new industrial regions?

Key Terms

Apparel (11.4.1) An article of clothing.

Break-of-bulk point (11.2.2) A location where transfer is possible from one mode of transportation to another.

Bulk-gaining industry (11.5.1) An industry in which the final product weighs more or comprises a greater volume than the inputs.

Bulk-reducing industry (11.5.1) An industry in which the final product weighs less or comprises a lower volume than the inputs.

Cottage industry (11.1.1) Manufacturing based in homes rather than in factories, commonly found prior to the Industrial Revolution.

Fordist production (11.3.1) A form of mass production in which each worker is assigned one specific task to perform repeatedly.

Just-in-time delivery (11.2.2) Shipment of parts and materials to arrive at a factory moments before they are needed.

Labor-intensive industry (11.3.1) An industry for which labor costs comprise a high percentage of total expenses.

New international division of labor (11.9.1) Transfer of some types of jobs, especially those requiring low-paid, less skilled workers, from developed to developing countries.

Outsourcing (11.9.1) A decision by a corporation to turn over much of the responsibility for production to independent suppliers.

Post-Fordist production (11.3.1) Adoption by companies of flexible work rules, such as the allocation of workers to teams that perform a variety of tasks.

Right-to-work law (11.8.1) A law in some U.S. states that prevents a union and a company from negotiating a contract that requires workers to join the union as a condition of employment.

Site factors (11.3.1) Costs related to manufacturing inside a plant, such as expenses associated with land, labor, and capital.

Situation factors (11.5.1) Location factors related to the transportation of materials into and from a factory.

Textile (11.4.1) A fabric made by weaving, used in making clothing.

Vertical integration (11.9.1) An approach typical of traditional mass production in which a company controls all phases of a highly complex production process.

Geospatial Analysis

Log in to the Mastering Geography Study Area to access MapMaster 2.0.

Railroads & Industries

U.S. manufacturers make heavy use of railroads.

Add the ***Major Manufacturing Regions (North America)*** layer. Select the *Settings* icon from the *Legend*, select *Change Map Projection*, select *Albers*, and select *Show City Labels*. Zoom to North America.

1. What major cities in the United States appear to be within major manufacturing regions?
2. Add the ***World Railroads*** layer. In which city do the largest number of railroads appear to converge? What industry discussed in this chapter is highly clustered near this city? What might account for this clustering?

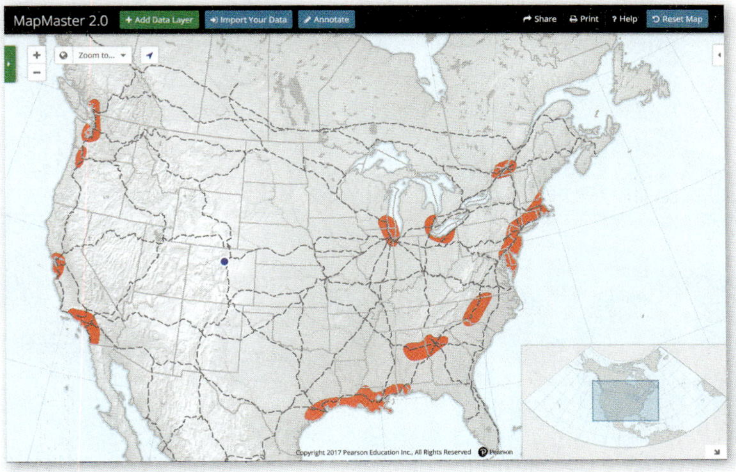

▲ **11.CR.5** MAJOR MANUFACTURING REGIONS AND RAILROADS OF NORTH AMERICA

Explore

Auto Assembly Plants

Auto assembly plants take up a lot of land. Let's compare one of the oldest and newest in the United States.

1. In Google Earth, ***search*** for **7777 Kia Parkway, West Point, Georgia**. This is a Kia assembly plant opened in 2009. Describe the use of land surrounding the factory.
2. ***Search*** for **12600 S. Torrence Ave., Chicago**. This is a Ford assembly plant opened in 1924. Zoom out to around 2 km. Describe the uses of land that surround the factory. How do the surroundings differ from those around the Kia plant?

▲ **11.CR.6** KIA MOTORS, WEST POINT, GEORGIA

CHAPTER 12

Services & Settlements

▲ Phuket, Thailand, an urban settlement.

Most employed people in the world work in jobs that provide services, such as in shops, offices, restaurants, universities, and hospitals. Services are closely linked with the geography of settlements because most services are located in settlements. Geography plays an especially important role in the provision of services. For example, service providers apply geographic principles to determine the best place to locate.

KEY ISSUES

1 Where Are Settlements & Services Distributed?

Services are clustered in urban settlements. Three types of services are consumer, business, and public.

2 Where Are Consumer Services Distributed?

Consumer services are distributed in a regular pattern, especially in developed countries. Central place theory helps explain the distribution of consumer services.

3 Where Are Business Services Distributed?

Business services tend to cluster in a relatively small number of large settlements, known as global cities. Developing countries specialize in some forms of business services.

4 Why Do Services Cluster in Settlements?

Services cluster in settlements, which could be rural or urban. An increasing number and an increasing percentage of people live in urban settlements.

LOCATIONS IN THIS CHAPTER

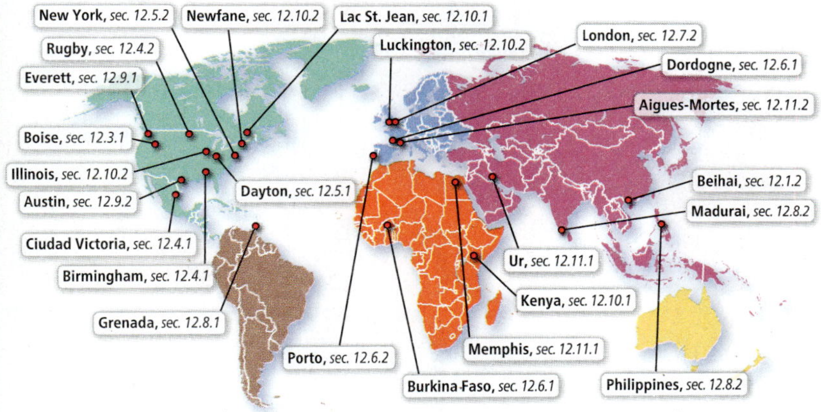

Urban Settlements

- **Describe the distribution of urban settlements.**

A **settlement** is a permanent collection of buildings where people reside, work, and obtain services. Settlements occupy a very small percentage of Earth's surface, well under 1 percent, but settlements are home to nearly all people because few people live in isolation.

12.1.1 Urban & Rural Settlements

The jobs that employ people are divided into the primary sector (primarily agriculture), the secondary sector (manufacturing), and the tertiary sector (services) (refer to Section 10.2.2). A **rural settlement** is a community where most people are engaged in agriculture. An **urban settlement** is a community where people are employed in secondary- and tertiary-sector jobs.

According to urban sociologist Louis Wirth, an urban settlement is distinguished from a rural settlement because the urban settlement is larger in area, of a higher population density, and inhabited by people with more diverse personal characteristics and modes of behavior compared with those living in a rural settlement.

▼ 12.1.1 GROWTH IN PERCENTAGE OF PEOPLE LIVING IN URBAN SETTLEMENTS

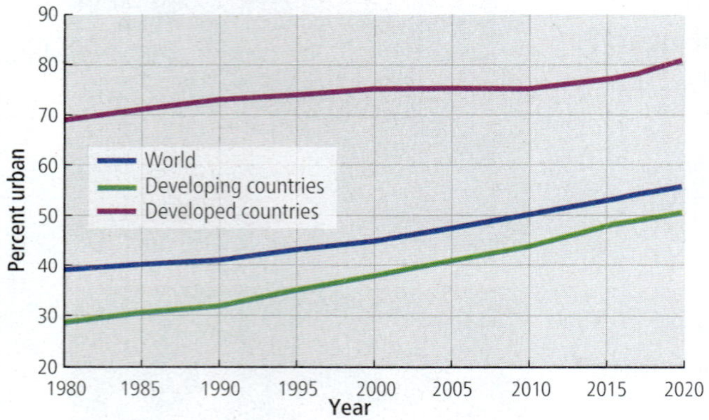

Growth in Urban Settlements

Urbanization is an increase in the percentage of people living in urban settlements. Worldwide, the percentage living in urban settlements increased rapidly during the twentieth century, from 14 percent in 1900 to 45 percent in 2000. The population of Earth's urban settlements exceeded that of rural settlements for the first time in human history around 2008. Approximately 56 percent of the world's population now live in an urban settlement (Figure 12.1.1).

The percentage of people living in urban settlements reflects a country's level of development. In developed countries, 79 percent live in urban areas, compared to 52 percent in developing countries (Figure 12.1.2). Only 41 percent live in urban areas in sub-Saharan Africa and 37 percent in South Asia. Not by coincidence, these two developing regions have the lowest Human Development Indexes (HDIs), according to the U.N. (refer to Section 10.1.1). On the other hand, in Latin America—the developing region with the highest HDI—the percentage living in urban areas is 79 percent, virtually the same as in developed countries.

The percentage living in urban settlements is relatively high in developed countries because over the past 200 years, residents of rural settlements immigrated to work in factories and services in urban settlements. The need for fewer farm workers pushed people out of rural settlements, and rising employment opportunities in manufacturing and services lured them into urban settlements.

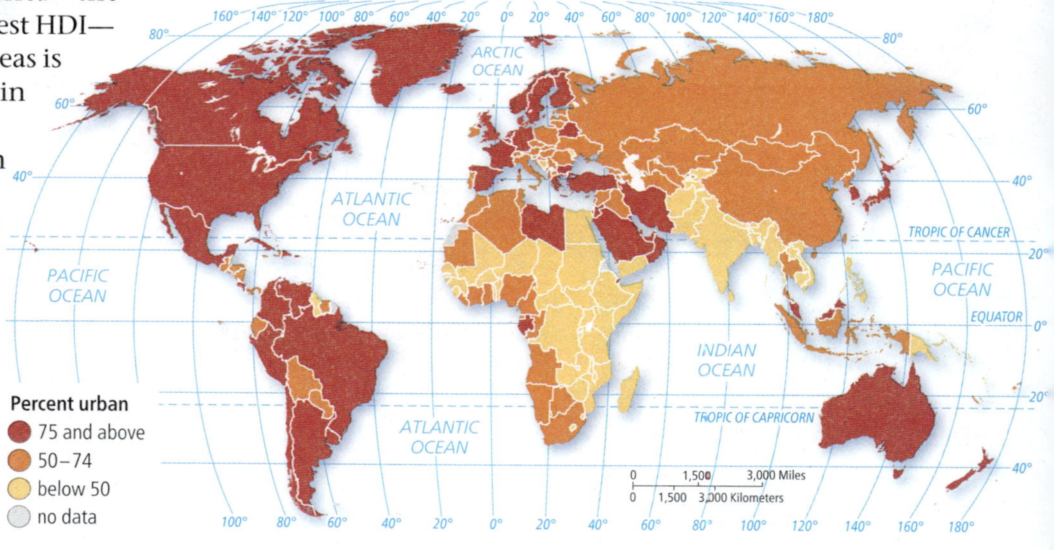

▲ 12.1.2 PERCENTAGE OF PEOPLE LIVING IN URBAN SETTLEMENTS

12.1.2 Megacities

Developed countries have a higher percentage of urban residents, but developing countries have most of the very large urban settlements. The U.N. defines a **megacity** as an urban settlement with a total population in excess of 10 million people and a **metacity** as an urban settlement with more than 20 million people.

The U.N. identifies 33 urban settlements in the world as megacities and further identifies 6 of those 33 as metacities. Only 6 of the 33 megacities (Tokyo, Osaka, New York, Los Angeles, Moscow, and Paris) are in developed countries; the other 27 are in developing countries (Figure 12.1.3). Other researchers identify nine more urban settlements that may qualify as megacities; all but two (London and Nagoya) are in developing countries.

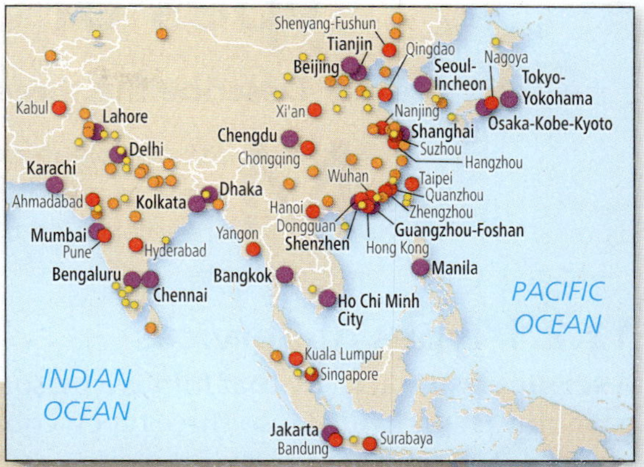

▼ **12.1.3 MEGACITIES AND METACITIES**

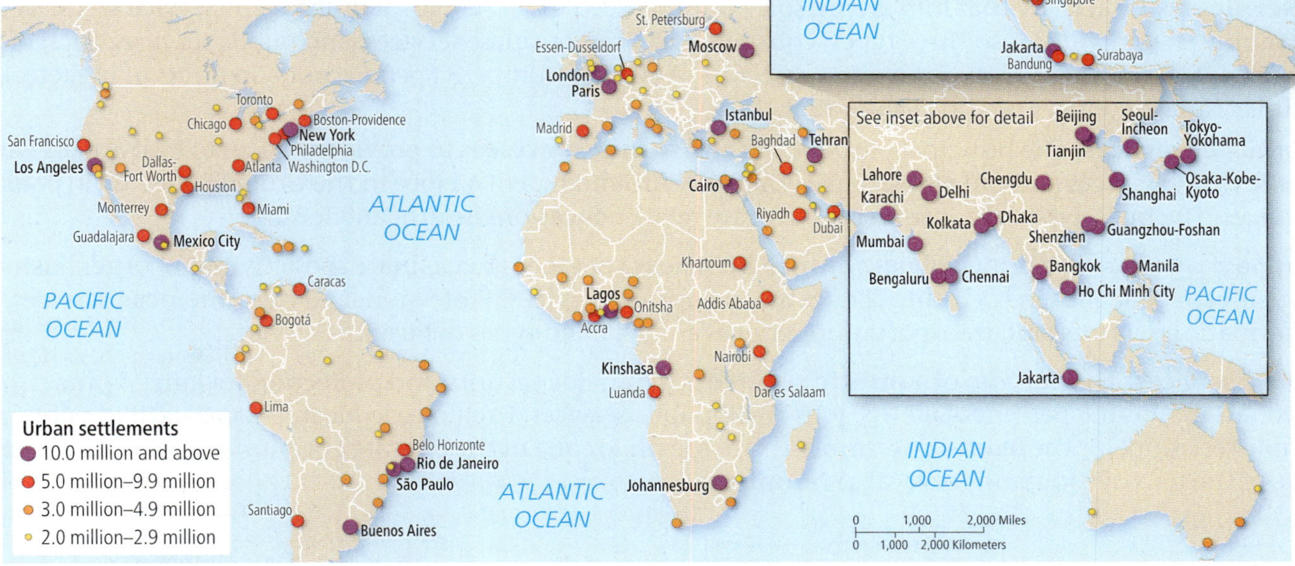

Rapidly Growing Urban Settlements

All but 4 of the 100 fastest-growing urban settlements are in developing countries (Figure 12.1.4). Beihai, China, is ranked as the fastest-growing (Figure 12.1.5). The fact that developing countries dominate the lists of largest and fastest-growing urban settlements is remarkable because urban settlements were once associated with economic development. In developing countries, migration from the countryside is fueling half the increase in population in urban settlements, even though job opportunities are not always available. The other half of that increase in population results from high natural increase rates, especially in Africa (refer to Section 2.3.1).

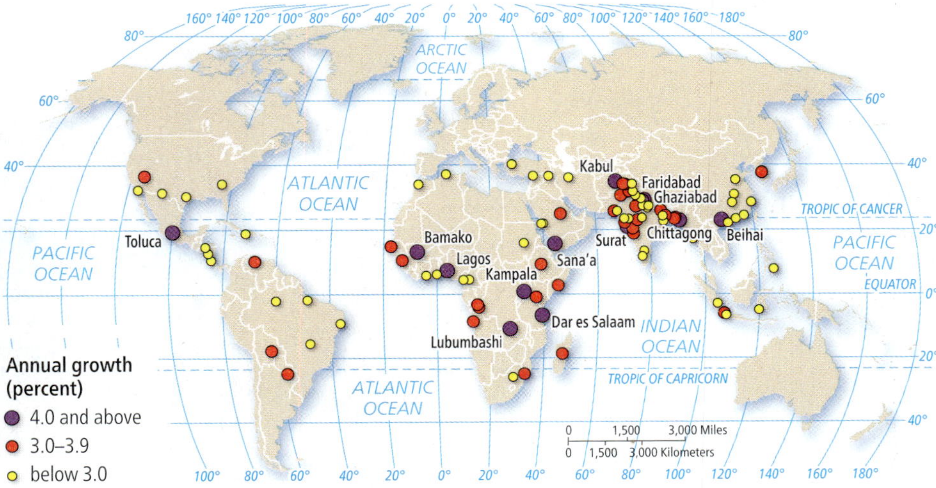

▲ **12.1.4 WORLD'S 100 FASTEST-GROWING URBAN SETTLEMENTS**
Settlements growing more than 4.0 percent per year are labeled.

▶ **12.1.5 BEIHAI, CHINA**
The world's fastest-growing city.

Types of Services

- **Describe the three types of services.**

Most people in developed countries work in urban settlements that provide services, such as shops, offices, restaurants, universities, and hospitals. These are examples of the tertiary—or service—sector of the economy.

12.2.1 Types of Services

A **service** is any activity that fulfills a person's want or need in exchange for money. The service sector of the economy is subdivided into three types: consumer services, business services, and public services. Each of these sectors is subdivided into several major subsectors. A smaller number of people work on farms or in factories, the primary and secondary sectors (Figure 12.2.1).

Consumer services. The principal purpose of a **consumer service** is to provide services to individuals who desire those services and can afford to pay for them. Around half of all jobs in the United States are in consumer services. The main types of consumer services are retail, education, health, and leisure.

Business services. The principal purpose of a **business service** is to facilitate the activities of other businesses. One-fourth of all jobs in the United States are in business services. Professional services, financial services, information services, and transportation services are the main types of business services.

Public services. The purpose of a **public service** is to provide security and protection for citizens and businesses. About 8 percent of all U.S. jobs are in public services. Excluding educators, one-sixth of public service employees work for the federal government, one-fourth for one of the 50 state governments, and three-fifths for one of the tens of thousands of local governments.

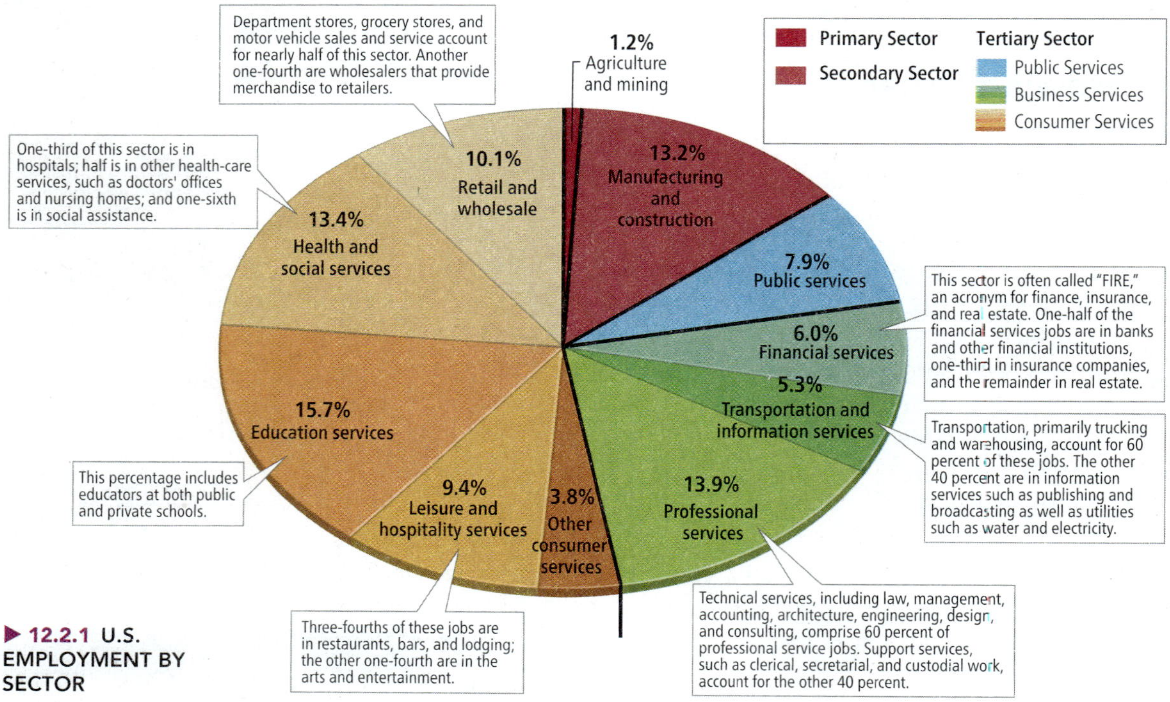

▶ **12.2.1** U.S. EMPLOYMENT BY SECTOR

The distinction among services is not absolute. For example, individual consumers use business services, such as legal services and banks, and businesses use consumer services, such as video conferencing software and hotels. Still, geographers find the services classifications useful because the various services differ in distributions and different factors influence where they are located.

12.2.2 Distribution of Services

Services generate more than two-thirds of GDP in most developed countries, compared to around one-half in developing countries (Figure 12.2.2 and refer to Figure 10.2.3). Logically, the distribution of service workers is opposite the distribution of primary workers (refer to Figure 9.4.2).

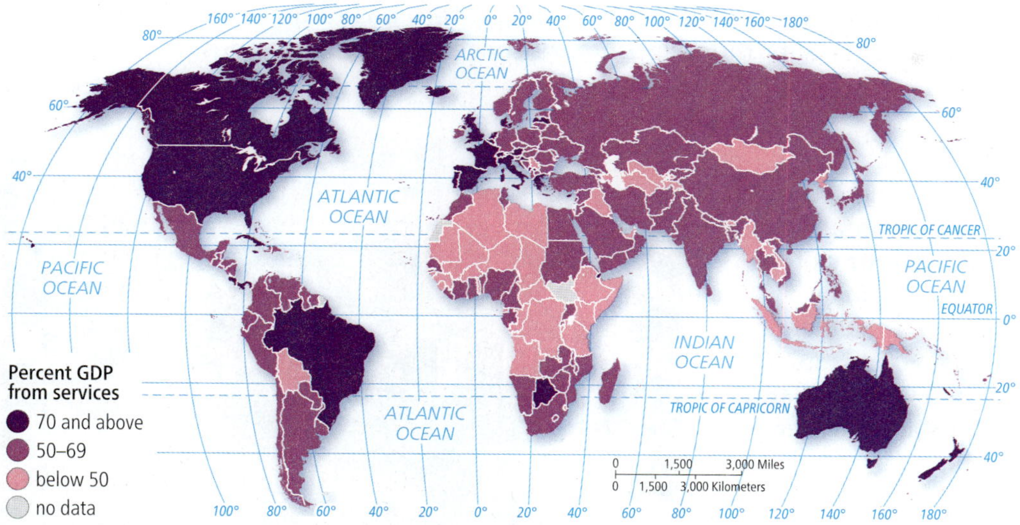

◀ **12.2.2 PERCENTAGE OF GDP FROM SERVICES**

If services were located merely where people lived, then China and India would have the most, rather than the United States and other developed countries. Services cluster in developed countries because more people who are able to buy services live there. Within developed countries, cities offer a larger scale of services than do small towns because more customers reside in cities.

Geographers see a close link between services and settlements because providing services is a principal reason that settlements exist. Thus, the world map showing the percentage of people engaged in services (Figure 12.2.2) looks very similar to the world map showing the percentage of people living in urban settlements (Figure 12.1.2).

On the one hand, explaining why services are clustered in settlements is straightforward for geographers. In geographic terms, only one locational factor is critical for a service: proximity to the market. The optimal location of industry, described in Chapter 11, requires balancing a number of site and situation factors, but the optimal location for a service is simply near its customers. Thus, a settlement is where customers are clustered.

On the other hand, locating a service calls for far more precise geographic skills than locating a factory. The optimal location for a factory may be an area of several hundred square kilometers, whereas the optimal location for a service may be a very specific place, such as a street corner.

Changes in Job Sectors

Figure 12.2.3 shows changes in U.S. employment. All the growth in employment in the United States has been in services, whereas employment in primary- and secondary-sector activities has declined. Within business services, jobs expanded most rapidly in professional services and more slowly in finance and transportation services because of improved efficiency; fewer workers are needed to run trains and answer phones, for example. On the consumer services side, the most rapid increase has been in the provision of health care, education, entertainment, and recreation.

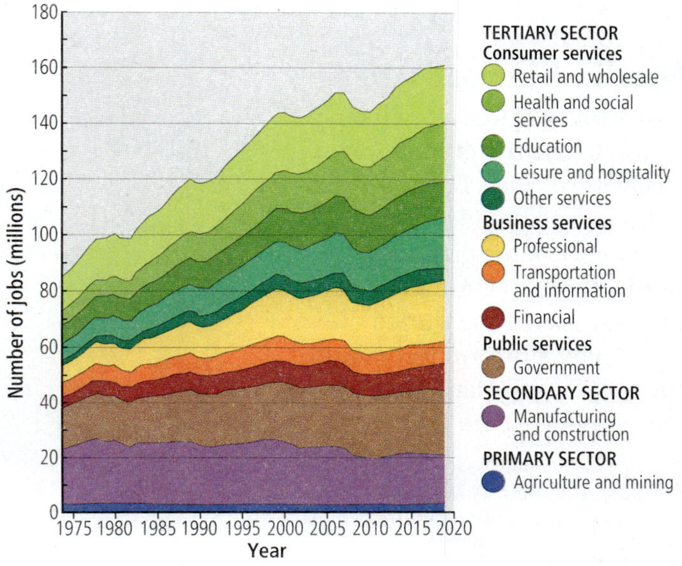

▲ **12.2.3 CHANGE IN U.S. EMPLOYMENT**

Central Place Theory

- **Explain the concepts of market area, range, and threshold.**

Consumer services generally follow a regular pattern based on size of settlements. Larger settlements offer more consumer services than smaller ones.

12.3.1 Central Place Theory

Selecting the right location for a new shop is probably the single most important factor in the profitability of a consumer service. **Central place theory** helps explain how the most profitable location can be identified. German geographer Walter Christaller first proposed the theory in the 1930s based on his studies of southern Germany. August Lösch in Germany and Brian Berry and others in the United States further developed the concept during the 1950s.

▲ 12.3.1 BOISE, IDAHO, THE CENTER OF A MARKET AREA

Market Area of a Service

A **central place** is a market center for the exchange of goods and services by people attracted from the surrounding area. The central place is so called because it is centrally located to maximize accessibility.

Businesses in central places compete against each other to serve as markets for goods and services for the surrounding region. According to central place theory, this competition creates a regular pattern of settlements.

The area surrounding a service from which customers are attracted is the **market area**, or **hinterland**. A market area is a good example of a nodal region—a region with a core where the characteristic is most intense. The hinterland around Boise, Idaho, is an example of a market area. To establish the market area, a circle is drawn around the node of service on a map. The territory inside the circle is its market area (Figure 12.3.1).

Because most people prefer to get services from the nearest location, consumers near the center of the circle obtain services from local establishments. The closer to the periphery of the circle, the greater the percentage of consumers who will choose to obtain services from other nodes. People on the circumference of the market-area circle are equally likely to use the service or go elsewhere.

Media Market Area

A **media market area** is a region where the population receives the same TV and radio station broadcasts and newspapers and consumes similar internet content. Media market areas in the United States typically form hinterlands surrounding the largest urban settlements (Figure 12.3.2).

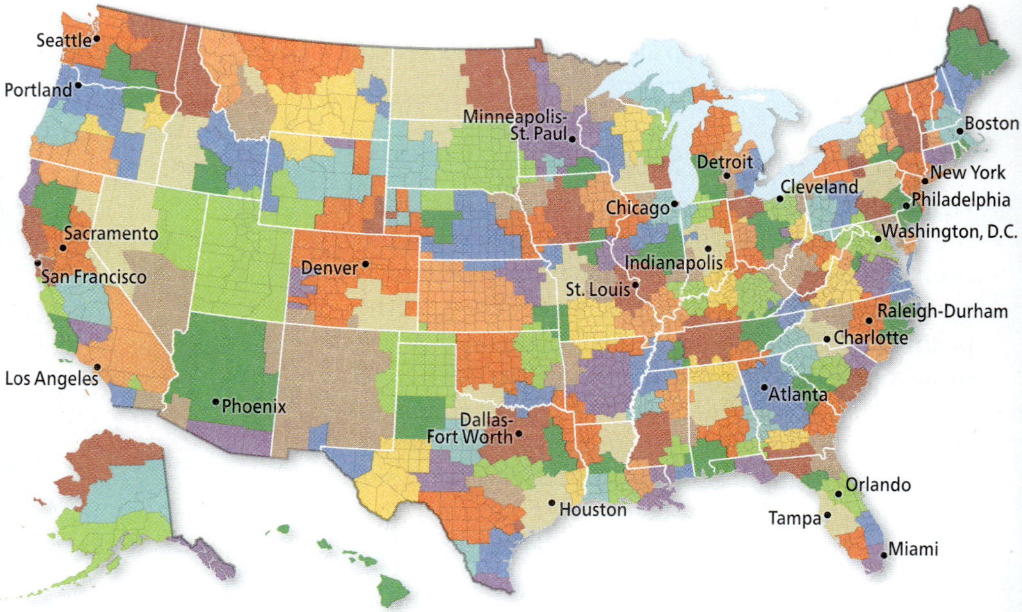

▶ 12.3.2 U.S. MEDIA MARKET AREAS
The Nielsen Corporation determined the market areas based on patterns of TV, radio, newspaper, and internet coverage. The 25 largest market areas are labeled.

12.3.2 Range of a Service

The market area of every service varies. To determine the extent of a market area, geographers need two pieces of information about a service: its range and its threshold.

How far are you willing to travel for a pizza? To see a doctor for a serious problem? To watch a ball game? The range is the maximum distance people are willing to travel to use a service. The **range** is the radius of the circle drawn to delineate a service's market area.

People are willing to go only a short distance for everyday consumer services, such as groceries and pharmacies. But they will travel longer distances for other services, such as a concert or professional ball game. Thus, a convenience store has a small range, whereas a stadium has a large range. As a rule, people tend to go to the nearest available service.

Most people think of the range in terms of time: "One hour" may translate into 100 kilometers (60 miles) of driving on an expressway but only 50 kilometers (30 miles) on city streets.

Threshold of a Service

The second piece of geographic information needed to compute a market area is the **threshold**, which is the minimum number of people needed to support the service. Every enterprise has a minimum number of customers required to generate enough sales to make a profit. So once the range has been determined, a service provider must determine whether a location is suitable by counting the potential customers inside the irregularly shaped circle—often estimated using census data.

How expected consumers inside the range are counted depends on the product. Convenience stores and fast-food restaurants appeal to nearly everyone, whereas other goods and services appeal primarily to certain consumer groups. For example, movie theaters attract younger people, and chiropractors attract older folks. Amusement parks attract families with children, whereas nightclubs appeal to singles.

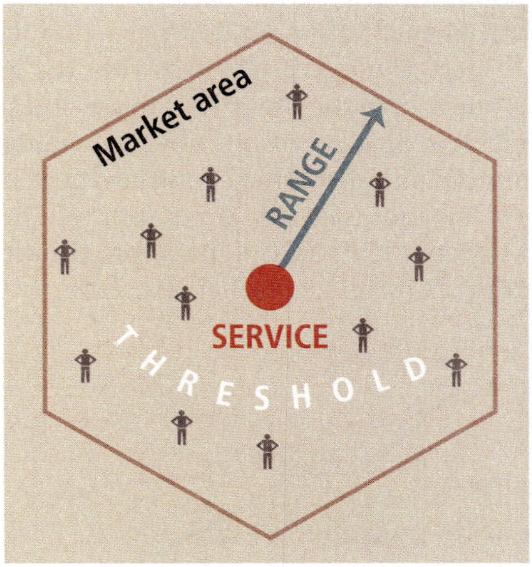

▲ **12.3.3 MARKET AREA, RANGE, AND THRESHOLD**
The market area is the area of the hexagon, the range is the radius, and the threshold is a sufficient number of people inside the area to support the services.

The Geometry of Market Areas

To represent market areas in central place theory, geographers draw hexagons around settlements (Figure 12.3.3). Hexagons represent a compromise between circles (Figure 12.3.4) and squares (Figure 12.3.5). Like squares, hexagons nest without gaps (Figure 12.3.6).

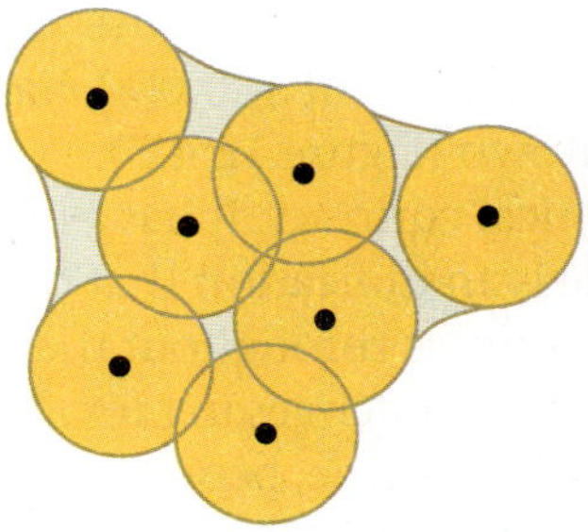

▲ **12.3.4 CENTRAL PLACE THEORY: THE PROBLEM WITH CIRCLES**
Circles are equidistant from center to edge, but they overlap or leave gaps. An arrangement of circles that leaves gaps indicates that people living in the gaps are outside the market area of any service, which is obviously not true. Overlapping circles are also unsatisfactory in determining market area for competing services because one service or another will be closer, and people will tend to patronize it.

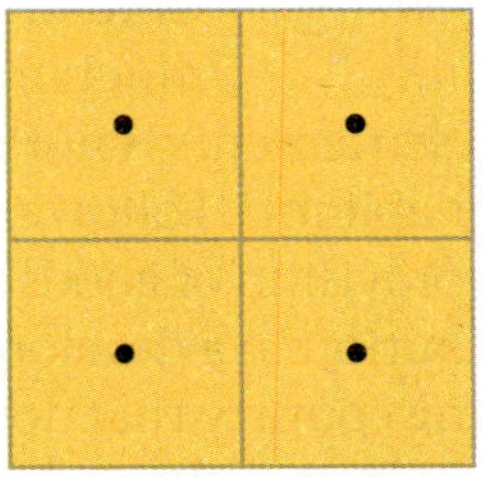

▲ **12.3.5 CENTRAL PLACE THEORY: THE PROBLEM WITH SQUARES**
Squares nest together without gaps, but their sides are not equidistant from the center. If the market area is a circle, the radius—the distance from the center to the edge—can be measured because every point around a circle is the same distance from the center. However, in a square, the distance from the center varies among points along a square.

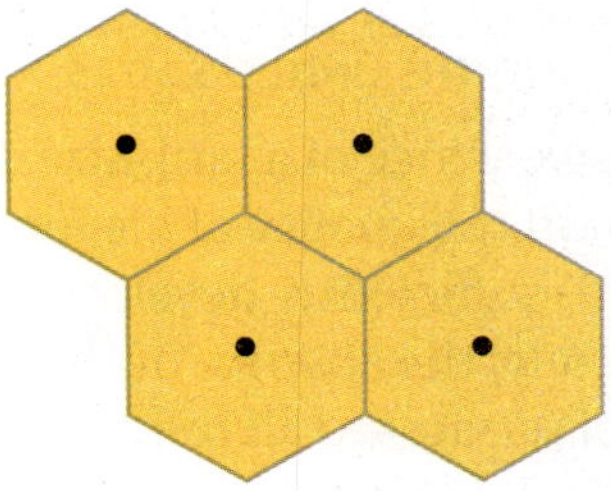

▲ **12.3.6 CENTRAL PLACE THEORY: THE HEXAGON COMPROMISE**
Geographers use hexagons to depict the market area of a service because hexagons offer a compromise between the geometric properties of circles and squares. Like squares, hexagons nest without gaps. Although all points along the hexagon are not the same distance from the center, the variation is less than with a square.

Hierarchy of Consumer Services

- **Describe the distribution of different-sized settlements.**

We tend to spend as little time and effort as possible in obtaining consumer services and thus go to the nearest place that fulfills our needs. There is no point in traveling to a distant store if the same merchandise is available at a nearby one. Typically, we travel a greater distance than to our local shops only if the price of a product we want is much lower elsewhere or if the product is unavailable locally.

12.4.1 Rank-Size Rule

In many developed countries, geographers observe that, when they rank settlements from largest to smallest population, a regular pattern emerges. This is the **rank-size rule**, in which the population of a country's *n*th-largest settlement is 1/*n* the population of the largest settlement.

According to the rank-size rule, the second-largest city is one-half the size of the largest, the fourth-largest city is one-fourth the size of the largest, and so on. When the population of settlements is plotted on a logarithmic scale graph in descending order by size, a fairly straight line appears if the rank-size rule is being followed. In the United States and a handful of other countries, the distribution of settlements follows the rank-size rule at least reasonably closely.

The rank-size rule is not followed in many developing countries, including Mexico. The largest settlement in Mexico (Mexico City) has approximately the same population as the largest settlement in the United States (New York). But the second-largest settlement in Mexico is much smaller than the second-largest settlement in the United States (Figure 12.4.1).

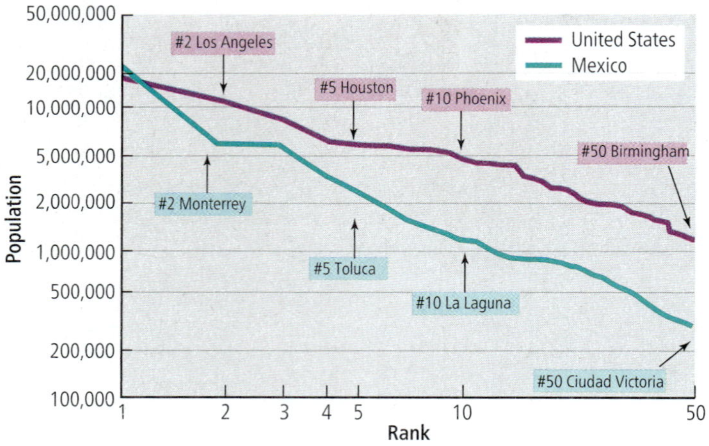

▲ 12.4.1 RANK-SIZE DISTRIBUTION OF SETTLEMENTS IN MEXICO AND THE UNITED STATES

Primate City Rule

In many countries, including, Mexico, the distribution of settlements follows the **primate city rule**, in which the largest settlement has more than twice as many people as the second-ranking settlement. In this distribution, the country's largest city is called the **primate city**.

The difference between the fiftieth-largest settlements of Mexico (Ciudad Victoria) and the United States (Birmingham, Alabama) is visible in the downtown areas. Birmingham, which has a population of more than 1,000,000, has a downtown area filled with high-rise buildings occupied by service providers (Figure 12.4.2). Ciudad Victoria, with a population of 350,000, has a much more modest downtown (Figure 12.4.3).

The existence of a rank-size distribution of settlements is not merely a mathematical curiosity. It has a real impact on the quality of life for a country's inhabitants. Following the rank-size rule is an indication that the society is sufficiently developed to justify the provision of goods and services to consumers throughout the country. Conversely, the absence of the rank-size distribution in a developing country indicates challenges in providing a full range of services to all residents. In a developing country that does not follow the rank-size rule, people may need to travel long distances to reach an urban settlement with shops, hospitals, and other important services.

▼ 12.4.2 DOWNTOWN BIRMINGHAM, ALABAMA

▼ 12.4.3 DOWNTOWN CIUDAD VICTORIA, MEXICO

12.4.2 Nesting of Services & Settlements

According to central place theory, market areas across a developed country would be a series of hexagons of various sizes, unless interrupted by physical features such as mountains and bodies of water. Developed countries have numerous small settlements with small thresholds and ranges and far fewer large settlements with large thresholds and ranges. In his original study, Walter Christaller showed that the distances between settlements in southern Germany followed a regular pattern.

Four different levels of market area—hamlet, village, town, and city—are shown in Figure 12.4.4. Overlapping hexagons of different sizes form a nesting pattern. Hamlets with very small market areas are represented by the smallest contiguous hexagons. Larger hexagons represent the market areas of larger settlements and are overlaid on the smaller hexagons because consumers from smaller settlements shop for some goods and services in larger settlements.

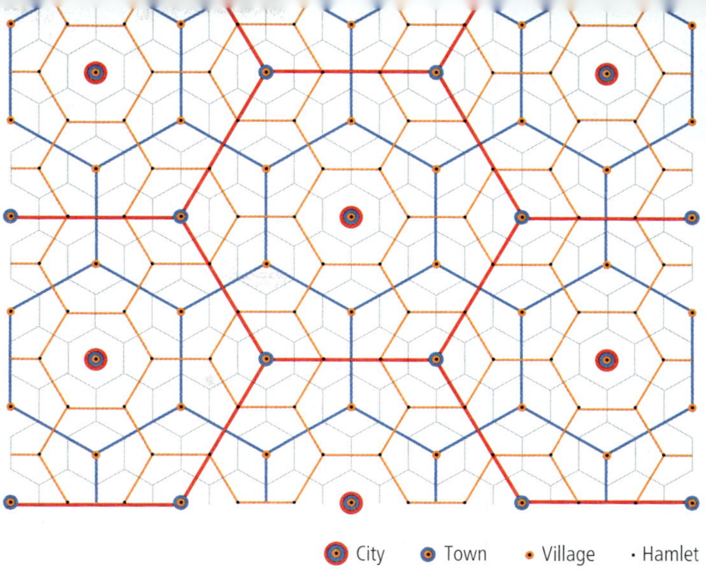

▲ **12.4.4** CENTRAL PLACE THEORY

Central Place Theory Applied to North Dakota

Businesses in central places compete against each other to serve as markets for goods and services for the surrounding region. According to central place theory, this competition creates a regular pattern of settlements.

Across much of the interior of the United States, a regular pattern of settlements can be observed, even if not precisely the same as the generalized model shown in Figure 12.4.4. North-central North Dakota is an example. Minot—the largest city in the area, with 48,000 inhabitants—is surrounded by:

- 13 small towns of between 1,000 and 3,000 inhabitants, with average ranges of 30 kilometers (20 miles) and market areas of around 2,800 square kilometers (1,200 square miles), including Rugby (Figure 12.4.5).
- 29 villages of between 100 and 999 inhabitants, with ranges of 20 kilometers (12 miles) and market areas of around 1,200 square kilometers (500 square miles).
- 34 hamlets of fewer than 100 inhabitants, with ranges of 15 kilometers (10 miles) and market areas of around 800 square kilometers (300 square miles) (Figure 12.4.6).

▲ **12.4.5** CENTRAL PLACE THEORY SMALL TOWN
Rugby, North Dakota.

Larger settlements provide consumer services that have larger thresholds, ranges, and market areas. Only consumer services that have small thresholds, short ranges, and small market areas are found in small settlements because too few people live in small settlements to support many services. A large store cannot survive in a small settlement because the threshold (the minimum number of consumers needed) exceeds the population within range of the settlement. For example, Minot and Bottineau are the only settlements in Figure 12.4.4 that have Walmarts.

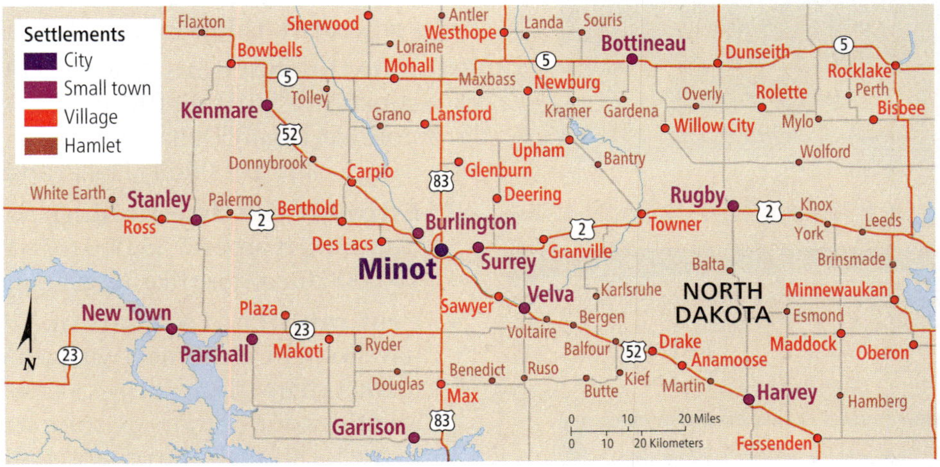

▲ **12.4.6** CENTRAL PLACE THEORY APPLIED TO NORTH DAKOTA

Market Area Analysis

- **Explain how to use threshold and range to find the optimal location for a service.**

Geographers apply central place theory to create market area studies that assist service providers with opening new businesses. And in a severe economic downturn, market area analysis helps determine where to close businesses.

12.5.1 Gravity Model Applied to Retail Geography

Geographers have adapted the gravity model from physics. In retailing geography, the gravity model measures the level of interaction between a person and a service. The **gravity model** predicts that the optimal location of a service is directly related to the number of people in the area and inversely related to the distance people must travel to access it. The best location will be the one that minimizes the distances that all potential customers must travel to reach the service.

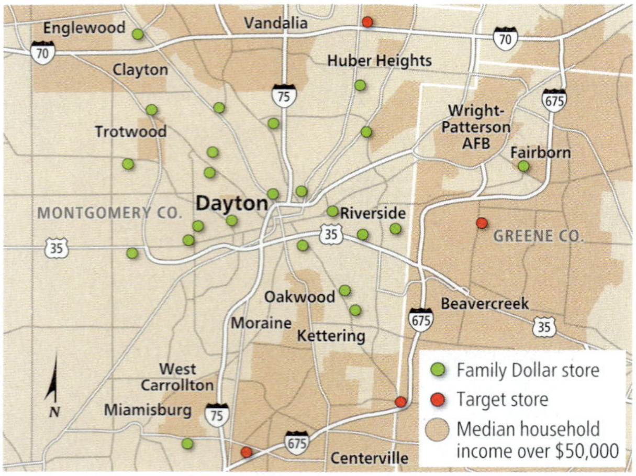

▲ 12.5.1 FAMILY DOLLAR AND TARGET STORES IN DAYTON, OHIO

According to the gravity model, consumer behavior reflects two patterns:

1. The greater the number of people living in a particular place, the greater the number of potential customers for a service. A neighborhood that contains 100 families will generate more customers than an individual house containing only 1 family.
2. The farther people are from a particular service, the less likely they are to use it. People who live 1 kilometer from a store are more likely to patronize it than people who live 10 kilometers away.

Profitability of a Location

The best location for a factory is typically described as a large region, such as auto alley. For service providers, the optimal location is much more precise. For example, one corner of an intersection can be profitable and another corner of the same intersection unprofitable.

Major U.S. consumer services, such as supermarkets and fast-food franchises, employ geographers to determine the best location to build new stores. Geographers use the two components of central place theory—range and threshold—to determine whether a location would be profitable. Here's how:

1. **Define the market area.** The first step in forecasting sales for a proposed new consumer service is to define the market area where the store would derive most of its sales. The market area of a department store is often defined as the ZIP codes where two-thirds to three-fourths of the customers live.
2. **Estimate the range.** Based on the ZIP codes of credit-card customers, geographers can estimate the range for the service. For example, the range is around 15 minutes for a store like Family Dollar and 30 minutes for a store like Target. Figure 12.5.1 shows the location of Family Dollar and Target stores in Dayton, Ohio.
3. **Estimate the threshold.** The threshold varies for each service. The threshold is around 25,000 for a store like Family Dollar (Figure 12.5.2) and 100,000 for a store like Target (Figure 12.5.3).
4. **Profitability of a location.** The proposed new consumer service will have to share customers with competitors. Geographers typically predict market share through the so-called analog method. The geographer identifies one or more existing stores in locations judged to be comparable to the location of the proposed store. The geographer then applies the market share of the comparable stores to the proposed store.

▼ 12.5.2 FAMILY DOLLAR STORE

▼ 12.5.3 TARGET STORE

12.5.2 E-commerce

E-commerce accounts for a rapidly increasing share of consumer services. At the height of the COVID-19 pandemic, when many in-person consumer services were closed, some shoppers found that e-commerce was the principal means of obtaining consumer services. In the United States, Amazon is the dominant e-commerce provider of retail services. Instead of going to stores, consumers order products online for delivery to their homes.

The distribution system established by Amazon in the United States relies for the most part on central place theory. Amazon's geography has three main elements: procurement & fulfillment, distribution, and last mile.

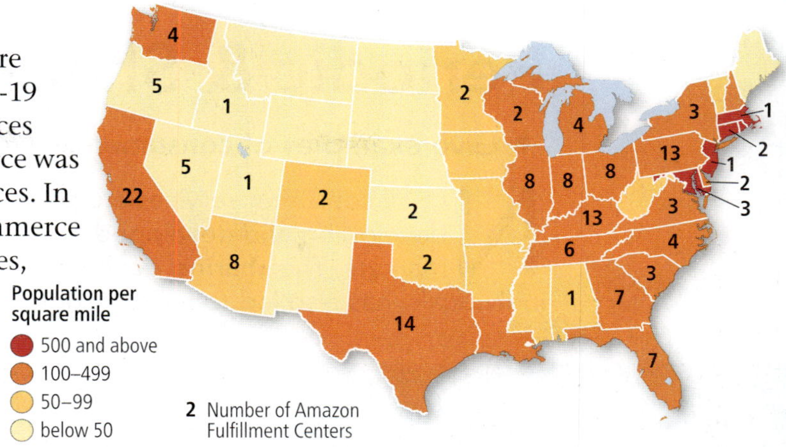

▲ 12.5.4 AMAZON FULFILLMENT CENTERS

Procurement & Fulfillment

Amazon operated 187 fulfillment centers as of 2020 (Figure 12.5.4). These are very large buildings where the various products ordered by an individual are assembled. For the most part, the fulfillment centers are located in more densely populated areas of the United States.

In terms of central place theory, these fulfillment centers have large market areas, offering products with very large ranges and very large thresholds. Some fulfillment centers fill orders that fit in a small box, whereas others fill orders that require a large box. Other fulfillment centers specialize in specific products, such as perishables like groceries or items like appliances too large to box.

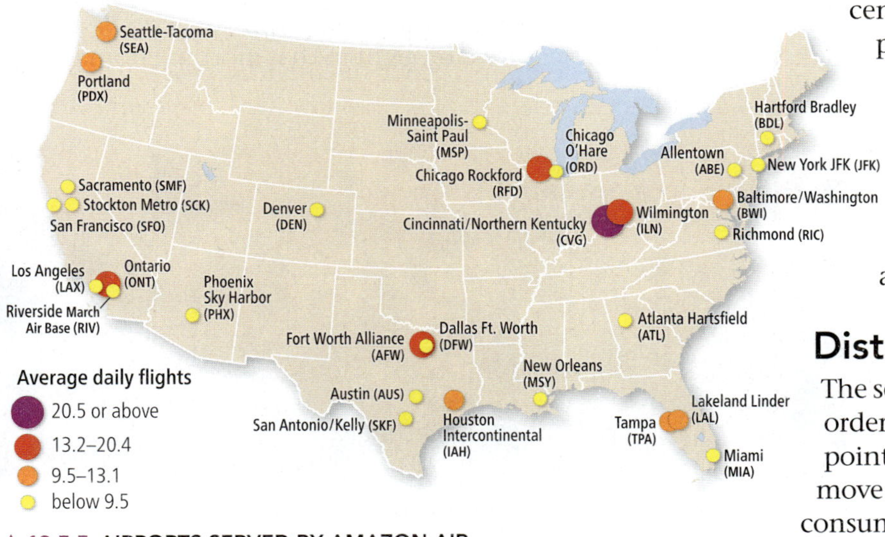

▲ 12.5.5 AIRPORTS SERVED BY AMAZON AIR

Distribution

The second stage is the distribution of parceled orders toward facilities near the final delivery point. Amazon operates its own airline to move goods into fulfillment centers and out to consumers. Amazon Air had landing rights at 28 airports as of 2020 (Figure 12.5.5). Most fulfillment centers are located near an airport served by Amazon Air.

Last Mile

The third stage, known as the last mile, brings parcels to their final destination, mainly through delivery routes from specialized facilities. Parcel hubs arrange packages arriving from various fulfillment centers according to their regional destinations. Sorting centers sort packages into local destinations, such as ZIP code. Packages are then sent to local post offices or other delivery services for delivery to the customer (Figure 12.5.6).

◀ 12.5.6 AMAZON LAST MILE DELIVERY
New York.

Periodic & Sharing Services

- **Give examples of consumer services that do not have a fixed location.**

Not all consumer services are provided in buildings in fixed locations. In both developing and developed countries, some groceries are sold in periodic markets. The **sharing economy**, in which services are shared between private individuals, has also led to more flexible consumer services.

▼ 12.6.1 WEEKLY PERIODIC MARKET, BURKINA FASO

12.6.1 Periodic Markets

Services at the lower end of the central place hierarchy may be provided at a periodic market. A **periodic market** is a collection of individual vendors who come together to offer goods and services in a location on specified days. It is typically set up in a street or other public space early in the morning, taken down at the end of the day, and set up in another location the next day (Figure 12.6.1).

A periodic market provides goods to residents of developing countries where sparse populations and low incomes produce purchasing power too low to support full-time retailing. A periodic market makes services available in more villages than would otherwise be possible, at least on a part-time basis. In urban areas, periodic markets offer residents fresh food brought in that morning from the countryside.

Many of the vendors in periodic markets are mobile, driving their trucks from farm to market, back to the farm to restock, and then to another market. Other vendors, especially local residents who cannot or prefer not to travel to other villages, operate on a part-time basis, perhaps only a few times a year. Other part-time vendors are individuals who are capable of producing only a small quantity of food or handicrafts.

▲ 12.6.2 PERIODIC MARKET, SARLAT-LA-CANÉDA, DORDOGNE, FRANCE

Farmers Markets

A farmers market is a form of periodic market found in many developed countries (Figure 12.6.2). Farmers appear at a specified time and place to sell fresh produce from farms in the surrounding region to consumers living in urban areas.

The frequency of periodic markets varies by culture. In the United States, a farmers market typically sets up once or twice a week, perhaps Wednesdays and Saturdays. In rural areas of Europe, the same farmers might go from town to town through the week; they all appear in town A on Mondays, town B on Tuesdays, town C on Wednesdays, and so forth (Figure 12.6.3).

▲ 12.6.3 PERIODIC MARKETS, DORDOGNE, FRANCE

12.6.2 House Sharing

Services that involve sharing have expanded rapidly, especially in transportation and lodging. These sharing services are challenging the traditional classification of services between consumer and business.

Airbnb started in 2007 in San Francisco to provide short-term visitors with an alternative to hotels. The alternatives can be an entire house or apartment or a room in a house also occupied by the owner. Airbnb house sharing is especially popular among people in their 20s, who make up one-third of Airbnb customers.

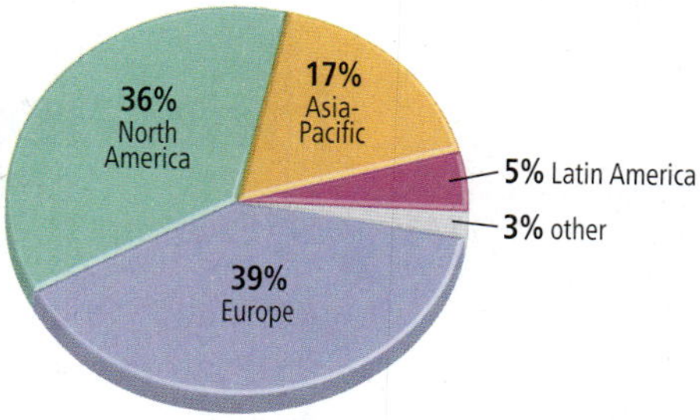

▲ 12.6.4 AIRBNB REVENUES BY REGION

Airbnb has grown to 6 million listings in 100,000 cities, primarily in Europe and North America (Figure 12.6.4). Paris has the largest number of listings, around 80,000. In the United States, Airbnb has an especially strong presence in Florida. Eight of the eleven cities with the most Airbnb listings per capita are in Florida (Figure 12.6.5).

Airbnb is primarily a business service, and most of its employees are computer operators. The owners of the housing pay Airbnb a service fee for the ability to list the property. Airbnb has no responsibility for the cleaning and management of the housing units.

Airbnb provides travelers with a wider variety of accommodations than are available from hotels and typically at a lower cost. However, the conversion of a large number of housing units to Airbnb usage has had negative impacts on some communities. Units are sometimes rented to host large noisy parties, a practice that is especially disruptive for neighbors.

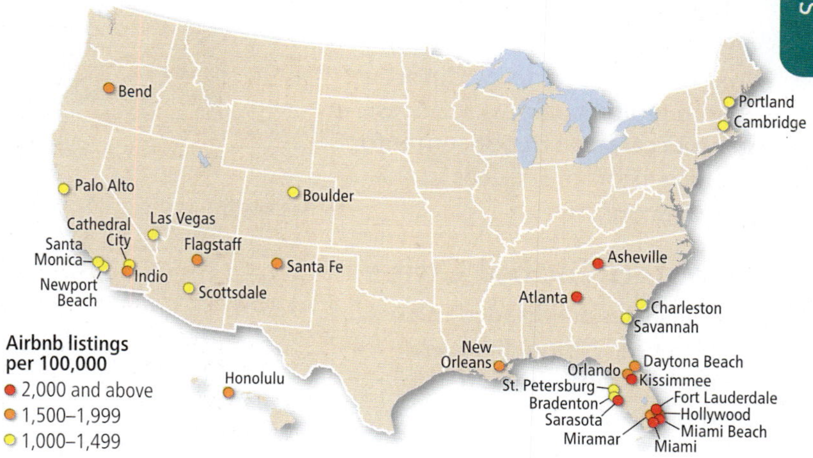

▲ 12.6.5 AIRBNB TOP 30 U.S. CITIES LISTINGS PER CAPITA

Of greater concern in many cities is the loss of affordable rental units for people with limited incomes. Property owners can make more money with short-term rentals through Airbnb than with long-term residents on annual leases. As a result, some cities attempt to limit the use of Airbnb. For example, an apartment in Paris can be used legally by Airbnb for only 120 days of the year.

Ridesharing

Ridesharing services, such as Uber and Lyft, match people looking for a ride with people who are willing for a fee to use their own cars to transport the ride seeker. Most ridesharing services compete with taxis that are company- or independent-contractor-owned cars typically summoned for hire through a phone call or a hit-or-miss process of flagging one down on the street. In contrast, ridesharing services are summoned by using a smartphone app. The driver's smartphone shows the passenger's current location and desired destination as well as the optimal route. It also confirms that the passenger has already paid for the ride.

Uber claims that its four million drivers are independent contractors, essentially one-person businesses providing a transportation business service. Some government regulators have ruled that ridesharing drivers should instead be classified as employees. The classification matters because it is the ridesharing company's responsibility to screen, train, and insure them if they are consumer service employees but not if they are independent business service contractors. Taxi drivers claim that ridesharing services are unfair competitors because they are not subject to the same regulations as taxis (Figure 12.6.6).

◀ 12.6.6 TAXI DRIVERS PROTEST UBER AS AN UNFAIR COMPETITOR
Porto, Portugal.

Hierarchy of Business Services

- **Explain the clustering of business services in global cities.**

Every urban settlement provides consumer services to people in a surrounding area, but not every settlement of a given size has the same number and types of business services. Business services disproportionately cluster in a handful of urban settlements, and individual settlements specialize in particular business services.

12.7.1 Global Cities

Geographers identify a handful of urban settlements known as global cities (also called world cities) that play an especially important role in global business services. A **global city** is a major center for the provision of services in the global economy.

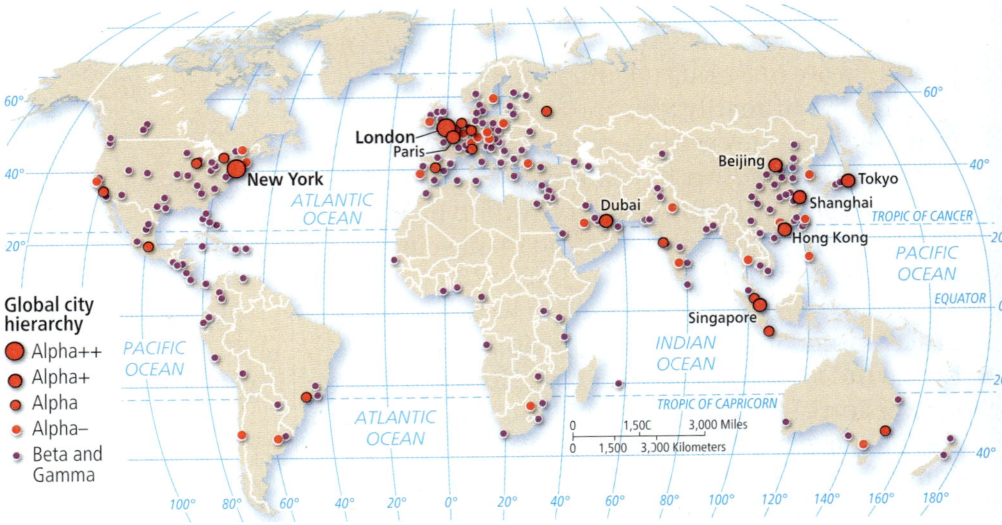

▲ 12.7.1 GLOBAL CITIES

Ranking Global Cities

Global cities are divided into three levels: Alpha, Beta, and Gamma. Alpha cities are the primary nodes in the global economic network. Beta and Gamma cities are smaller global cities that link economic regions into the global network. These three levels are further subdivided (Figure 12.7.1). A combination of factors is used to identify and rank global cities.

- **Economic factors.** Economic factors include the number of headquarters for multinational corporations, financial institutions, and law firms that influence the global economy.
- **Political factors.** Political factors include hosting headquarters for international organizations and capitals of countries that play a leading role in international events.
- **Cultural factors.** Cultural factors include the presence of renowned cultural institutions, influential media outlets, sports facilities, and educational institutions.
- **Infrastructure factors.** Infrastructure factors include rail terminals, interstate highway junctions, international airports, health-care facilities, and advanced communications systems. These factors facilitate communication with coworkers, clients, and customers around the world. Technology was expected to reduce the need for clustering of services in large cities, but it hasn't.

▲ 12.7.2 GLOBAL CITIES IN NORTH AMERICA

Global Cities in North America

The same hierarchy of business services used worldwide can be used within individual regions, including North America (Figure 12.7.2). New York is the only Alpha++ settlement. Chicago, Los Angeles, and Toronto are Alpha settlements.

12.7.2 Business Services in Global Cities

Global cities are most closely integrated into the global economic system because they are at the center of the flow of information and capital. Business services that concentrate in disproportionately large numbers in global cities include financial institutions, corporate headquarters, and professional services.

Financial institutions. As centers for finance, global cities attract the headquarters of the major banks, insurance companies, and specialized financial institutions where corporations obtain and store funds for expansion of production (Figure 12.7.3).

▲ **12.7.3 LONDON BUSINESS SERVICES**
London's financial district is known as The City.

Headquarters of large corporations. Shares of these corporations are bought and sold on stock exchanges located in global cities. Obtaining information in a timely manner is essential in order to buy and sell shares at attractive prices. Executives of manufacturing firms meeting far from the factories make key decisions concerning what to make, how much to produce, and what prices to charge. Support staff far from the factory accounts for the flow of money and materials to and from the factories. This work is done in offices in global cities.

Lawyers, accountants, & other professional services. Professional services cluster in global cities to provide advice to major corporations and financial institutions. Advertising agencies, marketing firms, and other services concerned with style and fashion locate in global cities to help corporations anticipate changes in taste and to help shape those changes.

Consumer Services in Global Cities

Because of their large size, global cities have consumer services with extensive market areas, but they may have even more consumer services than large size alone would predict. A disproportionately large number of people with high incomes live in global cities, so luxury and highly specialized products are especially likely to be sold there.

▲ **12.7.4 LONDON CONSUMER SERVICES**
Theater district.

Leisure services of national significance are especially likely to cluster in global cities, in part because they require large thresholds and large ranges and in part because of the presence of patrons with wealth. Global cities typically offer the most plays, concerts, operas, night clubs, restaurants, bars, and professional sporting events. They contain the largest libraries, museums, and theaters (Figure 12.7.4).

▼ **12.7.5 LONDON PUBLIC SERVICES**
Houses of Parliament.

Public Services in Global Cities

Global cities may be centers of national or international political power. Most are national capitals, and they contain mansions or palaces for the head of state, imposing structures for the national legislature and courts, foreign embassies, and offices for the government agencies (Figure 12.7.5). Also clustered in global cities are offices for groups having business with the government, such as representatives of foreign countries, trade associations, labor unions, and professional organizations.

▲ 12.8.1 OFFSHORE FINANCIAL SERVICE CENTER
Grenada.

Business Services in Developing Countries

- **Describe two types of business services in developing countries.**

In the global economy, developing countries specialize in two distinctive types of business services: offshore financial services and back-office functions. These businesses tend to locate in developing countries for a number of reasons, including the presence of supportive laws, weak regulations, and low-wage workers.

12.8.1 Offshore Financial Services

The International Monetary Fund (IMF) defines an **offshore financial center** as a country or jurisdiction that provides financial services to nonresidents on a scale that is much larger than the size of its domestic economy. Small countries, usually islands and microstates, exploit niches in the circulation of global capital by offering offshore financial services (Figure 12.8.1). Offshore financial centers provide two important functions in the global circulation of capital:

- **Taxes.** Taxes on income, profits, and capital gains are typically low or nonexistent. Companies incorporated in an offshore center also have tax-free status, regardless of the nationality of the owners. The United States loses an estimated $70 billion in tax revenue each year because companies operating in the country conceal their assets in offshore tax havens.
- **Privacy.** Bank secrecy laws can help individuals and businesses evade disclosure of income, profits, and capital gains to their home countries.

Corporations and people who may be accused of malpractice, such as a doctor or lawyer, or negligence, such as a developer of a collapsed building, can protect some of their assets from lawsuits by storing them in offshore centers. So can an individual who wants to protect assets in a divorce. Creditors cannot reach such assets in bankruptcy hearings. Short statutes of limitation protect offshore accounts from long-term investigation.

The privacy laws and low tax rates in offshore centers can also provide havens to tax dodges and other illegal schemes. By definition, the extent of illegal activities is unknown.

Distribution of Offshore Financial Service Centers

Figure 12.8.2 shows locations that the European Union defines as offshore financial service centers. Four types of locations are included: dependencies of the United Kingdom, dependencies of countries other than the United Kingdom, independent island countries, and independent countries that are not islands.

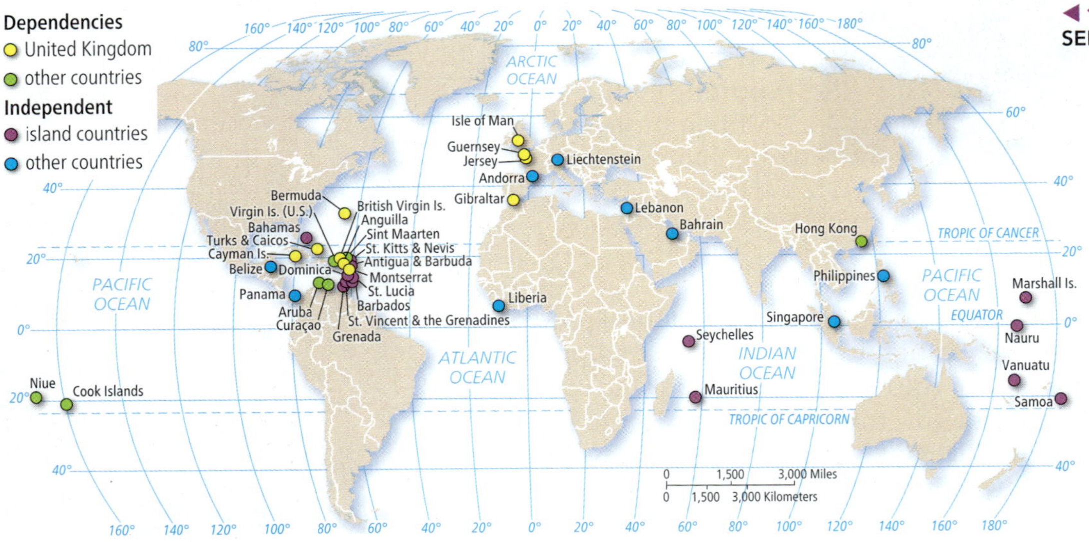

◄ 12.8.2 OFFSHORE FINANCIAL SERVICE CENTERS

12.8.2 Business-Process Outsourcing

A second distinctive type of business service found in peripheral regions is back-office functions, also known as business-process outsourcing (BPO). Typical BPO functions include insurance claims processing, payroll management, transcription work, and other routine clerical activities. BPO work also includes centers for responding to billing inquiries related to credit cards, shipments, and claims, or technical inquiries related to installation, operation, and repair.

Traditionally, companies housed their BPO in the same office building downtown as their management staff, or at least in nearby buildings. A large percentage of the employees in a downtown bank building, for example, would be responsible for sorting paper checks and deposit slips. Proximity was considered important to assure close supervision of routine office workers and rapid turnaround of information.

▲ **12.8.3 ADVERTISEMENT FOR CALL CENTER TRAINING INSTITUTE** Madurai, India.

For many business services, the most important factor in eliminating the need for spatial proximity is improved telecommunications. Better digital connectivity, as well as digitization of more products and services, is sometimes termed the *Fourth Industrial Revolution*. Investment in telecommunications has facilitated the relocation of more BPO services to developing countries (Figure 12.8.3).

Labor Needs in BPO Services

Selected developing countries have attracted BPO services for two reasons related to labor:

- **Low wages.** Most back-office workers earn a few thousand dollars per year—higher than wages paid in most other sectors of the economy but only one-tenth the wages paid for workers performing similar jobs in developed countries. As a result, what is regarded as menial and dead-end work in developed countries may be considered relatively high-status work in developing countries and therefore may attract better-educated, more-motivated employees in developing countries than would be the case in developed countries.
- **Ability to speak English.** Many developing countries offer lower wages than developed countries, but only a handful of developing countries possess a large labor force fluent in English. In Asia, countries such as India, Malaysia, and the Philippines have substantial numbers of workers with English-language skills, a legacy of British and American colonial rule (Figure 12.8.4).

The ability of people in some Asian countries to communicate in English over the telephone is a strategic advantage in competing for back offices with neighboring countries, such as Indonesia and Thailand, where English is less commonly used. Familiarity with English is an advantage not only for answering the telephone but also for gaining a better understanding of the preferences of American consumers through exposure to English-language music, movies, and television. Major multinational companies such as American Express and General Electric have extensive back-office facilities in Asian countries.

Workers in back offices often must work late at night, when it's daytime in the United States and demand for answers to inquiries is at its peak. Many employees must arrive at work early and stay late because they lack their own transportation and depend on public transportation, which typically does not operate late at night. Sleeping and entertainment rooms may be provided at work to fill the extra hours.

▲ **12.8.4 CALL CENTER IN THE PHILIPPINES FOR AN AUSTRALIAN MOVING COMPANY**

▲ 12.9.1 ECONOMIC BASE OF THE SEATTLE AREA Boeing Company, Everett, Washington.

Economic Specialization of Settlements

- **Explain the concept of economic base.**

Settlements can be classified by the distinctive types of economic activities that take place there. All sectors of the economy—be they the various types of agriculture, the various types of manufacturers, or the various types of services—have distinctive geographic distributions.

12.9.1 Economic Base

The economic activities in a settlement can be divided into two types: A **basic business** exports primarily to customers outside the settlement (Figure 12.9.1). A **nonbasic business** serves primarily customers living in the same settlement. The unique cluster of basic businesses in a settlement is its **economic base**.

A settlement's economic base is important because exporting by the basic businesses brings more money into the local economy, thus stimulating the provision of more nonbasic services for the settlement. It works like this: (1) New basic businesses attract new workers to a settlement. (2) The new basic business workers bring their families with them. (3) New nonbasic services are opened to meet the needs of the new workers and their families. For example, when a new motor vehicle factory opens, new consumer services like groceries and restaurants soon follow. But the opposite doesn't occur: A new restaurant does not induce construction of a new car plant.

Economic Base of U.S. Settlements

Settlements in the United States can be classified by their distinctive collection of basic businesses. The concept of basic businesses originally referred to manufacturing, but with the growth of the service sector of the economy, the basic businesses of many communities are in consumer, business, and public services (Figure 12.9.2). If basic businesses of a settlement are growing, they attract other basic and nonbasic businesses that can benefit from proximity. The result can be a cluster of businesses that reinforce each other's growth. For example, Boston's basic sector in biotechnology consists of a cluster of business sectors that complement each other. Conversely, if a settlement's basic businesses are shedding jobs—such as Detroit's auto industry—then other businesses in the cluster may also decline.

▶ 12.9.2 ECONOMIC BASE OF SELECTED U.S. COMMUNITIES

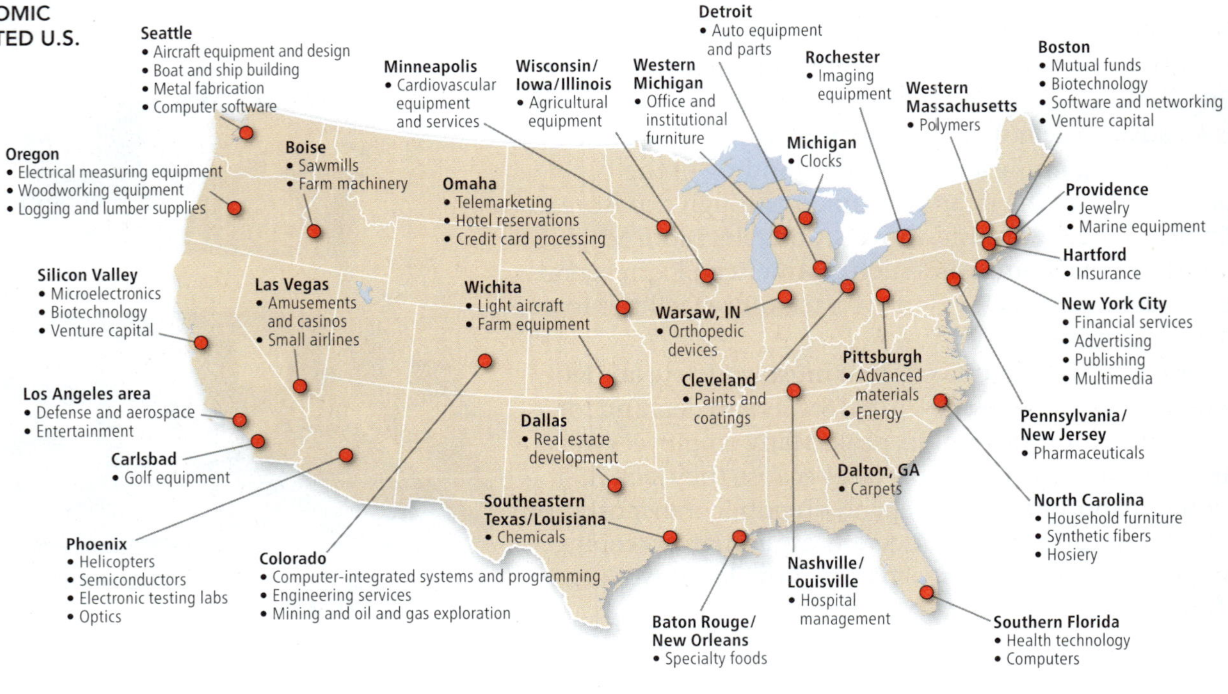

12.9.2 Distribution of Talent

Individuals possessing special talents are not distributed uniformly among settlements. Some settlements have more than their fair share of individuals with various talents. Attracting talented individuals is important because these individuals are responsible for promoting economic innovation in a settlement. They are likely to start new businesses and infuse the settlement's economy with fresh ideas.

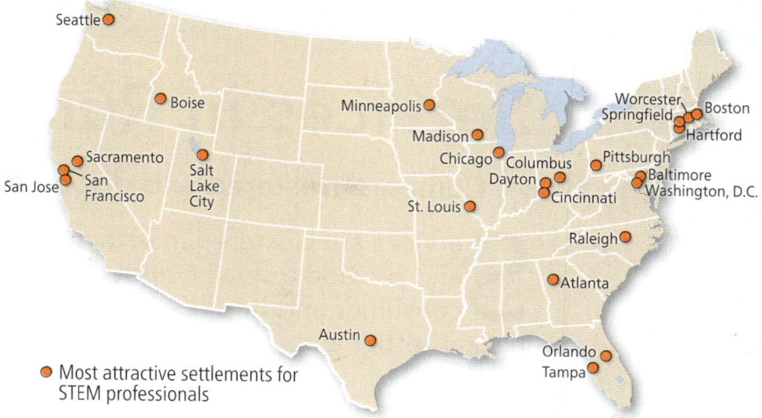

▲ **12.9.3 MOST ATTRACTIVE U.S. SETTLEMENTS FOR STEM PROFESSIONALS**

Settlements for Talented Scientists

For science, technology, engineering, and mathematics (STEM) professionals, the most attractive settlements are Seattle, Boston, Atlanta, Austin, and Minneapolis, according to wallethub.com (Figure 12.9.3). The rankings were based in part on job openings for STEM professionals as well as a judgment concerning the quality of life in a settlement that STEM professionals value.

Settlements for Talented Artists

The settlements most attractive for people in the creative arts differ from those for STEM professionals. New York, San Francisco, Seattle, Los Angeles, and Minneapolis were considered the most attractive in a Creative City Index assessment (Figure 12.9.4). The rankings were based on the numbers of creative jobs per capita, creative arts schools, performing arts companies, and jobs in the motion picture and video industries.

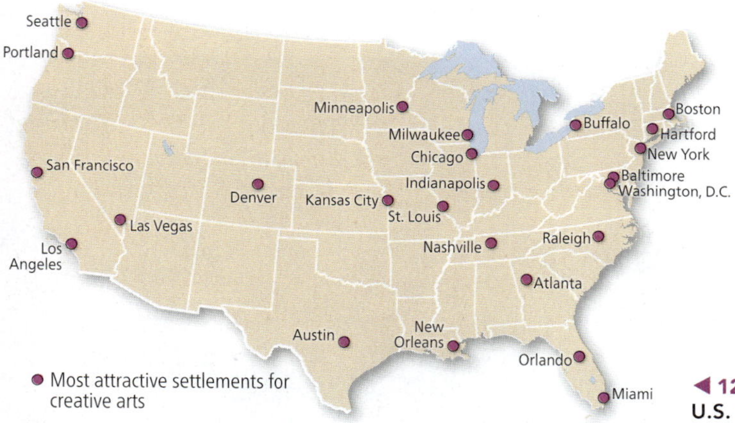

◀ **12.9.4 MOST ATTRACTIVE U.S. SETTLEMENTS FOR CREATIVE ARTS**

Settlements for Recent Graduates

For recent college graduates, the most attractive settlements in the United States are Austin (Figure 12.9.5), San Francisco, Houston, San Antonio, and Columbus, Ohio, according to onlinedegrees.com (Figure 12.9.6). The determination was based on several factors, including median earnings for bachelor's degree holders, housing costs, crime rate, commuting time to work, and the number of arts and entertainment centers.

◀ **12.9.6 MOST ATTRACTIVE U.S. SETTLEMENTS FOR RECENT COLLEGE GRADUATES**

▼ **12.9.5 AUSTIN, TEXAS**
Recreation in Barton Creek.

Services in Rural Settlements

- **Compare clustered and dispersed rural settlements.**

Services are clustered in settlements. Rural settlements are centers for agriculture and provide a small number of services. Urban settlements are centers for consumer and business services. Around one-half of the people in the world live in rural settlements and the other half in urban settlements.

12.10.1 Two Arrangements of Rural Settlements

Rural settlements are either dispersed or clustered. They are most commonly dispersed in the United States and clustered in most other parts of the world.

- A **dispersed rural settlement** is characterized by farmers living on individual farms isolated from neighbors rather than alongside other farmers in settlements.
- A **clustered rural settlement** is an agricultural-based community in which a number of families live in close proximity to each other, with fields surrounding the collection of houses and farm buildings.

Clustered Rural Settlements

A clustered rural settlement typically includes homes, barns, tool sheds, and other farm structures, plus consumer services, such as religious structures, schools, and shops. A handful of public and business services may also be present in a clustered rural settlement.

Each person living in a clustered rural settlement is allocated strips of land in the surrounding fields. Homes, public buildings, and fields in a clustered rural settlement are arranged according to local cultural and physical characteristics. Clustered rural settlements are often arranged in one of two types of patterns: circular or linear.

Circular clustered rural settlements. A circular clustered rural settlement consists of a central open space surrounded by structures. In sub-Saharan Africa, the Maasai people, who are pastoral nomads, build circular settlements known as kraal (Figure 12.10.1). Women have the principal responsibility for constructing them. The kraal villages have enclosures for livestock in the center, surrounded by a ring of houses. Von Thünen observed this circular pattern in Germany in his landmark agricultural studies in the early nineteenth century (refer to Figure 9.6.4).

▲ 12.10.1 CIRCULAR CLUSTERED RURAL SETTLEMENT
Kraal village, Kenya.

Linear clustered rural settlements. Linear rural settlements comprise buildings clustered along a road, river, or dike to facilitate communications. The fields extend behind the buildings in long, narrow strips. Long-lot farms can be seen today along the St. Lawrence River in Québec (Figure 12.10.2). Québec got the system from the French.

▲ 12.10.2 LINEAR CLUSTERED RURAL SETTLEMENT
Lac St. Jean, Québec.

12.10.2 Dispersed Rural Settlements

Isolated farms are typical of most of the rural United States. A dispersed pattern developed from the time of initial settlement of the Middle Atlantic colonies because most immigrants to these colonies arrived individually rather than as members of a cohesive group. As people moved westward from the Middle Atlantic region, they took with them their preference for isolated individual farms. Land was plentiful and cheap, so people bought as much as they could manage (Figure 12.10.3).

▲ **12.10.3 DISPERSED RURAL SETTLEMENT**
Illinois.

Conversion of Rural Settlements from Clustered to Dispersed

In Europe, some clustered settlements were converted to dispersed settlements in order to make agriculture more efficient. Clustered rural settlements worked when the population was low, but they had no spare land to meet the needs of a growing population. With the introduction of machinery, farms operated more efficiently at a larger scale.

▲ **12.10.4 RURAL SETTLEMENT: UNITED KINGDOM**
Luckington was originally a clustered rural settlement, but during the enclosure movement, the surrounding fields were consolidated into large farms.

For example, in the United Kingdom, the **enclosure movement** between 1750 and 1850 resulted in the consolidation of individually owned strips of land surrounding villages into large farms owned by single individuals (Figure 12.10.4). When necessary, the government forced people to give up their holdings.

As displaced farmers moved to urban settlements, the population of clustered rural settlements declined drastically. Because the enclosure movement coincided with the Industrial Revolution, villagers displaced from farming became workers in urban factories.

Clustered Rural Settlements in New England

Although most of the United States was developed with dispersed rural settlements, clustered rural settlements were characteristic of colonial New England. Colonists who settled in New England typically came from England in a group, and they wanted to live close together to reinforce common cultural and religious values. The contemporary New England landscape contains remnants of the old clustered rural settlement pattern. Many New England towns still have a central common surrounded by the church, school, and various houses (Figure 12.10.5).

▲ **12.10.5 CLUSTERED NEW ENGLAND RURAL SETTLEMENT**
Newfane, Vermont. The county courthouse is to the left and the Congregational Church to the right.

Services in Early Urban Settlements

- **Summarize important urban settlements through history.**

No one knows the precise sequence of events through which settlements were established to provide services. The first urban settlements were established prior to the beginning of recorded history around 5,000 years ago.

12.11.1 Prehistoric Urban Settlements

Based on archaeological research, settlements probably originated to provide consumer and public services. Business services probably came later.

Early consumer services. The earliest permanent settlements may have been established to offer consumer services, specifically places to bury the dead. Having established a permanent resting place for the dead, the group might then install officials at the site to watch over the deceased. This would have encouraged the building of structures—places for ceremonies and dwellings. By the time recorded history began, many settlements existed, and some of them featured places of worship (Figure 12.11.1).

Early public services. Early settlements housed political leaders as well as defense forces to guard the residents of the settlement and defend the surrounding hinterland from seizure by other groups.

Early business services. Early urban settlements were places where groups could store surplus food and trade with other groups. People brought plants, animals, and minerals as well as tools, clothing, and containers to the urban settlements and exchanged them for items brought by others. To facilitate this trade, officials in the settlement set fair prices, kept records, and created currency.

▲ 12.11.1 THE ZIGGURAT OF UR IN PRESENT-DAY IRAQ
Constructed as part of a temple complex around 2500 B.C.E.

▼ 12.11.2 MEMPHIS, EGYPT
Possibly the world's largest urban settlement around 5,000 years ago.

Ancient Urban Settlements

Urban settlements may have been first established in the eastern Mediterranean around 2500 B.C.E. in the area known as the Fertile Crescent (refer to Section 8.3.1 and Figure 8.3.1). Urban settlements may have diffused from the Fertile Crescent west to Egypt and east to China and to South Asia's Indus Valley. Or settlements may have originated independently in each of the four hearths. The largest urban settlement 5,000 years ago may have been Memphis, Egypt (Figure 12.11.2).

12.11.2 Medieval Urban Settlements

The rise of the Roman Empire encouraged urban settlement. With much of Europe and Southwest Asia & North Africa under Roman rule, settlements were established as centers of administrative, military, and other public services as well as retail and other consumer services. Trade was encouraged through transportation and utility services, notably construction of many roads and aqueducts, and the security the Roman legions provided. Rome—the empire's center for administration, commerce, culture, and all other services—was the world's most populous city 2,000 years ago and may have been the first city to reach half a million inhabitants.

With the collapse of the Roman Empire in the fifth century C.E., urban settlements declined. The empire's prosperity had rested on trading in the secure environment of imperial Rome. But with the empire fragmented under hundreds of rulers, trade diminished. Large urban settlements shrank or were abandoned. For several hundred years, Europe's cultural heritage, such as books and art, was preserved largely in monasteries and isolated rural areas.

The largest medieval European urban settlements served as power centers for the lords and church leaders as well as major market centers. The most important public services occupied palaces, churches, and other prominent buildings arranged around a central market square. The tallest and most elaborate structures were usually churches, many of which still dominate the landscape of smaller European towns. In medieval times, European urban settlements were often surrounded by walls even though by then cannonballs could destroy them (Figure 12.11.3).

▲ **12.11.3** MEDIEVAL WALLED URBAN SETTLEMENT
Aigues-Mortes, France.

World's Largest Urban Settlement Through History

In ancient times, the world's largest urban settlement was in the Fertile Crescent. After the collapse of the Roman Empire, most of the world's largest urban settlements were clustered in China. Several cities in China are estimated to have been the world's most populous at various times between 600 and 1500 C.E., including Beijing, Ch'ang-an, Hangzhou, Kaifeng, and Nanking. During the past 300 years, the world's largest urban settlement has been London, New York, or Tokyo (Figure 12.11.4).

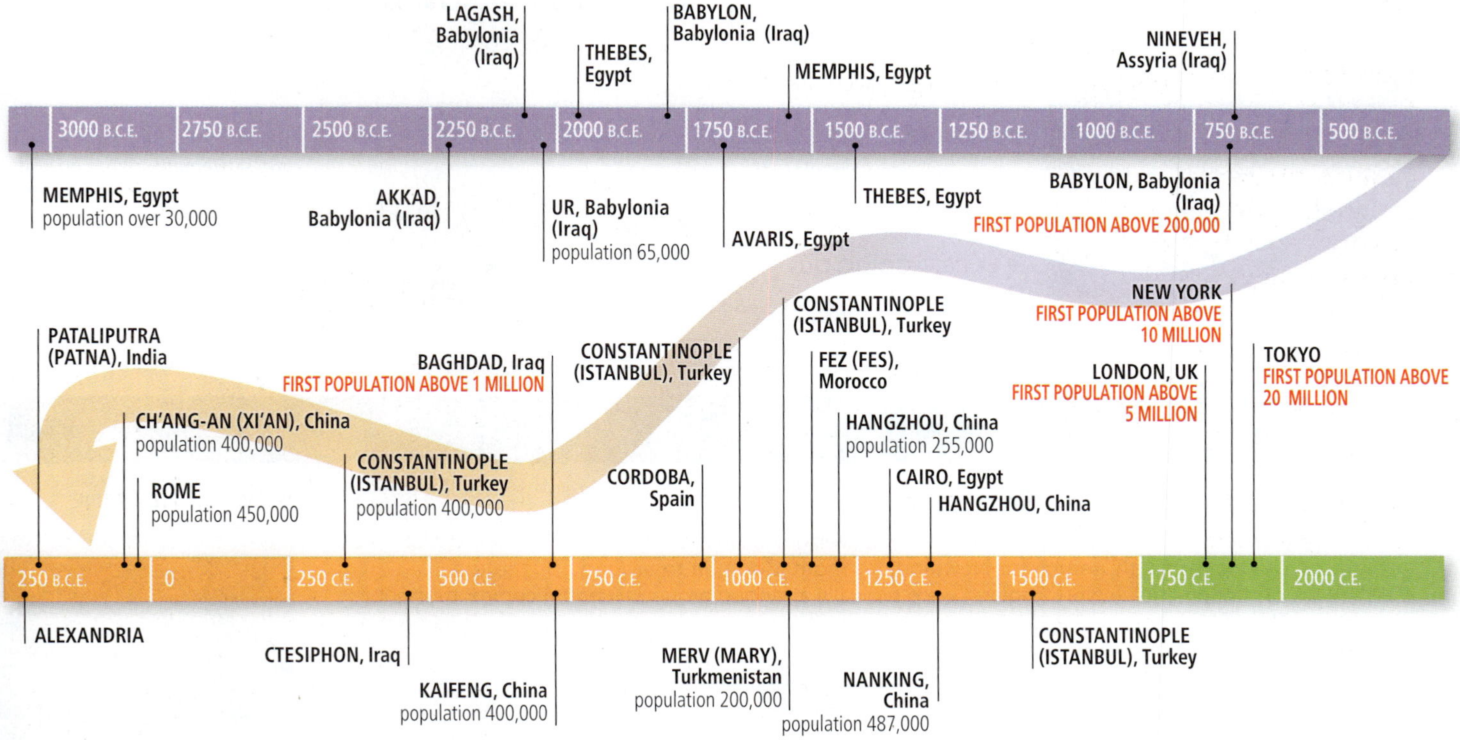

▲ **12.11.4** LARGEST URBAN SETTLEMENTS THROUGH HISTORY

CHAPTER 12

Review, Analyze, & Apply

KEY ISSUE 1 Where Are Settlements & Services Distributed?

- Settlements are either rural or urban.
- More than half of the people in the world now live in urban settlements.
- Settlements are centers for the provision of services.
- Three types of services are consumer, business, and public.

THINKING GEOGRAPHICALLY

An initiative called "Tulsa Remote" is offering $10,000 to tech workers to relocate to Oklahoma. Compared with a megacity, what advantages might be offered by a small settlement like Tulsa?

◀ 12.CR.1 YOGA IN THE PARK, TULSA

KEY ISSUE 2 Where Are Consumer Services Distributed?

- A central place is surrounded by a market area that has a range and a threshold.
- Market areas of varying sizes nest and overlap.
- Regular patterns of settlements that provide consumer services can be observed, especially in developed countries.

THINKING GEOGRAPHICALLY

Shopping malls and specialized retailers have been going out of business in many U.S. settlements. What changes in people's shopping habits have contributed to these closures?

◀ 12.CR.2 GOING OUT OF BUSINESS

KEY ISSUE 3 Where Are Business Services Distributed?

- Business services cluster in global cities.
- Developing countries provide offshore financial services and business-process outsourcing.
- A settlement may specialize in the provision of a particular service.
- A settlement's specialized services constitute its economic base.

THINKING GEOGRAPHICALLY

Liechtenstein is a developed country in the heart of Europe that is an offshore financial service center (refer to Figure 12.8.2). What advantages might Liechtenstein's location in Europe offer as an offshore financial service center?

◀ 12.CR.3 NATIONAL BANK OF LIECHTENSTEIN

KEY ISSUE 4 Why Do Services Cluster in Settlements?

- The first settlements were established prior to recorded history.
- Rural settlements are centers for agriculture and provide a small number of services.
- Urban settlements are centers for consumer, business, and public services.

THINKING GEOGRAPHICALLY

Memphis, Egypt, was the world's largest settlement 5,000 years ago. The largest structure in Memphis was a pyramid (refer to Figure 12.11.2). Based on our understanding of services in early settlements, what do you expect to be the purpose of the settlement's most prominent structure?

◀ 12.CR.4 MEMPHIS, EGYPT

Explore

New England Clustered Settlement

Most U.S. rural settlements are dispersed. Newfane, Vermont, is an example of a clustered rural settlement. In Google Earth, search for Newfane, Vermont. Zoom in to around 500 meters. Clustered rural settlements in New England were typically arranged around a central common or public open space. Newfane's common is divided by roads into three sections—a small triangular section to the west, a larger triangular section in the middle, and a semicircular section to the east.

1. Describe the uses and the building materials of the structures that surround the common in Newfane. Some are labeled. You will have to enter Street View to check out most of them.
2. Why do you think these structures were placed at the most prominent location in the rural village?

▼ 12.CR.5 NEWFANE

Key Terms

Basic business (12.9.1) A business that sells its products or services primarily to consumers outside the settlement.

Business service (12.2.1) A service that primarily meets the needs of other businesses, including professional, financial, and transportation services.

Central place (12.3.1) A market center for the exchange of services by people attracted from the surrounding area.

Central place theory (12.3.1) A theory that explains the distribution of services based on the fact that settlements serve as centers of market areas for services; larger settlements are fewer and farther apart than smaller settlements and provide services for a larger number of people who are willing to travel farther.

Clustered rural settlement (12.10.1) A rural settlement in which the houses and farm buildings of each family are situated close to each other, with fields surrounding the settlement.

Consumer service (12.2.1) A business that provides services primarily to individual consumers, including retail services and education, health, and leisure services.

Dispersed rural settlement (12.10.1) A rural settlement pattern characterized by isolated farms rather than clustered villages.

Economic base (12.9.1) A community's collection of basic businesses.

Enclosure movement (12.10.2) The process of consolidating small landholdings into a smaller number of larger farms in England during the eighteenth century.

Global city (12.7.1) A major center for the provision of services in the global economy.

Gravity model (12.5.1) A model that holds that the potential use of a service at a particular location is directly related to the number of people in a location and inversely related to the distance people must travel to reach the service.

Market area (or **hinterland**) (12.3.1) The region surrounding a central place of goods and services from which people travel to access those goods and services.

Media market area (12.3.1) A region where the population receives the same TV and radio station broadcast and newspapers and consumes similar internet content.

Megacity (12.1.2) An urban settlement with a total population in excess of 10 million people.

Metacity (12.1.2) An urban settlement with a total population in excess of 20 million people.

Nonbasic business (12.9.1) A business that sells its products primarily to consumers in the community.

Offshore financial center (12.8.1) A country or jurisdiction that provides financial services to nonresidents on a scale that is much larger than the size of its domestic economy.

Periodic market (12.6.1) A collection of individual vendors who come together to offer goods and services in a location on specified days.

Primate city (12.4.1) The largest settlement in a country, if it has more than twice as many people as the second-ranking settlement.

Primate city rule (12.4.1) A pattern of settlements in a country such that the largest settlement has more than twice as many people as the second-ranking settlement.

Public service (12.2.1) A service offered by the government to provide security and protection for citizens and businesses.

Range (of a service) (12.3.2) The maximum distance people are willing to travel to use a service.

Rank-size rule (12.4.1) A pattern of settlements in a country such that the *n*th largest settlement is 1/*n* the population of the largest settlement.

Rural settlement (12.1.1) A community where most people are engaged in agriculture.

Service (12.2.1) Any activity that fulfills a person's want or need in exchange for money.

Settlement (12.1.1) A permanent collection of buildings and inhabitants.

Sharing economy (12.6.1) Services that are shared between private individuals.

Threshold (12.3.2) The minimum number of people needed to support a service.

Urban settlement (12.1.1) A community where people are employed in secondary- and tertiary-sector jobs.

Urbanization (12.1.1) An increase in the percentage of people living in urban settlements.

Geospatial Analysis

Log in to the Mastering Geography Study Area to access MapMaster 2.0.

Urbanization & Environment in China

China's cities are not distributed uniformly across the country. How are China's cities related to environmental features?

Add the ***Cities*** layer. Select the *Settings* icon from the *Legend* and select *Show Physical Labels*. Zoom to East Asia.

1. Are China's cities more likely to be located near water or far away from water? What might account for this distribution?
2. Add the ***World Elevations*** layer. Are China's cities more likely to be located in regions of high elevation or low elevation? What might account for this distribution?
3. Add the ***Areas Affected by Acid Rain*** layer and replace the ***World Elevations*** layer. Are China's cities more likely to be located in regions adversely affected by acid rain? What might account for this distribution?

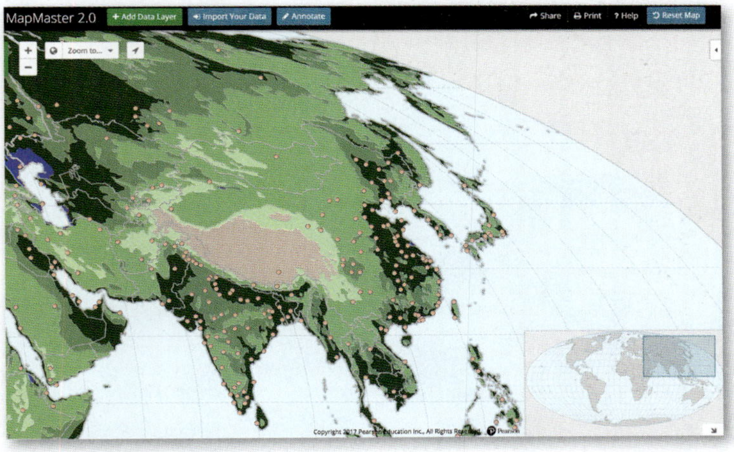

▲ **12.CR.6** ENVIRONMENTS IN CHINA

CHAPTER

13 Urban Patterns

House repairs, Chicago.

When you are staring at a skyscraper, you know you are in a city. When you are standing in a cornfield, you have no doubt that you are in the country. Geographers help explain why urban and rural settlements are different. The previous chapter examined the distribution of urban settlements at national and global scales. This chapter looks at where people and activities are distributed within urban areas.

KEY ISSUES

1 Why Are Cities Challenging to Define?

Services, especially business and public services, cluster downtown. The growth of cities surrounding downtown has resulted in the need to construct various definitions of their extent.

2 Where Are People Distributed Within a City?

Several models have been created to describe where different people tend to cluster within U.S. urban areas, depending on their age, income, and ethnicity, among other factors.

3 How Are Cities Outside North America Structured?

The models of urban structure developed in the United States can also describe—with modifications—urban areas elsewhere in the world.

4 How Might Cities Become More Sustainable?

The rapid growth of cities has resulted in development of extensive suburbs where residents depend exclusively on motor vehicles for transportation. Meanwhile, cities face a variety of challenges with perseverance and hope.

LOCATIONS IN THIS CHAPTER

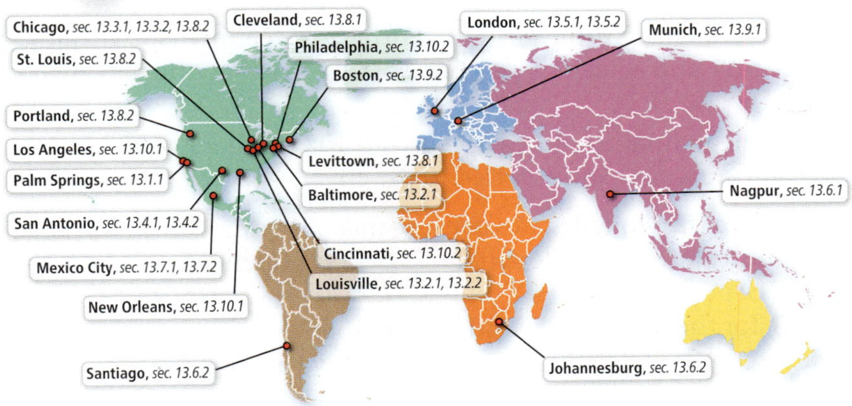

Defining Urban Settlements

- **Compare various definitions of urban settlements.**

Historically, urban settlements were small and compact. As settlements have grown, definitions have been created to characterize their different parts. Examples include municipality (Figure 13.1.1), urban area, and metropolitan area.

▲ 13.1.1 PALM SPRINGS, CALIFORNIA, AN EXAMPLE OF A MUNICIPALITY

13.1.1 Definitions of Urban Settlements in the United States

Figure 13.1.2 illustrates several ways that urban settlements are defined in the United States.

Municipality. A **municipality** or **central city** (or simply city) is an urban settlement that has been legally incorporated into an independent, self-governing unit.

A city has locally elected officials, the ability to raise taxes, and responsibility for providing essential services. The boundaries of the city define the geographic area within which the local government has legal authority, making it an excellent example of a formal region (refer to Section 1.6.1). The city limits are often clearly delineated, as in Figure 13.1.1.

Urban area. An **urban area** consists of a central city and its surrounding built-up suburbs. The U.S. census recognizes two types of urban areas:

- An **urbanized area** is an urban area with at least 50,000 inhabitants.
- An **urban cluster** is an urban area with between 2,500 and 50,000 inhabitants.

The census identified 486 urbanized areas and 3,087 urban clusters in the United States as of 2010. Approximately 70 percent of the U.S. population lived in one of the 486 urbanized areas, including about 30 percent in central cities and 40 percent in surrounding jurisdictions. Approximately 10 percent of the U.S. population lived in one of the 3,087 urban clusters.

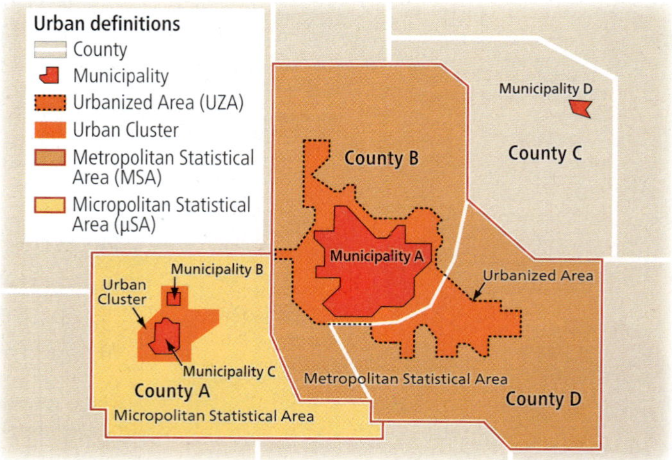

▲ 13.1.2 DEFINITIONS OF URBAN SETTLEMENTS IN THE UNITED STATES

Metropolitan area. The U.S. Office of Management and Budget (OMB) recognizes two types of metropolitan areas.

- A **metropolitan statistical area (MSA)** includes an urbanized area with a population of at least 50,000; the county within which the central city is located (in New England, towns are sometimes used instead of counties); and adjacent counties with a high population density and a large percentage of residents working in the central city's county (specifically, a county with a density of 25 persons per square mile and at least 50 percent working in the central city's county).
- A **micropolitan statistical area (µSA)** includes an urbanized area of between 10,000 and 50,000 inhabitants, the county in which it is located, and adjacent counties tied to the city.

MSAs are widely used to define urban settlements in the United States because many statistics are published for counties, the basic building block for MSAs in most states. OMB designated 392 MSAs (including 8 in Puerto Rico) as of 2020, encompassing 86 percent of the U.S. population.

The U.S. OMB has designed 543 µSAs as of 2020, for the most part found around southern and western communities previously considered rural in character. About 8 percent of Americans live in an µSA.

13.1.2 Megalopolis

A **megalopolis** is a collection of adjacent or overlapping metropolitan areas that merge into a continuous urban region. The word is derived from the Greek meaning "great city." Geographer Jean Gottmann applied the term *Megalopolis* (with an upper case *M*) specifically to a continuous urban complex in the northeastern United States that extends more than 700 kilometers (440 miles) between north of Boston and south of Washington (Figure 13.1.3). The Megalopolis region, also known as the Boswash corridor, contains one-fourth of the U.S. population on only 2 percent of the country's total land area.

▲ **13.1.3 MEGALOPOLIS**
The red lines show boundaries between MSAs and μSAs, and the white lines show boundaries between counties within a single area.

Other urban complexes exist in the United States, including the southern Great Lakes between Milwaukee and Pittsburgh and southern California between Los Angeles and Tijuana. Among important examples in other of the world's developed regions are the German Ruhr (including the cities of Dortmund, Düsseldorf, and Essen), Randstad in the Netherlands (including the cities of Amsterdam, The Hague, and Rotterdam), and Japan's Tokaido (including the cities of Tokyo and Yokohama).

Within Megalopolis, individual municipalities, such as Baltimore, New York, and Philadelphia, retain distinctive identities, and the urban areas are visibly separated from each other by open space used as parks, military bases, and farms. But at the periphery of the urban areas, the boundaries overlap.

Combining Metropolitan Areas

Because adjacent metropolitan areas often overlap, OMB combines MSAs and μSAs in two ways:

- A **core-based statistical area (CBSA)** is any one MSA or μSA. The United States had 935 CBSAs as of 2020, including the 392 MSAs and the 543 μSAs.
- A **combined statistical area (CSA)** is two or more contiguous CBSAs tied together by commuting patterns. The United States had 176 CSAs as of 2020.

Figure 13.1.4 includes two CBSAs (the MSA consisting of Counties B and D and the μSA consisting of County A) and one CSA (Counties A, B, and D).

▶ **13.1.4 CBSA AND CSA**

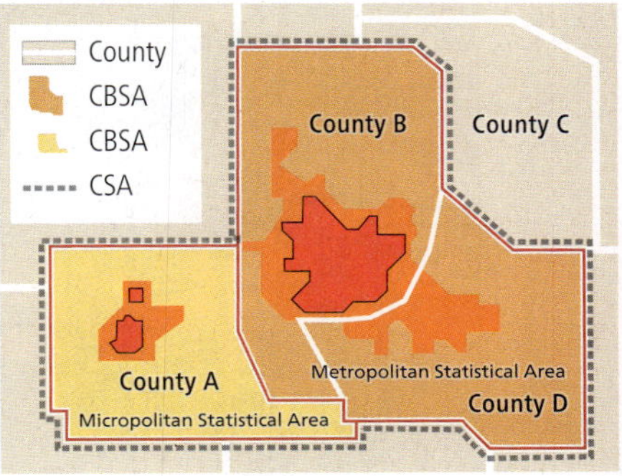

The Central Business District

- **Describe distinctive features of the central business district.**

Downtown is the best-known and the most visually distinctive area of most cities. Downtown is one of the oldest districts in a city, usually at or near the original site of the settlement.

▼ 13.2.1 BALTIMORE, MARYLAND, CBD
(a) In 1903, when industry was an important land use. (b) In recent years, consumer services and boats for recreation have replaced factories and ships transporting goods to and from factories.

(a)

(b)

13.2.1 Central Business District

Downtown is known to geographers by the more precise term **central business district (CBD)**. The CBD is compact—less than 1 percent of the urban land area—but contains a large percentage of the public, business, and consumer services. Services are attracted to the CBD because of its accessibility. The CBD is the focal point of the region's transportation network and therefore often the easiest part of the city to reach from the rest of the region.

A CBD's accessibility produces extreme competition for the limited available land. As a result, land values are very high in the CBD. In a rural area a hectare of land might cost several thousand dollars. In a suburb it might run tens of thousands of dollars. In the CBD of a global city like London, if a hectare of land were even available, it would cost more than $200 million.

As a result of intense competition for land, the CBD has distinctive features:

- Land uses commonly found elsewhere in the urban area, especially industrial and residential activities, are rare in the CBD because rents can be extremely high. Factories once located in the CBD, especially in a port city (Figure 13.2.1). Modern factories require large parcels of land to spread operations among one-story buildings. Suitable land is found in suburbs and rural areas rather than downtown (refer to Section 11.3.2).
- The CBD is three-dimensional, with more space used below and above ground level than elsewhere in the urban area. Demand for space makes construction of high-rise buildings economically feasible in the CBD, as Figures 13.2.1 and 13.2.2 illustrate.

▼ 13.2.2 LOUISVILLE, KENTUCKY, CBD

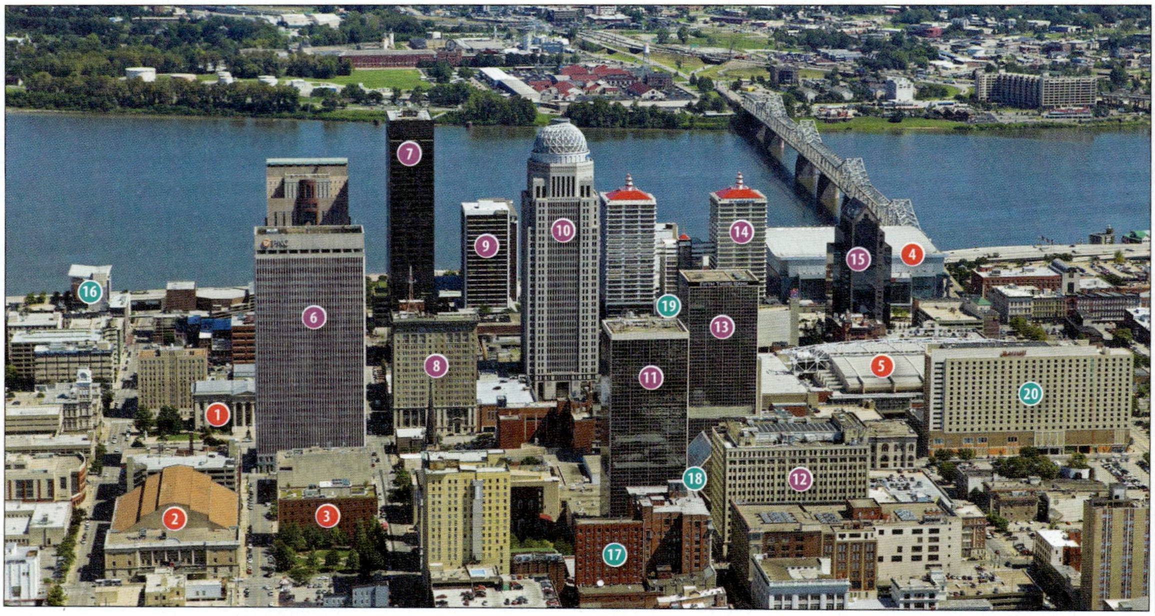

Public services
1 Metro Hall
2 Louisville Gardens
3 Louisville Metro Development Center
4 KFC Yum! Center
5 Kentucky International Convention Center

Business services
6 PNC Plaza
7 National City Tower
8 Kentucky Home Life Building
9 One Riverfront Plaza
10 400 West Market
11 Meidinger Tower
12 Starks Tower
13 B&W Tower
14 Waterfront Plaza
15 LG&E Center

Consumer services
16 Muhammad Ali Center
17 Seelbach Hilton Hotel
18 Fourth Street Live
19 Actors Theater
20 Marriott Hotel

13.2.2 Services in the CBD

Land in the CBD is devoted primarily to the service sector of the economy. All three types of services—public services, business services, and consumer services—are found in the CBD.

Public services in the CBD. Public services that typically locate in a CBD include government offices (Figure 13.2.3), libraries, and museums. These facilities historically clustered downtown, in many cases in substantial buildings. Today, many public services remain in the CBD to facilitate access for people living in all parts of the city. Similarly, semipublic services such as places of worship and social service agencies also cluster downtown in handsome historic structures.

▲ **13.2.3 PUBLIC SERVICE IN LOUISVILLE**
Statue of King Louis XIV in front of Jefferson County Court House.

Business services in the CBD. Offices cluster in a CBD for accessibility (Figure 13.2.4). People in business services such as advertising, banking, finance, journalism, and law particularly depend on proximity to professional colleagues. Lawyers, for example, choose locations near government offices and courts. Services such as temporary secretarial agencies and instant printers locate downtown to be near lawyers, forming a chain of interdependency that continues to draw offices to the center city.

Even with the diffusion of modern telecommunications, such as meetings on Zoom, many professionals prefer to exchange information with colleagues primarily through face-to-face contact. Financial analysts discuss attractive stocks or impending corporate takeovers. Lawyers meet to settle disputes out of court.

A central location also helps businesses that employ workers from a variety of neighborhoods. Top executives may live in one neighborhood, junior executives in another, secretaries in another, and custodians in still another. Only a central location is readily accessible to all groups.

When the COVID-19 pandemic forced most office buildings to close during 2020, most employees of business services were forced to work from home. Once restrictions on in-person work were lifted, some employees still worked from home, at least part time.

▲ **13.2.4 BUSINESS SERVICE IN LOUISVILLE**
400 West Market, the tallest building in Louisville.

Consumer services in the CBD. In the past, three types of consumer services clustered in a CBD because they required accessibility to everyone in the region—retailers with high thresholds (the minimum number of people needed to support a service), retailers with high range (the maximum distance people are willing to travel to use a service), and consumer services that served people working and living in the CBD.

A retailer with a high threshold traditionally preferred a CBD location in order to be accessible to many people. For example, high volumes of shoppers patronized large department stores downtown. A retailer with a high range was often a specialist, with customers who patronized them infrequently. For example, a jewelry or clothing store attracted shoppers from all over the urban area, but each customer visited infrequently.

Changing shopping habits and residential patterns have reduced the importance of consumer services with high thresholds and high ranges in the CBD. Most retailers with high thresholds and high ranges prefer suburban locations and have closed their downtown shops. Some retailers with high ranges have remained in CBDs because they are patronized by tourists.

In contrast to the other two types of retailers, shops that appeal to nearby office workers and residents have expanded in the CBD, in part because the number of downtown office workers has increased and in part because downtown offices require more services. Patrons of downtown shops tend increasingly to be downtown employees who shop during the lunch hour. These services sell primarily food and office supplies. Local residents also patronize shops and restaurants in the CBD as a leisure activity on evenings and weekends (Figure 13.2.5).

Most consumer services in the CBD patronized by office workers were forced to close in 2020 during the COVID-19 pandemic. Once consumer services were allowed to reopen, some remained permanently closed.

▶ **13.2.5 CONSUMER SERVICE IN LOUISVILLE**
Fourth Street entertainment district.

Models of Urban Structure

- **Describe four models of internal structure of urban areas.**

People are not distributed randomly within an urban area. They concentrate in particular neighborhoods, depending on the social characteristics of a neighborhood. Geographers describe where different groups of people are likely to live within an urban area, and they offer explanations for why these patterns occur.

13.3.1 Four Models of Urban Structure, Part 1

Geographers, sociologists, and economists developed four models to help explain where different groups of people tend to live in an urban area. Three of the models are concentric zone, sector, and multiple nuclei. A fourth model—galactic (or peripheral)—is a variation of the multiple nuclei model.

Concentric zone model. According to the **concentric zone model**, created in 1923 by sociologist Ernest Burgess, a city grows outward from a central area in a series of concentric rings, like the growth rings of a tree (Figure 13.3.1). Back in the 1920s, Burgess identified five rings, using labels widely used at the time (Figure 13.3.2). The precise number and width of the rings vary from one city to another, but the same basic types of rings appear in every city in roughly the same order.

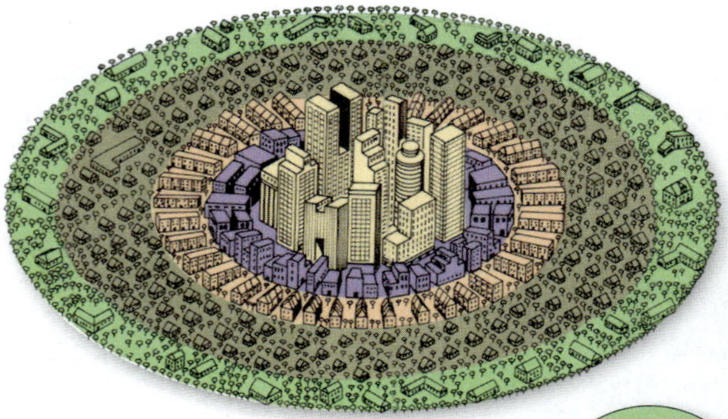

▶ **13.3.1 CONCENTRIC ZONE MODEL** Burgess gave names to each zone that are now obsolete.

1 Central business district
2 Zone of transition
3 Zone of independent workers' homes
4 Zone of better residences
5 Commuter's zone

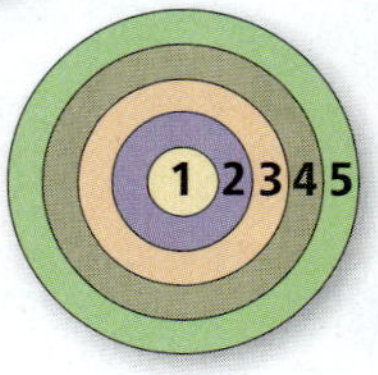

▶ **13.3.2 HOUSING IN CHICAGO'S ZONE OF INDEPENDENT WORKERS' HOMES**

Sector model. According to the **sector model**, developed in 1939 by land economist Homer Hoyt, a city develops in a series of sectors (Figure 13.3.3). Certain areas of the city are more attractive for various activities, originally because of an environmental factor or even by chance. As a city grows, activities expand outward in a wedge, or sector, from the center.

Once a district with expensive housing is established, the most expensive new housing is found in a corridor extending outward from that district (Figure 13.3.4). Industrial and retailing activities and less-expensive housing develop in other sectors.

▼ **13.3.3 SECTOR MODEL** Hoyt gave names to each sector that are now obsolete.

1 Central business district
2 Transportation and industry
3 Low-class residential
4 Middle-class residential
5 High-class residential

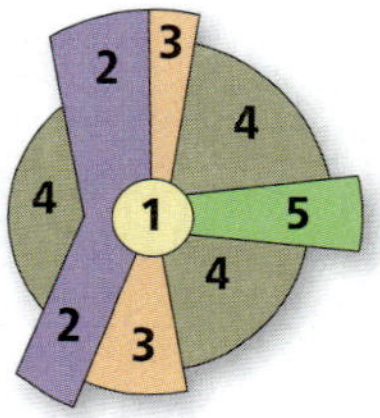

▼ **13.3.4 EXPENSIVE RESIDENTIAL SECTOR IN CHICAGO**

13.3.2 Four Models of Urban Structure, Part 2

In addition to concentric zone and sector, a third model of urban structure is multiple nuclei, developed by geographers Chauncey Harris and Edward Ullman in 1945. A fourth model, the galactic, was developed by Harris in 1960.

Multiple nuclei model. According to the **multiple nuclei model**, a city is a complex structure that includes many nodes (or nuclei)—a CBD as well as other centers—around which activities occur (Figure 13.3.5). Examples of nodes in a city include a port, a neighborhood business center, a university, an airport, and a park. The multiple nuclei theory states that some people and public, business, and consumer services are attracted to particular nodes, whereas others try to avoid them. For example, a university node may attract well-educated residents, pizzerias, and bookstores (Figure 13.3.6), whereas an airport node may attract hotels and warehouses. Furthermore, according to the multiple nuclei model, centers with incompatible land uses are located at a distance from each other. Heavy industry and expensive housing, for example, rarely exist in the same neighborhood.

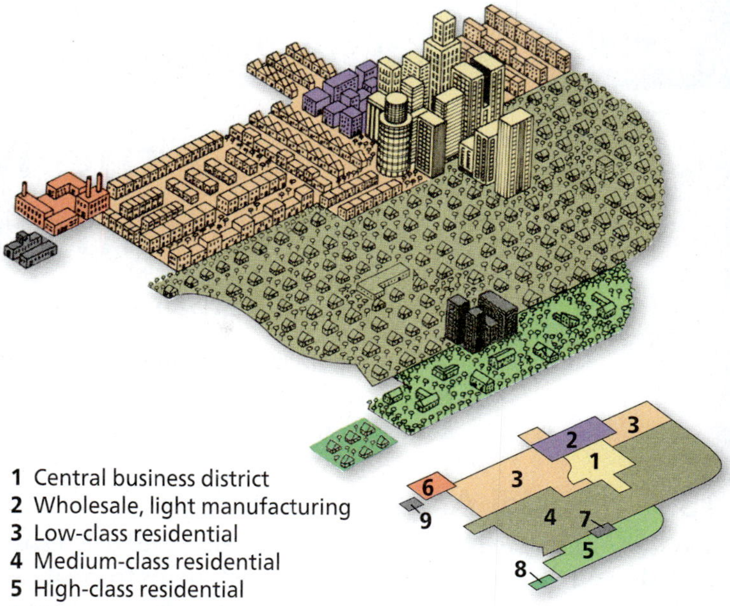

1 Central business district
2 Wholesale, light manufacturing
3 Low-class residential
4 Medium-class residential
5 High-class residential
6 Heavy manufacturing
7 Outlying business district
8 Residential suburb
9 Industrial suburb

▲ **13.3.5 MULTIPLE NUCLEI MODEL**
Harris and Ullman gave names to each node that are now obsolete.

▶ **13.3.6 THE UNIVERSITY OF CHICAGO, A NODE IN CHICAGO**

Galactic model. According to the **galactic model (**or **peripheral model)**, developed by Harris in 1960, an urban area consists of a city surrounded by large suburban residential and service nodes tied together by a beltway or ring road (Figure 13.3.7). Rather than an entirely new model, Harris considered the galactic model to be a modification of the multiple nuclei model (which he co-authored), reflecting the growth of suburbs. In the galactic model, a node of consumer and business services around the beltway is called an **edge city** (Figure 13.3.8).

▼ **13.3.7 GALACTIC (PERIPHERAL) MODEL**

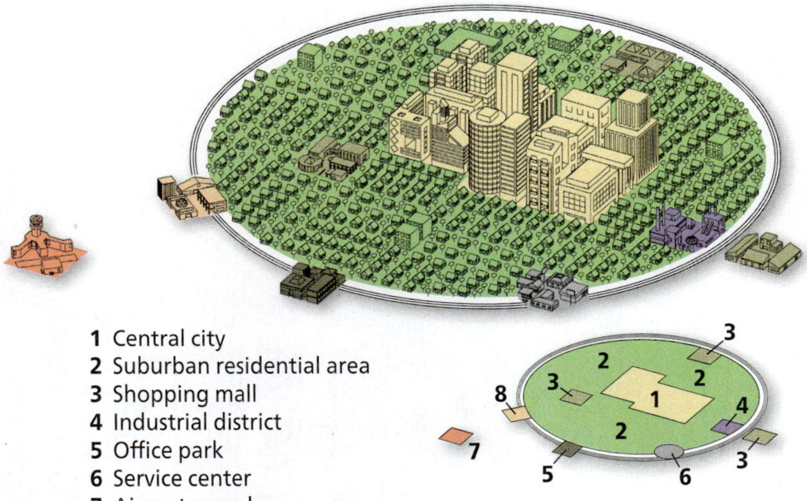

1 Central city
2 Suburban residential area
3 Shopping mall
4 Industrial district
5 Office park
6 Service center
7 Airport complex
8 Combined employment & shopping center

▼ **13.3.8 ROSEMONT, ILLINOIS, AN EDGE CITY ADJACENT TO CHICAGO O'HARE AIRPORT**

Applying the Models in North America

- **Analyze how urban structure models help to explain where people live.**

The models of urban structure help us understand where people with different personal characteristics tend to live within an urban area. They can also help explain why certain groups of people tend to live in particular places.

13.4.1 Social Area Analysis

The study of where people of varying living standards, ethnic background, and lifestyle live within an urban area is **social area analysis**. Social area analysis helps to create an overall picture of where various groups of people tend to live, depending on their particular personal characteristics.

Social area analysis depends on the availability of data at the scale of individual neighborhoods. In the United States and many other countries, that information comes from the census. Urban areas in the United States are divided into **census tracts** that each contain approximately 5,000 residents and correspond, where possible, to neighborhood boundaries. The census also divides the entire United States into blocks, which are typically a collection of several dozen houses; inside urban areas, blocks are typically bounded by four streets. A block group, as the name implies, is a collection of several neighboring blocks.

Every decade the U.S. Bureau of the Census publishes data summarizing the characteristics of the residents and the housing in each tract. Annual estimates are issued through the American Fact Finder service of the census's American Community Survey program. Examples of information the census provides at the tract level include the number of various races, the median household income, and the percentage of adults who finished high school.

The spatial distribution of any of the household and housing characteristics can be plotted on a map of the community's census tracts. Computers are invaluable in this task because they permit rapid creation of maps and storage of voluminous data about each census tract. Relatively little is available at the block level because the number of people in a block is so small that publishing information at the block level could violate the privacy of individuals.

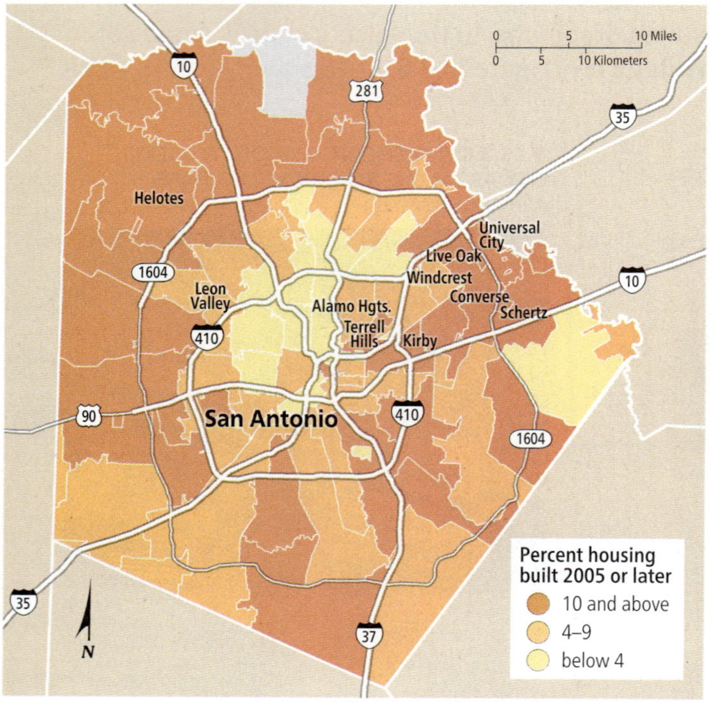

▲ 13.4.1 CONCENTRIC ZONES IN SAN ANTONIO
The outer ring has a higher percentage of recently built housing.

Social Area Analysis: Concentric Zones

Consider two households, one living in a newly constructed home and the other living in an older home. The concentric zone model suggests (Figure 13.4.1) that the household in the newer house is much more likely to live in an outer ring and the household in the older house in an inner ring (Figure 13.4.2).

▼ 13.4.2 OLDER INNER-RING HOUSING IN SAN ANTONIO

13.4.2 Social Area Analysis: Sector & Multiple Nuclei

Along with the concentric zone model, the sector and multiple nuclei models also help to explain where people of various characteristics tend to live within an urban area. Examples can be drawn from San Antonio.

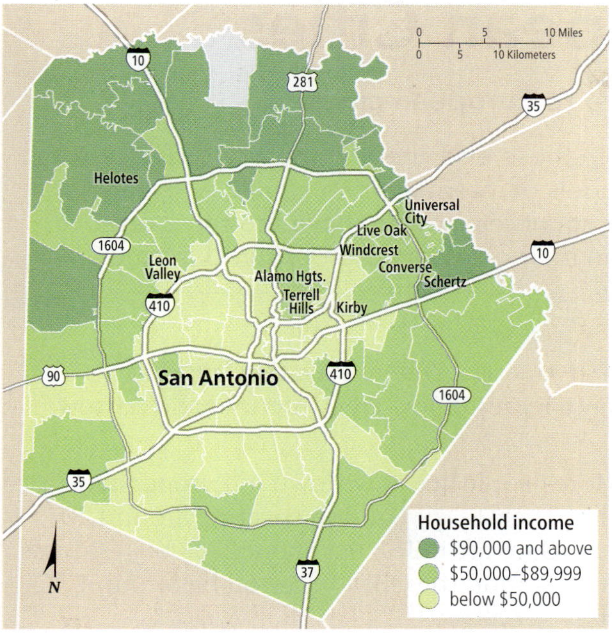

▲ **13.4.3 SECTORS IN SAN ANTONIO**

Sector model. Given two households who own their homes, the sector model suggests that a household of modest income and a household of high income are likely living in different sectors of the city (Figure 13.4.3). For example, in San Antonio, a household with a relatively high income is most likely to live in the northern part of the urban area (Figure 13.4.4).

▲ **13.4.4 SECTOR OF RELATIVELY HIGH-INCOME HOUSEHOLDS IN SAN ANTONIO**

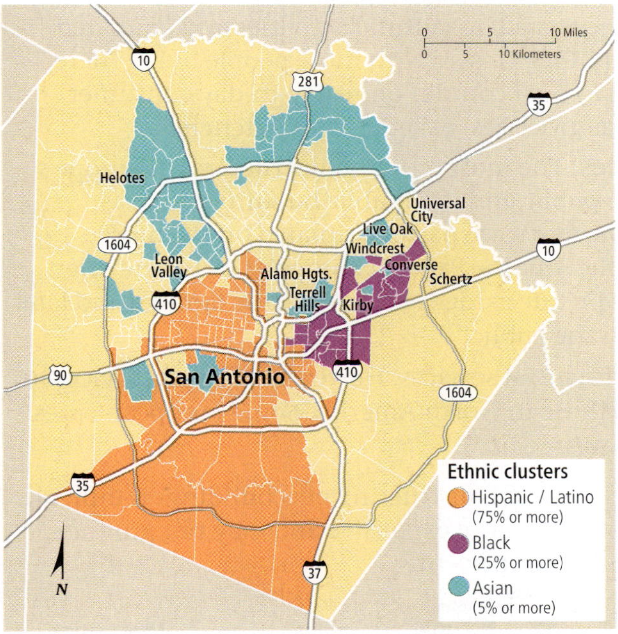

▲ **13.4.5 MULTIPLE NUCLEI IN SAN ANTONIO**

Multiple nuclei model. According to the multiple nuclei model, people with the same ethnic or racial background are likely to live near each other (Figure 13.4.5). For example, in San Antonio, Asian Americans are most likely to live to the north, Blacks to the east, and Hispanics to the south (Figure 13.4.6).

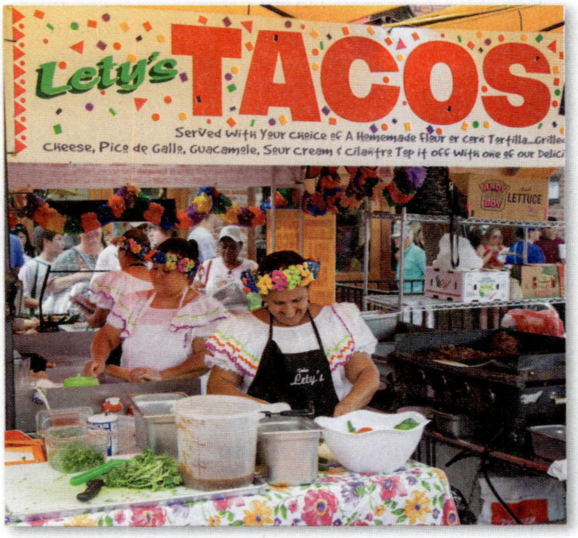

◀ **13.4.6 HISPANIC NODE IN SAN ANTONIO** Market Square.

Limitations of the Models

None of the urban structure models taken individually completely explains why different groups of people live in distinctive parts of a city. If the models are combined rather than considered independently, they help geographers explain where different groups of people live in a city. Putting the models together, we can identify, for example, areas where Asian Americans with relatively high incomes who own their homes are most likely to live.

Still, critics point out that the models are too simple and fail to consider the variety of reasons that lead people to select particular residential locations. Because the models are all based on conditions that existed in U.S. cities during the twentieth century, critics also question the relevance of the urban structure models to contemporary urban patterns in the United States or in other countries.

Structure of European Cities

- **Relate the models of urban structure to European cities.**

North American urban areas differ from those elsewhere in the world. These differences do not invalidate the models of internal urban structure, but they do point out that social groups in other countries may not have the same reasons for selecting particular neighborhoods within their cities as do groups in U.S. cities.

13.5.1 CBDs in Europe

Europe's CBDs have a different mix of land uses than those in North America. Differences stem from the medieval origins of many of Europe's CBDs. European cities display a legacy of low-rise structures and narrow streets, in many cases built several hundred years ago.

▲ 13.5.1 LONDON CBD
Expensive apartments are on the upper floors, above a pub frequented by local residents.

- **Residences.** More people live downtown in Europe than in North America. The CBD of London covering around 20 square kilometers (8 square miles) has about 270,000 residents. A comparable area around the CBD of Detroit has around 25,000 residents.
- **Consumer services.** More people live in Europe's CBDs in part because they are attracted to the concentration of consumer services, such as cultural activities and animated nightlife (Figure 13.5.1). And with more people living there, Europe's CBDs in turn contain more day-to-day consumer services, such as groceries, bakeries, and butchers.
- **Public services.** The most prominent structures in Europe's CBDs are often public and semipublic buildings, such as churches and former royal palaces, situated on the most important public squares. Parks in Europe's CBDs were often first laid out as private gardens for aristocratic families and later were opened to the public.
- **Business services.** Europe's CBDs contain professional and financial services. However, business services in Europe's CBDs are less likely to be housed in skyscrapers than those in North America. Some European cities try to preserve their historic CBDs by limiting high-rise buildings.

Many service providers wish to be in the center of European cities, but constructing new buildings is difficult. The alternative is renovation of older buildings. However, renovation is more expensive and may not produce enough space to meet the demand. As a result, rents are much higher in the center.

Concentric Zones in London

The urban structure within London can be used to illustrate similarities and differences in the distribution of people within U.S. and European cities. As in U.S. urban areas, newer housing in the London region is in outer rings and older housing is closer to the center (Figure 13.5.2). Unlike U.S. urban areas, though, much of the newer housing in Europe is in high-rise apartments rather than single-family homes.

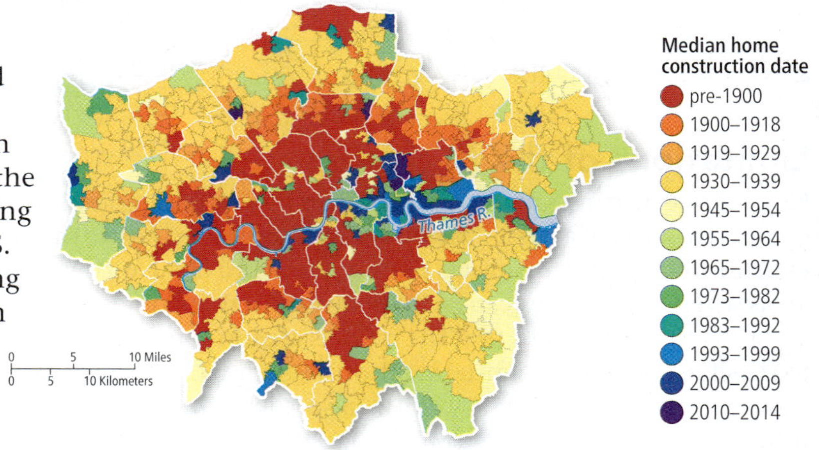

▶ 13.5.2 CONCENTRIC ZONES IN LONDON
The older housing is generally in an inner ring, and the newer housing is generally in an outer ring.

13.5.2 Sectors & Nodes in London

In the United States, people with different incomes tend to live in different sectors, and people having differing ethnic identities tend to live in different nodes. The sector and multiple nuclei models also help to explain where different groups of people live within London.

Sector model. Similar to U.S. urban areas, people with different incomes tend to cluster in distinctive sectors in the London region (Figure 13.5.3). People with higher incomes are more likely to live in sectors to the southwest and northwest, whereas people with lower incomes are more likely to live in a northeast sector.

In the past, Londoners with higher incomes built homes toward the west, near royal palaces. During the Industrial Revolution, most of London's factories were built toward the east, especially along the River Thames. People with higher incomes preferred to live toward the west, away from the factories.

Multiple nuclei model. European urban areas, including London, have experienced a large increase in immigration from other regions of the world. One-third of London's inhabitants were born outside the United Kingdom, including 18 percent in Asia, 7 percent in Africa, and 6 percent in the Caribbean. London's ethnic groups have formed nodes in various parts of the city. For example, Bangladeshis have clustered in the east (Figure 13.5.4), Africans in the northeast, Caribbean immigrants in the north and south, and Arabs in the west (Figure 13.5.5).

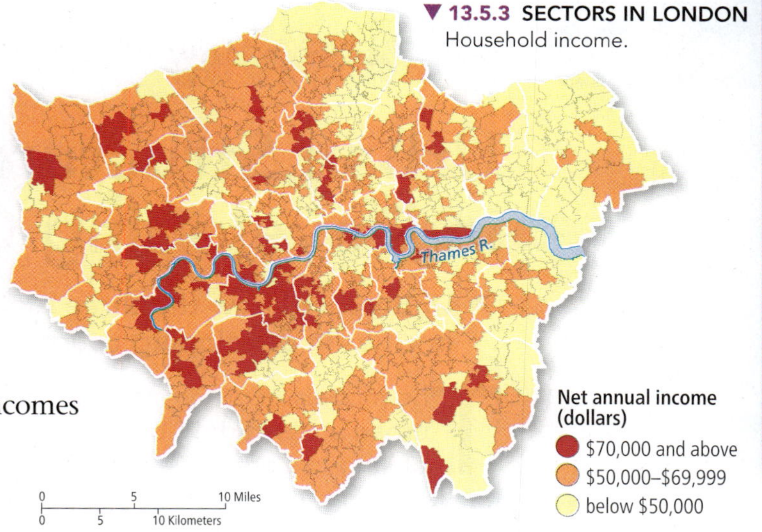

▼ **13.5.3 SECTORS IN LONDON**
Household income.

▼ **13.5.4 MULTIPLE NUCLEI NODES EXAMPLE IN LONDON**
Brick Lane, heart of London's Bangladeshi community.

▼ **13.5.5 MULTIPLE NUCLEI NODES IN LONDON: ETHNICITIES**
(a) Bangladeshi, (b) African, (c) Caribbean, (d) Arab.

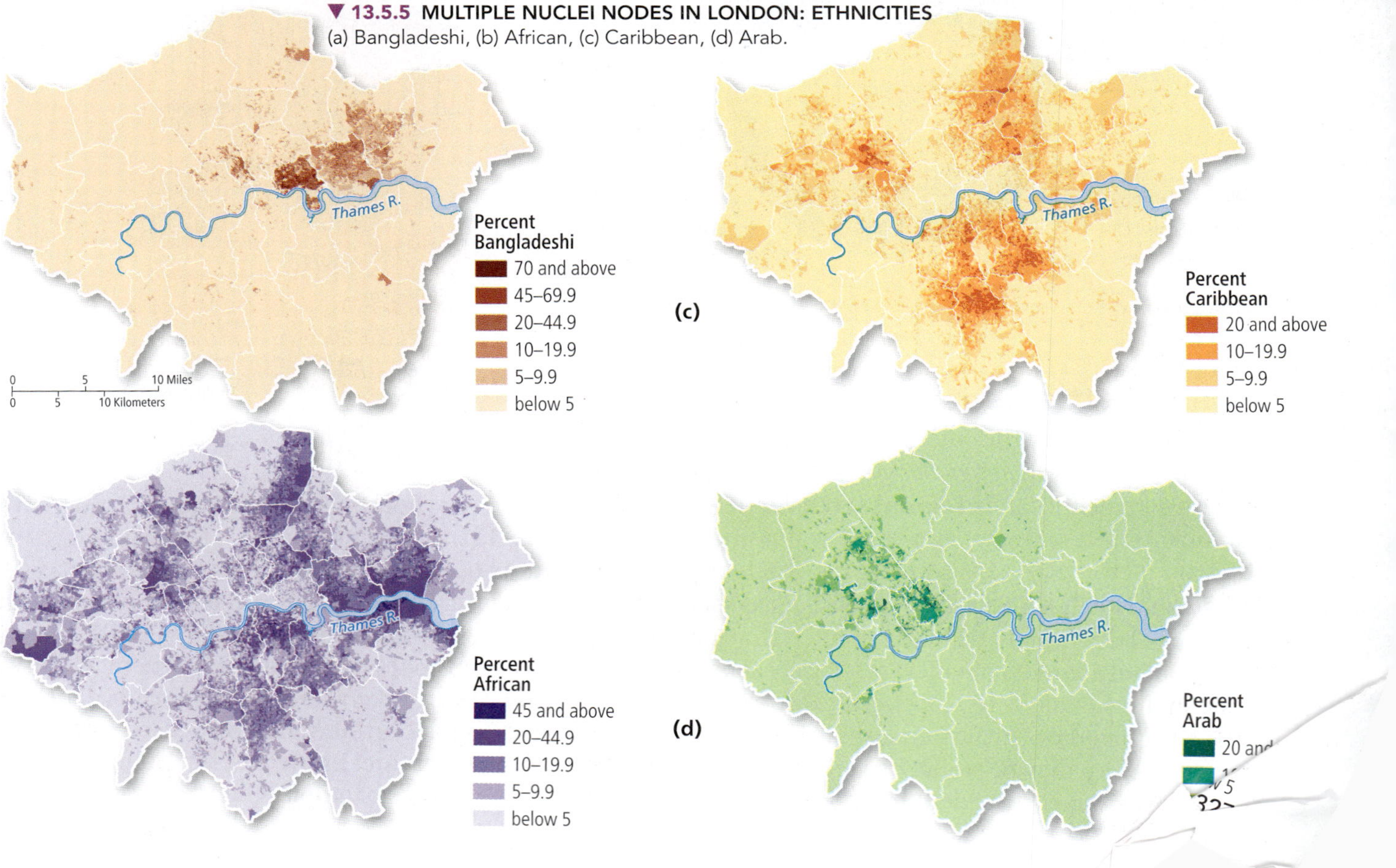

Urban Structure Models in Developing Countries

- **Understand how the three models of urban structure describe patterns in cities in developing countries.**

The models of urban structure help to explain contemporary patterns within the urban areas in developing countries. Rapid growth of population and expansion of land-use area has strengthened the applicability of the models in some cities but reduced their usefulness in other cases.

13.6.1 Concentric Zones in Cities of Developing Countries

The concentric zone model has been applied most frequently to cities in developing countries. Geographer Harm deBlij's model of sub-Saharan African cities is an example (Figure 13.6.1). The inner rings house higher-income people. Inner rings have the most attractive residential areas because they are near business and consumer services, and they offer such vital public services as water, electricity, paved roads, and garbage pickup.

As cities grow rapidly in developing countries, rings are constantly being added on the periphery to accommodate immigrants from rural areas attracted by job opportunities. For example, in Nagpur, India, an urban area of 3 million inhabitants, the highest-cost housing is in the center, and the cost of housing declines in outer rings (Figure 13.6.2).

▶ 13.6.1 CONCENTRIC ZONE MODEL APPLIED TO CITIES IN DEVELOPING COUNTRIES

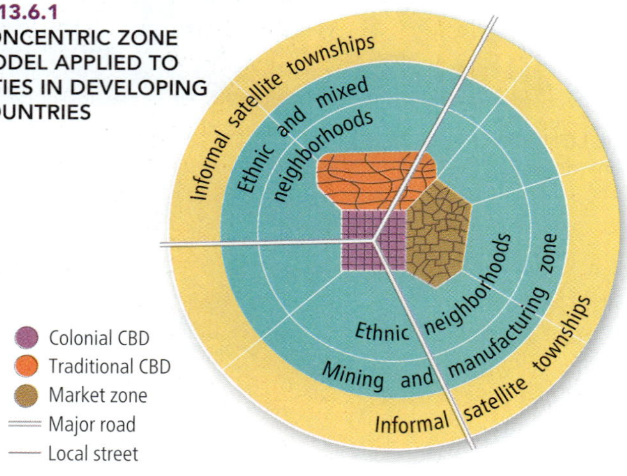

▼ 13.6.2 CONCENTRIC ZONE MODEL APPLIED TO NAGPUR, INDIA

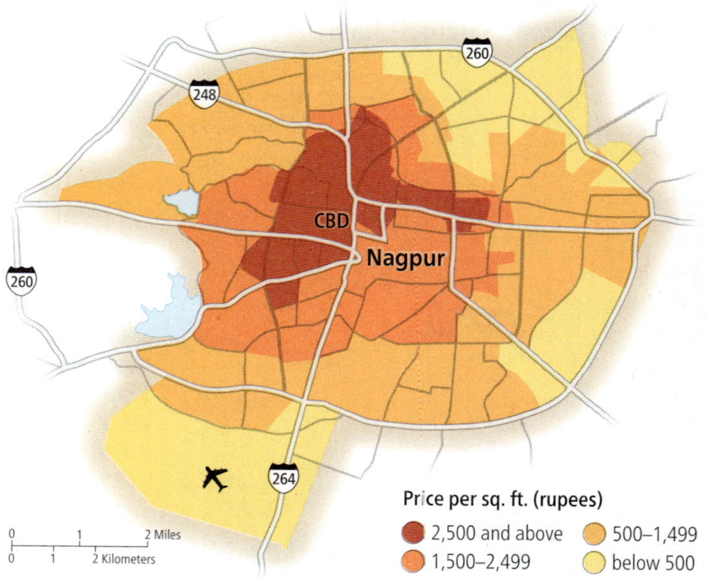

Informal Settlements

Much of the housing in the outer rings is in informal settlements, also known as *squatter settlements*. The United Nations defines an **informal settlement** as a residential area where housing has been built on land to which the occupants have no legal claim or where housing has not been built to the city's standards for legal buildings.

Informal settlements are known by a variety of names, including *barriadas* and *favelas* in Latin America, *bidonvilles* in North Africa, *bastees* in India, *gecekondu* in Turkey, *kampongs* in Malaysia, and *barong-barong* in the Philippines. The United Nations estimated that 1 billion people lived in informal settlements in 2018.

Informal settlements offer few services because neither the city nor the residents can afford them. Homes are in basic shelters made with scavenged cardboard, wood boxes, sackcloth, and crushed beverage cans. Latrines may be designated by the settlement's leaders, and water is carried from a central well or dispensed from a truck. Electricity service may be stolen by running a wire from the nearest power line. In the absence of bus service or affordable private cars, a resident may have to walk two hours to reach a place of employment.

13.6.2 Sectors in Cities of Developing Countries

Geographers Ernest Griffin and Larry Ford show that in Latin American cities, people with relatively high incomes push out from the center in a well-defined elite residential sector (Figure 13.6.3). The elite residential sector forms on either side of a narrow spine (the green commercial area in Figure 13.6.3) that contains offices, shops, and amenities, such as restaurants, theaters, parks, and zoos, attractive to people of wealth. For example, Santiago, Chile, has an elite residential sector extending northeast from the CBD (Figure 13.6.4).

People with relatively high incomes are also attracted to the center and spine because services such as water and electricity are more readily available and reliable there than elsewhere. Residents of the elite residential sector tend to avoid areas that may be noisy or polluting or that cater to people with limited incomes.

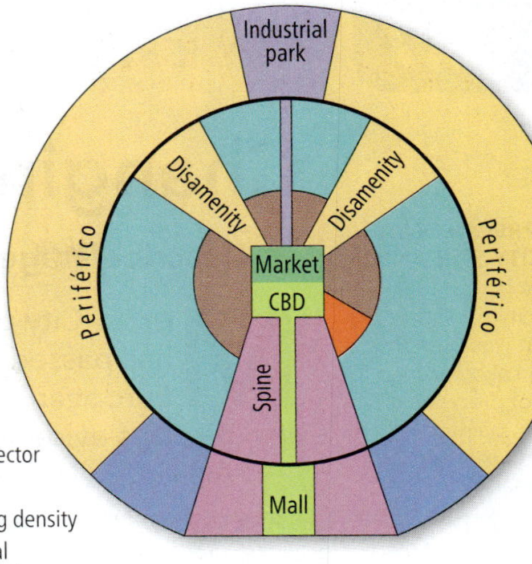

▲ 13.6.3 SECTOR MODEL APPLIED TO CITIES IN DEVELOPING COUNTRIES

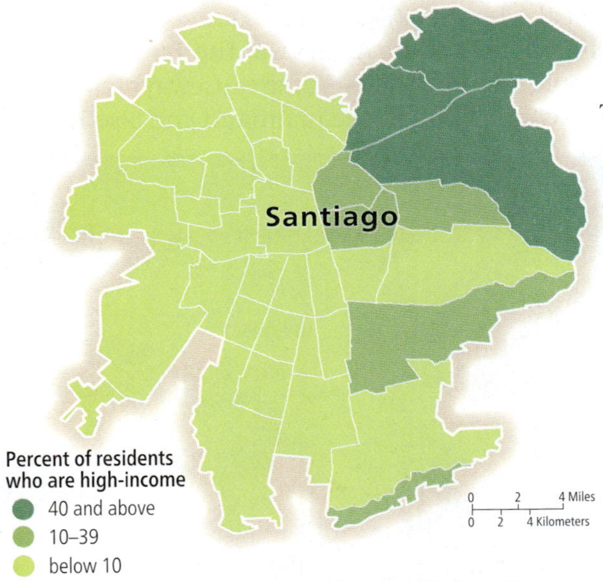

▲ 13.6.4 SECTOR MODEL APPLIED TO SANTIAGO, CHILE

Multiple Nuclei in Cities of Developing Countries

T. G. McGee's model of a Southeast Asian city superimposes on concentric zones several nodes of informal settlements and what he called "alien" zones, where newcomers live and work (Figure 13.6.5). McGee found that Southeast Asian cities do not typically have a strong CBD. Instead, the various functions of the CBD are dispersed to several nodes.

Cities in some developing countries show evidence of the multiple nuclei model by containing a complex mix of ethnic groups. During South Africa's apartheid era (refer to Section 7.4.2), its cities showed especially clear evidence of the multiple nuclei model because Black Africans, Colored, Whites, and Asians (primarily Indians) were forced by law to segregate into distinct neighborhoods (Figure 13.6.6).

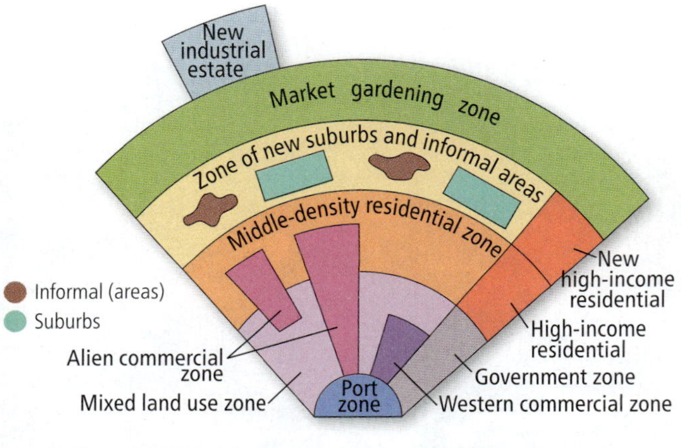

▲ 13.6.5 MULTIPLE NUCLEI MODEL APPLIED TO CITIES IN DEVELOPING COUNTRIES

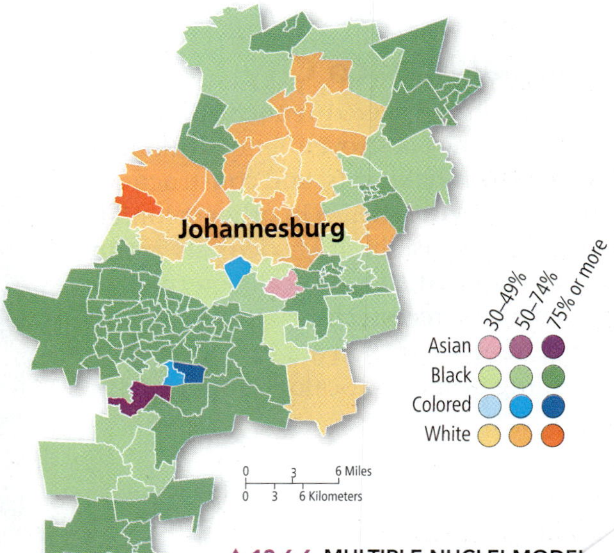

▲ 13.6.6 MULTIPLE NUCLEI MODE APPLIED TO JOHANNESBURG, AFRICA

Changing Structure of Mexico City

- **Describe stages of development and apply urban models to Mexico City.**

Mexico City provides a good example of a city in a developing country that has passed through three stages of development—pre-European origin, the European colonial period, and postcolonial independence. The modern city also displays evidence of the models of urban structure.

▼ **13.7.1 PRECOLONIAL MEXICO CITY SITE AND SITUATION**
The Aztec city of Tenochtitlán was bult on an island in Lake Texcoco. This mural of precolonial Tenochtitlán, painted by Diego Rivera, is in Mexico City's National Palace.

13.7.1 Precolonial Mexico City

The Aztecs founded Mexico City—which they called Tenochtitlán—on a hill known as Chapultepec ("the hill of the grasshopper"). When forced by others to leave the hill, the Aztecs migrated a few kilometers south, near the present-day site of the University of Mexico, and then in 1325 to a marshy 10-square-kilometer island in Lake Texcoco (Figure 13.7.1).

The node of religious life for the Aztecs was the Templo Mayor or Great Temple (Figure 13.7.2). Three causeways with drawbridges linked Tenochtitlán to the mainland and also helped control flooding. An aqueduct brought fresh water from Chapultepec. Most food, merchandise, and building materials crossed from the mainland to the island by boat, and the island was laced with canals to facilitate movement. Over the next two centuries, the Aztecs conquered the neighboring peoples and extended their control through much of present-day Mexico. As their wealth and power grew, Tenochtitlán grew to a population of around 200,000.

▲ **13.7.2 PRECOLONIAL MEXICO CITY CBD**
The center of Tenochtitlán was dominated by the Templo Mayor. The blue shrine on the top of the temple was dedicated to the Aztec god of rain and agriculture, and the red shrine was dedicated to the Aztec god of war. This is part of a reconstruction made in the 1990s.

Colonial Mexico City

The Spanish conquered Tenochtitlán in 1521 after a two-year siege. They destroyed Tenochtitlán, dispersed or killed most of the inhabitants, and constructed a new city on the site. As in other colonial cities, the Spanish built Mexico City around a main square, called the Zócalo, on the site of the Aztecs' sacred precinct in the center of the island. Streets were laid out in a grid pattern extending from the Zócalo (Figure 13.7.3).

▶ **13.7.3 COLONIAL MEXICO CITY CBD**
The main square in downtown Mexico, the Zócalo, was laid out by the Spanish. The Metropolitan Cathedral is to the left. City Hall is to the right. The National Palace is at the far end of the square. Excavations at the site of the Templo Mayor are to the left of the National Palace and behind the Metropolitan Cathedral.

13.7.2 Postcolonial Mexico City

At independence, Mexico City was a relatively small city. The population grew modestly during the nineteenth century, reaching around 345,000 in 1900. The population has grown rapidly since then. Currently, the population of the municipality of Mexico City is 9 million, and the population of the urban area is 21 million. Millions of people have migrated to the city in search of work. Rapid population growth has resulted in an expansion of the land area of Mexico City. In 1903, most of Lake Texcoco was drained by a gigantic canal and tunnel project, allowing the city to expand to the north and east.

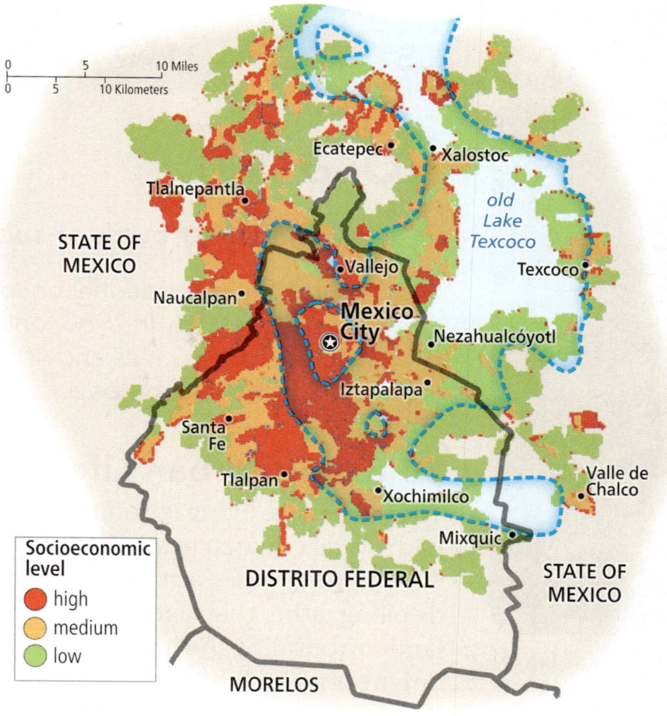

▲ 13.7.4 SOCIAL AREAS IN MEXICO CITY

Applying the Models of Urban Structure to Mexico City

The concentric zone model and the sector model help to explain the distribution of people with different income levels within the Mexico City urban area. Figure 13.7.4 divides Mexico City into areas inhabited primarily by people with high incomes, with medium incomes, and with low incomes.

▲ 13.7.5 CONCENTRIC ZONE IN MEXICO CITY
Housing for people with limited incomes is built into hillsides in outer rings.

Concentric zone model. Like many urban areas in developing countries, the character of Mexico City's concentric zones differs from that of North America. In Mexico City, the outer ring contains the most modest and least expensive housing.

The high cost of housing near the center of the city has forced people with limited incomes to live in informal settlements on the outskirts of the city. Many of these informal settlements are located on hillsides (Figure 13.7.5). Residents of the hillside informal settlements have difficulty reaching jobs that are concentrated in the CBD and inner rings of Mexico City.

Sector model. Mexico City's elite residential sector is toward the southwest, along the Paseo de la Reforma, a 14-lane tree-lined boulevard designed by Emperor Maximilian (1864–1867) and patterned after the Champs-Elysées in Paris (Figure 13.7.6). Reforma extends 3 kilometers southwest from the CBD to Chapultepec. During the late nineteenth century, people with wealth built palacios (palaces) along the boulevard.

Environmental factors influenced the movement of people with high incomes toward the southwest sector of Mexico City. Because elevation is higher than elsewhere in the city, wastewater once flowed freely toward the northeast, away from Chapultepec. As Mexico City's population grew rapidly during the twentieth century, people with wealth migrated further out along the southwestern sector anchored by the Reforma.

▶ 13.7.6 SECTOR IN MEXICO CITY
The Paseo de la Reforma is the main boulevard through the heart of the elite residential sector.

▲ **13.8.1 SUBURBAN GROWTH AFTER WORLD WAR II**
Levittown, Pennsylvania, 1957.

Sprawl

- **Describe the growth of suburbs and of suburban sprawl.**

A **suburb** is a residential or commercial area situated within an urban area but outside the central city. In 1950, only 20 percent of Americans lived in suburbs compared to 40 percent in cities and 40 percent in small towns and rural areas. Today, 60 percent of Americans live in suburbs and 20 percent each in cities and rural areas.

13.8.1 Suburban Lifestyle

The percentage living in suburbs climbed rapidly after World War II, when returning GIs married, had several children (the baby boomer generation), and needed decent housing (Figure 13.8.1). Suburbs offer varied attractions: a detached single-family dwelling rather than a row house or an apartment, private land surrounding the house, space to park cars, and a greater opportunity for home ownership. Families with children were especially attracted to suburbs, which offered more space for play and protection from the high crime rates and heavy traffic that characterized city life (Figure 13.8.2).

▲ **13.8.2 SUBURBAN LIFESTYLE**
Ice cream delivered by the Good Humor Man.

Density Gradient

The change in density within an urban area is called the **density gradient**. According to the density gradient, the number of houses per unit of land diminishes as distance from the center city increases.

Historically, suburbs were built at much lower densities than the central city. For example, Figure 13.8.3 shows a sharp difference in density between the central city and suburbs in Cleveland in 1900. Since then, as people have migrated from the city to suburbs, density has declined in the city and increased in suburbs.

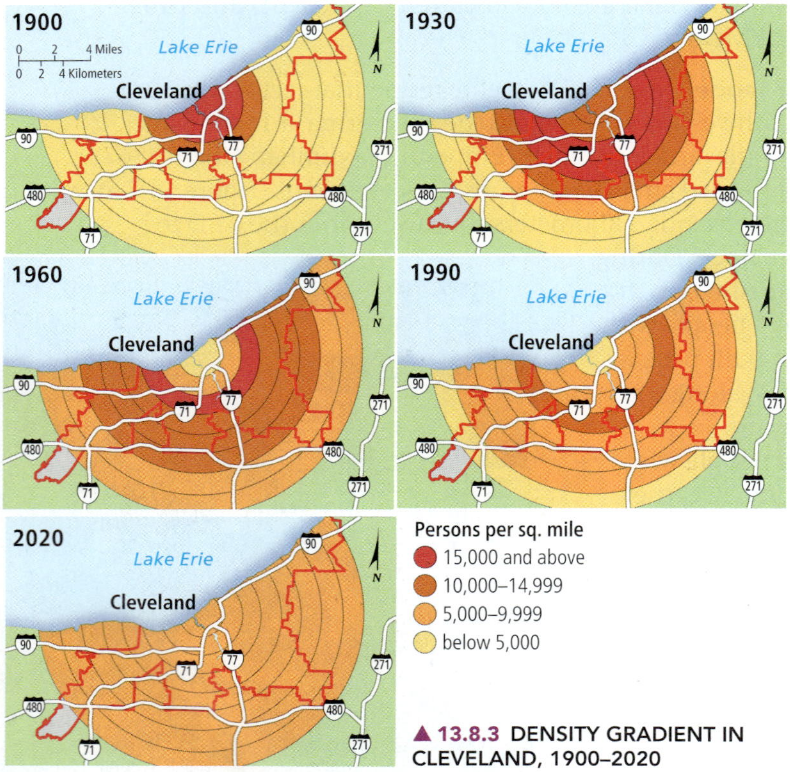

▲ **13.8.3 DENSITY GRADIENT IN CLEVELAND, 1900–2020**

Suburban Sprawl

Sprawl is the development of suburbs at relatively low density and at locations that are not contiguous to the existing built-up area. Sprawl is fostered in the United States by the desire of many families to own large tracts of land and by private developers who recognize that land for new housing sites is cheaper if not contiguous to the existing built-up area. In contrast, suburban growth is more carefully planned in the United Kingdom (Figure 13.8.4).

As a result of sprawl, roads and utilities must be extended to connect isolated new developments; motorists must drive longer distances, consuming more fuel; and agricultural land is lost.

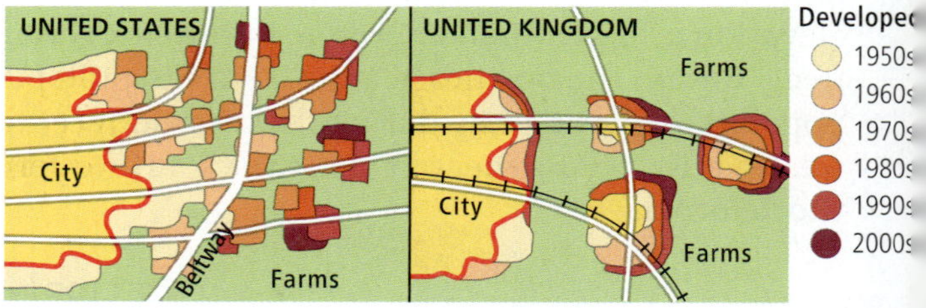

▲ **13.8.4 SUBURBAN DEVELOPMENT PATTERNS IN THE U.K. AND U.S.**

13.8.2 Annexation & Fragmentation

Until recently in the United States, as cities grew, they expanded by adding peripheral land (Figure 13.8.5). The process of legally adding land area to a city is **annexation**. Now cities are surrounded by a collection of suburban jurisdictions whose residents prefer to remain legally independent of the large city. Given the difficulty in annexing suburbs, local government in the United States is extremely fragmented (Figure 13.8.6).

According to the 2017 Census of Governments, the United States had 90,075 local governments, including 3,031 counties, 19,495 municipalities, 16,253 townships, 12,754 school districts, and 38,542 special districts (which provide such services as fire protection, water supply, libraries, and public transportation).

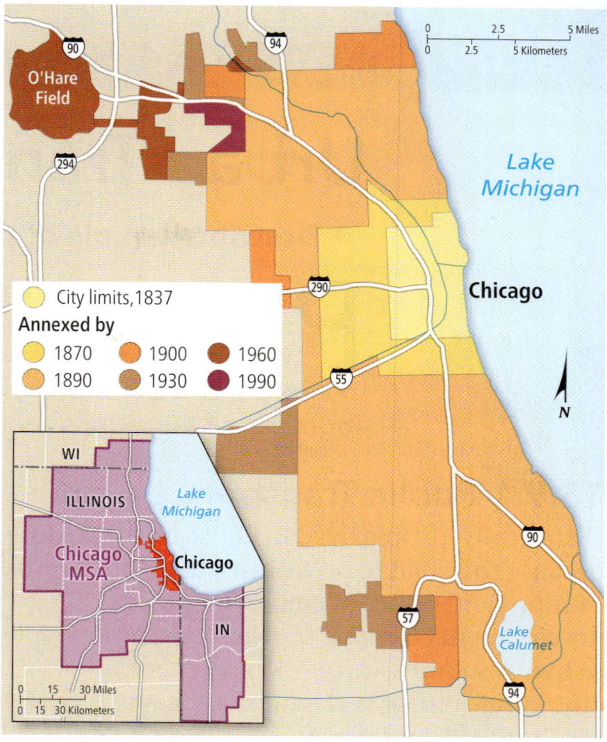

▲ 13.8.5 ANNEXATION IN CHICAGO

Suburban Segregation

Many suburbs display two types of segregation:

- **Social class segregation.** Suburban housing is often built for people of a single social class. Segregation by race and ethnicity also persists in some suburbs. People with relatively low incomes are less likely to live in many U.S. suburbs because of the high cost of the housing and unwelcoming attitudes of some suburbanites. Suburban communities may discourage the inclusion of Blacks, people of color, and people with relatively low incomes because of fear that property values will decline if the composition of the suburb is altered.
- **Land-use segregation.** Residents are separated from commercial and manufacturing activities that are confined to compact, distinct areas. Segregation of commercial and residential land uses is legally protected through zoning ordinances. A **zoning ordinance** is a law that limits the permitted uses of land and maximum density of development in a community. Zoning ordinances typically identify districts designed only for single-family houses, apartments, industry, or commerce.

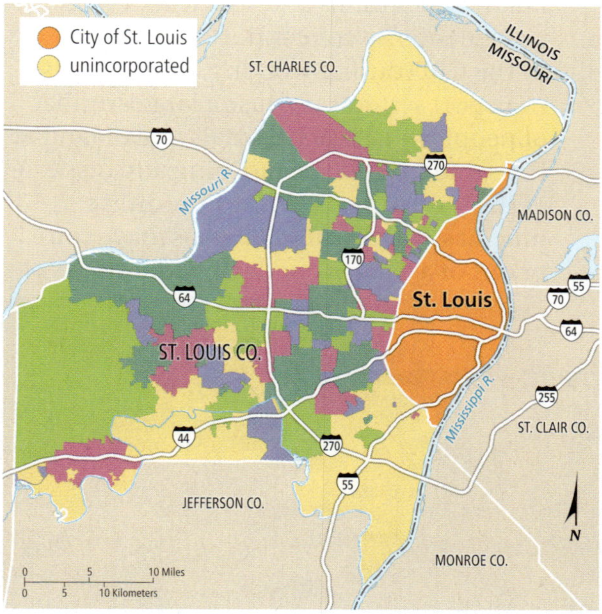

▲ 13.8.6 MUNICIPALITIES IN ST. LOUIS COUNTY

Smart Growth

Several U.S. states have taken steps to curb suburban growth. The goal is to produce a pattern of compact and contiguous development and protect rural land for agriculture, recreation, and wildlife. Legislation and regulations to limit suburban growth is called **smart growth**.

Oregon and Tennessee define growth boundaries within which new development must occur. Cities like Portland, Oregon, can annex only lands that are included in the urban growth areas (Figure 13.8.7). New Jersey, Rhode Island, and Washington were also early leaders in enacting strong state-level smart-growth initiatives. Maryland's smart-growth law discourages the state from funding new highways and other projects that would extend suburban sprawl and destroy farmland. State development money must be allocated to "fill in" already urbanized areas.

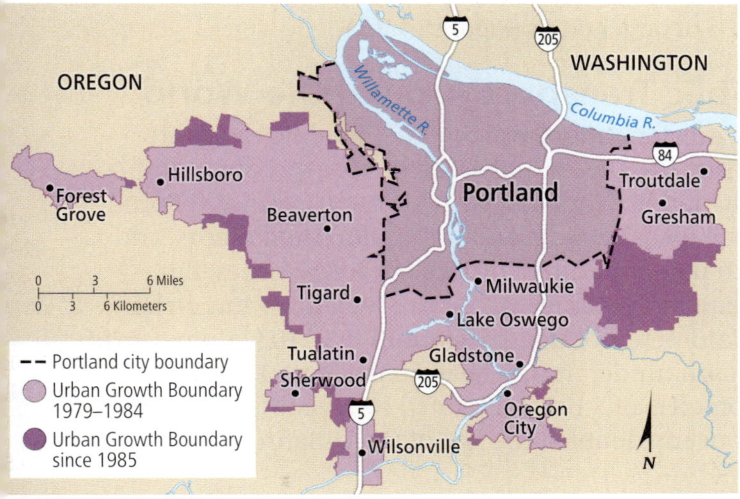

◀ 13.8.7 URBAN GROWTH BOUNDARY, PORTLAND, OREGON

Urban Transportation

- **Describe the role of public and private transportation in cities.**

People do not travel aimlessly; their trips have a precise point of origin, destination, and purpose. More than half of all trips are work related—commuting between work and home, business travel, or deliveries. Shopping accounts for one-fourth of all trips, and other personal business and social journeys account for the remainder. Together, all these trips produce congestion in urban areas.

13.9.1 Public Transport

Historically, people lived in dense central cities because they had to be within walking distance of shops and places of employment. The development of public transport enabled people to live in suburbs and work in the central city. Public transit is classified as heavy rail, light rail, and bus. Heavy rail includes underground subways and elevated trains. Light rail includes trams and streetcars.

The intense concentration of people in the CBD during working hours strains transport systems because a large number of people must reach a small area of land at the same time in the morning and disperse at the same time in the afternoon. As much as 40 percent of all trips made into or out of a CBD occur during four hours of the day—two in the morning and two in the afternoon. **Rush hour** is the four consecutive 15-minute periods that have the heaviest traffic.

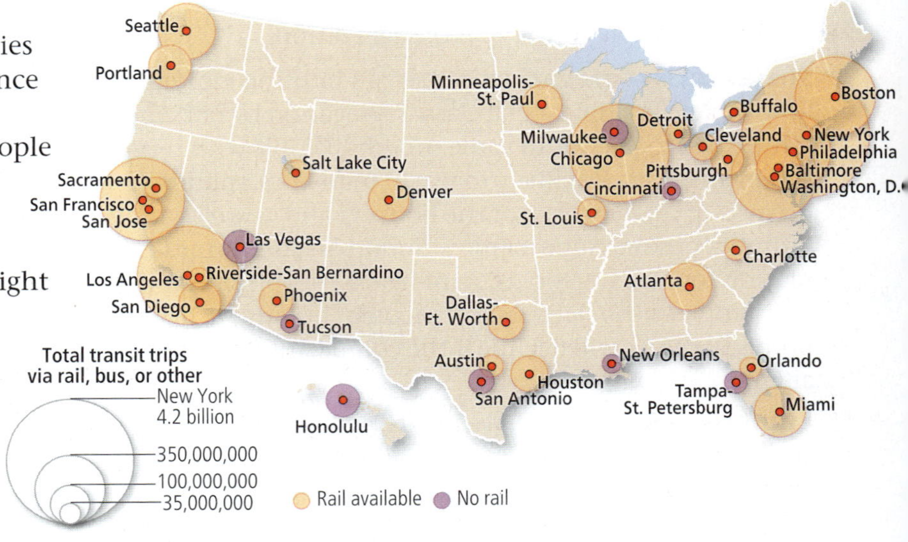

▲ 13.9.1 PUBLIC TRANSPORT PASSENGERS IN U.S. CITIES

In hundreds of cities around the world extensive networks of public transport lines have been maintained, and funds for new construction have been provided in recent years, including in a number of U.S. cities. But public transport ridership in the United States has declined from 23 billion per year in the 1940s to 10 billion in 2019. Service is minimal or nonexistent outside the CBDs of larger cities (Figure 13.9.1).

▲ 13.9.2 MUNICH TRAMS

▲ 13.9.3 MUNICH U-BAHN STATION

Public Transport Around the World

The greater importance placed on public transport outside the United States can be seen by comparing Indianapolis, Indiana, with Munich, Germany. Both urban areas have around 1.4 million inhabitants, and both have around 500 kilometers of bus lines. But Indianapolis has no rail service, whereas Munich has 83 kilometers of trams (Figure 13.9.2), 103 kilometers of U-Bahn (underground heavy rail, Figure 13.9.3), and 434 kilometers of S-Bahn (elevated heavy rail), and Munich has nearly doubled the size of the U-Bahn since 1990.

13.9.2 Benefits & Costs of Motor Vehicles

There are 1.4 billion motor vehicles in the world, including 280 million in the United States. The United States actually has more registered motor vehicles than licensed drivers. Motor vehicle ownership is nearly universal among American households, with the exception of some families with limited incomes, older individuals, and people living in the centers of large cities such as New York.

Motor vehicles offer two principal benefits:

- **Comfort, choice, & flexibility.** Motorists can live wherever they wish and travel whenever they wish. They are not constrained by the timetable of public transport service. A motor vehicle offers comfortable seats, choice of music, and isolation from strangers otherwise encountered on a bus or subway.
- **Perceived cost.** Motorists perceive that the cost of using vehicles is less than the cost of using public transport. Each time public transport is used a fare must be paid, and the fare is higher than the cost of fuel, at least in the United States. Most of the costs associated with motor vehicles, such as insurance and license, are paid on an annual basis, regardless of the amount of driving that is actually done.

The U.S. government has encouraged the use of cars and trucks by paying 90 percent of the cost of limited-access, high-speed interstate highways, which stretch for 75,000 kilometers (47,000 miles) across the country. However, some U.S. cities demolished freeways that once sliced through CBDs, including Boston, which has replaced the Central Artery (Figure 13.9.4) with a park (Figure 13.9.5).

▲ 13.9.4 BOSTON'S CENTRAL ARTERY FREEWAY, 1990

▲ 13.9.5 ROSE KENNEDY GREENWAY, WHICH REPLACED THE CENTRAL ARTERY

The freeway was placed in a tunnel.

Transportation Epochs

Transportation improvements played a key role in the changing structure of urban areas. Geographer John Borchert identified five eras of U.S. urban areas resulting from changing transportation systems (Figure 13.9.6).

Cities prospered or suffered during the various epochs, depending on their proximity to economically important resources and migration patterns. At the same time, cities retain physical features from earlier eras that may be assets or liabilities in subsequent eras.

1800 — **SAIL-WAGON EPOCH (1790–1830)**
Urban areas were clustered along the Atlantic Coast. Communication was primarily by wind-powered ships plying up and down the Atlantic Coast.

1850 — **IRON HORSE EPOCH (1830–1870)**
Steam-powered railroads provided transport from outlying areas into cities. Canals connected newly founded inland cities with existing urban centers on the East Coast.

1900 — **STEEL RAIL EPOCH (1870–1920)**
Long-haul rail lines connected urban areas around the country. Travel time between cities was greatly reduced.

1950 — **AUTO-AIR-AMENITY EPOCH (1920–1970)**
The gasoline engine made it possible for motor vehicles to become the dominant transport within and between urban areas. Gasoline-powered airplanes facilitated travel between distant urban centers.

2000 — **SATELLITE-ELECTRONIC-JET PROPULSION EPOCH (1970–?)**
The current era is characterized by the ability to communicate electronically, as well as to control transport systems electronically.

▲ 13.9.6 TRANSPORTATION EPOCHS

Challenges & Hopes for Cities

- **Summarize challenges and hopes for cities.**

Contemporary cities display ever-sharper contrasts. Some cities are thriving while others struggle. Some neighborhoods are animated, while others are deserted. The final section of this chapter examines some of these contrasts as well as the opportunities and challenges for the future.

13.10.1 The City Challenged

One hundred years ago, inner-city neighborhoods that surrounded the CBDs of U.S. cities teemed with throngs of recent immigrants from Europe. Such neighborhoods that housed perhaps 100,000 a century ago may contain fewer than 5,000 inhabitants today. Those remaining in these inner-city neighborhoods face a variety of distinctive social and physical challenges that are very different from those facing suburban residents.

▲ **13.10.1 WHEN HOME IS A STREET**
New Orleans.

Job skills. Residents of inner-city neighborhoods are increasingly disadvantaged in competing for jobs. Because fewer than half complete high school, they may lack technical skills needed for jobs in factories that now require skills in operating electronics. Businesses that are willing to hire people with relatively modest skills, such as cleaning services and fast-food restaurants, are mostly located in suburbs. But many people living in inner-city neighborhoods have difficulty reaching suburban jobs because they don't own cars.

Homelessness. Around 580,000 Americans live in homelessness, according to the U.S. Department of Housing and Urban Development (Figure 13.10.1). The principal cause of homelessness is insufficient income to afford housing. Affordable housing is increasingly difficult to find in inner-cities. Homelessness may have been sparked by family problems or job loss. One-fourth of those living without homes are children, according to government surveys.

Challenges for families. Unwed mothers give birth to two-thirds of the babies in U.S. inner-city neighborhoods, and 80 percent of children in the inner-city live with only one parent. Because of inadequate child-care services, single mothers may be forced to choose between working to generate income and staying at home to take care of the children. If the father or a partner moves back home, the mother may lose government assistance and other benefits, leaving the couple financially worse off together than apart.

Challenges to health & safety. Inner-city neighborhoods have a relatively high share of a metropolitan area's serious crimes, such as murder. Elevated crime rates in an inner-city neighborhood may reflect the neighborhood's high rates of joblessness and addiction. For example, some drug users obtain money through criminal activities. Gangs may form in inner-city neighborhoods to control lucrative drug distribution. Violence may erupt when two gangs fight over the boundaries between their drug distribution areas. Although drug use is increasing more rapidly in rural areas, inner-cities still have higher rates of use. Heroin and cocaine use is especially higher in cities compared to rural areas.

Services. Inner-city neighborhoods lack adequate police and fire protection, shops, hospitals, clinics, and other healthcare facilities. Access to food is an especially common challenge in the areas where residents with low incomes dwell (Figure 13.10.2). Residents with low incomes need the same range of public services as residents with moderate and high incomes. However, cities face a growing gap between the cost of needed services and the availability of funds to pay for them.

▲ **13.10.2 INNER-CITY FOOD INSECURITY**
Feeding those in need, Los Angeles.

13.10.2 The City Renewed

The process of converting an inner-city neighborhood from a predominantly renter-occupied area inhabited by people with modest incomes to a predominantly owner-occupied area inhabited by people with relatively high incomes is called **gentrification**. Most cities have at least one substantially renovated inner-city neighborhood that has attracted residents with higher incomes, especially unmarried people and couples without children (Figure 13.10.3).

A deteriorated inner-city neighborhood is attractive for several reasons:

- Inner-city houses may be larger and more substantially constructed yet less expensive than houses in the suburbs.
- Inner-city houses may possess attractive architectural details, such as ornate fireplaces, cornices, high ceilings, and wood trim.
- Inner-city living eliminates the strain of commuting on crowded freeways or public transport for people who work downtown.
- Inner-city neighborhoods are near theaters, bars, restaurants, stadiums, and other cultural and recreational facilities.

(a) (b)

▲ **13.10.3 GENTRIFICATION IN CINCINNATI**

(a) The neighborhood had many abandoned residences and shops. (b) Many of the abandoned structures have been renovated into attractive shops and higher-cost housing.

Because renovating an old inner-city house can be even more expensive than buying a new one in the suburbs, cities encourage the process by providing low-cost loans and tax breaks. Public expenditures for renovation have been criticized as subsidies for the middle class at the expense of people with lower incomes, who are sometimes forced to move out of the gentrified neighborhoods because the rents in the area suddenly become too high for them.

▼ **13.10.4 READING TERMINAL MARKET, PHILADELPHIA**

Reviving Consumer Services

Most consumer services have relocated from cities to suburbs to be near suburban residents. However, some consumer services are returning to the city, in part to meet day-to-day needs of residents of gentrified neighborhoods. Consumer services in the city are also attracting people looking for leisure activities, such as unusual shops in a dramatic downtown setting or view of a harbor. Several North American CBDs combine new retail services with leisure services. For example:

- Boston's Faneuil Hall Marketplace is located in renovated eighteenth-century buildings.
- Philadelphia's nineteenth-century Reading Terminal Market, which barely survived during the twentieth-century suburbanization movement, was renovated into a thriving marketplace with individual stalls operated by different merchants (Figure 13.10.4).
- Chicago's Navy Pier, a former cargo dock, was converted to shops and attractions.

CHAPTER 13

Review, Analyze, & Apply

KEY ISSUE 1 Why Are Cities Challenging to Define?

- Urban settlements are defined in various ways, including municipalities (or central cities), urban areas, and metropolitan areas.
- Business services, public services, and some consumer services cluster in the central business district (CBD).

THINKING GEOGRAPHICALLY

Some professional sports stadiums and arenas are located in the CBD, and some are located in suburbs. What are the advantages and drawbacks for the fans of each location?

◀ 13.CR.1 ORIOLE PARK AT CAMDEN YARDS, IN BALTIMORE'S CBD

KEY ISSUE 2 Where Are People Distributed Within a City?

- Four urban structure models help to explain where different groups of people live in an urban area.
- According to the concentric zone model, a city grows outward in rings.
- According to the sector model, a city grows along transportation corridors.
- According to the multiple nuclei model, a city grows around several nodes.
- The galactic or peripheral model is a variation of the multiple nuclei model to account for suburban nodes.

THINKING GEOGRAPHICALLY

Identify the ring, sector, and node in which you (or a friend or relation) live within an urban area. Do conditions in your place of residence fit the overall patterns expected of the urban structure models? Why or why not?

▲ 13.CR.2 LAS VEGAS
(Background) CBD, (foreground) residential area in the north sector and outer ring.

KEY ISSUE 3 How Are Cities Outside North America Structured?

- The models of urban structure can also be used to describe where people and activities are located within urban areas in Europe and in developing countries.
- As in North America, people with relatively high incomes in other regions of the world tend to cluster in particular sectors of cities.
- Unlike in North America, people with relatively low incomes are more likely to locate in the outer rings of cities elsewhere in the world.

THINKING GEOGRAPHICALLY

Efforts are underway to improve the condition of informal settlements, such as the free Medellín Metrocable, which connects informal settlements with Medellín's CBD. Why might the gondola be useful to residents of informal settlements?

▲ 13.CR.3 MEDELLÍN, COLOMBIA, METROCABLE

KEY ISSUE 4 How Might Cities Become More Sustainable?

- Cities have grown rapidly in land area, especially in North America, where suburban sprawl is common.
- Most trips in North America are taken in private motor vehicles.
- Public transport is designed to move a lot of people into the CBD in the morning and back to suburbs in the afternoon.

THINKING GEOGRAPHICALLY

Electric scooters and electric bikes are diffusing rapidly into cities. How might their diffusion alter transportation patterns in CBDs? What impact might they have in urban areas outside the CBD?

◀ 13.CR.4 ELECTRIC SCOOTERS
Washington, D.C.

Explore

London's Congestion Charge

To try to reduce congestion, London charges vehicles to enter the CBD during weekdays. In Google Earth, ***search*** for *1 Marble Arch, London*. ***Zoom*** in to around 100 meters. Note the white C in a red circle. Motorists must pay a charge to drive across the red and white C into the CBD.

1. Use Street View to see street conditions in front of 1 Marble Arch. Do you see evidence that the charge is reducing congestion? What might be done to further address congestion?
2. There are no toll booths at the entrance to the congestion charge zone. How do you think the toll is being collected?

▼ 13.CR.5 ENTERING LONDON'S CENTRAL ZONE

Key Terms

Annexation (13.8.2) Legally adding land area to a city in the United States.

Census tract (13.4.1) An area delineated by the U.S. Bureau of the Census for which statistics are published; in urban areas, census tracts correspond roughly to neighborhoods.

Central business district (CBD) (13.2.1) The area of a city where retail and office activities are clustered.

Central city (13.1.1) See municipality.

Combined statistical area (CSA) (13.1.2) In the United States, two or more contiguous core-based statistical areas tied together by commuting patterns.

Concentric zone model (13.3.1) A model of the internal structure of cities in which social groups are spatially arranged in a series of rings.

Core-based statistical area (CBSA) (13.1.2) In the United States, the collection of all metropolitan statistical areas and micropolitan statistical areas.

Density gradient (13.8.1) The change in density in an urban area from the center to the periphery.

Edge city (13.3.2) A node of office and retail activities on the edge of an urban area.

Galactic (or **peripheral) model** (13.3.2) A model of North American urban areas consisting of a city surrounded by large suburban residential and business areas tied together by a beltway or ring road.

Gentrification (13.10.2) A process of converting an urban neighborhood from an area of residents with predominantly low incomes who rent to an area of residents with middle incomes who own their homes.

Informal settlement (13.6.1) An area within a city in a less developed country in which people illegally establish residences on land they do not own or rent and erect homemade structures.

Megalopolis (13.1.2) A continuous urban complex in the northeastern United States.

Metropolitan statistical area (MSA) (13.1.1) In the United States, an urbanized area with a population of at least 50,000, the county within which the city is located, and adjacent counties meeting one of several tests indicating a functional connection to the central city.

Micropolitan statistical area (μSA) (13.1.1) An urbanized area of between 10,000 and 50,000 inhabitants, the county in which it is located, and adjacent counties tied to the city.

Multiple nuclei model (13.3.2) A model of the internal structure of cities in which social groups are arranged around a collection of nodes of activities.

Municipality (or **central city**) (13.1.1) An urban settlement that is legally incorporated into an independent, self-governing unit.

Rush hour (13.9.1) The four consecutive 15-minute periods in the morning and evening with the heaviest volumes of traffic.

Sector model (13.3.1) A model of the internal structure of cities in which social groups are arranged around a series of sectors, or wedges, radiating out from the central business district.

Smart growth (13.8.2) Legislation and regulations to limit suburban sprawl and preserve farmland.

Social area analysis (13.4.1) Statistical analysis used to identify where people of similar living standards, ethnic background, and lifestyle live within an urban area.

Sprawl (13.8.1) Development of new housing sites at relatively low density and at locations that are not contiguous to the existing built-up area.

Suburb (13.8.1) A residential or commercial area situated within an urban area but outside the central city.

Urban area (13.1.1) A dense core of census tracts, densely settled suburbs, and low-density land that links the dense suburbs with the core.

Urban cluster (13.1.1) In the United States, an urban area with between 2,500 and 50,000 inhabitants.

Urbanized area (13.1.1) In the United States, an urban area with at least 50,000 inhabitants.

Zoning ordinance (13.8.2) A law that limits the permitted uses of land and maximum density of development in a community.

Geospatial Analysis

Log in to the Mastering Geography Study Area to access MapMaster 2.0.

Urban Air Quality

Air quality varies among U.S. cities.

Add the *Urban Population (U.S.)* layer, and ***deselect*** all classes below *61% urban*. ***Add*** the *Poor Urban Air Quality* layer, and *join* the data. In *Settings*, ***click*** *Show City Labels*.

1. Are cities with poor air quality distributed uniformly across the United States? What might account for the distribution?

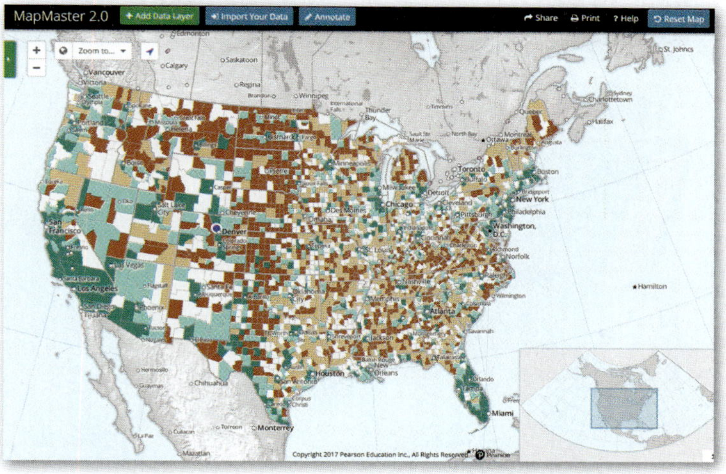

▲ **13.CR.6** URBAN AREAS WITH POOR AIR QUALITY

CHAPTER

14 Resource Issues

▲ Recycling plastic caps, Sofia, Bulgaria.

People have always transformed Earth's land, water, and air for their benefit. Humans once believed Earth's resources to be infinite, or at least so vast that human actions could never harm or deplete them. But human actions in recent years have consumed, polluted, and modified much of Earth's resources. Future generations will pay the price if we continue to rapidly consume and mismanage Earth's resources.

KEY ISSUES

1 Where Are Energy Resources Distributed?

Currently, energy is derived primarily from three fossil fuels: coal, natural gas, and petroleum. Neither the supply nor the demand for these resources is uniformly distributed.

2 What Is the Future for Energy Resources?

Energy resources can be divided between renewable and nonrenewable. The three fossil fuels are nonrenewable, whereas other increasingly important energy resources are renewable.

3 How Are Resources Polluted?

Human actions pollute Earth's air, water, and land. Pollution occurs when more waste is added to a resource than it can accommodate.

4 How Are Resources Protected?

Sustainable development can be achieved through preservation and conservation of resources. Humans can reduce the amount of pollution by either consuming less of a product that causes pollution or recycling more of the products that they use.

LOCATIONS IN THIS CHAPTER

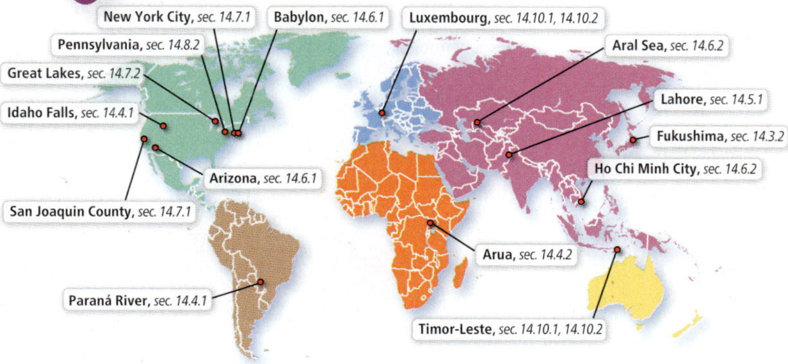

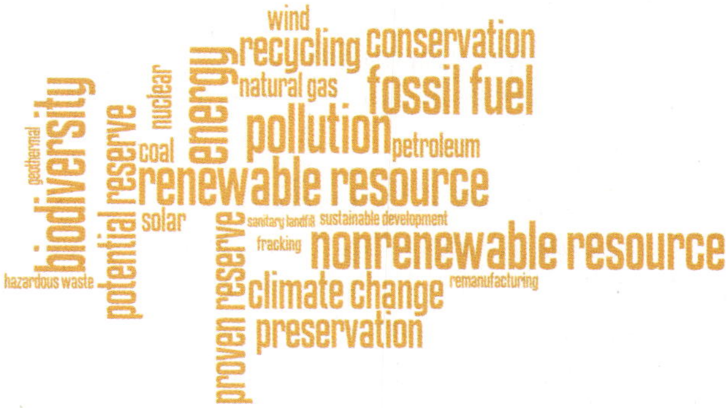

▲ 14.1.1 ENERGY FOR TRANSPORTATION

Energy Supply & Demand

- Identify the principal sources of supply and demand for energy.

Earth offers a large array of resources for people to use. A resource was defined in Chapter 1 as a substance in the environment that is useful to people, is economically and technologically feasible to access, and is socially acceptable to use. Energy resources are especially valuable. We depend on abundant, low-cost energy and minerals to run our industries, keep our homes comfortable, and transport ourselves (Figure 14.1.1). But we are depleting the supply of some energy resources.

14.1.1 Fossil Fuels

Historically, people relied primarily on **animate power**, which is power supplied by animals or by people themselves. Now five-sixths of the world's energy is supplied by three fossil fuels (Figure 14.1.2). A **fossil fuel** is an energy source formed from the residue of plants and animals buried millions of years ago. As sediment accumulated over these remains, intense pressure and chemical reactions slowly converted the remains into the fossil fuels that are currently used. When fossil fuels are burned, energy that was stored in plants and animals millions of years ago is released. The three principal fossil fuels are coal, petroleum, and natural gas.

A resource can be renewable or nonrenewable. A **renewable resource** is produced in nature more rapidly than it is consumed by humans. A **nonrenewable resource** is produced in nature more slowly than it is consumed by humans. The fossil fuels are examples of nonrenewable energy resources that have finite supplies capable of being exhausted.

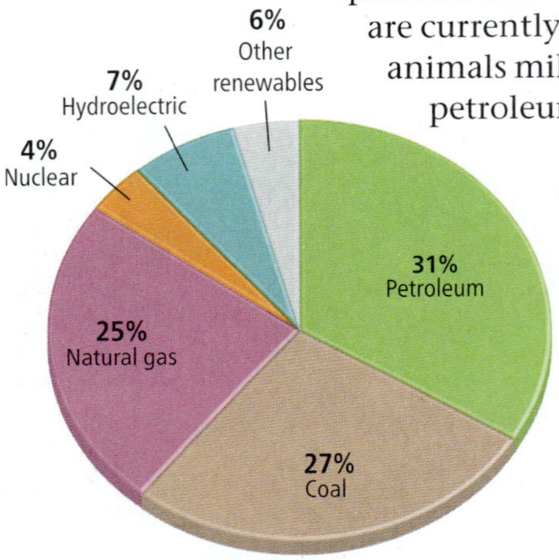

▲ 14.1.2 WORLD ENERGY SUPPLY

Supply of Energy

Geographers observe two important inequalities in the distribution of energy supply and demand:

- **Supply.** The quantity of something that producers have available for sale is **supply**. Some regions have abundant supply of energy resources, whereas others have little.
- **Demand.** The quantity that people wish to consume and are able to buy is **demand**. The heaviest consumers of energy are in developed countries, whereas most of the energy sources are currently found in developing countries.

Given the centrality of energy in contemporary economy and culture, the unequal distribution in the supply and demand of energy has been a major challenge for industry and a source of instability between developed and developing countries.

U.S. Energy Supply

The United States is highly dependent on the three fossil fuels (Figure 14.1.3). Wood was the principal energy source in the nineteenth century. Coal became most important during the late nineteenth century, and petroleum and natural gas gained prominence during the second half of the twentieth century.

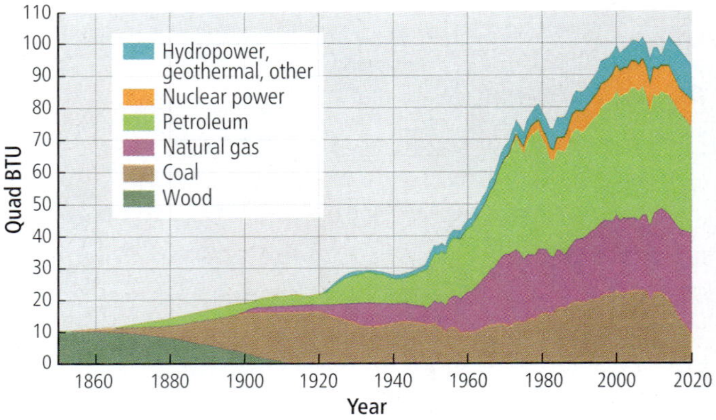

▲ 14.1.3 CHANGING U.S. ENERGY SUPPLY

14.1.2 Demand for Energy

Demand for energy in the United States comes from four principal sources:

- **Industry.** Factories are powered by roughly 41 percent natural gas, 33 percent petroleum, 6 percent coal, and 20 percent other sources. Most of the natural gas and petroleum is burned directly, whereas the other sources are consumed primarily through purchasing electricity.
- **Transportation.** Almost all transportation systems run on petroleum products.
- **Homes.** Natural gas provides the largest share of home energy needs. Natural gas is the principal source of home heating and air-conditioning, whereas coal and nuclear power are the principal sources for the generation of electricity.
- **Services.** Energy uses and sources for retailers and offices are similar to those for homes.

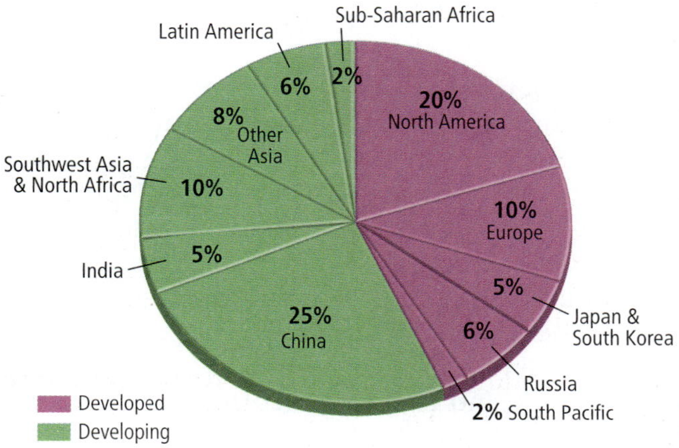

◀ **14.1.4** SHARE OF WORLD ENERGY DEMAND BY COUNTRY

Distribution of Demand for Energy

Demand for the world's energy is currently divided about equally between developed and developing countries (Figure 14.1.4). China has surpassed the United States as the country that uses the most energy. However, North America has the highest per capita consumption of energy. The region contains 5 percent of the world's people but consumes 20 percent of the world's energy (Figure 14.1.5).

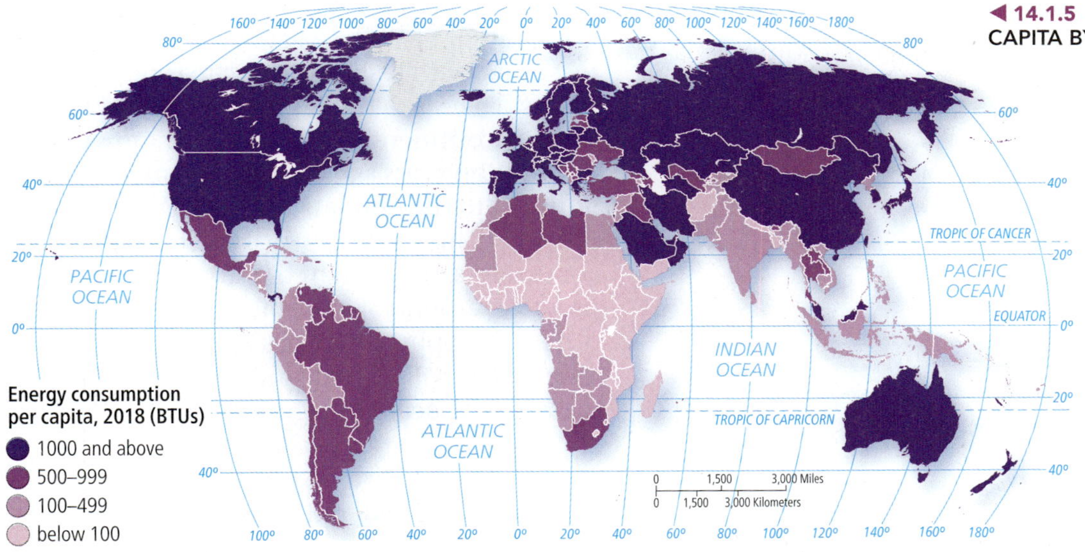

◀ **14.1.5** ENERGY DEMAND PER CAPITA BY COUNTRY

Future Demand for Energy

Developed countries are expected to continue to consume approximately the same amount of energy as they are now. However, demand for energy is expected to increase rapidly in developing countries. As a result, developing countries are expected to demand two-thirds of the world's energy in 2040 (Figure 14.1.6).

▼ **14.1.6** FUTURE ENERGY DEMAND

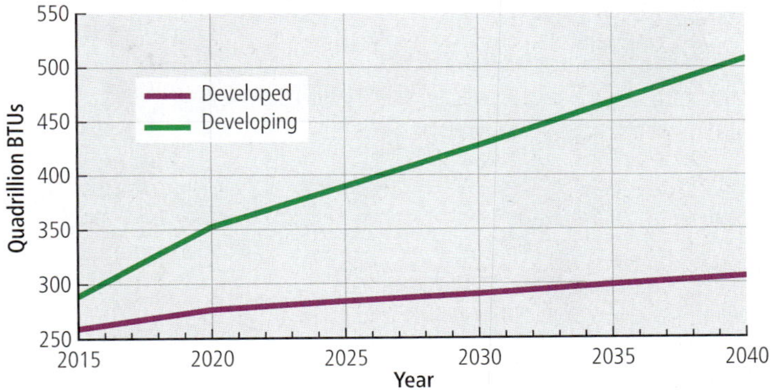

Energy Production

- **Describe the distribution of fossil fuel production.**

Earth's fossil fuel resources are not distributed evenly. Some regions are well endowed with one or more fossil fuels, whereas other regions have little. The uneven distribution of fossil fuels partly reflects how fossil fuels form.

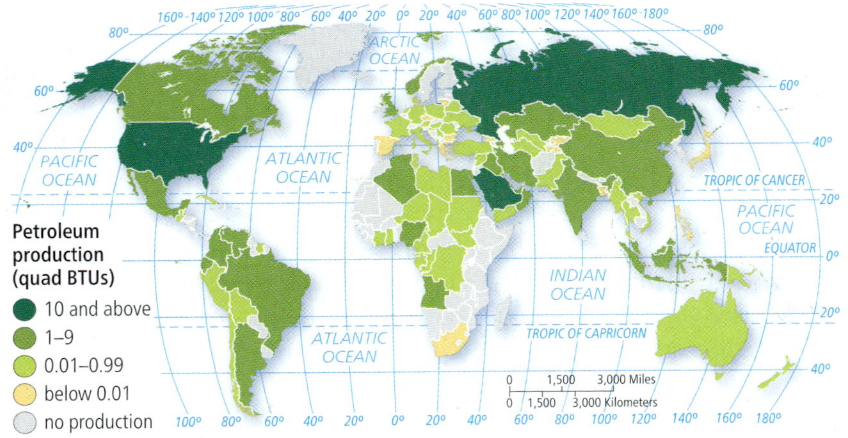

▲ **14.2.1 DISTRIBUTION OF PETROLEUM PRODUCTION**

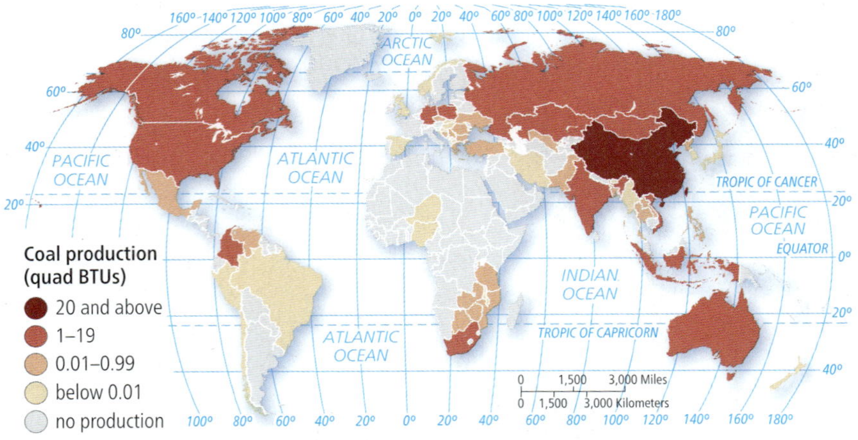

▲ **14.2.2 DISTRIBUTION OF COAL PRODUCTION**

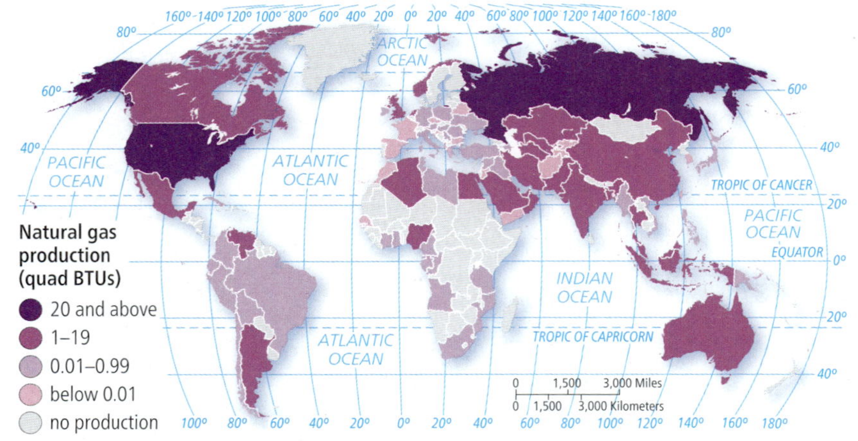

▲ **14.2.3 DISTRIBUTION OF NATURAL GAS PRODUCTION**

14.2.1 Distribution of Fossil Fuels

The three principal fossil fuels have distinctive distributions.

Distribution of petroleum. Petroleum formed millions of years ago from residue deposited on the seafloor. Some petroleum still lies beneath such seas as the Persian Gulf and the North Sea, but other reserves are located beneath land that was under water millions of years ago. Around 29 percent of the world's petroleum is produced in Southwest Asia, 29 percent in other developing regions, 26 percent in North America, and 16 percent in other developed regions (Figure 14.2.1).

Distribution of coal. Coal formed in tropical locations, with lush, swampy areas rich in plants. Thanks to the slow movement of Earth's drifting continents, the tropical swamps of 250 million years ago relocated from the low latitudes to the mid-latitudes. As a result, today's main reserves of coal are in mid-latitude countries rather than in the tropics. China produces around 52 percent of the world's coal, other developing countries 23 percent, the United States 8 percent, and other developed countries 17 percent (Figure 14.2.2).

Distribution of natural gas. Like petroleum, natural gas formed millions of years ago from sediment deposited on the seafloor. Around 33 percent of the world's natural gas is produced in North America, 20 percent in Russia, 8 percent in other developed countries, 22 percent in Southwest Asia, and 17 percent in other developing countries (Figure 14.2.3).

Figures 14.2.1, 14.2.2, and 14.2.3 all use the same units (quad BTUs). Quad is short for quadrillion (1 quadrillion = 1,000,000,000,000,000), and BTU is short for British thermal unit. One quad BTU equals 8 billion U.S. gallons of gasoline.

14.2.2 Trade in Energy

Maintaining a consistent supply of petroleum has been especially challenging in recent years. On the one hand, most of the world's petroleum is supplied by countries in Southwest Asia & North Africa and Central Asia, two regions strongly impacted by religious, ethnic, and political conflicts discussed in Chapters 6 through 8. On the other hand, developed countries in North America and Europe demand disproportionately large shares of energy.

Developed countries supply a large share of the world's fossil fuels, but they demand more energy than they produce, so they must import fossil fuels, especially petroleum. The largest flows of oil are from Russia to Europe and from Canada to the United States (Figure 14.2.4). Europe imports more than half their petroleum, and Japan imports more than 90 percent.

Several developing countries possessing substantial petroleum reserves, primarily in Southwest Asia & North Africa, created the Organization of the Petroleum Exporting Countries (OPEC) in 1960. OPEC was originally formed to enable oil-rich countries to gain more control over their resource. At the time, U.S. and European transnational companies, which had originally explored and exploited the oil fields, were selling the petroleum at low prices to consumers in developed countries and keeping most of the profits. With the formation of OPEC, countries possessing the oil reserves nationalized or more tightly controlled the fields, and prices were set by governments rather than by petroleum companies.

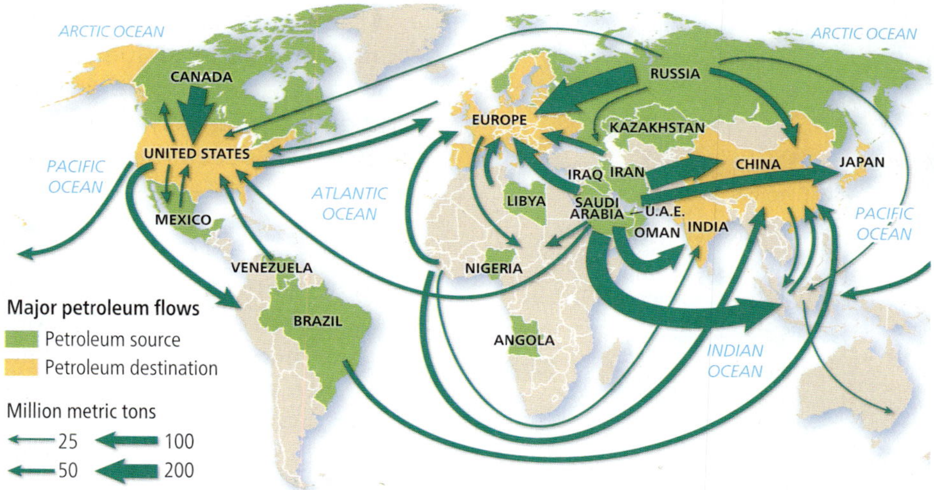

▲ **14.2.4 PETROLEUM TRADE**

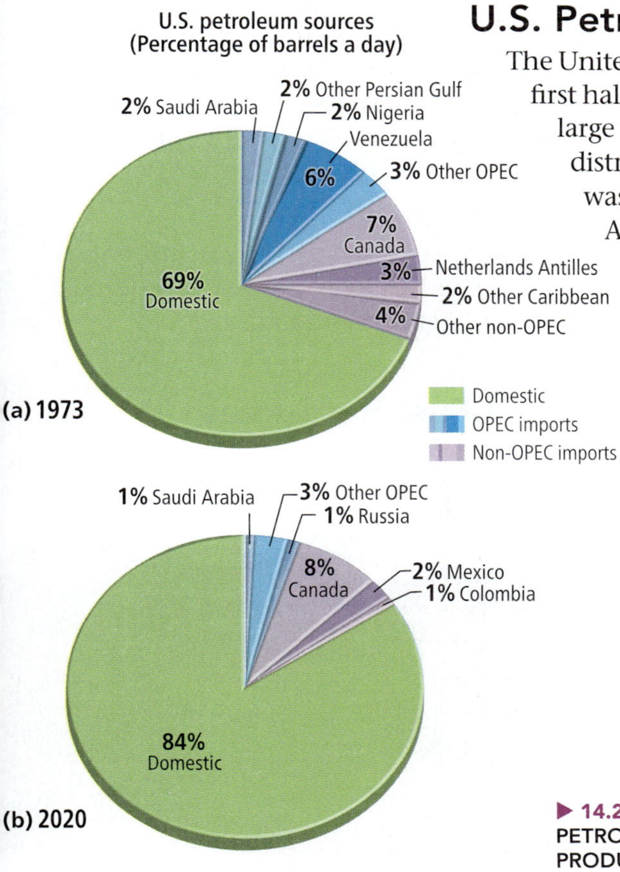

▲ **14.2.5 U.S. PETROLEUM SOURCES**

U.S. Petroleum Supply & Demand

The United States produced more petroleum than it consumed during the first half of the twentieth century. Beginning in the 1950s, the handful of large transnational companies then in control of international petroleum distribution determined that extracting petroleum in the United States was more expensive than importing it from Southwest and Central Asia. As a result, the United States became a net importer of petroleum.

Since 1973, petroleum imported into the United States has come from a different set of countries. For example, the share of U.S. petroleum imported from OPEC members declined from 15 percent in 1973 to 4 percent in 2020 (Figure 14.2.5).

In 2020, the United States exported more petroleum than it imported for the first time in at least 70 years (Figure 14.2.6). Domestic production has increased, and demand declined during the COVID-19 lockdown.

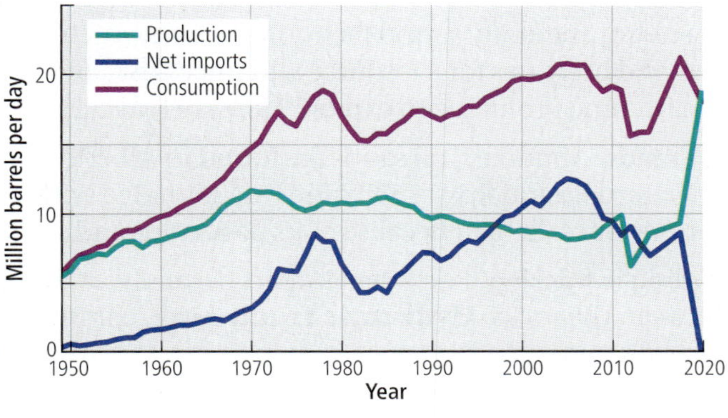

▶ **14.2.6 U.S. PETROLEUM IMPORTS, PRODUCTION, AND CONSUMPTION**

Fossil Fuel Futures

- **Describe the distribution of fossil fuel production.**

The world faces an energy challenge because of the depleting supply of the nonrenewable fossil fuels, petroleum, natural gas, and coal. Although we will not literally "run out" of nonrenewable fossil fuels during the twenty-first century, at some point, extracting the remaining reserves will prove so expensive and environmentally damaging that use of alternative energy sources will accelerate.

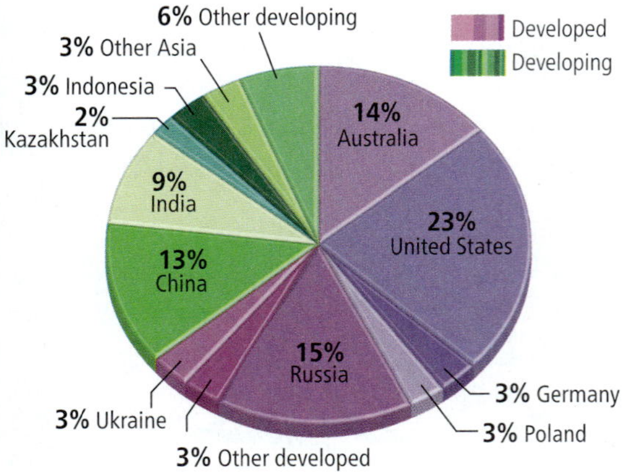

▲ 14.3.1 PROVEN RESERVES OF COAL

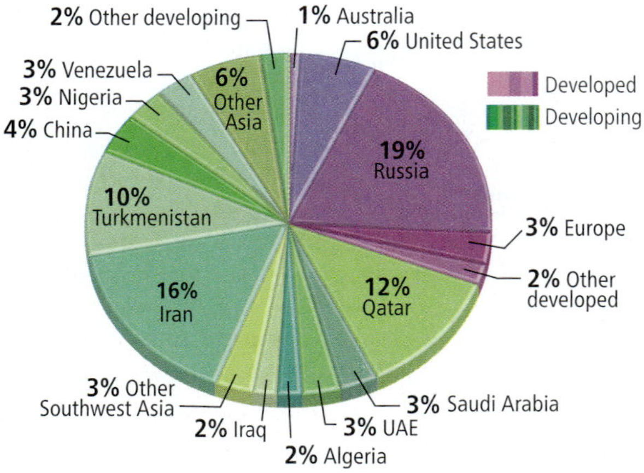

▲ 14.3.2 PROVEN RESERVES OF NATURAL GAS

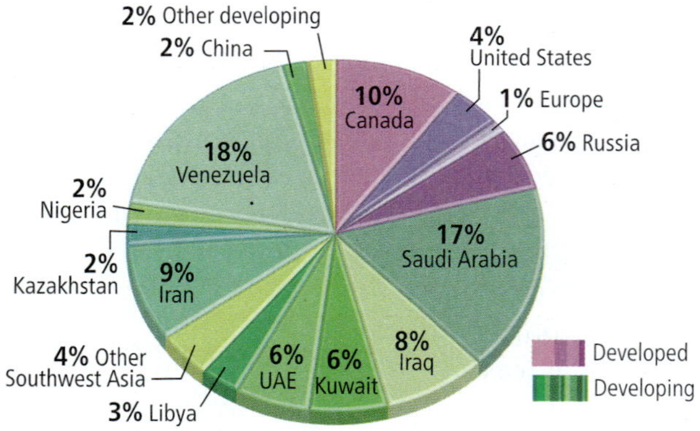

▲ 14.3.3 PROVEN RESERVES OF PETROLEUM

14.3.1 Proven & Potential Reserves

The supply of energy remaining in deposits that have been discovered is called a **proven reserve**. The supply in deposits that are undiscovered but thought to exist is a **potential reserve**. When a potential reserve is actually discovered, it is reclassified as a proven reserve.

Proven reserves can be measured with reasonable accuracy:

Coal reserves. World proven coal reserves total approximately 1 quadrillion metric tons (23 million quad BTUs). At current demand, proven coal reserves would last 110 years. The United States, Russia, China, and Australia have 65 percent of the world's proven coal reserves (Figure 14.3.1).

Natural gas reserves. World proven natural gas reserves are approximately 7,000 trillion cubic meters (7,200 quad BTUs). At current demand, proven natural gas reserves would last 54 years. Iran, Russia, and Qatar together have half of the world's proven natural gas reserves (Figure 14.3.2).

Petroleum reserves. World proven petroleum reserves are approximately 1.7 trillion barrels (9,000 quad BTUs). At current demand, proven petroleum reserves would last 43 years. Venezuela and Saudi Arabia are the two leading locations of proven petroleum reserves (Figure 14.3.3).

Unconventional Fossil Fuel Resources

Resources are considered unconventional if we lack economically feasible or environmentally sound technology with which to extract them. As demand increases for a resource and prices rise, an unconventional source can become profitable to exploit. Here are two current examples:

Oil sands. Abundant oil sands are found in Alberta, Canada, as well as in Venezuela and Russia. Oil sands are saturated with thick petroleum commonly called tar because of its dark color and strong odor.

Hydraulic fracking. Rocks break apart naturally, and gas can fill the space between the rocks. **Hydraulic fracturing**, commonly called *fracking*, involves pumping a lot of water at high pressure into the ground to further break apart rocks and thereby release more trapped gas (Figure 14.3.4).

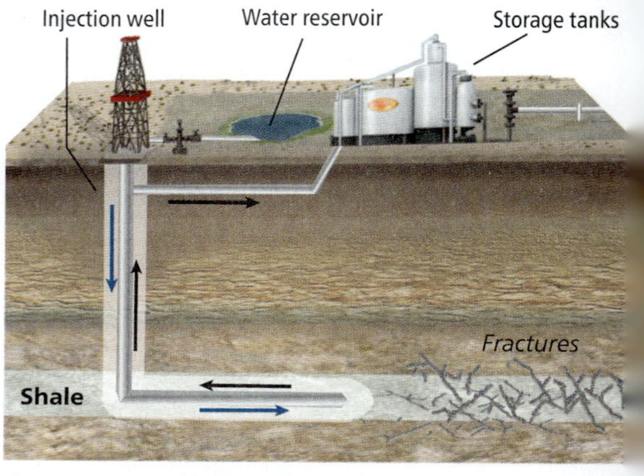

▲ 14.3.4 HYDRAULIC FRACKING

14.3.2 Nuclear Energy

Nuclear power supplies 10 percent of the world's electricity. Although nuclear power is nonrenewable, some view it as an alternative to fossil fuels. The big advantage of nuclear power is the large amount of energy released from a small amount of material. One kilogram of enriched nuclear fuel contains more than 2 million times the energy in 1 kilogram of coal.

Two-thirds of the world's nuclear power is generated in developed countries, including 34 percent in Europe and 27 percent in North America. Only 32 of the world's nearly 200 countries make some use of nuclear power, including 22 developed countries and 10 developing countries. The countries most highly dependent on nuclear power are clustered in Europe (Figure 14.3.5).

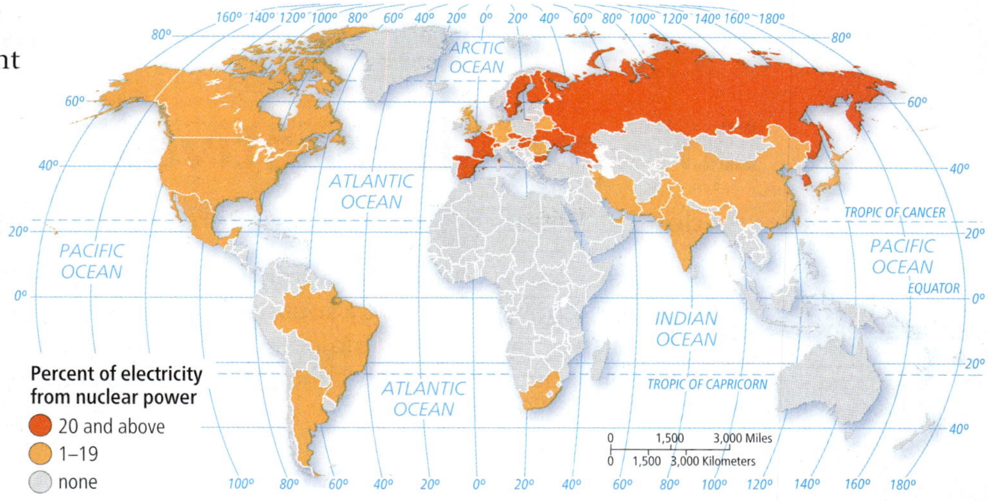

▲ **14.3.5 ELECTRICITY FROM NUCLEAR POWER**

Nuclear Power in the United States

Dependency on nuclear power varies widely among U.S. states (Figure 14.3.6). Nuclear power accounts for more than 50 percent of electricity in Illinois, New Hampshire, and South Carolina. At the other extreme, 22 states and the District of Columbia generate no electricity from nuclear power.

▼ **14.3.6 NUCLEAR POWER BY U.S. STATE**

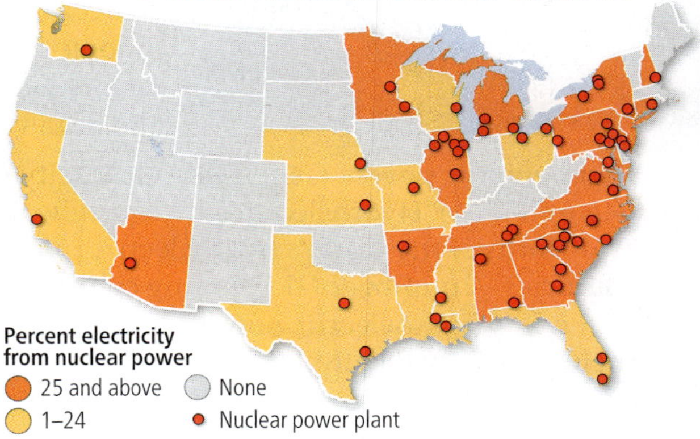

Distinctive Challenges with Using Nuclear Power

Dependency on nuclear power presents several serious challenges. These include a lack of plans to dispose of radioactive waste, the potential to create nuclear weapons, and the threat of serious accidents.

Radioactive waste. A nuclear power plant produces electricity from energy released by splitting uranium atoms in a controlled environment, a process called **fission**. One product of all nuclear reactions is radioactive waste, certain types of which are lethal to people exposed to it. The waste cannot be burned or chemically treated, and it must be isolated for several thousand years until it loses its radioactivity. No country has yet devised permanent storage for radioactive waste. Elaborate safety precautions are taken to prevent the leaking of nuclear fuel from a power plant.

Nuclear weapons. The world has eight acknowledged nuclear powers, as discussed in Section 8.11.2. The diffusion of nuclear programs to countries sympathetic to terrorists is particularly worrying to the rest of the world. For example, a major factor in long-time tensions between Iran and other countries is the possible pursuit by Iran to gain the capability of building a nuclear weapon.

Accidents. Nuclear power plants cannot explode, like a nuclear bomb, because the quantities of uranium are too small and cannot be brought together fast enough in a nuclear power plant to cause an explosion. However, it is possible to have a runaway reaction, which overheats the reactor, causing a meltdown, possible steam explosions, and scattering of radioactive material into the atmosphere. This happened in 1986 at Chernobyl, a nuclear power plant then in the Soviet Union and now in the north of Ukraine, near the Belarus border. The accident caused 43 deaths due to exposure to high radiation doses and an estimated 4,000 cancer-related deaths to people who lived near the plant.

Following an earthquake and tsunami in 2011, three of the six reactors at Japan's Fukushima Daiichi Nuclear Power Plant experienced full meltdown, resulting in release of radioactive materials. Three workers died; the death toll among nearby residents exposed to high levels of radioactivity won't be known for years.

Some nuclear power issues might be addressed through nuclear **fusion**, which is the fusing of hydrogen atoms to form helium. Fusion can occur only at very high temperatures (millions of degrees). Scientists have not yet discovered how to manage this process on a sustained basis for energy production.

Energy Alternatives

- Identify alternative sources of energy and challenges to using them.

Whereas fossil fuels that currently supply most of the world's energy needs are nonrenewable, energy sources of theoretically unlimited supply are renewable. Tapping into renewable energy sources for personal, agricultural, and industrial use is age-old, but advanced technologies that help us harness renewable energy sources are relatively new.

▼ 14.4.1 HYDROELECTRIC DAM
Itaipu Dam, Paraná River, between Brazil and Paraguay.

14.4.1 Renewable Energy Sources

Several sources of renewable energy are currently utilized to varying extent. These include biomass, hydroelectric, wind, geothermal, and solar.

Biomass. Before the widespread use of fossil fuels, biomass was the principal energy source that supplemented animate power. **Biomass fuel** is derived from burning wood, plant material, and animal waste directly or converting them to charcoal, alcohol, or methane gas. Biomass remains an important source of fuel in some developing countries. When carefully harvested in forests, wood is a renewable resource that can be used to generate electricity and heat. Wood can also be salvaged from surplus construction material and demolition of buildings.

Hydroelectric power. Generating electricity from the movement of water is called **hydroelectric power** (Figure 14.4.1). Hydroelectric power is now the world's second-most-popular source of electricity after coal. Two-thirds of the world's hydroelectric power is generated in developing countries and one-third in developed countries. A number of developing countries depend on hydroelectric power for most of their electricity (Figure 14.4.2).

Wind power. The benefits of wind-generated power seem irresistible. Construction of a wind turbine modifies the environment less severely than does construction of a dam. And wind power has greater potential for increased use because only a small portion of the resource has been harnessed. Hundreds of wind "farms" consisting of large numbers of wind turbines have been constructed across the United States (Figure 14.4.3). But wind power has divided the environmental community. Some oppose construction of wind turbines because they can be noisy and lethal for birds and bats. Some also find them a visual blight when they're constructed in places of outstanding beauty.

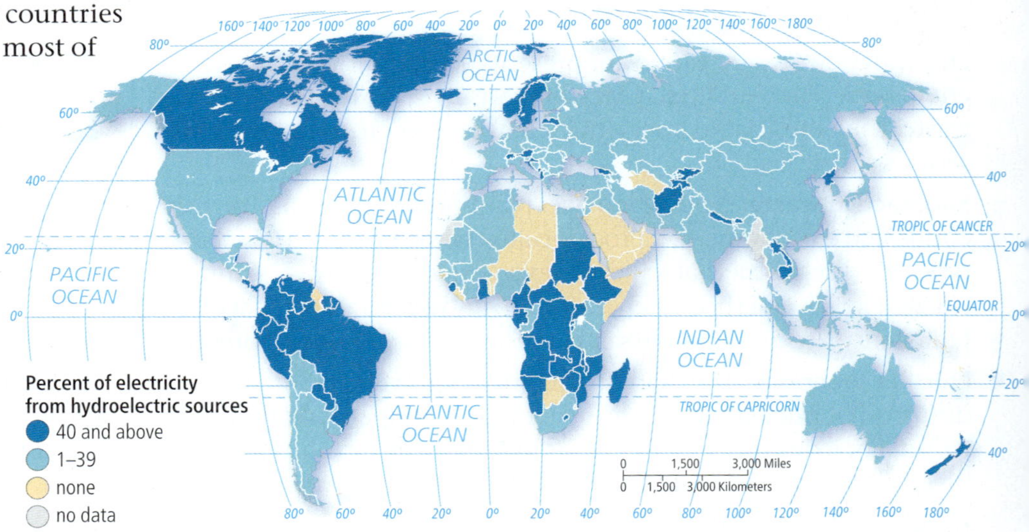

▲ 14.4.2 ELECTRICITY FROM HYDROELECTRIC POWER

▼ 14.4.3 WIND FARM
Idaho Falls, Idaho.

Geothermal energy. Energy from hot water or steam is called **geothermal energy**. Natural nuclear reactions make Earth's interior hot. Toward the surface, in volcanic areas, this heat is especially pronounced. The hot rocks can encounter groundwater and produce heated water or steam that can be tapped by wells. Harnessing geothermal energy is most feasible at sites along Earth's surface where crustal plates meet, which are also the sites of many earthquakes and volcanoes.

14.4.2 Solar Energy

The ultimate renewable energy resource is solar energy supplied by the Sun. Solar energy offers the possibility for countries at low levels of development to pursue sustainable economic growth. Through solar energy, people, businesses, schools, and hospitals in developing countries unable to obtain electricity through conventional power plant sources can generate their own energy.

The potential for growth is limitless. The Sun's remaining life is estimated at 5 billion years, and humans appear to be incapable of destroying or depleting that resource. The Sun's energy is free and ubiquitous and cannot be exclusively owned, bought, or sold by any particular individual or enterprise. Utilizing the Sun as a resource does not damage the environment or cause pollution, as do the extraction and burning of nonrenewable fossil fuels. Despite its promise, solar energy currently generates only 2 percent of the world's electricity. China accounts for 32 percent of the world's solar energy generation (Figure 14.4.4). The United States is second in solar energy generation, but that total is responsible for only 2 percent of total U.S. electricity.

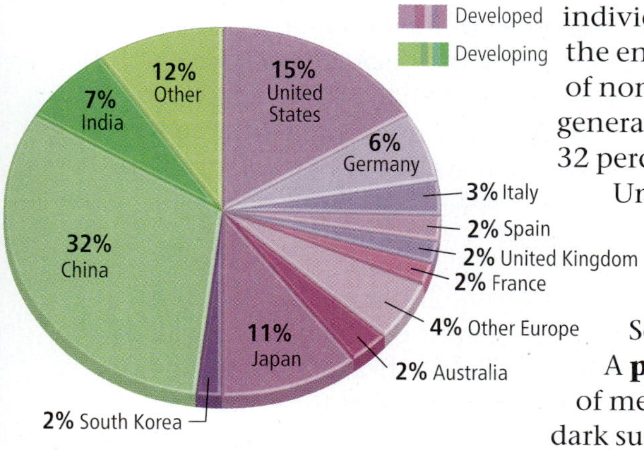

▲ 14.4.4 SOLAR POWER

Passive Solar Power

Solar energy is harnessed through either passive or active means. A **passive solar energy system** collects energy without the use of mechanical devices. Passive systems use south-facing windows and dark surfaces to heat and light buildings on sunny days. The Sun's rays penetrate the windows and are converted to heat.

Active Solar Power

An **active solar energy system** collects solar radiation through the use of mechanical devices and converts the solar radiation either to heat energy or to electricity. The conversion can be accomplished either directly or indirectly.

In a direct active solar energy system, solar radiation is captured through **photovoltaic cells**. A photovoltaic cell is a device made primarily of silicon (the element also contained in computer chips) that converts light energy to electrical energy. Silicon is the second-most-abundant element in Earth's crust. When silicon is combined with one or more other materials, it exhibits distinctive electrical properties in the presence of sunlight, known as the photovoltaic effect. Electrons excited by the sunlight move through the silicon, producing direct current (DC) electricity.

In indirect active solar energy power, the Sun's rays are concentrated by reflectors onto a pipe filled with synthetic oil. The heat from the oil-filled pipe generates steam to run turbines that, in turn, generate electricity.

Solar-Powered Electricity

Solar power can be produced at a central station and distributed by an electric company, as are coal-generated electricity and nuclear-generated electricity. However, because coal and natural gas are still relatively cheap and investment in nuclear facilities is already substantial, public and private utility companies show little interest in investing in solar technology.

▼ 14.4.5 SOLAR-POWERED ELECTRICITY IN A DEVELOPING COUNTRY
Arua, Uganda.

The interest in solar power that does exist in developed countries comes in the form of solar-generated electricity used in spacecraft, light-powered calculators, and solar power plants at remote sites where conventional power is unavailable, such as California's Mojave Desert. Solar energy in developed countries is used primarily in homes as a substitute for electricity in heating water. Rooftop devices collect, heat, and store water for apartment buildings in Israel and Japan and individual homes in the United States.

In developing countries, the largest and fastest-growing market for photovoltaic cells includes the 2 billion people who lack electricity, especially residents of remote villages (Figure 14.4.5). For example, in sub-Saharan Africa, more homes have been electrified in recent years using photovoltaic cells than by hooking up to the central power grid.

The cost of photovoltaic cells must drop and their efficiency must improve for solar power to expand rapidly throughout the United States and the world at large, with or without government support. Solar energy will become more attractive as other energy sources become more expensive.

Air Pollution

- **Describe causes and effects of air pollution at local, regional, and global scales.**

In our consideration of resources, consumption is one-half of the equation—disposal of waste after consuming resources is the other half. All of the resources we use are eventually returned to the atmosphere, bodies of water, or the land surface through burning, rinsing, or discarding. **Pollution** occurs when more of the resource is burned, rinsed, or discarded than the atmosphere, bodies of water, and land surface can accommodate.

14.5.1 Distribution of Air Pollution

At ground level, Earth's average atmosphere is made up of about 78 percent nitrogen, 21 percent oxygen, and less than 1 percent argon. The remaining 0.04 percent includes several trace gases, some of which are critical.

Air pollution is a concentration of trace substances at a greater level than occurs in average air. Concentrations of these trace gases in the air can damage property and adversely affect the health of people, other animals, and plants.

Most air pollution is generated from factories and power plants, as well as from motor vehicles. Factories and power plants produce sulfur dioxides and solid particulates, primarily from burning coal. Burning petroleum in motor vehicles produces carbon monoxide, hydrocarbons, and nitrogen oxides. The countries with the most polluted air are clustered in Asia and Africa (Figure 14.5.1).

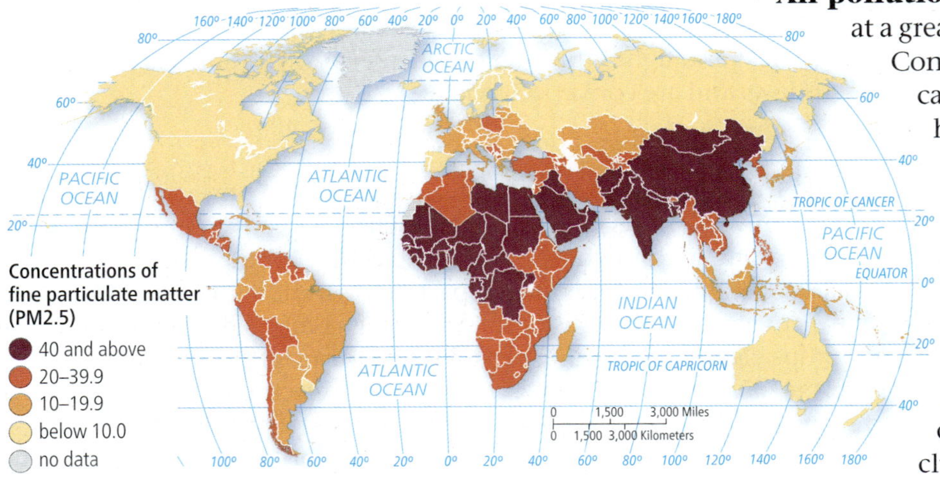

▲ 14.5.1 AIR POLLUTION BY COUNTRY

Local-Scale Air Pollution

At the local scale, air pollution is especially severe in places where emission sources are concentrated, such as in urban areas. The air above urban areas may be polluted because a large number of factories, motor vehicles, and other polluters emit residuals in a concentrated area. Most of the cities with the worst air pollution are in Asia (Figure 14.5.2). Urban air pollution has three basic components:

▲ 14.5.2 AIR POLLUTION, LAHORE, PAKISTAN

- **Carbon monoxide.** Breathing carbon monoxide reduces the oxygen level in blood, impairs vision and alertness, and threatens those with breathing problems.
- **Hydrocarbons.** In the presence of sunlight, hydrocarbons, as well as nitrogen oxides, form **photochemical smog**, which causes respiratory problems, stinging in the eyes, and an ugly haze over cities.
- **Particulates.** A combination of dust and smoke particles results in particulates. The dark plume of smoke from a factory stack and the exhaust of a diesel truck are examples of particulate emissions.

The worst urban air pollution occurs when winds are slight, skies are clear, and a temperature inversion exists. When the wind blows, it disperses pollutants; when the wind is calm, pollutants build. Sunlight combined with particulates lead to the formation of photochemical smog. Air is normally cooler at higher elevations, but when air is warmer at higher elevations—a phenomenon known as *temperature inversion*—pollutants are trapped near the ground.

14.5.2 Regional-Scale Air Pollution

At the regional scale, air pollution may damage vegetation and water supply through **acid deposition**, which is the accumulation of acids, including sulfuric acid and nitric acid, on Earth's surface. Especially affected by acid deposition are the world's principal industrial regions (Figure 14.5.3).

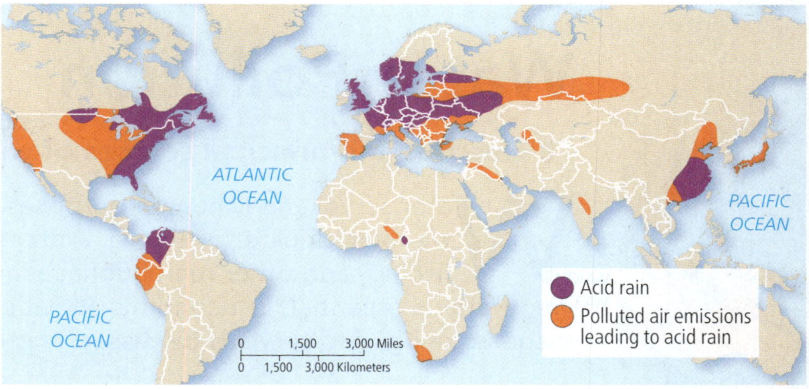

▲ 14.5.3 ACID PRECIPITATION

Sulfur oxides and nitrogen oxides emitted by burning fossil fuels enter the atmosphere, where they combine with oxygen and water. When dissolved in water, the acids may fall as **acid precipitation** (rain, snow, or fog containing pollutants) or be deposited in dust. Before they reach the surface, acid precipitation droplets might be carried hundreds of kilometers.

Acid precipitation damages lakes, killing fish and plants. On land, concentrations of acid in the soil deprive it of nutrients, injuring plants, worms, and insects. Buildings and monuments made of marble and limestone corrode from acid rain.

Geographers are particularly interested in the effects of acid precipitation because the worst damage from the emission of its pollutants occurs at a distance from where it is generated. Within the United States, the major generators of acid deposition are in industrial states along the southern Great Lakes; however, the severest effects of acid rain are felt in several areas farther east. Attempting to reduce the effects of acid deposition and acid rain, the United States reduced sulfur dioxide emissions significantly during the late twentieth century.

Global-Scale Air Pollution: Ozone Damage

Ozone is a gas that absorbs ultraviolet radiation in the stratosphere, a zone between 10 and 50 kilometers (6 to 30 miles) above Earth's surface. Were it not for the ozone in the stratosphere, UV rays would severely damage plants, cause rampant skin cancer, and seriously disrupt food chains.

Earth's protective ozone layer is threatened by manufactured chemicals, such as **chlorofluorocarbons (CFCs)**. CFCs such as Freon were once widely used as coolants in refrigerators and air conditioners. When they leak from these appliances, the CFCs are carried into the stratosphere, where they break down Earth's protective layer of ozone gas.

Global-Scale Air Pollution: Climate Change

At the global scale, the average temperature of Earth's surface has increased by 1°C (2°F) since 1880 (Figure 14.5.4). This temperature increase is directly linked to human actions, especially the burning of fossil fuels in factories and vehicles, according to an international team of U.N. scientists.

When fossil fuels are burned, carbon dioxide is discharged into the atmosphere. Plants and oceans absorb much of the discharge, but increased fossil fuel burning during the past 200 years has caused the level of carbon dioxide in the atmosphere to rise by 45 percent, according to the U.S. National Aeronautics and Space Administration (NASA). The concentration of trace gases in the atmosphere can delay the heat leaving Earth's surface heading for space, thereby raising Earth's temperatures. The increase in Earth's temperature, caused by carbon dioxide trapping some of the radiation emitted by the surface, is called the **greenhouse effect**.

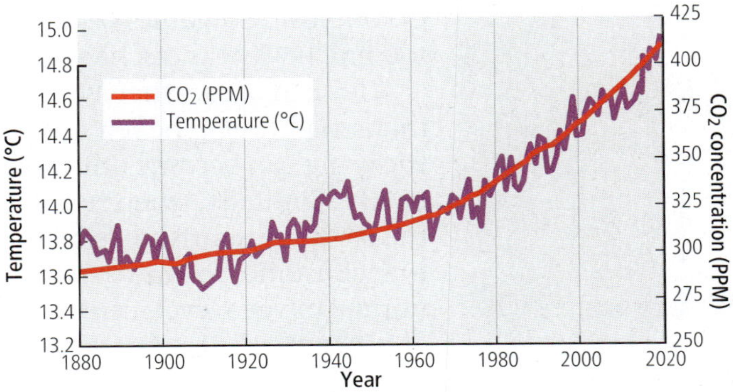

▶ 14.5.4 GLOBAL WARMING AND CARBON DIOXIDE CONCENTRATIONS

▲ 14.6.1 IRRIGATING CROPS, ARIZONA

Water Pollution

- **Describe the principal sources of water pollution.**

Water serves many human purposes. We use it for cooking and bathing, and we must drink it to survive. Water provides a location for boating, swimming, fishing, and other recreational activities. It is home to fish and other aquatic life. Water is essential for economic activities, including agriculture (Figure 14.6.1), manufacturing, and services discussed in previous chapters.

14.6.1 Demand for Water

We depend on fresh, clean, unpolluted water for many purposes. But that is not always available because people also use water for purposes that pollute it. Pollution is widespread because it is easy to dump waste into a river and let the water carry it downstream where it becomes someone else's problem. Water can decompose some waste without adversely impacting other activities, but the volume of waste in waterways exceeds the capacity of many rivers and lakes to accommodate it.

Water usage is either nonconsumptive or consumptive. Water usage is considered **nonconsumptive** if the water is returned to nature as a liquid and **consumptive** if the water is evaporated. Most industrial and municipal uses of water are nonconsumptive because the wastewater is primarily discharged into lakes and streams. Most agricultural uses are consumptive because the water is primarily taken up by plants that transpire it.

Humans use around 4,000 cubic kilometers of water per year. The heaviest water demand is for agriculture, followed by industry and municipal sewage systems (Figure 14.6.2). China, India, and the United States use more than 500 cubic kilometers each of water. The United States has one of the world's highest per capita daily consumption of water, at 3,800 liters (1,000 gallons) per person, compared with 1,700 liters (450 gallons) in India and 1,165 (300 gallons) in China.

Water usage is extremely high in the United States primarily because of agriculture. U.S. farmers raise a large number of animals to meet the high demand for meat that the average American consumes, as discussed in Chapter 9. A large amount of water is also needed in U.S. agriculture to irrigate fields of crops.

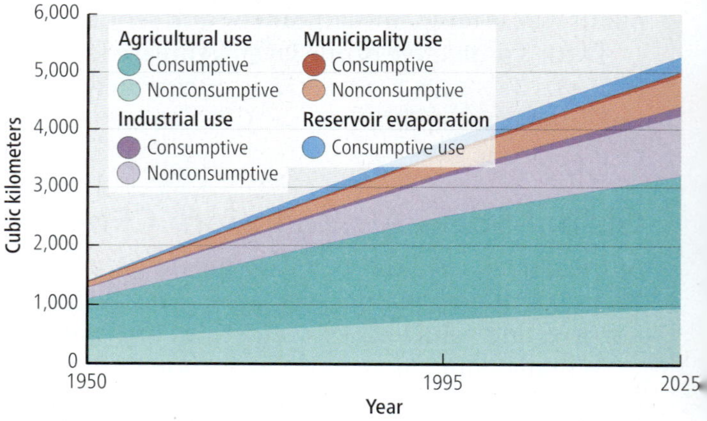

▲ 14.6.2 DEMAND FOR WATER

Impact on Aquatic Life

Polluted water can harm aquatic life. Aquatic plants and animals consume oxygen and so does the decomposing organic waste that humans dump in the water. The oxygen consumed by the decomposing organic waste constitutes the **biochemical oxygen demand (BOD)**. If too much waste is discharged into water, the water becomes oxygen starved, and fish die.

This can occur when bodies of water become loaded with municipal sewage or industrial waste. The sewage and industrial pollutants consume so much oxygen that the water can become unlivable for normal plants and animals, creating a "dead" stream or lake. Similarly, when runoff carries fertilizer from farm fields into streams or lakes, the fertilizer nourishes excessive aquatic plant production—a "pond scum" of algae—that consumes too much oxygen (Figure 14.6.3). Either type of pollution reduces the normal oxygen level, threatening aquatic plants and animals. The result is a reduction of **biological diversity**, or **biodiversity** for short, which is the variety of species in a specific place or on Earth as a whole.

▲ 14.6.3 EXCESSIVE ALGAE, BABYLON, NEW YORK

14.6.2 Point-Source Water Pollution

The sources of pollution can be divided into point sources and nonpoint sources. **Point-source pollution** enters a body of water at a specific location, whereas **nonpoint-source pollution** comes from a large, diffuse area. Point-source pollutants are usually smaller in quantity and much easier to control than nonpoint-source pollutants.

Point-source water pollution originates from a specific point, such as a pipe from a wastewater treatment plant. The two main point sources of water pollution are manufacturers and municipal treatment plants. Many factories use water for cooling and then discharge the warm water back into the river or lake. Fish adapted to cold water, such as salmon and trout, might not be able to survive in the warmer water. The manufacturing processes that produce steel, paper, and food generate chemicals that are major polluters of water.

In developed countries, sewers carry wastewater from sinks, bathtubs, and toilets to a municipal treatment plant, where most—but not all—of the pollutants are removed. The treated wastewater is then typically dumped back into a river or lake. Since passage of the U.S. Clean Water Act and equivalent laws in other developed countries, most treatment plants meet high water-quality standards.

In developing countries, sewer systems are rare, and wastewater usually drains, untreated, into rivers and lakes (Figure 14.6.4). The drinking water, usually removed from the same rivers, may be inadequately treated as well. The combination of untreated water and poor sanitation makes drinking water deadly in developing countries. Waterborne diseases such as cholera, typhoid, and dysentery are major causes of death in developing countries.

▲ **14.6.4** WATER POLLUTION, VIETNAM
In the absence of sewers, waste in Ho Chi Minh City, Vietnam, is dumped into the Saigon River. Sanitation workers on boats try to remove the trash from the river.

Nonpoint-Source Water Pollution

Nonpoint sources usually pollute in greater quantities and are much harder to control than point sources of pollution. The principal nonpoint source is agriculture. Fertilizers and pesticides spread on fields to increase agricultural productivity are carried into rivers and lakes by irrigation systems or natural runoff. Expanded use of fertilizers and pesticides may help prevent a global food crisis in the short term, but it can destroy aquatic life by polluting rivers and lakes.

One of the world's most extreme instances of nonpoint water pollution is the Aral Sea in the former Soviet Union, now divided between the countries of Kazakhstan and Uzbekistan. The Aral Sea was once the world's fourth-largest lake, but it has shrunk to less than 10 percent of its former size (Figure 14.6.5). When it had control of the Aral Sea, the Soviet Union diverted the waters into canals to irrigate nearby cotton fields.

As the lake dried, wind carried the dust generated from the exposed lake bed into the nearby farm fields, as seen in Figure 14.6.5c. The dust, enriched with fertilizers and pesticides from the nearby fields, as well as the increasingly salty seawater, poses health hazards for nearby residents.

▼ **14.6.5** DISAPPEARING ARAL SEA

(a) 1990 (b) 2000 (c) 2020

Solid Waste Pollution

- **Describe the principal strategies for managing solid waste pollution.**

As people acquire more material goods, they generate more solid waste. The average American generates about 2.2 kilograms (5 pounds) of solid waste per day (Figure 14.7.1).

▲ **14.7.1 SOLID WASTE**
New York City.

14.7.1 Solid Waste Generation

People in developed countries generate more than twice as much solid waste than people in developing countries (Figure 14.7.2). Per capita solid waste disposal in the United States is three times the worldwide average. Overall in developed countries, residences generate around 60 percent of the solid waste and businesses 40 percent. Paper products, such as corrugated cardboard and newspapers, account for the largest share of solid waste in the United States, especially among residences and retailers. Manufacturers discard large quantities of metals as well as paper.

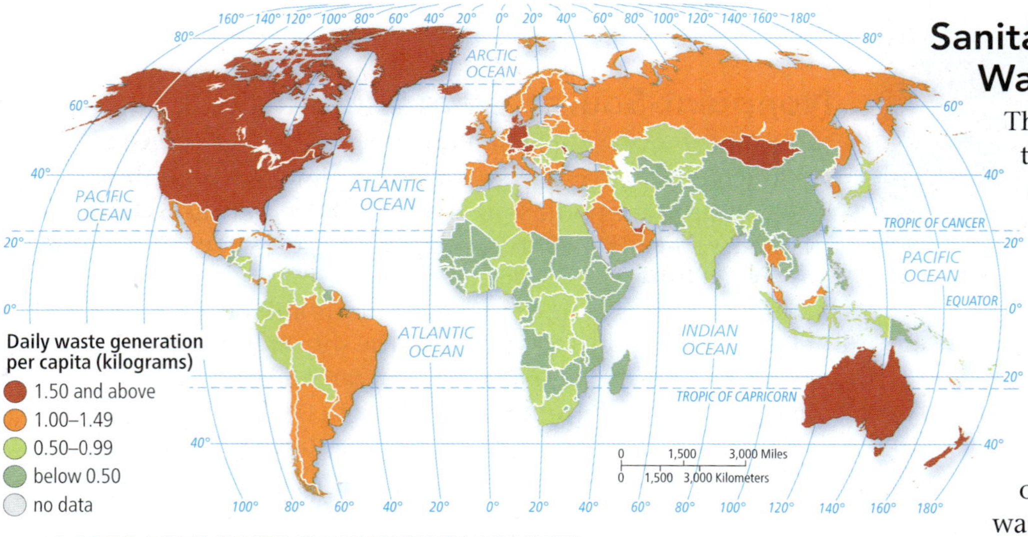

▲ **14.7.2 SOLID WASTE GENERATION BY COUNTRY**

Sanitary Landfills for Solid Waste Disposal

The **sanitary landfill** is by far the most common strategy for disposal of solid waste in the United States. More than half of the country's waste is trucked to landfills and buried under soil.

This strategy is the opposite of our disposal of gaseous and liquid wastes: We disperse air and water pollutants into the atmosphere, rivers, and eventually the ocean, but we concentrate solid waste in thousands of landfills. Concentration would seem to eliminate solid-waste pollution, but it may only hide it temporarily. Chemicals released by the decomposing solid waste can leak from landfills into groundwater and contaminate nearby water wells, soil, and streams.

The number of landfills in the United States has declined by three-fourths since 1990. Thousands of small-town "dumps" have been closed and replaced by a small number of large regional ones (Figure 14.7.3). Better compaction methods combined with expansion in the land area of some of the large regional dumps have resulted in expanded landfill capacity.

The United States sends some of its waste to other countries. For example, in 2019 around 350 million kilograms (770 million pounds) of plastic waste was sent to a number of countries in Asia, including around 120 million kilograms (265 million pounds) to Malaysia.

▲ **14.7.3 SANITARY LANDFILL**
San Joaquin County, California.

14.7.2 Plastic Waste

Most plastics are used in manufacturing a wide variety of products, from cars to computers. Plastics are also used for packaging of other products, notably beverages. All too often, discarded plastic packages flow into bodies of water and wash up on shore.

A major source of plastic waste is water bottles (Figure 14.7.4). The average American consumes around 150 plastic water bottles per year (3 per week). It takes around 15 liters (4 gallons) of petroleum to produce the 150 plastic bottles consumed by a typical American. That equates to the gasoline burned during a 160-kilometer (100-mile) drive in a typical American vehicle. The average American recycles only 1 out of 6 water bottles. The rest are "thrown away" into landfills or into rivers, lakes, and oceans, where they take hundreds of years to disintegrate.

▲ **14.7.4 DISCARDED WATER BOTTLE POLLUTES A RIVER**

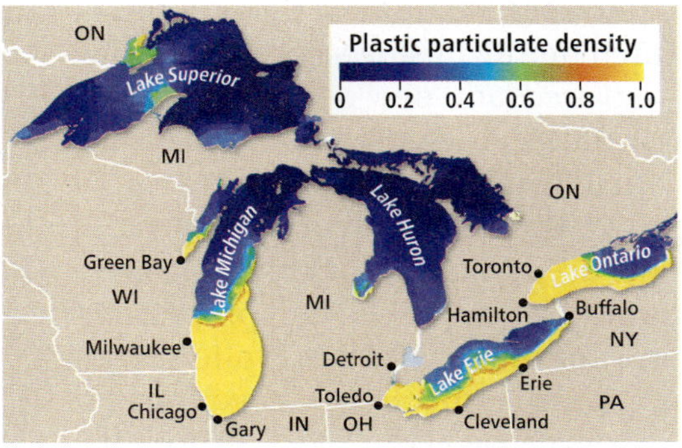

▲ **14.7.5 DENSITY OF PLASTIC IN THE GREAT LAKES**

Rochester Institute of Technology scientists Matthew Hoffman and Eric Hittinger measured the amount and spatial distribution of plastic debris dumped into the Great Lakes. Using mathematical modeling and GIS, they calculated that 10,000 metric tons (22 million pounds) of plastic end up in the Great Lakes annually.

The plastic waste is not distributed uniformly through the Great Lakes. Lake Michigan accounted for 52 percent of the plastic debris, Lake Erie 26 percent, Lake Ontario 16 percent, Lake Huron 6 percent, and Lake Superior 0.3 percent (Figure 14.7.5). The distribution is related to population concentrations as well as distinctive water circulation patterns. Plastic waste discarded in Chicago is trapped at the southern end of Lake Michigan. Lake Erie's plastic debris is clustered near the Detroit River, as well as a southern strip originating from Cleveland.

Hazardous Waste

Disposing of hazardous waste is especially difficult. Hazardous wastes include heavy metals (including mercury, cadmium, and zinc), PCB oils from electrical equipment, cyanides, strong solvents, acids, and caustics. These may be unwanted by-products generated in manufacturing or waste to be discarded after usage.

According to the toxics release inventory published by the U.S. Environmental Protection Agency (EPA), 3.4 billion pounds of toxic chemicals were released into the land in 2019 at 21,393 facilities in the United States. Most production-related toxics were recycled, treated, or discharged into receptacles other than into land. Mining operations were the largest polluters. The Midwest and Southeast had the largest share of facilities generating toxic releases (Figure 14.7.6).

If poisonous industrial residuals are not carefully placed in protective containers, the chemicals may leak into the soil and contaminate groundwater or escape into the atmosphere. Breathing air or consuming water contaminated with toxic wastes can cause cancer, genetic mutations, and chronic ailments.

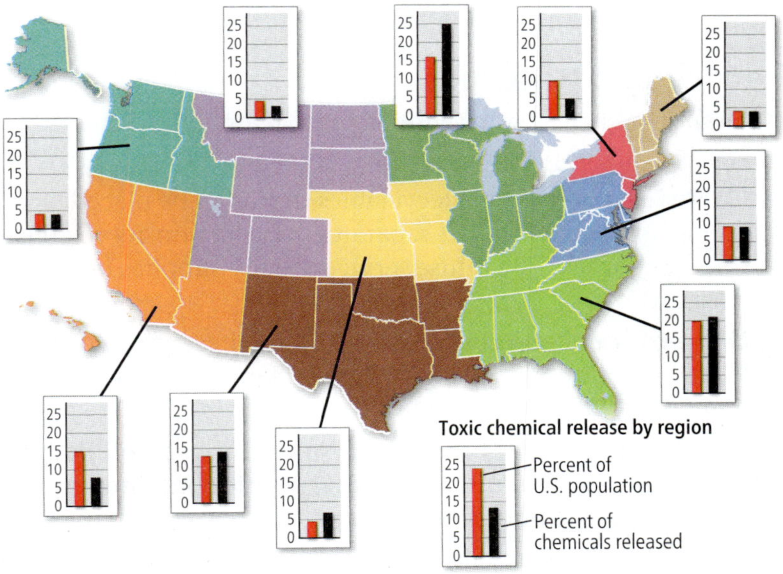

▲ **14.7.6 TOXIC CHEMICAL RELEASE SITES**

Recycling & Remanufacturing

- **Describe alternative strategies for recycling.**

We can reduce the amount of pollution we generate in two basic ways: We can consume less of a product that causes pollution (such as petroleum), or we can recycle more of the products we do use.

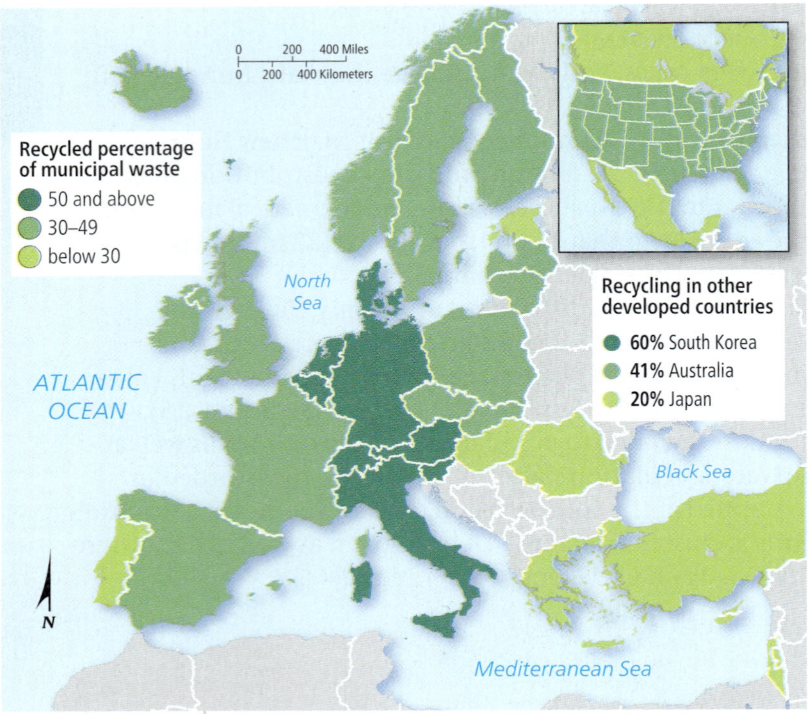

▲ 14.8.1 RECYCLING IN EUROPE AND OTHER DEVELOPED COUNTRIES

14.8.1 Recycling

Recycling is the separation, collection, processing, marketing, and reuse of the unwanted material. **Remanufacturing** is the rebuilding of a product to specifications of the original manufactured product using a combination of reused, repaired, and new parts. Both are increasingly used in industry as ways to promote more sustainable industrial processes.

Recycling involves the breaking down of the components used in the creation of plastics, papers, aluminum, and glass and preparing the components for use in a future application. Formal programs to encourage recycling are common in developed countries, although rates vary widely (Figure 14.8.1). Around 67 percent of solid waste in Germany is recycled, but only 20 percent in Japan. Recycling in the United States increased from 10 percent in 1985 to around 32 percent now (Figure 14.8.2).

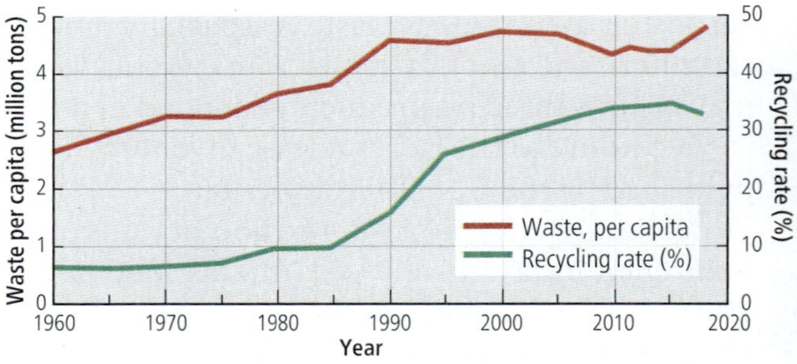

▲ 14.8.2 RECYCLING IN THE UNITED STATES

Formal recycling programs are not in place in most developing countries, though informal recycling is common. Pickers comb through landfills and resell what they find.

Recycling Strategies

Materials that would otherwise be thrown away are collected and sorted in four principal ways:

- **Curbside programs.** Recyclables can often be placed at the curb in a container separate from the non-recyclable trash for pickup by a refuse service at a specified time each week. The refuse service usually supplies homes with specially marked containers for the recyclable items.
- **Drop-off centers.** Drop-off centers are sites, typically with several large containers placed at a central location, for individuals to leave recyclable materials. The containers are periodically emptied by a processor or recycler but are otherwise left unattended.
- **Buy-back centers.** Commercial operations sometimes pay consumers for recyclable materials, especially aluminum cans but also sometimes plastic containers and glass bottles. These materials are usually not processed at the buy-back center.
- **Deposit programs.** Glass and aluminum containers can sometimes be returned to retailers. The price a consumer pays for a beverage may include a deposit fee of 5¢ or 10¢ that the retailer refunds when the container is returned.

14.8.2 Remanufacturing

The percentage of materials recovered by recycling varies widely by product (Figure 14.8.3). Common household items that contain recycled materials include newspapers and paper towels; aluminum, plastic, and glass soft-drink containers; steel cans; and plastic laundry detergent bottles. Materials are manufactured into new products for which a market exists.

The principal inputs into manufacturing include recycled paper, plastic, glass, and aluminum. Recycled materials are also used in such industrial applications as recovered glass in roadway asphalt ("glassphalt") and recovered plastic in carpet, park benches, and pedestrian bridges.

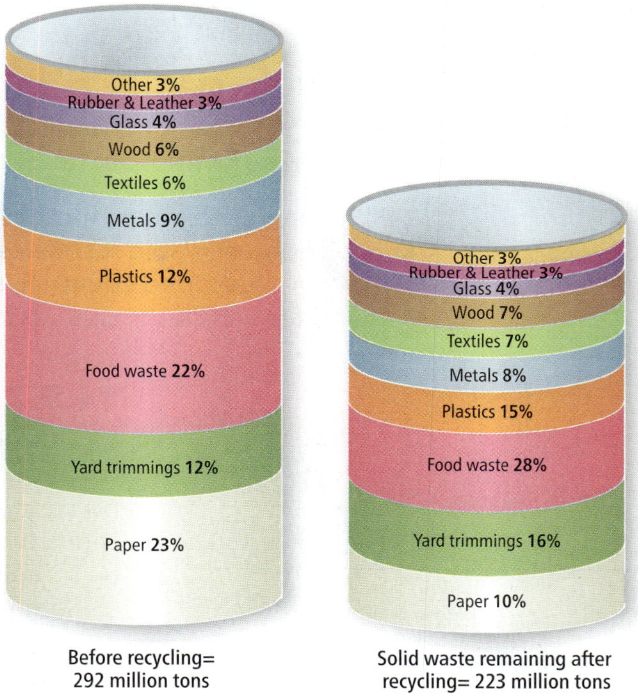

▲ **14.8.3 SOURCES OF U.S. SOLID WASTE BEFORE AND AFTER RECYCLING**
The left side shows the share of different sources of waste. The right side shows the share remaining after some of the waste has been recycled.

Recycled Inputs into Remanufacturing

Four major manufacturing sectors account for more than half the recycling activity—paper mills, steel mills, plastic converters, and iron and steel foundries.

- **Paper.** Most types of paper can be recycled. Newspapers have been recycled profitably for decades, and recycling of other paper, especially computer paper, is growing. The key to recycling is collecting large quantities of clean, well-sorted, uncontaminated, and dry paper. Rapid increases in the cost of virgin paper pulp are stimulating construction of more plants capable of using waste paper.
- **Plastic.** Plastics are of many types made from a variety of materials that undergo various chemical manipulations. To successfully recycle plastic, different plastic types must not be mixed, as even a small amount of the wrong type of plastic can ruin the melt. Because it is impossible to tell one type of plastic from another by sight or touch, the plastic industry has developed a system of numbers marked inside triangles on the bottom of containers. Types 1 and 2 are commonly recycled, and the others generally less frequently.
- **Glass.** Glass can be used repeatedly with no loss in quality and is 100 percent recyclable (Figure 14.8.4). The process of creating new glass from old is extremely efficient, producing virtually no waste or unwanted by-products; however, there are drawbacks in recycling glass. Whereas unbroken clear glass is valuable in recycling, mixed-color glass is nearly worthless, and broken glass is hard to sort.
- **Aluminum.** The principal source of recycled aluminum is beverage containers. Aluminum cans began to replace glass bottles for beer during the 1950s and for soft drinks during the 1960s. Aluminum scrap is readily accepted for recycling, in contrast to the other metals that are rarely accepted.

▲ **14.8.4 RECYCLING GLASS**
Pennsylvania.

The Car of the Future

- **Describe the importance of reducing vehicle emissions.**

One of the greatest challenges to reducing pollution and conserving nonrenewable resources is reliance on petroleum as automotive fuel. Consumers in developed countries are reluctant to give up their motor vehicles, and demand for vehicles is soaring in developing countries.

▼ **14.9.1 CORRELATION BETWEEN GNI PER CAPITA AND CO_2 EMISSIONS**

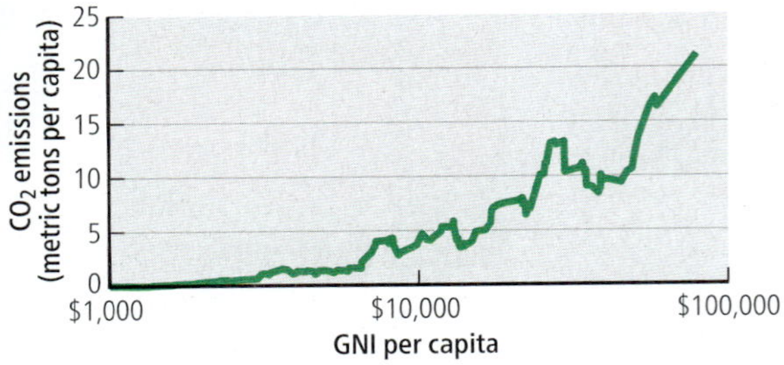

14.9.1 Carbon Dioxide Emissions

What will it take for the world to reduce air pollution and fossil fuel dependency in the years ahead? The challenge is especially great because, as a country improves its level of development and its per capita income increases, the country's per capita carbon dioxide (CO_2) emissions generally increase (Figure 14.9.1).

Some developed countries in Europe have made progress in reducing CO_2 emissions (Figure 14.9.2). Overall, though, developed countries with 17 percent of the world's population are responsible for generating 40 percent of CO_2 emissions (Figure 14.9.3).

◀ **14.9.2 SHARE OF CO_2 EMISSIONS AND POPULATION**

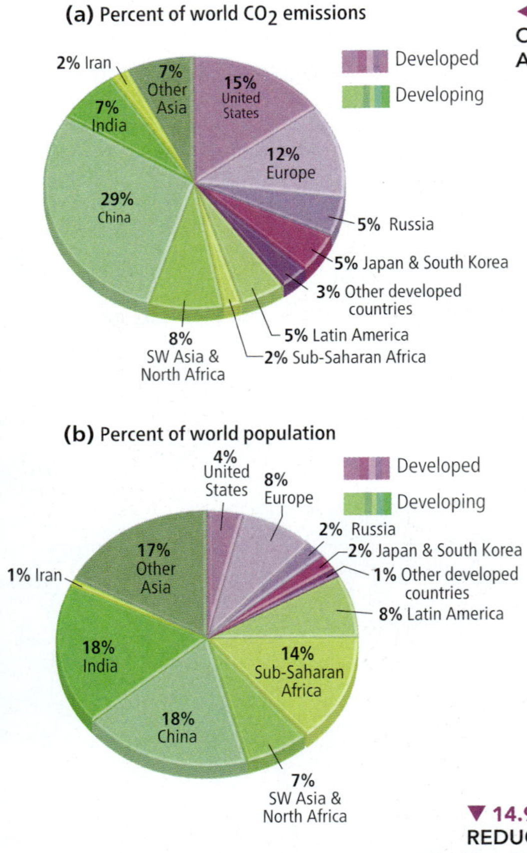

▶ **14.9.3 CO_2 EMISSIONS PER CAPITA**

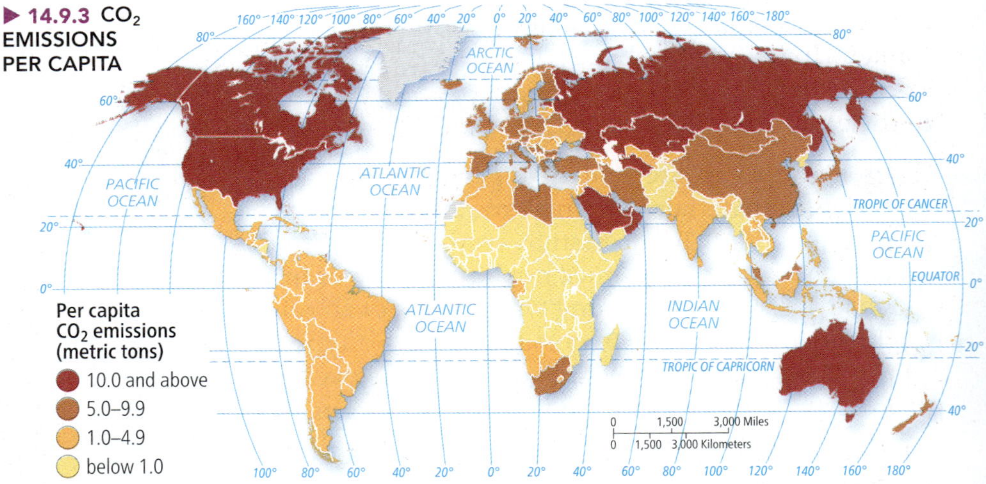

U.S. Emissions Reduction Strategies

According to the U.N., strategies to reduce CO_2 emissions vary among countries. The U.S. scientists working with the United Nations offer a strategy with three key elements (Figure 14.9.4):

- Sharply decrease the use of the three fossil fuels.
- Increase the use of renewable energy.
- Use **carbon capture and storage (CCS)**, a process that involves capturing waste CO_2, transporting it to a storage site, and depositing it where it will not enter the atmosphere, which is normally underground.

▼ **14.9.4 U.N. TARGET FOR CO_2 EMISSIONS REDUCTION IN THE UNITED STATES**

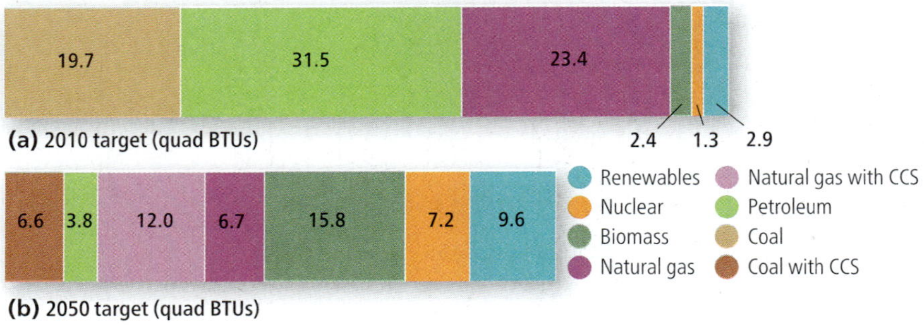

14.9.2 Alternative Fuel Vehicles

A principal strategy for reducing CO_2 emissions is increased reliance for transportation powered by electricity generated from sources other than the three fossil fuels.

Electric-powered vehicles (EVs) comprised only 2 percent of the world's motor vehicles and only 7 percent of total vehicle sales in 2021. However, by 2040, half of new vehicle sales and one-fourth of all vehicles on the road are forecast to be EVs.

Consumer surveys suggest that the most important drawback to driving an EV is the need for frequent charges (Figure 14.9.5). Recharging takes much longer than refueling with gasoline, and charging stations are not always conveniently located. In countries where EV use is increasing, more charging stations are being installed, and improved batteries will reduce the amount of time needed for recharging.

▲ 14.9.5 ELECTRIC VEHICLE (EV) CHARGING STATION

Ready or not, EVs are set to be the only option for consumers in some countries in the years ahead. Around two dozen countries, mostly in Europe, are phasing out the sale of new gasoline-powered vehicles. Canada, India, and Japan are examples of other countries that are phasing out the sale of new gas cars (Figure 14.9.6). Most of the phase-outs are effective between 2030 and 2040.

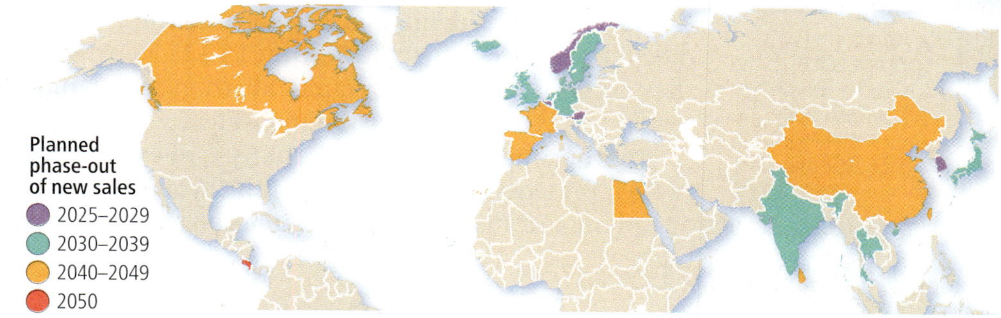

▲ 14.9.6 PHASE-OUT OF GASOLINE-POWERED VEHICLES

Regional Variations in Electricity

The source of power for EVs may or may not be sustainable. Although fossil fuel is not being pumped directly into the tank of the EV, fossil fuel may still be consumed by it indirectly if the electricity at the EV charging station comes from a fossil fuel power plant.

The United States as a whole generated around 40 percent of its electricity from natural gas, 19 percent from coal-burning power plants, 20 percent from nuclear power plants, and 20 percent from renewable energy sources. An EV does reduce consumption of an increasingly scarce and expensive resource—petroleum. If the electricity is generated by coal, an EV may cause more air pollution than a gas car. However, the share of electricity generated by coal has decreased rapidly in the United States, from 33 percent in 2015 to 19 percent in 2020.

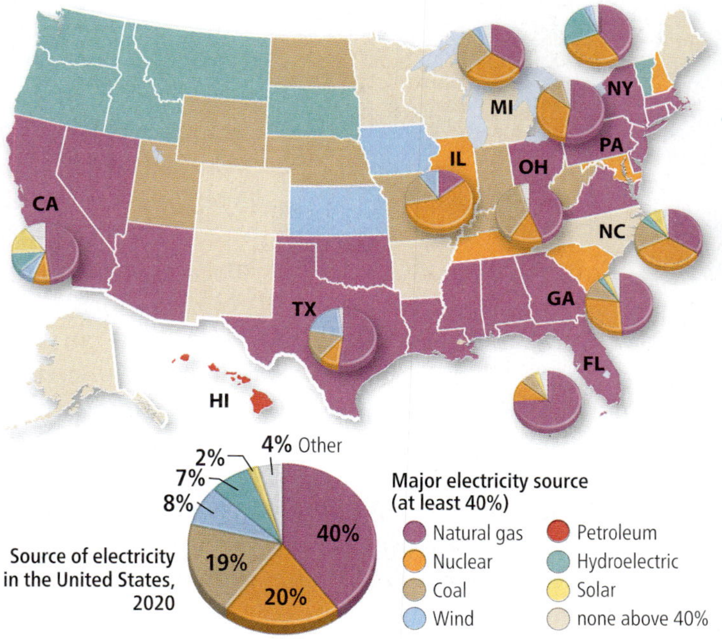

▲ 14.9.7 ELECTRICITY BY U.S. STATE

Electricity is generated differently among the 50 U.S. states. Coal was the leading source of electricity in 8 states in 2020, compared with 20 states in 2015. In Oregon and Washington State, where hydroelectric power is the leading source of electricity, recharging EVs has a much lower impact on air quality than is the case in Kentucky, Missouri, West Virginia, and Wyoming, where coal is the leading source of electricity (Figure 14.9.7). Thus, the "greenest" vehicle varies by location. But the rapid decline in the use of coal to generate electricity means that in most states EVs generate fewer emissions than gas cars.

14.10 KEY ISSUE 4 How are resources protected?

Revisiting Geography's Key Concepts

- **Apply geography's five basic concepts to two countries.**

▲ **14.10.1 PLACE: TIMOR-LESTE**
A massive statue called Cristo Rei (Christ the King) towers over the coast.

Sustainable development is defined by the U.N. as development that meets the needs of the present without compromising the ability of future generations to meet their own needs. Sustainable development is implemented in two distinct ways: through conservation and through preservation.

Sustainability: Conservation or Preservation?

Conservation was defined in Chapter 1 as the sustainable management of natural resources to meet human needs (refer to Section 1.12.1). Renewable resources such as trees are conserved if they are consumed at a less rapid rate than they can be replaced. Nonrenewable resources such as fossil fuels are conserved if remaining reserves are retained for future generations.

Another approach to sustainability is **preservation**, which is the maintenance of resources in their present condition with as little human impact on resources as possible. Preservation is a perspective that every plant and animal is valuable in its own right and should be preserved regardless of the cost.

14.10.1 Every Location Is Unique

To conclude the chapter on resources as well as the entire study of human geography, we return to the five key concepts introduced at the beginning of the book—place, region, scale, space, and connection. We can apply the five concepts to understand resource challenges facing two small countries—Luxembourg and Timor-Leste. Geographers refer to place and region to explain why every location on Earth is in some ways unique. A place is a specific point on Earth, distinguished by a particular characteristic. A region is an area of Earth defined by one or more distinctive characteristics.

▲ **14.10.2 PLACE: LUXEMBOURG**
The Alzette River flows through the center of Luxembourg.

▼ **14.10.3 REGION: TIMOR-LESTE**
Timor-Leste is in Southeast Asia.

Place. Luxembourg and Timor-Leste are small countries located on bodies of water that face pollution challenges. Timor-Leste, an island situated between the Savu Sea and the Timor Sea (Figure 14.10.1), struggles with assuring its population access to clean water and sanitation. According to the U.N., 70 percent of Timor-Leste's water sources are contaminated with microbiological entities, often holding potential for spreading diseases such as typhoid and cholera. Contaminations such as these are partly the result of a lack of effective sewage systems. Offshore protected areas have been established in recent years to preserve the biodiversity of aquatic life, which support local fisheries and tourism. The Alzette River, which flows through Luxembourg (Figure 14.10.2), has been polluted by spillages of waste water and chemicals from farms. But the country has elaborate policing and cleanup procedures that minimize the damage.

Region. Timor-Leste is part of the region of Southeast Asia (Figure 14.10.3). The region's many low-lying islands are exposed to a range of impacts from global climate change, especially rising seas and increased flooding (refer to Figure 3.5.5). Luxembourg is part of the region of Europe (Figure 14.10.4). The region's resource problems include acid precipitation from the burning of fossil fuels (refer to Figure 14.5.4), dependency on foreign sources of fossil fuels (refer to Figure 14.2.4), and reliance on nuclear fuel (refer to Figure 14.3.5).

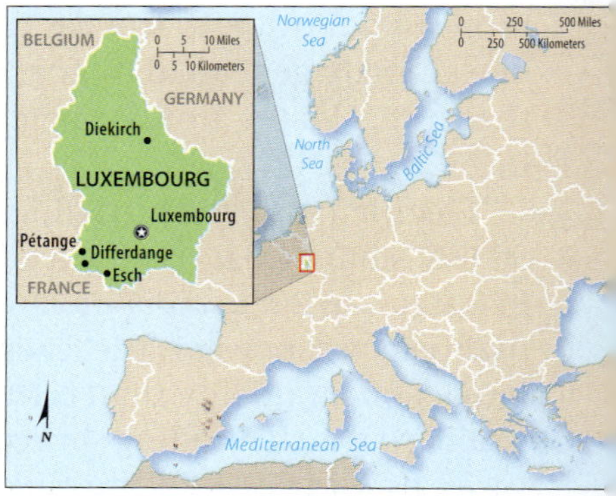

▲ **14.10.4 REGION: LUXEMBOURG**
Luxembourg is in Europe.

14.10.2 Every Location Is Connected

Geographers refer to scale, space, and connection to explain why all locations are interrelated. Scale is the relationship between the portion of Earth being studied and Earth as a whole. Geographers observe that many objects are distributed across space in a regular manner. Connection refers to relationships among people across the barrier of space.

▲ **14.10.5 SCALE: TIMOR-LESTE**
Market in Timor-Leste.

Scale. Climate change, including increasing temperatures, rainfall, and rising sea levels, poses challenges for Timor-Leste, a developing country that relies on climate stability for its agricultural-based economy (Figure 14.10.5). Timor-Leste must find ways to minimize the loss of agricultural land from the extreme flooding and landslides that come with climate change and attempt to reduce poverty and hunger through sustainable management practices that conserve agricultural land.

Luxembourg is a developed country, with one of the world's highest incomes per capita. One of the world's most prominent international organizations, the European Union, has many of its offices in Luxembourg (Figure 14.10.6).

▲ **14.10.6 SCALE: LUXEMBOURG**
European Union offices in Luxembourg City, the capital of Luxembourg.

Space. The space, or gap, between places can be minimal or substantial. In Timor-Leste, the gap is only 70 kilometers (45 miles) between its capital Dili (refer to Figure 14.10.1) at sea level and the country's highest elevation, Mount Ramelau (Figure 14.10.7), at 3,000 meters (10,000 feet). In Luxembourg, the gap is also only 70 kilometers between the capital Luxembourg City (refer to Figure 14.10.2) at an elevation of 300 meters (1,000 feet) and the country's highest elevation of 700 meters (2,200 feet) near the town of Clervaux (Figure 14.10.8).

▲ **14.10.7 SPACE: TIMOR-LESTE**
Mount Ramelau, Timor-Leste's highest elevation.

In reality, the 70-kilometer gap is more substantial in Timor-Leste than in Luxembourg. In a developed country like Luxembourg, with good roads and train service, Clervaux can be reached in less than an hour from the capital, whereas to travel from Dili to Mount Ramelau requires a 5-hour drive followed by a 5-hour hike. In both countries, the mountains are a resource to be preserved rather than utilized.

Connection. Although Luxembourg has one of the world's highest incomes per capita and Timor-Leste has one of the world's lowest incomes per capita, they share a similar form of connection to the rest of the world. Both countries possess international airports (Figures 14.10.9 and 14.10.10). And their international airports mean that both countries contribute more to global air pollution than they would otherwise. Air travel is responsible for 5 percent of the world's CO_2 emissions.

▼ **14.10.8 SPACE: LUXEMBOURG**
Clervaux, near Luxembourg's highest elevation.

▼ **14.10.9 CONNECTION: TIMOR-LESTE**
Plane flies over Dili.

▼ **14.10.10 CONNECTION: LUXEMBOURG**
Plane flies over Luxembourg City.

CHAPTER 14 Review, Analyze, & Apply

KEY ISSUE 1 Where Are Energy Resources Distributed?

- Most energy comes from three fossil fuels—petroleum, coal, and natural gas.
- Fossil fuels are not distributed uniformly across Earth.
- Demand for fossil fuels is relatively high in developed countries.

THINKING GEOGRAPHICALLY

Much of the remaining petroleum is located either offshore or in shale or oil sands that need to be mined. What are some of the pollution challenges to increased use of these sources?

▲ **14.CR.1 OIL SANDS MINING**
Alberta.

KEY ISSUE 2 What Is the Future for Energy Resources?

- Fossil fuels are nonrenewable.
- Renewable energy sources are increasingly important, especially hydroelectric, wind, and solar.

THINKING GEOGRAPHICALLY

What initiatives have your community or state taken to promote alternative sources of energy? What are some of the positive outcomes and some of the challenges in using alternative energy sources?

◀ **14.CR.2 WIND TURBINES**
Palm Springs, California.

KEY ISSUE 3 How Are Resources Polluted?

- Waste—the unwanted by-products of using resources—are discharged to air, water, and land.
- Pollution occurs when more waste is discharged than can be accommodated in the air, water, or land.

THINKING GEOGRAPHICALLY

When you throw something away in a trash can, where does it go in your community? What challenges does your community face in handling your trash?

◀ **14.CR.3 STREET TRASH**
New York City.

KEY ISSUE 4 How Are Resources Protected?

- Sustainable development conserves resources for future generations and preserves biodiversity.
- Sustainable development is promoted through consuming fewer resources and recycling more of those that are used.

THINKING GEOGRAPHICALLY

What initiatives have been undertaken at your university or school to encourage more sustainable development? In what ways might students, teachers, and staff be encouraged to adopt more sustainable practices?

◀ **14.CR.4 CLEAN WATER FROM THE FAUCET**
A resource that most Americans take for granted.

Explore

Alternative Transportation in Paris

Paris has one of the world's most extensive projects to support alternative vehicles.

In Google Earth, ***search*** for *192 Boulevard Saint-Germain, Paris, France.* Use Street View to ***drop*** to 192 Boulevard Saint-Germain. Look at the cars parked along the street.

1. What is distinctive about the cars along the street?
2. What is written on the side of the cars parked there?
3. Use your search engine to learn about Car2Go. Use a translation app if necessary.

▲ **14.CR.5 PARIS EV STATION**

Key Terms

Acid deposition (14.5.2) The accumulation of acids on Earth's surface.

Acid precipitation (14.5.2) Conversion of sulfur oxides and nitrogen oxides to acids that return to Earth as rain, snow, or fog.

Active solar energy system (14.4.2) A solar energy system that collects energy through the use of mechanical devices such as photovoltaic cells or flat-plate collectors.

Air pollution (14.5.1) Concentration of trace substances, such as carbon monoxide, sulfur dioxide, nitrogen oxides, hydrocarbons, and solid particulates, at a greater level than occurs in average air.

Animate power (14.1.1) Power supplied by animals or by people.

Biochemical oxygen demand (BOD) (14.6.1) The amount of oxygen required by aquatic bacteria to decompose a given load of organic waste; a measure of water pollution.

Biological diversity (or **biodiversity)** (14.6.1) The number of species within a specific habitat.

Biomass fuel (14.4.1) Fuel derived from wood, plant material, or animal waste.

Carbon capture and storage (CCS) (14.9.1) The process of capturing waste CO_2, transporting it to a storage site, and depositing it where it will not enter the atmosphere (normally underground).

Chlorofluorocarbon (CFC) (14.5.2) A gas used as a solvent, a propellant in aerosols, and a refrigerant.

Consumptive water usage (14.6.1) The use of water that causes it to evaporate rather than return to nature as a liquid.

Demand (14.1.1) The quantity of something that consumers are willing and able to buy.

Fission (14.3.2) The splitting of an atomic nucleus to release energy.

Fossil fuel (14.1.1) An energy source formed from the residue of plants and animals buried millions of years ago.

Fusion (14.3.2) Creation of energy by joining the nuclei of two hydrogen atoms to form helium.

Geothermal energy (14.4.1) Energy from steam or hot water produced from hot or molten underground rocks.

Greenhouse effect (14.5.2) The anticipated increase in global temperature caused by carbon dioxide (emitted by burning fossil fuels) trapping some of the radiation emitted by Earth's surface.

Hydraulic fracturing (14.3.1) Pumping water at high pressure into the ground to break apart rocks and thereby release natural gas.

Hydroelectric power (14.4.1) Power generated from moving water.

Nonconsumptive water usage (14.6.1) The use of water that is returned to nature as a liquid.

Nonpoint-source pollution (14.6.2) Pollution that originates from a large, diffuse area.

Nonrenewable resource (14.1.1) Something produced in nature more slowly than it is consumed by humans.

Ozone (14.5.2) A gas that absorbs ultraviolet solar radiation in the stratosphere, a zone between 10 and 50 kilometers (6 and 30 miles) above Earth's surface.

Passive solar energy systems (14.4.2) Solar energy systems that collect energy without the use of mechanical devices.

Photochemical smog (14.5.1) An atmospheric condition formed through a combination of weather conditions and pollution, especially from motor vehicle emissions.

Photovoltaic cell (14.4.2) A solar energy cell, usually made from silicon, that collects solar rays to generate electricity.

Point-source pollution (14.6.2) Pollution that enters a body of water from a specific source.

Pollution (14.5.1) Concentration of waste added to air, water, or land at a greater level than occurs in average air, water, or land.

Potential reserve (14.3.1) The amount of energy in deposits not yet identified but thought to exist.

Preservation (14.10.1) The maintenance of resources in their present condition, with as little human impact as possible.

Proven reserve (14.3.1) The amount of a resource remaining in discovered deposits.

Recycling (14.8.1) The separation, collection, processing, marketing, and reuse of unwanted material.

Remanufacturing (14.8.1) The rebuilding of a product to specifications of the original manufactured product using a combination of reused, repaired, and new parts.

Renewable resource (14.1.1) Something produced in nature more rapidly than it is consumed by humans.

Sanitary landfill (14.7.1) A place to deposit solid waste, where a layer of earth is bulldozed over garbage each day to reduce emissions of gases and odors from the decaying trash, to minimize fires, and to discourage vermin.

Supply (14.1.1) The quantity of something that producers have available for sale.

Sustainable development (14.10.1) Development that meets the needs of the present without compromising the ability of future generations to meet their own needs.

Geospatial Analysis

Log in to the Mastering Geography Study Area to access MapMaster 2.0.

Emissions & Energy Consumption

Carbon dioxide emissions and energy consumption have both increased. At the national scale, are the two related?

Add the *Energy Consumption* layer.

1. What world regions have the highest energy consumption per capita?
2. ***Add*** the *Carbon Dioxide Emissions* layer, and ***select*** *Join with data layer*. Probe the map. Are countries with high carbon dioxide emissions per capita those with relatively high energy consumption per capita or relatively low? What might account for this relationship?

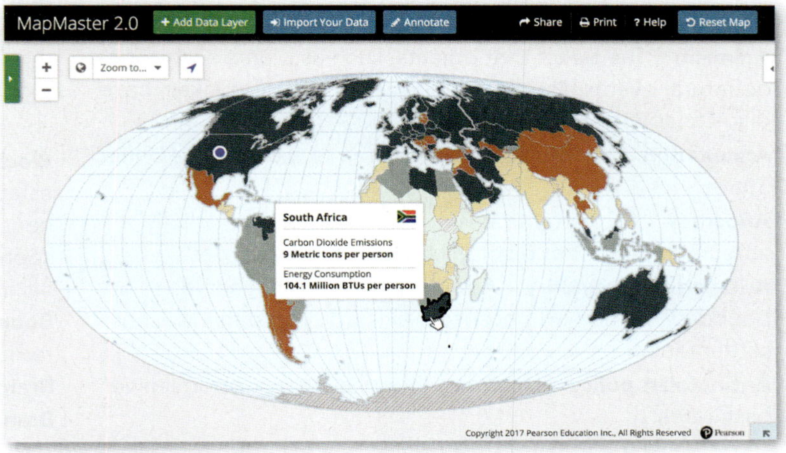

▲ **14.CR.6** ENERGY CONSUMPTION & CARBON DIOXIDE EMISSIONS

Glossary

A

Abiotic Composed of nonliving or inorganic matter.

Acculturation The process of changes in culture that result from the meeting of two groups, each of which retains distinct cultural features.

Acid deposition The accumulation of acids on Earth's surface.

Acid precipitation Conversion of sulfur oxides and nitrogen oxides to acids that return to Earth as rain, snow, or fog.

Active solar energy system A solar energy system that collects energy through the use of mechanical devices such as photovoltaic cells or flat-plate collectors.

Adolescent fertility rate The number of births per 1,000 women ages 15 to 19.

African American Vernacular English (AAVE) A dialect used by some Blacks.

Agnosticism The belief that the existence of God cannot be proven or disproven empirically.

Agribusiness Commercial agriculture characterized by the integration of different steps in the food-processing industry, usually through ownership by large corporations.

Agricultural density The ratio of the number of farmers to the total amount of arable land (land suitable for agriculture).

Agriculture The deliberate effort to modify a portion of Earth's surface through the cultivation of crops and the raising of livestock for sustenance or economic gain.

Air pollution Concentration of trace substances, such as carbon monoxide, sulfur dioxide, nitrogen oxides, hydrocarbons, and solid particulates, at a greater level than occurs in average air.

Alternative fact (alt fact) A statement that can be proved to be false.

Animate power Power supplied by animals or by people.

Animism The belief that objects, such as plants and stones, or natural events, like thunderstorms and earthquakes, have discrete spirits and conscious life.

Annexation Legally adding land area to a city in the United States.

Anocracy A country that is not fully democratic or fully autocratic but rather displays a mix of the two types.

Antecedent boundary A physical feature that serves as a boundary, so named because it existed before the present settlement.

Antinatalist policy Government policy that supports lower birth rates.

Apartheid Laws (no longer in effect) in South Africa that physically separated different races into different geographic areas.

Apparel An article of clothing.

Aquaculture The cultivation of seafood under controlled conditions.

Arable land Land suited for agriculture.

Arithmetic density The total number of people divided by the total land area.

Assimilation The process by which a group's cultural features are altered to resemble those of another group.

Asylum seeker Someone who has migrated to another country in the hope of being recognized as a refugee.

Atheism The belief that God does not exist.

Atmosphere The thin layer of gases surrounding Earth.

Autocracy A country that is run according to the interests of the ruler rather than the people.

Autonomous religion A religion that does not have a central authority but shares ideas and cooperates informally.

B

Balance of power A condition of roughly equal strength between opposing countries or alliances of countries.

Basic business A business that sells its products or services primarily to consumers outside the settlement.

Biochemical oxygen demand (BOD) The amount of oxygen required by aquatic bacteria to decompose a given load of organic waste; a measure of water pollution.

Biological diversity (or **biodiversity)** The number of species within a specific habitat.

Biomass fuel Fuel derived from wood, plant material, or animal waste.

Biosphere All living organisms on Earth, including plants and animals, as well as microorganisms.

Biotic Composed of living organisms.

Black Lives Matter (BLM) Movement that campaigns against violence and perceived racism toward Blacks and educates others about the challenges that Black Americans continue to face in the United States.

Blockbusting A deceptive business tactic whereby real estate agents convince White property owners to sell their houses at low prices by warning that persons of color will soon move into the neighborhood and cause a significant fall in housing prices.

Boundary An invisible line that marks the extent of a state's territory.

Brain drain Large-scale emigration by talented people.

Branch A large and fundamental division within a religion.

Break-of-bulk point A location where transfer is possible from one mode of transportation to another.
Bulk-gaining industry An industry in which the final product weighs more or comprises a greater volume than the inputs.
Bulk-reducing industry An industry in which the final product weighs less or comprises a lower volume than the inputs.
Business service A service that primarily meets the needs of other businesses, including professional, financial, and transportation services.

C

Carbon capture and storage (CCS) The process of capturing waste CO_2, transporting it to a storage site, and depositing it where it will not enter the atmosphere (normally underground).
Cartogram A map in which the projection and scale are distorted in order to convey the information of a variable.
Cartography The science of making maps.
Caste system The class or distinct hereditary order into which a Hindu is assigned, according to religious law.
Census tract An area delineated by the U.S. Bureau of the Census for which statistics are published; in urban areas, census tracts correspond roughly to neighborhoods.
Central business district (CBD) The area of a city where retail and office activities are clustered.
Central city See municipality.
Central place A market center for the exchange of services by people attracted from the surrounding area.
Central place theory A theory that explains the distribution of services based on the fact that settlements serve as centers of market areas for services; larger settlements are fewer and farther apart than smaller settlements and provide services for a larger number of people who are willing to travel farther.
Centrifugal force A cultural value that tends to pull people apart.
Centripetal force A cultural value that tends to unify people.
Cereal grain A grass that yields grain for food.
Chlorofluorocarbon (CFC) A gas used as a solvent, a propellant in aerosols, and a refrigerant.
Choropleth map A map where recognizable areas are shaded or patterned in proportion to the data collected.
Circular migration The temporary movement of a migrant worker between home and host countries to seek employment.
Circulation Short-term, repetitive, or cyclical movements that recur on a regular basis.
Citizen science Scientific research by amateur scientists.
City-state A sovereign state composed of a city and its immediately surrounding countryside.
Clustered rural settlement A rural settlement in which the houses and farm buildings of each family are situated close to each other, with fields surrounding the settlement.
Colonialism An attempt by one country to establish settlements and to impose its political, economic, and cultural principles in another territory.
Colony A territory that is legally tied to a sovereign state rather than completely independent.
Columbian Exchange The transfer of plants and animals, as well as people, culture, and technology, between the Western Hemisphere and Europe, as a result of European colonization and trade.
Combined statistical area (CSA) In the United States, two or more contiguous core-based statistical areas tied together by commuting patterns.
Commercial agriculture Agriculture undertaken primarily to generate products for sale off the farm.
Commercial gardening See truck farming.
Compact state A state in which the distance from the center to any boundary does not vary significantly.
Concentration The extent of a feature's spread over a given area.
Concentric zone model A model of the internal structure of cities in which social groups are spatially arranged in a series of rings.
Congregation A local assembly of persons brought together for common religious worship.
Connection The relationships among people and objects across the barrier of space.
Consequent boundary A boundary drawn in order to separate groups based on ethnic, linguistic, religious, or economic differences.
Conservation The sustainable management of a natural resource to meet human needs.
Conservation tillage A method of soil cultivation that reduces soil erosion and runoff.
Consumer service A business that provides services primarily to individual consumers, including retail services and education, health, and leisure services.
Consumptive water usage The use of water that causes it to evaporate rather than return to nature as a liquid.
Contagious diffusion The rapid, widespread diffusion of a feature or trend throughout a population.
Cooperative An association of persons united voluntarily to meet their common economic, social, and cultural needs and aspirations through a jointly-owned enterprise.
Coordinated Universal Time (UTC) Informally Greenwich Mean Time (GMT). The time in the zone encompassing the prime meridian, or 0° longitude.
Core-based statistical area (CBSA) In the United States, the collection of all metropolitan statistical areas and micropolitan statistical areas.
Cosmogony A set of beliefs concerning the origin of the universe.
Cottage industry Manufacturing based in homes rather than in factories, commonly found prior to the Industrial Revolution.

Counterurbanization Net migration from urban to rural areas in developed countries.
Creole (or **creolized) language** A language that results from the mixing of a colonizer's language with the indigenous language of the people being dominated.
Crop Any plant gathered from a field as a harvest during a particular season.
Crop rotation The practice of rotating use of different fields from crop to crop each year to avoid exhausting the soil.
Crude birth rate (CBR) The total number of live births in a year for every 1,000 people alive in the society.
Crude death rate (CDR) The total number of deaths in a year for every 1,000 people alive in the society.
Cultural ecology A geographic approach that emphasizes human–environment relationships.
Cultural homogenization The process of reduction in cultural diversity through the diffusion of popular culture.
Cultural landscape An approach to geography that emphasizes the relationships among social and physical phenomena in a particular study area.
Culture The body of material possessions, customary beliefs, and social norms that together constitute the distinct tradition of a group of people.
Custom The frequent repetition of an act, to the extent that it becomes characteristic of the group of people performing the act.
Cyber espionage The unauthorized and clandestine deployment of a computer virus to observe or destroy data in the computer systems of government agencies and large corporations.

D

Dairy farming A form of commercial agriculture that specializes in the production of milk and other dairy products.
Demand The quantity of something that consumers are willing and able to buy.
Democracy A country in which citizens elect leaders and can run for office.
Demographic transition The process of change in a society's population from a condition of high crude birth and death rates and low natural increase rate to a condition of low crude birth and death rates, low natural increase rate, and higher total population.
Denglish A combination of Deutsch (the German word for German) and English.
Denomination A division of a branch that unites a number of local congregations into a single legal and administrative body.
Density The frequency with which something exists within a given unit of area.
Density gradient The change in density in an urban area from the center to the periphery.
Dependency ratio The number of people under age 15 and over age 64 compared to the number of people active in the labor force.
Desertification Degradation of land, especially in semiarid areas, primarily because of human actions such as excessive crop planting, animal grazing, and tree cutting. Also known as semiarid land degradation.
Developed country A country that has progressed relatively far along a continuum of development.
Developing country A country that is at a relatively early stage in the process of development.
Development A process of improvement in the conditions of people through diffusion of knowledge and technology.
Devolution The transfer or delegation of power to a lower level, especially by a national government to local or regional administration.
Dialect A regional variety of a language distinguished by vocabulary, spelling, and pronunciation.
Dietary energy consumption The amount of food that an individual consumes, measured in kilocalories (Calories in the United States).
Diffusion The process of spread of a feature or trend from one place to another over time.
Dispersed rural settlement A rural settlement pattern characterized by isolated farms rather than clustered villages.
Distance decay The diminished importance and eventual disappearance of a phenomenon with increasing distance from its origin.
Distribution The arrangement of something across Earth's surface.
Dot distribution map A map that depicts data consisting of discrete observations. Each dot represents a predetermined number of observations, which could be one or many.
Double cropping Harvesting twice a year from the same field.
Doubling time The number of years needed to double a population, assuming a constant natural increase rate.

E

Ecology The scientific study of ecosystems.
Economic base A community's collection of basic businesses.
Ecosystem A group of living organisms and the abiotic spheres with which they interact.
Ecumene The portion of Earth's surface occupied by permanent human settlement.
Edge city A node of office and retail activities on the edge of an urban area.
Elongated state A state with a long, narrow shape.
Emigration Migration from a location.
Enclosure movement The process of consolidating small landholdings into a smaller number of larger farms in England during the eighteenth century.
Endangered language A language that is endangered or dying because it is losing users and may not be retained by the younger generation.
Endemic A widespread occurrence of an infectious disease that is considered seasonal and/or predictable.

Environmental determinism A nineteenth- and early twentieth-century approach to the study of geography that argued that the general laws sought by human geographers could be found in the physical sciences. Geography was therefore the study of how the physical environment caused human activities.

Epidemic A widespread occurrence of an infectious disease in a community at a particular time.

Epidemiologic transition The process of change in the distinctive causes of death in each stage of the demographic transition.

Epidemiology The branch of medical science concerned with the incidence, distribution, and control of diseases that are prevalent among a population at a certain time and are produced by some special causes not generally present in the affected locality.

Esports Sports competitions using video games.

Ethnic cleansing A purposeful policy designed by one ethnic or religious group to remove by violent and terror-inspiring means the civilian population of another ethnic or religious group from certain geographic areas.

Ethnic religion A religion with a relatively concentrated spatial distribution whose principles are likely to be based on the physical characteristics of the particular location in which its adherents are concentrated.

Ethnicity Identity with a group of people who share the cultural traditions of a particular homeland or hearth.

Ethnophobia Fear of people of a particular ethnicity.

Expansion diffusion The spread of a feature or trend among people from one area to another in an additive process.

Extinct language A language that was once used by people in daily activities but is no longer used.

F

Fair trade A variation of international trade that provides greater equity to workers, small businesses, and consumers, focusing primarily on products exported from developing countries to developed countries.

Fake news A false report disseminated under the guise of an authentic news report, created to maliciously spread misinformation and mislead consumers of the content.

Federal state An internal organization of a state that allocates most powers to units of local government.

Female labor force participation rate The percentage of women holding full-time jobs outside the home.

First agricultural revolution The time when human beings first domesticated plants and animals and no longer relied entirely on hunting and gathering.

Fishing The capture of wild fish and other seafood living in the waters.

Fission The splitting of an atomic nucleus to release energy.

Floodplain The area subject to flooding during a given number of years, according to historical trends.

Folk culture Culture traditionally practiced by a small, homogeneous, rural group living in relative isolation from other groups.

Food security Physical, social, and economic access at all times to safe and nutritious food sufficient to meet dietary needs and food preferences for an active and healthy life.

Forced migration Permanent movement, compelled by cultural or environmental factors.

Fordist production A form of mass production in which each worker is assigned one specific task to perform repeatedly.

Foreign direct investment (FDI) Investment made by a foreign company in the economy of another country.

Formal region (or **uniform region)** An area in which most people share in one or more distinctive characteristics.

Fossil fuel An energy source formed from the residue of plants and animals buried millions of years ago.

Fragmented state A state that includes several discontinuous pieces of territory.

Franchise An agreement between a corporation and businesspeople to market that corporation's products in a local area.

Franglais A combination of français and anglais (the French words for French and English, respectively).

Frontier A zone separating two states in which neither state exercises political control.

Functional region (or **nodal region)** An area organized around a node or focal point.

Fundamentalism Literal interpretation and strict adherence to basic principles of a religion (or a religious branch, denomination, or congregation).

Fusion Creation of energy by joining the nuclei of two hydrogen atoms to form helium.

G

Galactic (or **peripheral) model** A model of North American urban areas consisting of a city surrounded by large suburban residential and business areas tied together by a beltway or ring road.

Gender Development Index (GDI) An index constructed by the United Nations that modifies the Human Development Index to measure the gender gap in level of achievement in terms of income, education, and life expectancy.

Gender Inequality Index (GII) An index constructed by the United Nations to measure the extent of each country's gender inequality in terms of reproductive health, empowerment, and the labor market.

Gender Social Norms Index (GSNI) An index constructed by the United Nations to measure the strength of attitudes biased against women.

Genetically modified organism (GMO) A living organism that possesses a novel combination of genetic material obtained through the use of modern biotechnology.

Genocide The mass killing of a group of people in an attempt to eliminate the entire group from existence.

Gentrification A process of converting an urban neighborhood from an area of residents with predominantly low incomes who rent to an area of residents with middle incomes who own their homes.

Geographic information science (GIScience) Analysis of data about Earth acquired through satellite and other electronic information technologies.

Geographic information system (GIS) A computer system that captures, stores, queries, and displays geographic data.

Geometric boundary A boundary based on a human construct such as a straight line or a parallel of latitude.

Geotagging Identification and storage of a piece of information by its precise latitude and longitude coordinates.

Geothermal energy Energy from steam or hot water produced from hot or molten underground rocks.

Gerrymandering The process of redrawing legislative boundaries within a U.S. state for the purpose of benefiting the party in power.

Ghetto During the Middle Ages, a neighborhood in a city set up by law to be inhabited only by Jews; now used to denote a section of a city in which members of any minority group live because of social, legal, or economic discrimination.

Global city A major center for the provision of services in the global economy.

Global Positioning System (GPS) A system that determines the precise position of something on Earth through a series of satellites, tracking stations, and receivers.

Globalization Actions or processes that involve the entire world and result in making something worldwide in scope.

Graduated symbol map A map that displays symbols that change in size according to the value of the variable.

Grain Seed of a cereal grass.

Gravity model A model that holds that the potential use of a service at a particular location is directly related to the number of people in a location and inversely related to the distance people must travel to reach the service.

Green revolution Rapid diffusion of new agricultural technology, especially new high-yield seeds and fertilizers.

Greenhouse effect The anticipated increase in global temperature caused by carbon dioxide (emitted by burning fossil fuels) trapping some of the radiation emitted by Earth's surface.

Gross domestic product (GDP) The value of the total output of goods and services produced in a country in a year, not accounting for money that leaves and enters the country.

Gross national income (GNI) The value of the output of goods and services produced in a country in a year, including money that leaves and enters the country.

Gross national income per capita at purchasing power parity The value of the output of goods and services produced in a country in a year, including money that leaves and enters the country, adjusted to account for differences among countries in the cost of goods.

Guest worker A term once used for a worker who migrated to the developed countries of Northern and Western Europe, usually from Southern and Eastern Europe or from North Africa, in search of a higher-paying job.

H

Habit A repetitive act performed by a particular individual.

Hearth A place from which an innovation originates.

Hierarchical diffusion The spread of a feature or trend from one key person or node of authority or power to other persons or places.

Hierarchical religion A religion in which a central authority exercises a high degree of control.

Human Development Index (HDI) An index constructed by the United Nations to measure the level of development for a country through a combination of income, education, and life expectancy.

Hydraulic fracturing Pumping water at high pressure into the ground to break apart rocks and thereby release natural gas.

Hydroelectric power Power generated from moving water.

Hydrosphere All of the water on and near Earth's surface.

I

Immigration Migration to a new location.

Industrial Revolution A series of improvements in industrial technology that transformed the process of manufacturing goods.

Inequality-adjusted Human Development Index (IHDI) An index constructed by the United Nations that modifies the Human Development Index to account for inequality.

Infant mortality rate (IMR) The total number of deaths in a year among infants under 1 year of age for every 1,000 live births in a society.

Informal settlement An area within a city in a less developed country in which people illegally establish residences on land they do not own or rent and erect homemade structures.

Institutional language A language used in education, work, mass media, and government.

Intensive subsistence agriculture A form of subsistence agriculture characteristic of Asia's major population concentrations in which farmers must expend a relatively large amount of effort to produce the maximum feasible yield from a parcel of land.

Internal migration Permanent movement within a particular country.

Internally displaced person (IDP) Someone who has been forced to migrate for similar political reasons as a refugee but has not migrated across an international border.

International Date Line An arc that for the most part follows 180° longitude. When the International Date Line is crossed heading east (toward America), the clock moves back 24 hours, or one entire day. When it is crossed heading west (toward Asia), the calendar moves ahead one day.

International migration Permanent movement from one country to another.
Interregional migration Permanent movement from one region of a country to another.
Intervening obstacle An environmental or cultural feature of the landscape that hinders migration.
Intraregional migration Permanent movement within one region of a country.
Isogloss A boundary that separates regions in which different language usages predominate.
Isolated language A language that is unrelated to any other languages and therefore not attached to any language family.
Isoline map A map that connects places of a particular value by lines.

J

Just-in-time delivery Shipment of parts and materials to arrive at a factory moments before they are needed.

L

Labor-intensive industry An industry for which labor costs comprise a high percentage of total expenses.
Landlocked state A state that does not have a direct outlet to the sea.
Language A system of communication through speech, movement, sounds, or symbols that a group of people understands to have the same meaning.
Language branch A collection of languages related through a common ancestor that can be confirmed through archaeological evidence.
Language family A collection of languages related to each other through a common ancestor long before recorded history.
Language group A collection of languages within a branch that share a common origin in the relatively recent past and display relatively few differences in grammar and vocabulary.
Latitude The numbering system used to indicate the location of parallels drawn on a globe and measuring distance north and south of the equator (0°).
Life expectancy The average number of years an individual can be expected to live, given current social, economic, and medical conditions.
Lingua franca A language mutually understood and commonly used in trade by people who have different native languages.
Literacy rate The percentage of a country's people who can read and write.
Literary tradition A language that is written as well as spoken.
Lithosphere Earth's crust and a portion of upper mantle directly below the crust.
Location The position of anything on Earth's surface.
Logogram A symbol that represents a word rather than a sound.
Longitude The numbering system used to indicate the location of meridians drawn on a globe and measuring distance east and west of the prime meridian (0°).
Lunar calendar A calendar with months that correspond to cycles of moon phases.
Lunisolar calendar A calendar with lunar months that are brought into alignment with the solar year through some process.

M

Malware (malicious software) Hostile or intrusive computer software designed to cause intentional harm.
Map A two-dimensional, or flat, representation of Earth's surface or a portion of it.
Map scale The relationship between the size of an object on a map and the size of the actual feature on Earth's surface.
Market area (or **hinterland**) The region surrounding a central place of goods and services from which people travel to access those goods and services.
Mashup A map that overlays data from one source on top of a map provided by a mapping service.
Maternal mortality rate The annual number of female deaths per 100,000 live births from any cause related to or aggravated by pregnancy or its management (excluding accidental or incidental causes).
Media market area A region where the population receives the same TV and radio station broadcast and newspapers, and consumes similar internet content.
Medical revolution Medical technology invented in Europe and North America that has diffused to the developing countries in Latin America, Asia, and Africa. Improved medical practices have eliminated many of the traditional causes of death in developing countries and enabled more people to live longer and healthier lives.
Megacity An urban settlement with a total population in excess of 10 million people.
Megalopolis A continuous urban complex in the northeastern United States.
Mental map A representation of a portion of Earth's surface based on what an individual knows about a place that contains personal impressions of what is in the place and where the place is located.
Meridian An arc drawn on a map between the North and South poles.
Metacity An urban settlement with a total population in excess of 20 million people.
Metropolitan statistical area (MSA) In the United States, an urbanized area with a population of at least 50,000, the county within which the city is located, and adjacent counties meeting one of several tests indicating a functional connection to the central city.
Microfinance Provision of small loans and financial services to individuals and small businesses in developing countries.
Micropolitan statistical area (μSA) An urbanized area of between 10,000 and 50,000 inhabitants, the county in which it is located, and adjacent counties tied to the city.

Microstate A state that encompasses a very small land area.
Migration A form of relocation diffusion involving a permanent move to a new location.
Migration transition A change in the migration pattern in a society that results from industrialization, population growth, and other social and economic changes that also produce the demographic transition.
Milkshed The area surrounding a city from which milk is supplied.
Millennium Development Goals Eight goals adopted by the United Nations in 2002 to reduce disparities between developed and developing countries by 2015.
Missionary An individual who helps to diffuse a universalizing religion.
Mixed crop and livestock farming Commercial farming characterized by integration of crops and livestock; most of the crops are fed to animals rather than consumed directly by humans.
Mobility All types of movements between locations.
Monotheism The doctrine of or belief in the existence of only one God.
Multidimensional Poverty Index (MPI) An index constructed by the United Nations that modifies the Human Development Index to measure the complexities of the lives of people living in poverty.
Multinational state A state that contains two or more cultural groups with traditions of self-determination that agree to coexist peacefully by recognizing each other as distinct nationalities.
Multiple nuclei model A model of the internal structure of cities in which social groups are arranged around a collection of nodes of activities.
Municipality (or **central city**) An urban settlement that is legally incorporated into an independent, self-governing unit.
Mutual intelligibility The ability of people communicating in two ways to readily understand each other without prior familiarity or special effort.

N

Nation A large group of people who are united by common cultural characteristics, such as language and ethnicity, or by shared history.
Nation-state A state whose territory corresponds to that occupied by a particular nation.
Nationalism Loyalty and devotion to a particular nationality.
Nationality Identity with a group of people who share legal attachment to a particular country.
Natural increase rate (NIR) The percentage growth of a population in a year, computed as the crude birth rate minus the crude death rate.
Net migration The difference between the level of immigration and the level of emigration.
Network A chain of communication that connects places.
New international division of labor Transfer of some types of jobs, especially those requiring low-paid, less skilled workers, from developed to developing countries.
No tillage A farming practice that leaves all of the soil undisturbed and the entire residue of the previous year's harvest left untouched on the fields.
Nonbasic business A business that sells its products primarily to consumers in the community.
Nonconsumptive water usage The use of water that is returned to nature as a liquid.
Nonpoint-source pollution Pollution that originates from a large, diffuse area.
Nonrenewable resource Something produced in nature more slowly than it is consumed by humans.

O

Official language The language adopted for use by a government for the conduct of business and publication of documents.
Offshore financial center A country or jurisdiction that provides financial services to nonresidents on a scale that is much larger than the size of its domestic economy.
Outsourcing A decision by a corporation to turn over much of the responsibility for production to independent suppliers.
Overfishing Capturing fish faster than they can reproduce.
Overpopulation A situation in which the number of people in an area exceeds the capacity of the environment to support them at a decent standard of living.
Ozone A gas that absorbs ultraviolet solar radiation in the stratosphere, a zone between 10 and 50 kilometers (6 and 30 miles) above Earth's surface.

P

Paddy The Malay word for wet rice, increasingly used to describe a flooded field.
Pandemic Disease that occurs over a wide geographic area and affects a very high proportion of the population.
Parallel A circle drawn around the globe parallel to the equator and at right angles to the meridians.
Participatory GIS (PGIS) Community-based mapping, representing local knowledge and information.
Passive solar energy systems Solar energy systems that collect energy without the use of mechanical devices.
Pastoral nomadism A form of subsistence agriculture based on herding domesticated animals.
Pattern The geometric or regular arrangement of something in a particular area.
Perforated state A state that completely surrounds another one.
Periodic market A collection of individual vendors who come together to offer goods and services in a location on specified days.
Pestilence An especially virulent epidemic.

Photochemical smog An atmospheric condition formed through a combination of weather conditions and pollution, especially from motor vehicle emissions.
Photogrammetry The science of taking measurements of Earth's surface from photographs.
Photovoltaic cell A solar energy cell, usually made from silicon, that collects solar rays to generate electricity.
Physiological density The number of people per unit area of arable land, which is land suitable for agriculture.
Pidgin language A form of language that adopts a simplified grammar and limited vocabulary of a lingua franca; used for communications among speakers of two different languages.
Pilgrimage A journey to a place considered sacred for religious purposes.
Place A specific point on Earth, distinguished by a particular characteristic.
Planetary pressures-adjusted Human Development Index (PHDI) An index constructed by the United Nations that modifies the Human Development Index to account for a country's carbon dioxide emissions per person and the amount of raw materials extracted per person to meet the country's demand for goods and services.
Plantation A large farm in tropical and subtropical climates that specializes in the production of one or two crops for sale, usually to a more developed country.
Point-source pollution Pollution that enters a body of water from a specific source.
Polder Land that the Dutch have created by draining water from an area.
Pollution Concentration of waste added to air, water, or land at a greater level than occurs in average air, water, or land.
Polytheism Belief in or worship of more than one god.
Popular culture Culture found in a large, heterogeneous society that shares certain habits and seeks similar material goods despite differences in other personal characteristics.
Population pyramid A bar graph that represents the distribution of population by age and sex.
Possibilism The theory that the physical environment may set limits on human actions, but people have the ability to adjust to the physical environment and choose a course of action from many alternatives.
Post-Fordist production Adoption by companies of flexible work rules, such as the allocation of workers to teams that perform a variety of tasks.
Potential reserve The amount of energy in deposits not yet identified but thought to exist.
Potential support ratio (or **elderly support ratio)** The number of working-age people (ages 15 to 64) divided by the number of persons 65 and older.
Preservation The maintenance of resources in their present condition, with as little human impact as possible.
Primary sector The portion of the economy concerned with the direct extraction of materials from Earth, generally through agriculture.
Primate city The largest settlement in a country, if it has more than twice as many people as the second-ranking settlement.
Primate city rule A pattern of settlements in a country such that the largest settlement has more than twice as many people as the second-ranking settlement.
Prime agricultural land The most productive farmland.
Prime meridian The meridian, designated as 0° longitude, that passes through the Royal Observatory at Greenwich, England.
Productivity The value of a particular product compared to the amount of labor needed to make it.
Projection A system used to transfer locations from Earth's surface to a flat map.
Pronatalist policy Government policy that supports higher birth rates.
Prorupted state An otherwise compact state with a large projecting extension.
Proven reserve The amount of a resource remaining in discovered deposits.
Public service A service offered by the government to provide security and protection for citizens and businesses.
Pull factor A factor that induces people to move to a new location.
Pupil/teacher ratio The number of enrolled students divided by the number of teachers.
Purchasing power parity (PPP) The amount of money needed in one country to purchase the same goods and services in another country.
Push factor A factor that induces people to leave old locations.

Q

Quota In reference to migration, a law that places maximum limits on the number of people who can immigrate to a country each year.

R

Race Identity with a group of people who are perceived to share a physical trait, such as skin color.
Racism The belief that race is the primary determinant of human traits and capacities and that racial differences produce an inherent superiority of a particular race.
Racist A person who subscribes to the beliefs of racism.
Ranching A form of commercial agriculture in which livestock graze over an extensive area.
Range (of a service) The maximum distance people are willing to travel to use a service.
Rank-size rule A pattern of settlements in a country such that the nth largest settlement is 1/n the population of the largest settlement.
Ransomware (cryptoviral extortion) A form of malware that encrypts the victim's computer files, making them inaccessible, until a ransom is paid to decrypt them.

Received Pronunciation (RP) The dialect of English commonly used by politicians, broadcasters, and actors in the United Kingdom.
Recycling The separation, collection, processing, marketing, and reuse of unwanted material.
Redlining A process by which financial institutions draw red-colored lines on a map and refuse to lend money for people to purchase or improve property within the lines.
Refugees People who are forced to migrate from their home country and cannot return for fear of persecution because of their race, religion, nationality, membership in a social group, or political opinion.
Region An area distinguished by one or more distinctive characteristics.
Relic boundary A boundary that no longer functions but can still be detected on the cultural landscape.
Relocation diffusion The spread of a feature or trend through bodily movement of people from one place to another.
Remanufacturing The rebuilding of a product to specifications of the original manufactured product using a combination of reused, repaired, and new parts.
Remittance Transfer of money by workers to people in the country from which they emigrated.
Remote sensing The acquisition of data about Earth's surface from a satellite orbiting the planet or from other long-distance methods.
Renewable resource Something produced in nature more rapidly than it is consumed by humans.
Resource A substance in the environment that is useful to people, is economically and technologically feasible to access, and is socially acceptable to use.
Ridge tillage A system of planting crops on ridge tops in order to reduce farm production costs and promote greater soil conservation.
Right-to-work law A law in some U.S. states that prevents a union and a company from negotiating a contract that requires workers to join the union as a condition of employment.
Rural settlement A community where most people are engaged in agriculture.
Rush hour The four consecutive 15-minute periods in the morning and evening with the heaviest volumes of traffic.

S

Sanitary landfill A place to deposit solid waste, where a layer of earth is bulldozed over garbage each day to reduce emissions of gases and odors from the decaying trash, to minimize fires, and to discourage vermin.
Sawah A flooded field for growing rice.
Scale The relationship between the portion of Earth being studied and Earth as a whole.
Second agricultural revolution The increase in productivity through improvement of crop rotation, breeding of livestock, and other practices, beginning in the United Kingdom in the seventeenth century.
Secondary sector The portion of the economy concerned with manufacturing useful products through processing, transforming, and assembling raw materials.
Sector model A model of the internal structure of cities in which social groups are arranged around a series of sectors, or wedges, radiating out from the central business district.
Self-determination The concept that ethnicities have the right to govern themselves.
Service Any activity that fulfills a person's want or need in exchange for money.
Settlement A permanent collection of buildings and inhabitants.
Sex ratio The number of males per 100 females in the population.
Sharecropper A person who works fields rented from a landowner and pays the rent and repays the loans by turning over to the landowner a share of the crops.
Sharing economy Services that are shared between private individuals.
Shifting cultivation A form of subsistence agriculture in which people shift activity from one field to another; each field is used for crops for a few years and left fallow for a relatively long period.
Site The physical character of a place.
Site factors Costs related to manufacturing inside a plant, such as expenses associated with land, labor, and capital.
Situation The location of a place relative to another place.
Situation factors Location factors related to the transportation of materials into and from a factory.
Smart growth Legislation and regulations to limit suburban sprawl and preserve farmland.
Social area analysis Statistical analysis used to identify where people of similar living standards, ethnic background, and lifestyle live within an urban area.
Social construction An idea or a meaning that is widely accepted as natural by a society but may not represent a reality shared by those outside the society.
Solar calendar A calendar that relates to the season or the apparent position of the sun in relation to the stars.
Solstice An astronomical event that happens twice each year, when the tilt of Earth's axis is most inclined toward or away from the Sun, causing the Sun's apparent position in the sky to reach its northernmost or southernmost extreme, and resulting in the shortest and longest days of the year.
Sovereignty Ability of a state to govern its territory free from control of its internal affairs by other states.
Space The physical gap or interval between two objects.
Space–time compression The reduction in the time it takes to diffuse something to a distant place as a result of improved communications and transportation systems.
Spanglish A combination of Spanish and English spoken by Hispanic Americans.
Spatial association The relationship between the distribution of one feature and the distribution of another feature.

Spatial interaction The deliberate movement of goods, people, or information from one place to another place.
Sprawl Development of new housing sites at relatively low density and at locations that are not contiguous to the existing built-up area.
Stable language A language that is used for face-to-face communication within all generations, including children.
Standard language The form of a language used for official government, business, education, and mass communication.
State An area organized into a political unit and ruled by an established government that has control over its internal and foreign affairs.
Step migration Migration that follows a path of a series of stages or steps toward a final destination.
Stimulus diffusion The spread of an underlying principle even though a specific characteristic is rejected.
Structural adjustment program Economic policies imposed on less developed countries by international agencies to create conditions that encourage international trade.
Subdialect A subdivision of a dialect.
Subsequent boundary A boundary that is established after the settlement of an area and is changed to reflect developments such as war.
Subsistence agriculture Agriculture designed primarily to provide food for direct consumption by the farmer and the farmer's family.
Suburb A residential or commercial area situated within an urban area but outside the central city.
Superimposed boundary A boundary drawn on an area by a conquering or colonizing power that ignores existing cultural patterns.
Supply The quantity of something that producers have available for sale.
Sustainability The use of Earth's renewable and nonrenewable natural resources in ways that do not constrain resource use in the future.
Sustainable development Development that meets the needs of the present without compromising the ability of future generations to meet their own needs.
Sustainable Development Goals Seventeen goals adopted by the United Nations in 2015 to reduce disparities between developed and developing countries by 2030.
Syncretic Combining several religious traditions.
Syncretism The combining of elements of two groups into a new cultural feature.

T

Taboo A restriction on behavior imposed by social custom.
Terroir The contribution of a location's distinctive environment to the way food produced there tastes.
Terrorism The threatened or actual use of illegal force and violence by a non-state actor to attain a political, economic, religious, or social goal through fear, coercion, or intimidation.
Tertiary sector The portion of the economy concerned with transportation, communications, and utilities, sometimes extended to the provision of all goods and services to people in exchange for payment.
Textile A fabric made by weaving, used in making clothing.
Third agricultural revolution See Green revolution.
Threshold The minimum number of people needed to support a service.
Toponym The name given to a portion of Earth's surface.
Total fertility rate (TFR) The average number of children a woman will have throughout her childbearing years.
Transhumance Seasonal migration of livestock between mountains and lowland pasture areas.
Transnational corporation A company that conducts research, operates factories, and sells products in many countries, not just where its headquarters or shareholders are located.
Triangular slave trade A practice, primarily during the eighteenth century, in which European ships transported enslaved people from Africa to Caribbean islands, molasses from the Caribbean to Europe, and trade goods from Europe to Africa.
Trolling The practice of posting deliberately inflammatory, extraneous, or off-topic messages in social media in order to provoke quarrels or otherwise upset people.
Truck farming Relatively small-scale production of fruits, vegetables, and other horticulture, so named for the Middle English word *truck*, meaning *barter* or "exchange of commodities."

U

Unauthorized immigrant A person who enters a country without proper documents to do so.
Undernourishment Dietary energy consumption that is continuously below the minimum requirement for maintaining a healthy life and carrying out light physical activity.
Uneven development The increasing gap in economic conditions between core and peripheral regions as a result of the globalization of the economy.
Unitary state An internal organization of a state that places most power in the hands of central government officials.
Universalizing religion A religion that attempts to appeal to all people, not just those living in a particular location.
Urban area A dense core of census tracts, densely settled suburbs, and low-density land that links the dense suburbs with the core.
Urban cluster In the United States, an urban area with between 2,500 and 50,000 inhabitants.
Urban settlement A community where people are employed in secondary- and tertiary-sector jobs.
Urbanization An increase in the percentage of people living in urban settlements.
Urbanized area In the United States, an urban area with at least 50,000 inhabitants.
Utopian settlement A community built around an ideal way of life, often based on a religion.

V

Vernacular region (or **perceptual region**) An area that people believe exists as part of their cultural identity.

Vertical integration An approach typical of traditional mass production in which a company controls all phases of a highly complex production process.

Voluntary migration Permanent movement undertaken by choice.

Volunteered geographic information (VGI) Creation and dissemination of geographic data contributed voluntarily and for free by individuals.

W

Weapon of mass destruction A nuclear, biological, chemical, or other weapon that can kill and bring significant harm to a large number of humans or cause great damage to human-made structures, natural structures, or the biosphere.

Wet rice Rice planted on dry land in a nursery and then moved to a deliberately flooded field to promote growth.

World-systems theory Theory developed by Immanuel Wallerstein that in an increasingly unified world economy developed countries form an inner core area whereas developing countries are found on the periphery and semi-periphery areas.

X

Xenophobia Fear of people who are from other countries.

Z

Zero population growth (ZPG) A decline of the total fertility rate to the point where the natural increase rate equals zero.

Zoning ordinance A law that limits the permitted uses of land and maximum density of development in a community.

Credits

Note: Uncredited maps and figures, International Mapping/Pearson Education, Inc.

Image Credits

Front Matter p. xviii: Courtesy of James M. Rubenstein

Chapter 1 CO-01, p. 2: Bloomberg/Getty Images; UNFIG01-01, p. 2: Chip Chipman/Bloomberg/Getty Images; UNFIG01-02, p. 3: Google Maps; UNFIG01-03, p. 3: V. Dorosz/Alamy Stock Photo; UNFIG01-04, p. 3: Wilfredo Lee/AP Images; 01-06-05, p. 15: Canva Pty Ltd/Alamy Stock Photo; 01-07-01, p. 16: V. Dorosz/Alamy Stock Photo; 01-07-03, p. 17: ONOKY Photononstop/Alamy Stock Photo; 01-08-01, p. 18: ableimages/Alamy Stock Photo; 01-09-01, p. 20: Michael Kemp/Alamy Stock Photo; 01-09-01, p. 20: Michael B. Thomas/Getty Images; 01-10-01, p. 22: Shutterstock; 01-11-02, p. 24: Album/Alamy Stock Photo; 01-12-02, p. 26: Dennis MacDonald/Alamy Stock Photo; 01-12-03, p. 26: John Anderson/Alamy Stock Photo; 01-12-04, p. 26: incamerastock/Alamy Stock Photo; 01-12-05, p. 27: Yola Watrucka/Alamy Stock Photo; 01-13-01a, p. 28: Colin Underhill/Alamy Stock Photo; 01-13-01b, p. 28: Rosemarie Stennull/Alamy Stock Photo; 01-13-02a, p. 28: Xavi Ruiz (@xruiztru), twitter.com, November 3, 2020; 01-13-02b, p. 28: frans lemmens/Alamy Stock Photo; 01-13-03, p. 28: Climate Central; 01-13-05, p. 29: Climate Central; 01-13-06, p. 30: Wilfredo Lee/AP Images; 01-CR-01, p. 30: Aleksey Boldin/Alamy Stock Photo; 01-CR-01, p. 30: TY Lim/Shutterstock; 01-CR-02, p. 30: Kip Evans/Alamy Stock Photo 01; 01-CR-03, p. 30: Sorbis/Shuttestock; 01-CR-04, p. 30: Wavebreak Media Ltd/123RF; 01-CR-06, p. 33: Michael Runkel/Alamy Stock Photo; 01-CR-07, p. 33: Friedrich Stark/Alamy Stock Photo

Chapter 2 CO-02, p. 35: zakir hossain chowdhury zakir/Alamy Stock Photo; UNFIG02-01, p. 35: Anadolu Agency/Getty Images; UNFIG02-02, p. 35: Geoff Smith/Alamy Stock Photo; UNFIG02-03, p. 35: blickwinkel/A. Laule/Alamy Stock Photo; UNFIG02.04, p. 35: Jake Lyell/Alamy Stock Photo; 02-01-01, p. 36: frans lemmens/Alamy Stock Photo; 02-02-01, p. 38: BMD Images/Alamy Stock Photo; 02-02-03, p. 38: robertharding/Alamy Stock Photo; 02-03-06, p. 41: E.D. Torial/Alamy Stock Photo; 02-04-03, p. 42: Godong/Getty Images; 02-04-05, p. 43: Familia/Alamy Stock Photo; 02-04-07, p. 43: Geoff Smith/Alamy Stock Photo; 02-05-01, p. 44: blickwinkel/A. Laule/Alamy Stock Photo; 02-06-02, p. 46: KidStock/Getty Images; 02-06-02, p. 46: KidStock/Getty Images; 02-06-04, p. 47: Homer Sykes/Alamy Stock Photo; 02-07-03, p. 48: Phanie/Alamy Stock Photo; 02-09-05, p. 53: Lou Linwei/Alamy Stock Photo; 02-10-01, p. 54: Dinodia Photos/Alamy Stock Photo; 02-10-03, p. 54: Jake Lyell/Alamy Stock Photo; 02-12-01, p. 58: Joseph Odor/AP/Shutterstock; 02-CR-01, p. 60: Vinicius Bacarin/Alamy Stock Photo; 02-CR-02, p. 60: DGLimages/Alamy Stock Photo; 02-CR-03, p. 60: Jeffrey Isaac Greenberg 20+/Alamy Stock Photo; 02-CR-04, p. 60: Yusuke Harada/NurPhoto/Shutterstock

Chapter 3 CO-03, p. 62: Eyepix/NurPhoto/Shutterstock; UNFIG03-01, p. 63: imageBROKER/Alamy Stock Photo; UNFIG03-02, p. 63: AGF Srl/Alamy Stock Photo; UNFIG03-03, p. 63: Benedicte Desrus/Alamy Stock Photo; UNFIG03-04, p. 63: John Lamparski/NurPhoto/Shutterstock; 03-10-06p. : John Lamparski/NurPhoto/Shutterstock; 03-01-01, p. 64: Friedrich Stark/Alamy Stock Photo; 03-01-02, p. 64: XiXinXing/Shutterstock; 03-02-01, p. 66: Xinhua/Alamy Stock Photo; 03-02-02, p. 66: imageBROKER/Alamy Stock Photo; 03-03-01, p. 68: American Stock Archive/Getty Images; 03-04-01p. 70: AGF Srl/Alamy Stock Photo; 03-04-05, p. 71: Mike Shipman/Alamy Stock Photo; 03-05-02p. 72: Benedicte Desrus/Alamy Stock Photo; 03-05-05, p. 73: jeremy sutton-hibbert/Alamy Stock Photo; 03-06-02, p. 74: ALI HAIDER/EPA-EFE/Shutterstock; 03-06-04, p. 75: Bloomberg/Getty Images; 03-07-02, p. 76: Andre Jenny/Alamy Stock Photo 03-07-03, p. 76: Benedicte Desrus/Alamy Stock Photo; 03-07-06, p. 77: Jan Sochor/Alamy Stock Photo; 03-08-01, p. 78: 1000 Words/Shutterstock; 03-08-03, p. 79: olrat/Shutterstock; 03-08-04, p. 79: Vladimir Pomortzeff/Alamy Stock Photo; 03-08-05, p. 79: Oleg Anisimov/Shutterstock; 03-08-06, p. 79: ZUMA Press Inc/Alamy Stock Photo; 03-09-01, p. 80: Frame Stock Footage/Shutterstock; 03-09-02, p. 81: Vic Hinterlang/Shutterstock; 03-10-01, p. 82: LUIS TORRES/EPA-EFE/Shutterstock; 03-10-03, p. 83: Mauro Ladu/Alamy Stock Photo; 03-10-04, p. 83: Jess Merrill/Alamy Stock Photo; 03-10-05, p. 83: Robert Shantz/Alamy Stock Photo; 03-10-06, p. 83: John Lamparski/NurPhoto/Shutterstock; 03-CR-01, p. 83: Vineyard Perspective/Shutterstock; 03-CR-02, p. 84: Andriy Popov/Alamy Stock Photo; 03-CR-03, p. 84: Prasit Rodphan/Alamy Stock Photo; 03-CR-04, p. 84: cesar fernandez/Alamy Stock Photo

Chapter 4 CO-04, p. 86: SasinTipchai/Shutterstock; UNFIG04-01, p. 87: Sharkshock/Shutterstock; UNFIG04-02, p. 87: Steve Vidler/mauritius images GmbH/Alamy Stock Photo; UNFIG04-03, p. 87: blickwinkel/Alamy Stock Photo; UNFIG04-04, p. 87: Yawar Nazir/Getty Images; 04-01-01a, p. 88: Michael Grant Travel/Alamy Stock Photo; 04-01-01b, p. 88: yamatao/Getty Images; 04-01-01c, p. 88: MoreISO/Getty Images; 04-01-01d, p. 88: Chris Fredriksson/Alamy Stock Photo; 04-01-02a, p. 89: Chris Bradshaw/Shutterstock; 04-01-02b, p. 89: Jnix/Shutterstock; 04-01-03c, p. 89: Ron Buskirk/Alamy Stock Photo; 04-02-01, p. 90: Gina Power/Shutterstock; 04-02-04, p. 90: Hemis/Alamy Stock Photo; 04-02-05, p. 91: Sharkshock/Shutterstock; 04-03-02, p. 92: Fairfax Media/Getty Images; 04-04-01, p. 94: Steve Vidler/mauritius images GmbH/Alamy Stock Photo; 04-04-02, p. 94: H. Mark Weidman Photography/Alamy Stock Photo; 04-04-05, p. 95: Parmorama/Alamy Stock Photo; 04-04-06, p. 95: Peter Cavanagh/Alamy Stock Photo; 04-05-01, p. 96: HOANG DINH NAM/Getty Images; 04-06-02, p. 99: ZUMA Press, Inc./Alamy Stock Photo; 04-06-04, p. 99: Imaginechina Limited/Alamy Stock Photo; 04-07-04, p. 101: blickwinkel/Alamy Stock Photo; 04-08-02, p. 102: Vlad Deep/Alamy Stock Photo; 04-08-03, p. 103: STEPHEN SHAVER/UPI/Shutterstock; 04-08-04, p. 103: ALEX OGLE/Getty Images; 04-09-06, p. 105: CBS Photo Archive/Getty Images; 04-10-01, p. 106: Jake Lyell/Alamy Stock Photo; 04-10-02, p. 106: Yawar Nazir/Getty Images 04-10-03, p. 107: Peter Eastland/Alamy Stock Photo; 04-10-04, p. 107: mRGB/Shutterstock; 04-11-01, p. 108: andresr/Getty Images; 04-11-03, p. 108: Jaro68/Shutterstock; 04-11-04, p. 108: wideonet/Shutterstock; 04-11-05, p. 109: F.Bettex - Mysterra.org/Alamy Stock Photo; 04-11-06, p. 109: Hoberman Collection/Getty Images; 04-CR-01, p. 110: Guy Bell/Alamy Stock Photo; 04-CR-03 p. 110: ZUMA Press, Inc./Alamy Stock Photo; 04-CR-04, p. 100: PIUS UTOMI EKPEI/Getty Images 04-CR-06, p. 111: Dennis Cox/Alamy Stock Photo

Chapter 5 CO-05, p. 112: Tatyana Tomsickova/Alamy Stock Photo; UNFIG05-01, p. 113: Sean Sprague/Alamy Stock Photo; UNFIG05-02, p. 113: Wozzie/Shutterstock; UNFIG05-02, p. 113: jeffreychin/Shutterstock; UNFIG05-03, p. 113: Christopher Canty Photography/Alamy Stock Photo; UNFIG05-04, p. 113: Pearson Education Inc.; 05-01-01, p. 114: Sean Sprague/Alamy Stock Photo; 05-02-02, p. 116: Israel by Blake-Ezra Cole/Alamy Stock Photo; 05-04-01, p. 120:

jeffreychin/Shutterstock; 05-05-01, p. 121: Wozzie/Shutterstock; 05-05-05, p. 123: Charles O. Cecil/Alamy Stock Photo; 05-06-01, p. 124: Ian M Butterfield (Yorkshire)/Alamy Stock Photo; 05-07-01, p. 126: REUTERS/Alamy Stock Photo; 05-07-02, p. 126: Christopher Canty Photography/Alamy Stock Photo; 05-08-05, p. 129: Lusica/Shutterstock; 05-10-02, p. 133: paul prescott/Alamy Stock Photo; 05-10-03, p. 133: Pearson Education Inc.; 05-CR-04, p. 134: Wolfgang Kaehler/Alamy StockPhoto; 05-CR-06, p. 135: Skyfish/Alamy Stock Photo

Chapter 6 CO-06, p. 163: Paul Christian Gordon/Alamy Stock Photo; UNFIG06-01, p. 137: ABCDstock/Shutterstock; UNFIG06-02, p. 137: halil ibrahim kurucan/AlamyStock Photo; UNFIG06-03, p. 137: Ira Berger/Alamy Stock Photo; UNFIG06-04, p. 137: Baturina Yuliya/Shutterstock; 06-01-01, p. 138: Iain Masterton/Alamy Stock Photo; 06-02-01, p. 140: All Canada Photos/Alamy Stock Photo; 06-03-01, p. 142: Vera Volodkina/Alamy Stock Photo; 06-03-02, p. 142: Ata Giri/Shutterstock; 06-04-02, p. 144: Tibor Bognar/Alamy Stock Photo; 06-04-03, p. 144: Art Phaneuf/Alamy Stock Photo; 06-04-04, p. 144: Art Directors & TRIP/Alamy Stock Photo; 06-04-05, p. 145: ABCDstock/Shutterstock; 06-05-03, p. 147: angelo giampiccolo/Alamy Stock Photo; 06-05-04, p. 147: SB Stock/Shutterstock; 06-06-01, p. 148: DANIEL GREENHOUSE/Alamy Stock Photo; 06-06-03, p. 149: ATEF SAFADI/EPA-EFE/Shutterstock; 06-06-04, p. 149: halil ibrahim kurucan/Alamy Stock Photo; 06-07-03, p. 151: Horizon Images/Motion/Alamy Stock Photo; 06-09-0, p. 154: Oleksandr Rupeta/Alamy Stock Photo; 06-09-02, p. 154: Purepix/Alamy Stock Photo; 06-09-03, p. 155: Citizen59/Alamy Stock Photo; 06-09-04, p. 155: Ric Perin/Shutterstock; 06-10-01, p. 156: Don Smetzer/Alamy Stock Photo; 06-10-02, p. 156: Gary Whitton/Alamy Stock Photo; 06-11-01, p. 158: Jim Keir/Alamy Stock Photo; 06-11-02, p. 158: Michal Sikorski/Alamy Stock Photo; 06-11-03, p. 159: Ira Berger/Alamy Stock Photo; 06-11-04, p. 159: Martin Newman/Alamy Stock Photo; 06-12-01, p. 160: Raj Singh/Alamy Stock Photo; 06-12-02, p. 161: Baturina Yuliya/Shutterstock; 06-12-03, p. 161: Rajesh KUMAR/Alamy Stock Photo; 06-13-03, p. 163: StockStudio/Alamy Stock Photo; 06-CR-01, p. 164: Vehbi Koca/Alamy Stock Photo; 06-CR-02, p. 164: Valentin Sama-Rojo/Alamy Stock Photo; 06-CR-04, p. 164: Wai Kit Wong/Alamy Stock Photo; 06-CR-05, p. 165: Mawardi Bahar/Alamy Stock Photo

Chapter 7 CO-07, p. 166: Marcin Rogozinski/Alamy Stock Photo; UNFIG07-01, p. 167: Megapress/Alamy Stock Photo; UNFIG07-02, p. 167: Tobias Rostlund/Alamy Stock Photo; UNFIG07-03, p. 167: Steve Liss/Getty Images; UNFIG07-04, p. 167: Jeremy Horner/Alamy Stock Photo; 07-01-01, p. 168: GEORGE NIKITIN/Associated Press; 07-01-05, p. 169: Megapress/Alamy Stock Photo; 07-03-03, p. 172: Science History Images/Alamy Stock Photo; 07-04-01, p. 174: Elliott Erwitt/Magnum Photos; 07-04-02, p. 174: National Archives/Getty Images; 07-04-03, p. 175: Alain Nogues/Getty Images; 07-04-04, p. 175: Tobias Rostlund/Alamy Stock Photo; 07-05-01, p. 176: BrianPlrwin/Shutterstock; 07-05-03, p. 177: Jim West/Alamy Stock Photo; 07-05-04, p. 177: Allison C Bailey/Shutterstock; 07-05-06, p. 177: Jed Leicester/BPI/Shutterstock; 07-05-05, p. 177: Steph Chambers/Getty Images 07-05-07, p. 178: Rich Kane Photography/Alamy Stock Photo; 07-06-01, p. 179: Steve Liss/Getty Images; 07-06-04, p. 182: MLBARIONA/Alamy Stock Photo; 07-08-01, p. 182: Peter Stuckings/Shutterstock; 07-08-02, p. 182: Azamat Imanaliev/Shutterstock; 07-08-03, p. 183: Pearson Education Inc.; 07-08-04, p. 185: Rehman Asad/Alamy Stock Photo; 07-09-04, p. 186: AfriPics.com/Alamy Stock Photo; 07-10-02, p. 187: Jeremy Horner/Alamy Stock Photo; 07-10-04a, p. 187: Martin Mayer/Alamy Stock Photo; 07-10-04b, p. 187: Agencja Fotograficzna Caro/Alamy Stock Photo; 07-10-04c, 188p. : Aleksandar Todorovic/Shutterstock; 07-CR-02, p. 188: Diriye Amey/Alamy Stock Photo; 07-CR-03, p. 188: Pacific Press Media Production Corp./Alamy Stock Photo; 07-CR-04, p. 188: Eric Nathan/Alamy Stock Photo

Chapter 8 CO-08, p. 190: REUTERS/Alamy Stock Photo; UNFIG08-01, p. 191: Apix1/Shutterstock; UNFIG08-02, p. 191 : Iain Masterton/Alamy Stock Photo; UNFIG08-03, p. 191: Kyodo/Newscom; UNFIG08-04, p. 191: PlanetObserver/Science Source; 08-01-01, p. 192: Zoonar GmbH/Alamy Stock Photo; 08-02-01, p. 194: Julian Elliott News Photography/Alamy Stock Photo; 08-02-02, p. 194: Sandy Macys/Alamy Stock Photo; 08-02-03, p. 194: Apix1/Shutterstock; 08-02-06, p. 194: imageBROKER/Alamy Stock Photo; 08-03-07, p. 195: Iain Masterton/Alamy Stock Photo; 08-04-02, p. 197: Jayne Lloyd/Stockimo/Alamy Stock Photo; 08-05-02, p. 198: Ralph Rozema/Alamy Stock Photo; 08-05-04, p. 200: Chronicle/Alamy Stock Photo; 08-05-05, p. 201: Roger Sedres/Alamy Stock Photo; 08-06-01, p. 201: Patricia Burilli Fencz/Alamy Stock Photo; 08-06-02, p. 202: Jan Schneckenhaus/Alamy Stock Photo; 08-06-05, p. 203: Hercules Milas/Alamy Stock Photo; 08-06-06, p. 203: Hercules Milas/Alamy Stock Photo; 08-07-03, p. 204: Kyodo/Newscom; 08-07-08, p. 205: National Geographic Image Collection/Alamy Stock Photo; 08-10-01, p. 210: Laperruque/Alamy Stock Photo; 08-11-06, p. 213: PlanetObserver/Science Source; 08-CR-01, p. 214: dpa picture alliance/Alamy Stock Photo; 08-CR-02, p. 214: Gemma Fletcher/Shutterstock; 08-CR-04, p. 214: Diego Montoya/Alamy Stock Photo

Chapter 9 CO-09, p. 216: Thomas Dutour/Shutterstock; UNFIG09-01, 217p. : UDAZKENA/Alamy Stock Photo; UNFIG09-02, p. 217: PanyaStudio/Shutterstock; UNFIG09-03, p. 217: Bambang Sutiyoso/Shutterstock; UNFIG09-04, p. 217: FiledIMAGE/Shutterstock; 09-01-01, p. 218: CECIL BO DZWOWA/Shutterstock; 09-01-02A, p. 218: Robert Fried/Alamy Stock Photo; 09-01-02B, p. 218: UDAZKENA/Alamy Stock Photo; 09-02-02, p. 220: Abdullah Durman/Shutterstock; 09-02-03, p. 220: Nature Picture Library/Alamy Stock Photo; 09-02-04, p. 220: Paolo Bona/Shuttestock; 09-02-06, p. 221: MBI/Alamy Stock Photo; 09-02-07, p. 221: Michal Urbanek/Shutterstock; 09-02-08, p. 221: Betty LaRue/Alamy Stock Photo; 09-03-01, p. 222: franco lucato/Shutterstock; 09-04-01, p. 224: PanyaStudio/Shutterstock; 09-04-03, p. 224: Jake Lyell/Alamy Stock Photo; 09-05-02, p. 226: india view/Alamy Stock Photo; 09-05-03, p. 226: dbimages/Alamy Stock Photo; 09-05-04p. : Perry van Munster/Alamy Stock Photo; 09-05-05p. : Keren Su/China Span/Alamy Stock Photo; 09-06-01, p. 226: Pearson Education Inc.; 09-06-02, p. 226: REUTERS/Alamy Stock Photo; 09-07-03, p. 228: Flystock/Shutterstock; 09-07-09, p. 228: National Geographic Image Collection/Alamy Stock Photo; 09-08-02, p. 232: photoff/Alamy Stock Photo; 09-08-03, p. 232: rujithai/Shutterstock; 09-08-04, p. 233: Bambang Sutiyoso/Shutterstock; 09-08-05, p. 234: mark reinstein/Alamy Stock Photo; 09-09-01a, p. 234: Ivoha/Alamy Stock Photo; 09-09-01b, p. 234: Danita Delimont/Alamy Stock Photo; 09-09-01c, p. 234: Danita Delimont/Alamy Stock Photo; 09-09-01d, p. 234: Mike Goldwater/Alamy Stock Photo; 09-09-01e, p. 235: Adam Jan Figel/Shutterstock; 09-09-02, p. 235: Djohan Shahrin/Shutterstock; 09-09-03, p. 236: Marevan/Shutterstock; 09-10-03, p. 239: Jim West/Alamy Stock Photo; 09-11-04, p. 239: FiledIMAGE/Shutterstock; 09-11-05, p. 239: Lyroky/Alamy Stock Photo; 09-CR-01, p. 239: bodnar.photo/Shutterstock; 09-CR-02, p. 240: eFesenko/Alamy Stock Photo; 09-CR-03, p. 240: Jim West/Alamy Stock Photo; 09-CR-04, p. 240: Justin Sullivan/Getty Images; 09-CR-05, p. 240: judyfromjuly/Stockimo/Alamy Stock Photo

Chapter 10 CO-10, p. 242: Agencja Fotograficzna Caro/Alamy Stock Photo; UNFIG10-01, p. 243: BSIP SA/Alamy Stock Photo; UNFIG10-02, p. 243: Riccardo Lennart Niels Mayer/Alamy Stock Photo; UNFIG10-03, p. 243: Paolo Romiti/Alamy Stock Photo; UNFIG10-04, p. 243: MobileSaint/Shutterstock; 10-01-01, p. 244: Chad Case/Alamy Stock Photo; 10-03a, p. 245: robertharding/Alamy Stock Photo; 10-03b, p. 245: eye35/Alamy Stock Photo; 10-03c, p. 245. : dpa picture alliance/Alamy Stock Photo; 10-03d, p. 245: robertharding/Alamy Stock Photo; 10-03e, p. 245: Sergi Reboredo/Alamy Stock Photo; 10-03f, p. 245: Sean Pavone/Alamy Stock Photo; 10-03g, p. 245: galit seligmann/Alamy Stock Photo; 10-03h, p. 245: GRANT ROONEY PREMIUM/Alamy Stock Photo; 10-03i, p. 245: philipus/Alamy Stock Photo; 10-02-01, p. 246: LapaiIrKrapai/Shutterstock;

10-02-06, p. 247: india view/Alamy Stock Photo; 10-03-02, p. 248: Cultura Creative RF/Alamy Stock Photo; 10-03-03, p. 248: BSIP SA/ Alamy Stock Photo; 10-04-01, p. 250: Mike Goldwater/Alamy Stock Photo; 10-05-01, p. 252: Riccardo Lennart Niels Mayer/Alamy Stock Photo; 10-05-03, p. 252: Barbara Sauder/Shutterstock; 10-06-05, p. 255: Xinhua/Alamy Stock Photo; 10-07-01, p. 256: robert mangan/ Alamy Stock Photo; 10-07-02, p. 256: Maciej Dakowicz/Alamy Stock Photo; 10-07-03, p. 257: Paolo Romiti/Alamy Stock Photo; 10-08-05, p. 259: Joerg Boethling/Alamy Stock Photo; 10-09-05, p. 261: Friedrich Stark/Alamy Stock Photo; 10-11-01, p. 264: Joerg Boethling/ Alamy Stock Photo; 10-11-02, p. 264: Jenlo8/Shutterstock; 10-11-03, p. 265: MobileSaint/Shutterstock; 10-11-04, p. 265: Frances Roberts/ Alamy Stock Photo; 10-CR-01, p. 266: BSIP SA/Alamy Stock Photo; 10-CR-03, p. 266: Vibrant Pictures/Alamy Stock Photo; 10-CR-04, p. 266: Zute Lightfoot/Alamy Stock Photo

Chapter 11 CO-11, p. 268: Ulrich Baumgarten/Getty Images; UNFIG11-01, p. 269: ZUMA Press Inc/Alamy Stock Photo; UNFIG11-02, p. 269: The Photo Works/Alamy Stock Photo; UNFIG11-03, p. 269: Cavan Images/Alamy Stock Photo; UNFIG11-04, p. 269: ZUMA Press, Inc./Alamy Stock Photo; 11-02-04, p. 272: ZUMA Press Inc/Alamy Stock Photo; 11-02-05, p. 273: Steve Vidler/Alamy Stock Photo; 11-02-06, p. 273: DZMITRY PALUBIATKA/Alamy Stock Photo; 11-03-02, p. 274: RBM Vintage Images/Alamy Stock Photo; 11-03-03, p. 275: Uladzik Kryhin/Shutterstock; 11-03-04, p. 275: Archimage-Aerial/ Alamy Stock Photo; 11-04-01, p. 276: Joerg Boethling/Alamy Stock Photo; 11-04-02, p. 277: Annie Owen/robertharding/Alamy Stock Photo; 11-04-05, p. 277: The Photo Works/Alamy Stock Photo; 11-05-03, p. 278: Jim West/Alamy Stock Photo; 11-05-02, p. 278: Steven Milne/Alamy Stock Photo; 11-06-02, p. 280: Cavan Images/Alamy Stock Photo; 11-08-06, p. 285: Michele Naccari/EPA/Shutterstock; 11-09-05, p. 287: ZUMA Press, Inc./Alamy Stock Photo; 11-09-06, p. 287: Ognjen Stevanovic/Alamy Stock Photo; 11-CR-01, p. 288: MiniStocker/Shutterstock; 11-CR-02, p. 288: Stephen Plaster/Shutterstock; 11-CR-03, p. 288: RoClickMag/Shutterstock; 11-CR-04, p. 288: Universal Images Group North America LLC/DeAgostini/Alamy Stock Photo; 11-CR-05, p. 289: ZUMA Press, Inc./Alamy Stock Photo

Chapter 12 CO-12, p. 290: Angelo Cordeschi/Alamy Stock Photo; UNFIG12-01, p. 291: Weiming Xie/Shutterstock; UNFIG12-02, p. 291: Chad Ehlers/Alamy Stock Photo; UNFIG12-03, p. 291: inga spence/ Alamy Stock Photo; UNFIG12-04, p. 291: AfriPics.com/Alamy Stock Photo; 12-01-05, p. 293: Weiming Xie/Shutterstock; 12-03-01, p. 296: William Mullins/Alamy Stock Photo; 12-04-02, p. 298: ESB Professional/ Shutterstock; 12-04-03, p. 298: roberto galan/Alamy Stock Photo; 12-04-05, p. 299: Todd Strand/Alamy Stock Photo; 12-05-02, p. 300: Jonathan Weiss/Alamy Stock Photo; 12-05-03, p. 300: Mira/ Alamy Stock Photo; 12-05-06, p. 301: Tada Images/Shutterstock; 12-06-01, p. 302: Jake Lyell/Alamy Stock Photo; 12-06-02, p. 302: Chad Ehlers/Alamy Stock Photo; 12-06-06, p. 303: Diogo Baptista/ Alamy Stock Photo; 12-07-03, p. 305: agsaz/shutterstock; 12-07-04, p. 305: Steven Charles/Alamy Stock Photo; 12-07-05, p. 305: Tomas Marek/Shutterstock; 12-08-01, p. 306: Greg Balfour Evans/Alamy Stock Photo; 12-08-03, p. 307: dbimages/Alamy Stock Photo; 12-08-04, p. 307: ZUMA Press, Inc./Alamy Stock Photo; 12-09-01p. : First Class Photography/Shutterstock; 12-09-05p. : inga spence/Alamy Stock Photo; 12-10-01, p. 308: AfriPics.com/Alamy Stock Photo; 12-10-02, p. 309: All Canada Photos/Alamy Stock Photo; 12-10-03, p. 310: Aerial Archives/Alamy Stock Photo; 12-10-04, p. 310: Commission Air/Alamy Stock Photo; 12-10-05, p. 311: Pegaz/Alamy Stock Photo; 12-11-01, p. 311: arabianEye FZ LLC/Alamy Stock Photo; 12-11-02, p. 311: Geof Kirby/Alamy Stock Photo; 12-11-03, p. 312: IURII BURIAK/Shutterstock; 12-CR-01, p. 314: Susan Vineyard/Alamy Stock Photo; 12-CR-02, p. 314: ThreeRivers11/Shutterstock; 12-CR-03, p. 314: Zoonar GmbH/Alamy Stock Photo; 12-CR-04, p. 314: Vlad Siaber/Shutterstock; 12-CR-06, p. 314: Pegaz/Alamy Stock Photo

Chapter 13 CO-13, p. 316: Danita Delimont/Alamy Stock Photo; UNFIG13-01, p. 317: Rob Crandall/Alamy Stock Photo; UNFIG13-02, p. 317: Jim West/Alamy Stock Photo; UNFIG13-03, p. 317: Alena Veasey/Shutterstock; UNFIG13-04, p. 317: Jon Lovette/Alamy Stock Photo; 13-01-01, p. 318: Virrage Images/Shutterstock; 13-02-01, p. 320: Rob Crandall/Alamy Stock Photo; 13-02-02, p. 320: Aerial Archives/Alamy Stock Photo; 13-02-03, p. 320: Dennis MacDonald/ Alamy Stock Photo; 13-02-04, p. 321: Aerial Archives/Alamy Stock Photo; 13-02-05, p. 321: Thomas Kelley/Shutterstock; 13-03-02, p. 322: Serhii Chrucky/Alamy Stock Photo; 13-03-04, p. 322: Image Professionals GmbH/Alamy Stock Photo; 13-03-06, p. 323: dbimages/ Alamy Stock Photo; 13-03-08, p. 323: Jim West/Alamy Stock Photo; 13-04-02, p. 324: simon leigh/Alamy Stock Photo; 13-04-04, p. 325: Don Despain/Alamy Stock Photo; 13-04-06, p. 325: Jim West/Alamy Stock Photo; 13-05-01, p. 326: cktravels.com/Shutterstock; 13-05-04, p. 327: cktravels.com/Shutterstock; 13-11-EOC-01, p. 328: Patrick Smith/Getty Images; 13-11-EOC-02, p. 328: Yvette Cardozo/Alamy Stock Photo; 13-11-EOC-03, p. 328: Bella Falk/Alamy Stock Photo; 13-11-EOC-04, p. 328: Bob Korn/Alamy Stock Photo; 13-11-EOC-06, p. 328: SUNG KUK KIM/Alamy Stock Photo; 13-07-01, p. 330: DEA/G DAGLI ORTI/AgeFotostock; 13-07-02, p. 330: DEA PICTURE LIBRARY/Getty Images; 13-07-03, p. 330: Wangkun Jia/Alamy Stock Photo; 13-07-05, p. 331: wendy connett/Alamy Stock Photo; 13-07-06, p. 331: Diego Grandi/Alamy Stock Photo; 13-08-01, p. 332: Classic-Stock/Alamy Stock Photo; 13-08-02, p. 332: Glasshouse Images/ Alamy Stock Photo; 13-09-02, p. 334: reallifephotos/Alamy Stock Photo; 13-09-03, p. 334: Goran Jakuš/Alamy Stock Photo; 13-09-04, p. 335: Michael Dwyer/Alamy Stock Photo; 13-09-05, p. 335: Andrew Haliburton/Alamy Stock Photo; 13-10-01, p. 336: Images-USA/Alamy Stock Photo; 13-10-02, p. 336: Xinhua/Alamy Stock Photo; 13-10-03a, p. 337: Gijsbert Hanekroot/Alamy Stock Photo; 13-10-03b, p. 337: Gijsbert Hanekroot/Alamy Stock Photo; 13-10-04, p. 337: Jon Lovette/ Alamy Stock Photo

Chapter 14 CO-14, p. 340: Ognyan Trifonov/Alamy Stock Photo; UNFIG14-01, p. 341: Peter W Titmuss/Alamy Stock Photo; UNFIG14-02, p. 341: Joerg Boethling/Alamy Stock Photo; UNFIG14-03, p. 341: David Grossman/Alamy Stock Photo; UNFIG14-04, p. 341: ADEK BERRY/Getty Images; 14-01-01, p. 342: Peter W Titmuss/Alamy Stock Photo; 14-03-04, p. 343: Milesy/Alamy Stock Photo; 14-04-01, p. 348: Idealink Photography/Alamy Stock Photo; 14-04-03, p. 348: Brian Brown/Alamy Stock Photo; 14-04-05, p. 349: Joerg Boethling/ Alamy Stock Photo; 14-05-02, p. 350: CrowdSpark/Alamy Stock Photo; 14-06-01, p. 352: Idealink Photography/Alamy Stock Photo; 14-06-03, p. 352: WoodysPhotos/Shutterstock; 14-06-04, p. 353: Hong Hanh Mac Thi/Alamy Stock Photo; 14-06-05a, p. 353: Universal Images Group North America LLC/Alamy Stock Photo; 14-06-05b, p. 353: World History Archive/Alamy Stock Photo; 14-06-05c, p. 353: NASA; 14-07-01, p. 354: David Grossman/Alamy Stock Photo; 14-07-03, p. 354: inga spence/Alamy Stock Photo; 14-07-04, p. 355: Stu/ Alamy Stock Photo; 14-08-04, p. 357: Philip Scalia/Alamy Stock Photo; 14-09-05, p. 359: Ellen Isaacs/Alamy Stock Photo; 14-10-01, p. 360: SerrNovik/Shutterstock; 14-10-02, p. 360: ADEK BERRY/Getty Image; 14-10-05, p. 361: AGF Srl/Alamy Stock Photo; 14-10-06, p. 361: Joan Wozniak/Shutterstock; 14-10-07, p. 361: John Crux Photography/ Getty Image; 14-10-10, p. 361: Sergey Izotov/Shutterstock; 14-10-09, p. 361: Friedrich Stark/Alamy Stock Photo; 14-CR-01, p. 362: Design Pics Inc/Alamy Stock Photo; 14-CR-02, p. 362: Ruth Grimes/Alamy Stock Photo; 14-CR-03, p. 362: David Grossman/Alamy Stock Photo; 14-CR-04, p. 362: Jochen Tack/Alamy Stock Photo; 14-CR-05, p. 362: sebastien bonaime/Alamy Stock Photo

Front insert International Mapping, Inc.

Back insert Defense Meteorological Satellite Program (DMSP) NASA/GSFC.

Index

Note: Page numbers in bold indicate definitions. Page numbers with *f* indicate figures.

Countries of the World

The Caribbean

ARCTIC OCEAN
Barents Sea
Svalbard
North Cape
Lapland
ICELAND
Faroe Is.
Scandinavia
See inset below
Northern European Plain
ALPS
Iberian Peninsula
Balkan Peninsula
Anatolia
CAUCASUS MTS.
Caspian Depression
URAL MTS.
West Siberian Plain
SIBERIA
Taymyr Peninsula
Central Siberian Plateau
VERKHOYANSK RANGE
KOLYMA RANGE
RUSSIA
Sea of Okhotsk
Kamchatka Peninsula
Kuril Is.
KAZAKHSTAN
Aral Sea
ALTAY MTS.
MONGOLIA
YABLONOVY RANGE
GOBI (DESERT)
TIAN SHAN
Taklimakan Desert
UZBEKISTAN
KYRGYZSTAN
TAJIKISTAN
TURKMENISTAN
AZERBAIJAN
GEORGIA
ARMENIA
TURKEY
SYRIA
LEBANON
ISRAEL
JORDAN
GAZA STRIP
WEST BANK
IRAQ
IRAN
ELBURZ MTS.
ZAGROS MTS.
Iranian Plateau
KUWAIT
BAHRAIN
QATAR
SAUDI ARABIA
UNITED ARAB EMIRATES
OMAN
YEMEN
AFGHANISTAN
HINDU KUSH
PAKISTAN
HIMALAYAS
NEPAL
BHUTAN
BANGLADESH
INDIA
Thar Desert
Deccan Plateau
WESTERN GHATS
EASTERN GHATS
SRI LANKA
MALDIVES
CHINA
NORTH KOREA
SOUTH KOREA
JAPAN
Sea of Japan (East Sea)
East China Sea
Ryukyu Trench
TAIWAN
BURMA (MYANMAR)
LAOS
THAILAND
VIETNAM
CAMBODIA
PHILIPPINES
South China Sea
Philippine Trench
BRUNEI
MALAYSIA
SINGAPORE
INDONESIA
TIMOR-LESTE
PAPUA NEW GUINEA
Bay of Bengal
Arabian Sea
Emperor Seamounts
Kuril Tr.
Northwest Pacific Basin
PACIFIC OCEAN
Northern Mariana Is.
Wake Island
Guam
Mariana Trench
PALAU
FEDERATED STATES OF MICRONESIA
MARSHALL ISLANDS
Central Pacific Basin
Equator
NAURU
KIRIBATI
SOLOMON ISLANDS
TUVALU
VANUATU
FIJI
New Hebrides Tr.
New Caledonia
Norfolk Island
Cape York
GREAT DIVIDING RANGE
AUSTRALIA
Great Sandy Desert
Western Plateau
Great Victoria Desert
Simpson Desert
Tasman Sea
NEW ZEALAND
Perth Basin
South Australian Basin
Christmas Island
Cocos (Keeling) Islands
Ninetyeast Ridge
Mid-Indian Basin
Mid-Indian Ridge
INDIAN OCEAN
Broken Ridge
Southwest Indian Ridge
MAURITIUS
Réunion
MADAGASCAR
COMOROS
Mayotte
SEYCHELLES
Somali Basin
TUNISIA
ATLAS MTS.
ALGERIA
LIBYA
EGYPT
SAHARA
Ahaggar
Tibesti
Air
SAHEL
MALI
NIGER
CHAD
SUDAN
ERITREA
DJIBOUTI
Ethiopian Highlands
Horn of Africa
ETHIOPIA
SOMALIA
SOUTH SUDAN
CENTRAL AFRICAN REP.
CAMEROON
NIGERIA
BENIN
TOGO
GHANA
BURKINA FASO
GUINEA
LIBERIA
CÔTE D'IVOIRE
SÃO TOMÉ AND PRÍNCIPE
EQUATORIAL GUINEA
GABON
REP. OF THE CONGO
Congo Basin
DEM. REP. OF THE CONGO
UGANDA
RWANDA
BURUNDI
KENYA
TANZANIA
Great Rift Valley
Katanga Plateau
ANGOLA
ZAMBIA
MALAWI
MOZAMBIQUE
ZIMBABWE
NAMIBIA
Namib Desert
BOTSWANA
Kalahari Desert
SWAZILAND
LESOTHO
SOUTH AFRICA
Cape of Good Hope
Agulhas Plateau
Cape Plain
Walvis Ridge
St. Helena
Mid-Atlantic Ridge
ATLANTIC OCEAN
Weddell Plain
ANTARCTICA
0 1,000 2,000 Miles
0 1,000 2,000 Kilometers

Europe

0 250 500 Miles
0 250 500 Kilometers
NORWAY
FINLAND
SWEDEN
ESTONIA
LATVIA
LITHUANIA
RUSSIA
BELARUS
North Sea
Baltic Sea
DENMARK
IRELAND
UNITED KINGDOM
NETHERLANDS
BELGIUM
GERMANY
POLAND
Channel Islands
LUXEMBOURG
FRANCE
LIECHTENSTEIN
CZECHIA
SLOVAKIA
UKRAINE
MOLDOVA
SWITZERLAND
AUSTRIA
HUNGARY
SLOVENIA
CROATIA
ROMANIA
SAN MARINO
MONACO
BOSNIA AND HERZEGOVINA
SERBIA
BULGARIA
KOSOVO
MONTENEGRO
MACEDONIA
ALBANIA
GREECE
Black Sea
GEORGIA
TURKEY
ATLANTIC OCEAN
PORTUGAL
SPAIN
ANDORRA
Balearic Is.
Corsica
Sardinia
ITALY
VATICAN CITY
Mediterranean Sea
Sicily
MALTA
Crete
CYPRUS
SYRIA
LEBANON
MOROCCO
ALGERIA
TUNISIA

Earth at Night, City Lights

The Americas

These images of Earth at night from NASA's Suomi-NPP "Marble" series use a collection of satellite-based observations, stitched together in a seamless mosaic of our planet. This view is based on instrumentation that observes light emanating from the ground. Notice how strongly major population clusters show up in the image.

Africa, Europe, and the Middle East

Defense Meteorological Satellite Program (DMSP) NASA/GSFC